Assisted Reproductive Technologies in Animals Volume 1

Juan Carlos Gardón · Katy Satué Ambrojo

Editors

Assisted Reproductive Technologies in Animals Volume 1

Current Trends for Reproductive Management

🐎 Springer

Editors
Juan Carlos Gardón (iD)
Medicine and Animal Surgery
Catholic University of Valencia, Faculty of
Veterinary and Experimental Sciences
Valencia, Spain

Katy Satué Ambrojo (iD)
Medicine and Animal Surgery Department
CEU-Cardenal Herrera University
Alfara del Patriarca, Valencia, Spain

ISBN 978-3-031-73078-8   ISBN 978-3-031-73079-5   (eBook)
https://doi.org/10.1007/978-3-031-73079-5

This Springer imprint is published by the registered company Springer Nature Switzerland AG
The registered company address is: Gewerbestrasse 11, 6330 Cham, Switzerland

If disposing of this product, please recycle the paper.

Foreword

The continuous advancement of science and technology has led to a profound reshaping of our understanding of the biological world. This has resulted in a series of remarkable breakthroughs that were previously thought to be beyond reach. Among these remarkable strides, Assisted Reproductive Technologies (ART) have revolutionized not only human medicine but also the field of animal reproduction.

"Assisted Reproductive Technologies in Animals: Current Trends for Reproductive Management" is a testament to the profound impact these technologies have had on the management, conservation, and improvement of animal species across the globe. The journey of ART in animals began with a simple yet revolutionary idea: to overcome the natural barriers of reproduction and enhance the genetic potential of animals. Today, this field encompasses a wide array of techniques, including artificial insemination, in vitro fertilization, embryo transfer, and cloning as well as basic concepts about physiological reproductive events. These methods are not only critical for agricultural productivity but also play a pivotal role in wildlife conservation and the preservation of endangered species.

This book provides an extensive exploration of the current trends and practices in animal ART, offering invaluable insights for veterinarians, researchers, and students. It highlights the latest advancements, from the refinement of existing techniques to the development of novel approaches that push the boundaries of what is possible. The contributors to this volume are leading experts in their respective fields, each bringing a wealth of knowledge and experience to the table. This ensures that readers are equipped with the most up-to-date information and practical guidance.

One of the most significant contributions of the book is its emphasis on the ethical and practical implications of ART. As we endeavor to reconcile technological expertise with ethical considerations, it is imperative to comprehend the influence of these technologies on animal welfare, genetic diversity, and ecosystem stability. These technologies must recognize the responsibility that comes with such powerful tools, and we have to use them ethically and prudently.

Furthermore, the current trends in reproductive management also emphasize the collaborative nature of this field. The integration of knowledge from various disciplines, including genetics, endocrinology, veterinary medicine, and beyond, demonstrates that progress is often the result of interdisciplinary efforts. Such collaboration not only accelerates innovation but also ensures that the benefits of ART are maximized across different sectors and species.

As we stand on the threshold of discoveries and innovations, this book serves as both a reflection on the progress that has been made and a beacon guiding us toward future possibilities. This book is an essential resource for anyone involved in animal reproduction, providing a comprehensive foundation for understanding and applying ART in various contexts. In conclusion, we would like to express our profound gratitude to the authors and contributors who have dedicated their expertise and passion to this volume. Their dedication to the advancement of the field of animal reproductive technologies is evident in each chapter, and their work will undoubtedly serve as an inspiration to future generations of scientists and practitioners.

It is hoped that this book will serve as a cornerstone for continued progress and a source of inspiration for all those who seek to harness the power of ART for the betterment of animal health, conservation, and productivity.

Faculty of Veterinary and Experimental Sciences   Juan Carlos Gardón
Catholic University of Valencia-San Vicente Mártir,
Valencia, Spain

Faculty of Veterinary   Katy Satué
CEU Cardenal Herrera University,
Valencia, Spain

Preface

The field of Assisted Reproductive Technologies (ART) in animals has undergone a profound evolution over the past few decades, progressing from experimental concepts to indispensable tools in modern animal husbandry, conservation biology, and biomedical research. This book, *Assisted Reproductive Technologies in Animals: Current Trends for Reproductive Management*, aims to provide a comprehensive overview of the latest advancements, methodologies, and practical applications of ART in various animal species.

The impetus for compiling this book stems from the recognition of the profound impact that ART has on animal production, genetic improvement, and biodiversity conservation. As global demand for food increases, efficient and sustainable livestock production becomes more critical. ART plays a crucial role in meeting these demands by enhancing reproductive efficiency, accelerating genetic gain, and mitigating the impacts of disease and environmental stressors. Moreover, the application of ART in wildlife conservation offers hope for endangered species, providing tools to preserve genetic diversity and prevent extinction.

This book is structured to guide readers through the multifaceted world of animal ART. We begin with foundational principles, exploring the biology of reproduction and the historical context of ART development. Subsequent chapters delve into specific technologies such as artificial insemination, in vitro fertilization, embryo transfer, and cloning. Each chapter is crafted to not only explain the technical aspects but also to discuss the practical implications, challenges, and prospects of these technologies.

A notable aspect of this volume is its interdisciplinary approach. The volume is enriched by contributions from leading experts in veterinary medicine, reproductive biology, genetics, and conservation science, which collectively provide a well-rounded perspective. The inclusion of case studies and real-world applications provides readers with practical insights and demonstrates the versatility of ART across different contexts and species.

The text incorporates ethical considerations, reflecting the growing importance of responsible and humane use of reproductive technologies. As we endeavor to expand the boundaries of what is feasible, it is of the utmost importance to consider the well-being of the animals involved, the ecological implications, and the ethical dimensions of genetic manipulation. The objective of this book is to encourage a balanced view of the advancements in ART while promoting ethical stewardship.

Furthermore, the text addresses the latest research trends and emerging technologies that promise to shape the future of animal reproduction. The field of animal reproduction technology (ART) is undergoing rapid expansion, encompassing genomic selection and gene editing, advanced cryopreservation techniques, and biobanking. These innovations have the potential to revolutionize animal breeding, conservation efforts, and biomedical research, opening new avenues for exploration and application.

In preparing this book, we have been privileged to work with an outstanding group of contributors who have generously shared their knowledge and expertise. Their dedication to advancing the science and practice of animal ART is evident in every chapter, and their contributions are what make this book an authoritative resource.

We hope that *"Assisted Reproductive Technologies in Animals: Current Trends for Reproductive Management"* will serve as a valuable reference for researchers, practitioners, and students alike. Whether you are engaged in livestock production, wildlife conservation, or academic research, we believe that this book will provide the information and inspiration needed to advance the field of animal reproduction.

We would like to express our gratitude to all those who have embarked on this journey with us. We look forward to the discoveries and innovations that lie ahead, and we are confident that this book will play a part in shaping the future of reproductive management in animals.

Valencia, Spain   Juan Carlos Gardón
Valencia, Spain   Katy Satué

Contents

Abbreviations

11HSDB2	hydroxysteroid 11-beta dehydrogenase 2
2D	two-dimensional
3' Utrs	3' Untranslated Regions
3D	three-dimensional
4D	four-dimensional
A	area
ACTH	adrenocorticotropic hormone
ADG	average daily gain
AI	artificial insemination
AI	artificial intelligence
ALH	amplitude of lateral head displacement
ALT	alanine transaminase
amEVs	amniotic-derived EVs
ARMC12	armadillo repeat containing 12
ART	assisted reproductive technologies
ASMA	automatized sperm morphometry analysis
AST	aspartate aminotransferase
ATP	adenosine triphosphate
AV	artificial vagina
AVP	arginine vasopressin
b2M	b-2 microglobulin
BBSE	bull breeding season evaluation
BCECF AM	2',7'-bis-(2-carboxyethyl)-5-(and-6)-carboxyfluorescein, acetoxymethyl ester
BCF	beating frequency
BoHV-1	bovine herpesvirus-1
BP	before present
BPD	biparietal diameter
BRD	bovine respiratory disease
BS	bovine serum
BTV	bluetongue virus
BVDV	bovine viral diarrhea virus
BW	birth weight
BWC	body weight composite

cAMP	cyclic adenosine monophosphate
CASA	computerized analysis system for sperm motility
CAT	catalase
CATSPERB	catSper channel auxiliary subunit beta
CATSPERG	catSper channel auxiliary subunit gamma
CCR	cow conception rate
Cd	cadmium
CED/CEM	calving ease direct/maternal
CFDA	6-carboxymethylfluorescein diacetate
cGMP	cyclic guanosine monophosphate
CHO	carbohydrates
CL	corpus luteum
Claw/Angle	structural soundness
CM	conditioned medium
CMA	cow's milk allergy
CMPA	cow's milk protein allergy
CO2	carbon dioxide
COC	cumulus-oocyte complex
COX-1	cytochrome c oxidase subunit 1
COX-4	cytochrome c oxidase subunit 4
CRH	corticotropin-releasing hormone
CRISPR	clustered regularly interspaced short palindromic repeats
CRISPR-Cas9	CRISPR-associated nuclease (Cas) 9
CRL	crown-rump length
CSF2	colony-stimulating factor 2
CTUP	combined utero placental thickness
CV	coefficient of variation
CW	carcass weight
CYP19	cytochrome P450 family 19
D loop	displacement loop
DAF-2 DA	4,5-diaminofluorescein diacetate
DAZL	deleted in azoospermia-like
DFI	daughter fertility index
DHA	docohexanoic acid
DHE	dihydroethidium
DHR 123	dihydrorhodamine 123
DMSO	dimethylsulfoxide
DNA	deoxyribonucleic acid
DNM1L	dynamin-1-like protein
Doc	docility
DOHaD	developmental origins of health and disease
DPB	days before birth
DPR	daughter pregnancy rate
DSB	double-stranded break
E2	estradiol-17β

EBLV	enzootic bovine leukosis virus
EBV	estimated breeding value
eCG	equine chorionic gonadotropin
EDV	end-diastolic velocity
EE	electroejaculation
EEC	endometrial epithelial cell
EEL	early embryonic loss
eEV	embryonic EV
EF	embryo flushing
eFSH	equine follicle stimulating hormone
EG	egg yolk
EG	ethylene glycol
ELISA	enzyme-linked immunosorbent assay
ELSPBP	epididymal sperm binding protein 1
EMC	extracellular matrix components
ENJSRVs	endogenous beta-retroviruses
EPC	embryos recovered/cycle
EPD	expected progeny difference
EPE	equine pituitary extracts
EPO	embryos per ovulation
EqIFN-delta 1	equine interferon-delta 1
EqIFN-delta 2	equine interferon-delta 2
ER	embryo recovery
ERR	embryo recovery rate
ESCRT	endosomal sorting complex responsible for transport
ET	embryo transfer
EV	embryonic vesicle
EV	extracellular vesicle
FADH$_2$	flavin adenine dinucleotide
FAO	Food and Agriculture Organization of the United Nations
Fat	back fat thickness
FDA	Food and Drug Administration
FF	follicular fluid
FHR	fetal heart rate
FIS1	mitochondrial fission 1 protein
FITC	fluorescein isothiocyanate
FKBP4	progesterone co-factor
FLC	feet and legs composite
FMLP	formyl-methionyl-leucyl-phenylalanine
FR	caliper size
FSH	follicle stimulating hormone
GAPDH	glyceraldehyde-3-phosphate dehydrogenase
GAPDH	glyceraldehyde-3-phosphate dehydrogenase
GE-EPDs	genomic-enhanced EPDs
GHR	growth hormone receptor gene

GMO	genetically modified organism
GnIH	gonadotropin-inhibitory hormone
GnRH	gonadotropin-releasing hormone
GPx	glutathione peroxidase
GPX5	glutathione peroxidase 5
GSR	glutathione reductase
GST	glutathione s-transferase
GT	genital tubercle
GTPase	guanosine triphosphatase
GWA	genome-wide association
H2DCFDA	2',7'-dichlorodihydrofluorescein diacetate
H_2O_2	hydrogen peroxide
H3K27me3	histone 3 lysine 27 three methylation
H3K4me3	histone 3 lysine 4 three methylation
HCG/hCG	human chorionic gonadotropin
HCR	heifer conception rate
HDR	homology-directed repair
HMOX1	heme oxygenase inducible
HMOX2	heme oxygenase constitutive
HP/HPG	heifer pregnancy rate
HPA	hypothalamic-pituitary-adrenal
HPG	hypothalamic-pituitary-gonadal
HPO	hypothalamic-pituitary-ovarian
HPRT1	hypoxanthine phosphoribosyl transferase 1
HSP70	heat shock protein family A member 70
HSPA8	heat shock protein family A member 8
ICAM	intercellular adhesion molecules
ICSI	intracytoplasmic sperm injection
IETS	International Embryo Technology Society
IFN-delta	interferon-delta
IGF-1	insulin-like growth factor 1
IGFBP3	insulin-like growth factor binding protein 3
IM	intramuscular
INHBA	inhibin A/B
INHBB	inhibin A/B
ITO	interval between treatment with PGF2α and ovulation
IU	international unit
IU	intrauterine
IUD	intrauterine device
IV	intravenous
IVC	*in vitro* culture
IVD	*in vivo*-derived embryos
IVF	*in vitro* fertilization
IVM	*in vitro* maturation
IVP	*in vitro* embryo production

IVPE	*in vitro* production of embryos
KISS1	kisspeptin
KLC3	kinesin light chain 3
L	length
LAMA5	laminin subunit alpha 5
LEP	leptin
LH	luteinizing hormone
LIN	linearity index
LIV	cow livability
lncRNA	long non-coding RNA
LPI	Lifetime Performance Index
M-540	merocyanine 540
MegN	meganuclease
Mff	mitochondrial fission factor
MFN1/2	mitofusion-1/2
MGEF8	milk fat globule epidermal growth factor (EGF) 8 protein
MHC Class II	mayor histocompatibility complex class II
MHz	megahertz
MIA	maternal immune activation
MIF	macrophage migration inhibitory factor
Milk	milk production
miRNA	MicroRNA
MMP	mitochondrial membrane potential
MO	multiple ovulation
MOET	multiple ovulation and embryo transfer
MPC1	mitochondrial pyruvate carrier 1
Mrb	marbling score
mRNA	messenger ribonucleic acid
MRP	maternal recognition of pregnancy
MS	mass spectrometry
mSOF	modified synthetic oviduct fluid
mtDNA	mitochondrial DNA
MV	mean velocity
MV	microvesicle
MVB	multivesicular bodies
MWW	maternal weaning weight
MYH9	myosin heavy chain 9
N2	nitrogen
NADH	nicotinamide adenine dinucleotide
NET	neutrophil extracellular trap
NEU2	enzyme neuraminidase 2
NHEJ	non-homologous end-joining
NMR	nuclear magnetic resonance
NO	nitric oxide
NRF1/2	nuclear respiratory factor ½

O2	dioxygen
O_2^-	superoxide anion
°C	degree Celsius
OCDE-FAO	Organization for Economic Co-Operation and Development and the Food and Agriculture Organization of the United Nations
OD	orbit diameter
OEC	oviductal epithelial cells
oEV	oviductal EV
OF	oviductal fluid
OPU	ovum pick-up
OVGP1	oviduct-specific glycoprotein 1
OXT	oxytocin
P/AI	pregnancy per artificial insemination
P	perimeter
P4	progesterone
PAG	pregnancy-specific proteinase
PAWP	post-acrosomal WW domain-binding protein
Pb	lead
PBS	buffer phosphate saline
Pc	Celtic polled allele
PCR	polymerase chain reaction
PG H2	prostaglandin H2
PG	prostaglandin
PGC-1α	peroxisome proliferator-activated receptor gamma coactivator-1
PGD	pre-implantation genetic diagnosis
PGE	prostaglandin E
PGE2	prostaglandin E2
PGF	prostaglandin F
PGF2α	prostaglandin F2 α
PGFM	13,14-dihydro-15-keto-PGF2a
PHE	penicillamine, hypotaurine, and epinephrine
PI	propidium iodide
PI	pulsatility index
piRNA	PIWI-interacting RNA
PKA	protein kinase A
PL	productive life
PLCZ	phospholipase C zeta
PMCA4	plasma membrane calcium ATPase 4
PMIE	post-mating induced endometritis
PMN	polymorphonuclear neutrophil
PMSG	pregnant mare serum gonadotropin
PNA	peanut agglutinin
PNC	parvocellular neurosecretory cells
PO	post ovulation
POMC	proopiomelanocortin

PON	paraoxonase
PP1A	peptidylprolyl isomerase A
PR	pregnancy rate
PRM1	protamine 1
PS	phosphatidylserine
PSA	pisum sativum agglutinin
PSF	pre-spermatic fraction
PSV	peak systolic velocity
PTGFRN	prostaglandin receptor inhibiting protein
PUFA	polyunsaturated fatty acid
PVN	paraventricular nucleus of the hypothalamus
PVS	perivitelline space
QTL	quantitative trait loci
REA/RE	rib eye area
RI	resistance index
RNA	ribonucleic acid
RNS	reactive nitrogen species
ROS	reactive oxygen species
RRADG/RFI	residual feed intake
rRNA	ribosomal ribonucleic acid
S/D	systolic velocity/diastolic velocity ratio
SAM	sympathetic adrenomedullary
SC/SCR	scrotal circumference
SC	serum cortisol
SCE/DCE	calving ease sire/daughter
SCNT	somatic cell nuclear transfer
SCS	somatic cell score
SEM	standard error of mean
SHR	tenderness
siRNA	small interfering RNA
SLA	swine leukocyte antigen complex
SNARE	soluble NSF attachment protein receptor
SncRNA	small non-coding RNA
sncRNA	small non-coding RNA
SNP	single nucleotide polymorphism
SOD	superoxide dismutase
SP	seminal plasma
SPAM1	sperm adhesion molecule 1
SR	sperm reservoir
SRF	sperm-rich fraction
STAY	stayability
T	testosterone
TAI	timed artificial insemination
TALEN	translation activator-like effector nuclease
TD	trunk diameter

TDN	TSKS-derived nuage
TFAM	mitochondrial transcriptional factor A
TGF-B	transforming growth factor beta
THI	temperature-humidity index
TP	transition protein
TSKS	testis-specific serine kinase substrate
tsRNA	transfer RNA-derived small RNA
Tyr-P	tyrosine phosphorylation
UC	udder composite
UEC	uterine epithelial cell
uEV	uterine extracellular vesicle
UF	uterine fluid
UPS	ubiquitin-proteasome system
UTJ	utero-tubal junction
VEGF	vascular endothelial growth factor
vEV	vaginal EV
VLF	vaginal luminal fluid
WCVM	Western College of Veterinary Medicine
WW	weaning weight
YG	lean yield
YW	yearling weight
ZFN	zinc finger nucleases
ZP	zona pellucida

Part I

Ultrasonography in Animal Reproduction

Ultrasonography in Goats During Pregnancy

Salvador Ruiz López (ID)

Abstract

The aim of this review is to further study and analyze the use of ultrasonography in small ruminants and in the goats during gestation. Thus, the underlying principles of ultrasonographic techniques at the level of the genital tract of the goat are described, together with an indication of the current methods being applied. A historical analysis of the development and importance of ultrasonography at the reproductive level is presented and, in aspects such as the early diagnosis of pregnancy, the determination of the number and sex of fetuses, and the research in fetometry that has been carried out in recent years and its correlation with the gestational age of the fetus are considered. Finally, the importance of new ultrasound techniques, such as Doppler and three-dimensional ultrasound, in the field of goat reproduction is discussed. In addition, a series of ultrasound images are presented, which illustrate, among other aspects, for example, the formation of embryonic vesicles, the heartbeat in the early stages of gestation, the development of the embryo and/or fetus during gestation, various measurements used to estimate gestational age throughout gestation, and lastly, images of the applications of Doppler and three-dimensional ultrasound in pregnant goats.

S. R. López (✉)
Department of Physiology, Faculty of Veterinary Medicine, Espinardo Campus, University of Murcia, Campus Mare Nostrum (CMN), Murcia, Spain

Murcian Institute for Biosanitary Research Pascual Parrilla (IMIB), Campus of Health Sciences, Murcia, Spain
e-mail: sruiz@um.es

© The Author(s), under exclusive license to Springer Nature Switzerland AG 2024
J. C. Gardón, K. Satué Ambrojo (eds.), *Assisted Reproductive Technologies in Animals Volume 1*, https://doi.org/10.1007/978-3-031-73079-5_1

3

Keywords

Small ruminants · Goats · Ultrasonography · Gestation · Pregnancy diagnosis ·
Fetal number · Fetal sexing · Gestation age · Fetal growth parameters ·
Fetometry · Doppler · 3D ultrasonography

1   Introduction

Ultrasonography, also known as echography, is a diagnostic imaging technique that
employs high-frequency sound waves, or ultrasound, to examine the internal organs
of the body. These are sound waves with a frequency higher than that which is
audible to the human ear, namely above 20,000 Hz (Galián et al. 2021a).

The advent of ultrasonography has had a profound impact on the field of veterinary medicine over the past several decades. The development and refinement of
ultrasonography techniques for examining reproductive organs in animals have significantly enhanced reproductive management and health monitoring. Currently,
ultrasonography constitutes an indispensable component of reproductive management in small ruminants (Ginther 2014).

The early detection of pregnancy in goats is of paramount importance for the
effective management of reproduction, the improvement of productivity, and the
assurance of the health of both the doe and the developing fetus. Ultrasonography
has become a primary tool for early pregnancy detection due to its non invasive
nature, accuracy, and ability to provide real-time results. Transrectal ultrasound can
be employed to detect pregnancy in goats as early as 17–28 days postbreeding or
transabdominally between days 25 and 28 of gestation (Padilla-Rivas et al. 2005;
Jones et al. 2016). From 30–40 days of gestation, transabdominal ultrasound is the
preferred approach.

Ultrasound remains the gold standard for estimating and determining fetal numbers in pregnant goats. Transabdominal ultrasound is the most used method for fetal
counting in goats. The optimal time for fetal counting is between 40 and 70 days of
gestation when the fetuses are large enough to be distinguishable but not so large
that they overlap significantly (Abdelghafar et al. 2007; Andrabi and Gulavane 2015).

In addition to confirming pregnancy and counting the fetuses, one of the advanced
applications of ultrasonography is fetal sex determination. This capability is of
value to breeders and farmers for several reasons, including those related to management and economics. The sex of the fetus can be determined by visualizing the
external genitalia and/or localizing the genital tubercle. Between 46 and 48 days of
gestation, the process of genital tubercle (GT) migration has concluded, and fetal
sexing can be performed with equal accuracy in male and female goat fetuses
(Santos et al. 2007a, b; Azevedo et al. 2009).

Ultrasonography represents a valuable tool for determining fetal age in goats,
providing essential information for reproductive management, monitoring fetal
development, and improving pregnancy outcomes. Ultrasonographic fetal aging has
significant applications in veterinary practice and research. Techniques such as

transrectal and transabdominal ultrasonography, when combined with measurements of crown-rump length, trunk diameter, biparietal diameter, orbit diameter, and placentome size, among others, offer reliable estimates of gestational age throughout pregnancy (reviews by Erdogan 2012; Jones and Reed 2017).

Doppler ultrasonography has transformed reproductive imaging in goats, providing detailed insights into the blood flow and vascular health of reproductive structures. Its applications in assessing ovarian function, pregnancy diagnosis, fetal monitoring, and detecting reproductive pathologies make it an invaluable tool in veterinary practice. Despite challenges such as cost and the need for technical expertise, the benefits of enhanced diagnostic accuracy and improved reproductive management underscore the importance of Doppler ultrasonography in modern veterinary reproductive care (Fasulkov et al. 2021; Ramírez-González et al. 2023).

Three-dimensional (3D) ultrasonography represents a revolutionary advancement in the field of veterinary reproductive imaging, offering unparalleled detail and accuracy in assessing fetal development, placental health, and overall reproductive status in goats. Its applications in diagnosing abnormalities, monitoring growth, and managing reproductive health make it an invaluable tool for veterinarians. Despite the challenges of cost and the need for technical expertise, the benefits of enhanced diagnostic accuracy and improved reproductive management outweigh the importance of adopting 3D ultrasonography in veterinary practices (Kumar et al. 2015b; Karadaev et al. 2019).

In this review, the author presents a study on the historical development and importance of ultrasonography at the reproductive level in small ruminants, with special reference to the goat. In particular, the author considers aspects of interest such as early diagnosis of gestation, determination of the number and sex of fetuses, and research in fetometry that has been carried out in recent years and its correlation with fetal gestational age. Finally, the significance of novel ultrasound techniques, such as Doppler and 3D ultrasound, in the field of reproduction in goats is discussed.

2 A Historical Overview of the Development of Ultrasonography in Animals

The advent of ultrasonography has transformed the field of veterinary medicine over the past several decades. The development and refinement of ultrasonography techniques for examining reproductive organs in animals have significantly enhanced reproductive management and health monitoring (Medan and Abd El-Aty (2010).

The historical development of ultrasonography in animals exemplifies a remarkable journey of technological advancement and practical application (Ginther 2014). From the early, rudimentary machines to today's sophisticated, portable, and artificial intelligence (AI)-enhanced devices, ultrasonography has become a cornerstone of modern veterinary reproductive management. These advancements have not only improved animal health and welfare but also enhanced the economic viability of small ruminant farming through better reproductive efficiency and management.

The inception of ultrasonography dates to the early twentieth century, with the initial applications focused primarily on human medicine. The potential of the technology for veterinary use was first recognized in the 1960s. Early veterinary ultrasound machines were adapted from human models and were large, cumbersome, and limited in their imaging capabilities. The first veterinary applications of ultrasonography were primarily in large animals such as cattle and horses. These early machines were primarily used for pregnancy diagnosis and basic reproductive assessments. The images produced were rudimentary, yet they offered insights that were previously unattainable without invasive procedures (Galián et al. 2021a, b).

The 1980s saw a significant advance in ultrasonography technology, with machines becoming more compact, affordable, and user-friendly. This decade saw the introduction of real-time B-mode (brightness mode) ultrasonography, which allowed for dynamic imaging of reproductive structures (reviewed by Ginther 2014). The use of ultrasonography in small ruminants, including goats, began to gain traction. Researchers and veterinarians started to explore its potential for early pregnancy diagnosis and monitoring of fetal development, and the capacity to visualize embryonic vesicles as early as 18–20 days post-breeding represented a significant breakthrough.

The 1990s witnessed further enhancements in image resolution and the portability of ultrasound machines. These improvements facilitated greater accessibility and practicality for field use in veterinary medicine. The introduction of transrectal and transabdominal ultrasound probes provided more detailed and accurate imaging of the reproductive organs in small ruminants. These probes facilitated enhanced visualization of the ovaries, uterus, and developing fetuses. In the late 1990s, Doppler ultrasonography emerged as a significant advancement in reproductive medicine, enabling the assessment of blood flow within the reproductive organs and providing new insights into ovarian function, placental health, fetal well-being, and other aspects of animal reproduction (Schmidt et al. 1991; Serin et al. 2010; Velasco and Ruiz 2021).

The advent of the new millennium has witnessed a remarkable acceleration in the development of ultrasonography technology, largely attributable to the convergence of three key factors: improvements in digital imaging, exponential growth in computer processing power, and the evolution of sophisticated software algorithms.

In the early 2000s, portable and high-resolution ultrasound machines became widely available, revolutionizing on-farm reproductive management. The advent of these devices has made it possible to conduct comprehensive reproductive examinations in a variety of settings, thereby enhancing the precision of breeding programs and health monitoring.

The integration of advanced features, such as color, power, and spectral Doppler, has enhanced the ability to assess vascularization and blood flow dynamics (Elmetwally and Meinecke-Tillmann 2018; Fasulkov et al. 2021). 3D ultrasonography commenced its application in veterinary medicine and provided volumetric imaging of the reproductive organs, offering detailed views of fetal structures and complex pathologies that were previously difficult to diagnose with traditional two-dimensional (2D) imaging (Markov and Dimitrova 2006; Kumar et al. 2015b). All these new capabilities are of great importance for the evaluation of ovarian activity, the diagnosis of placental insufficiencies, and the monitoring of fetal health.

3   Pregnancy Diagnosis During Early Gestation

Early pregnancy detection in goats is essential for effective reproductive management, improving productivity, and ensuring the health of both the doe and the developing fetus. Ultrasonography has become a primary tool for early pregnancy detection due to its noninvasive nature, accuracy, and ability to provide real-time results. A portable ultrasound machine with a high-frequency transducer, typically 5–7.5 MHz, is used. These devices are ideal for examining the reproductive organs of small ruminants because of their high resolution (Erdogan 2012).

Ultrasound can detect pregnancy in goats as early as 17–28 days postbreeding. During this period, transrectal ultrasound techniques are used to diagnose very early pregnancy. The transducer is inserted into the goat's rectum, providing a clear view of the uterus and gestational sac.

The use of a transvaginal or endovaginal approach for early detection of pregnancy and ultrasound monitoring in goats has also been described with good results in different goat breeds, such as Saanen, Attappady Black, and Murciano-Granadina (Koker et al. 2012; Philip et al. 2017; Ramírez-González et al. 2023, respectively).

The first detectable feature of pregnancy is the development and enlargement of circular and elongated uterine cross-sections, which distinguish the gravid uterus from the bladder on ultrasound (Jones and Reed 2017). In does, increasing uterine fluid can be seen transrectally as early as days 17–19 of gestation (Padilla-Rivas et al. 2005; Ramírez-González et al. 2023) (Fig. 1a–c) or transabdominally between days 25 and 28 of gestation (Jones et al. 2016). However, Anya et al. (2017) reported that in a West African Dwarf goat, using a transabdominal approach, they observed as the earliest ultrasound evidence of pregnancy the image of a circumscribed anechoic fluid in the uterus on day 19 and the embryo on day 20.

Uterine fluid is not the only indicator of pregnancy, as it is also associated with estrus and uterine abnormalities. Accompaniment of the fluid by the amniotic vesicle and the presence of a fetus with a heartbeat is essential to confirm a viable pregnancy (Amer 2010). In does, pregnancy can be diagnosed transrectally by detecting the fetal heartbeat as early as 22–23 days (Amer 2008; Karen et al. 2009), and with >90% sensitivity achieved between 24 and 36 days (Karen et al. 2014) (Fig. 1b).

Mali et al. (2019), in Kokan Kanyal goats, compared the efficacy of the transrectal and transabdominal approach. Comparatively, the transrectal probe was found more efficient as it detected the early embryonic vesicles and embryonic heartbeat at 20–23 and 25–28 days postmating as compared to 25–28 and 35 days onwards, respectively, by a transabdominal transducer. Transrectal scanning diagnosed the pregnancy with 100% accuracy as early as 25 days of gestation in goats.

From 30–40 days of gestation, transabdominal ultrasound becomes the preferred technique. The transducer is placed on the lower abdomen of the goat, facilitating

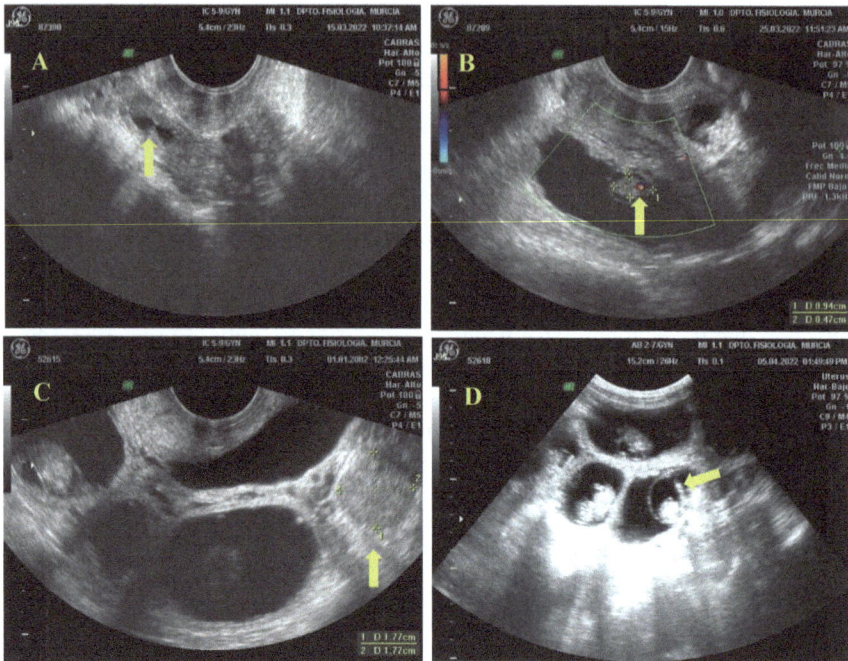

Fig. 1 (a) Ultrasonographic image obtained with an endovaginal probe at day 18 postbreeding in Murciano-Granadina goat. The presence of uterine fluid is clearly visible in the image, with two echogenic areas in the uterus (yellow arrow). (b) Ultrasonographic image obtained with an endovaginal probe at day 25 of gestation in Murciano-Granadina goat. The embryo is visible in the amniotic sac, and the fetal heartbeat is identifiable by the color Doppler (yellow arrow). (c) Ultrasonographic image obtained with an endovaginal probe at day 25 of gestation in Murciano-Granadina goat. Some embryonic vesicles and one embryo are visible. On the right, the gestational corpus luteum (CL) of the ovary can be seen in its full size (yellow arrow). (d) Ultrasonographic image obtained with a transabdominal probe at day 40 of gestation in a multiple gestation in Murciano-Granadina goat. Three embryos can be observed in their gestational sacs within the respective embryonic vesicles. The umbilical cord can be seen in the embryo located at the bottom right (yellow arrow)

the detection of fetuses and gestational structures without invasive methods (Fig. 1d). Using a transabdominal approach, the fetal heartbeat can be detected between days 27 and 30 (Karen et al. 2009; Amer 2010).

Fetal detection is accompanied by the development of placentome structures that support the pregnancy and establish the maternal–fetal interface. Transrectal scanning in goats allows imaging of placentomes as small nodules at day 35, which initially appear as irregular shapes on the uterine wall and mature into hollow hemispherical structures by day 42 (Kumar et al. 2015a).

During early pregnancy, the development of the fetal limbs, genitalia, and umbilical cord can be observed sequentially by ultrasound. In goats, the umbilicus can be visualized from day 32 (Ramírez-González et al. 2023); the limbs extend from the abdomen by day 42, and skeletal structures, including the rib cage, spinal cord, and skull, can be visualized from day 48 (Kumar et al. 2015c).

4 Fetal Number Determination

Accurately estimating and determining the number of fetuses in goats is essential for effective reproductive management and optimizing health outcomes for both dam and progeny, contributing to better reproductive management and overall herd health (Abdelghafar et al. 2007).

There are several practical advantages to determining the number of fetuses in goats, including the fact that knowing the number of fetuses allows more accurate nutritional planning, ensuring that the dam is receiving adequate nutrition to support multiple fetuses. Also, early detection of multiple pregnancies can lead to more frequent monitoring, reducing the risk of complications such as pregnancy toxemia, and lastly, accurate fetal counts help to plan subsequent breeding cycles and manage herd population dynamics (Vinoles-Gil et al. 2010).

Ultrasound remains the gold standard for estimating and determining fetal numbers in pregnant goats, with transabdominal ultrasound being the most used method for fetal counting in goats. Studies have shown that transabdominal ultrasonography can accurately determine the number of fetuses in goats with a high degree of reliability, especially when performed by experienced operators (Galián et al. 2021b).

The number of fetuses can be determined up to 100 days of gestation, but the optimal time is between 40 and 70 days of gestation, when the fetuses are large enough to be distinguishable but not so large that they overlap significantly (Medan and Abd El-Aty 2010). At 70 days and beyond, additional fetuses may lie beyond the depth of the 5 MHz linear array transducer (Andrabi and Gulavane 2015).

During fetal counting, the operator should slowly scan from right to left at the same level to localize the pregnancy. Thereafter, the direction of the scanning movements should be perpendicular to the first set of movements at different levels to systematically cover the entire scanning area and avoid scanning the same fetus twice (Vinoles-Gil et al. 2010; Erdogan 2012).

Although there are some reports of a very high percentage of correct and sensitive diagnoses of multiple pregnancies (Medan et al. 2004; Vinoles-Gil et al. 2010), it is well-known that it is difficult to differentiate between twins, triplets, or quadruplets at any stage of pregnancy (Padilla-Rivas et al. 2005; Abdelghafar et al. 2007).

Although less commonly used than the transabdominal approach, transrectal ultrasound can provide detailed images early in pregnancy. It is particularly useful between the 18–35 days of pregnancy. However, it requires more skill and experience to perform effectively and can be more invasive for the animal. In addition, Doppler ultrasound can help identify fetal heartbeats and differentiate between overlapping fetuses by identifying individual heartbeats (Ramírez-González et al. 2023).

Several systems and protocols have been developed to standardize the process of estimating the number of fetuses in goats by ultrasound. Some protocols use a scoring system based on visual criteria to estimate fetal numbers. These systems can help less experienced operators to achieve more consistent results. Modern ultrasound scanners may be equipped with software to assist in the counting and measurement of fetuses. These systems use algorithms to improve image clarity and automate parts of the counting process. To improve accuracy and consistency, various training programs and certifications for veterinarians and technicians focus on the specific techniques of small ruminant fetal counting (Galián et al. 2021a).

However, there are challenges associated with ultrasound fetal counting. The accuracy of ultrasound techniques is highly dependent on the skill and experience of the operator. Inexperienced operators may miscount or miss fetuses. High-quality, high-resolution ultrasound equipment is required for accurate fetal counting, but it can be expensive and may not be available in all settings. As pregnancy progresses, fetuses may overlap or change position, making accurate counting more difficult.

5   Fetal Sexing

Beyond pregnancy confirmation and fetal counting, one of the advanced applications of ultrasonography is fetal sex determination. This capability is valuable for breeders and farmers for various management and economic reasons. Thus, knowing the sex of the offspring can assist in the planning of future breeding strategies, the management of genetic lines, and the selection of replacement stock. Farmers can plan for market demands by knowing the sex ratio of the upcoming kid crop and optimizing sales strategies and pricing. Finally, sex determination allows for tailored nutritional and health management plans for pregnant goats carrying multiple fetuses of known sex, potentially improving offspring survival and health (Haibel 1990).

The sex of the fetus can be determined by visualizing the external genitalia (penis, prepuce, scrotal sac, nipples, and genital swelling) and/or localizing the GT. In male fetuses, the GT is located near the umbilical cord and will develop into the penis. This is seen as a hyperechoic structure between the hind limbs and the umbilical cord. In female fetuses, the GT is positioned under the tail and develops into the clitoris. This structure appears as a hyperechoic area located closer to the tail and away from the umbilical cord (Santos et al. 2007a, b).

The timing of the ultrasound examination is critical for accurate fetal sex determination. Between 46 and 48 days of gestation, GT migration has concluded, and fetal sexing can be performed with equal accuracy in male and female goat fetuses (Santos et al. 2007a, b; Azevedo et al. 2009). The optimal window for sexing goat fetuses using ultrasonography, based on the final location of GT and the identification of the external genitalia, is between 55 and 70 days of gestation.

Studies have demonstrated high accuracy rates for fetal sex determination when the examination is conducted within the optimal window by experienced practitioners. However, the accuracy of fetal sexing has been reported to decline as gestation advances due to the difficulty of positioning the fetus to visualize the necessary external genitalia. In goats, fetal sexing indices did not differ between litters of a singleton or multiples (Amer 2008).

Despite the advantages, several challenges are associated with fetal sex determination using ultrasonography. The accuracy of fetal sex determination is highly dependent on the skill and experience of the operator. Training and experience are crucial for reliable results, and poor-quality equipment can lead to inaccurate sex determination (Santos et al. 2007a). The position of the fetus during the ultrasound also can affect the visibility of the GT, leading to potential errors in sex determination. Finally, determining the sex outside the optimal window (55–70 days) can be challenging due to the development stage and positioning of the fetuses.

6 Determination of Fetal Age: Measures of Fetal Growth During Gestation

Ultrasonography represents a valuable tool for determining fetal age in goats, providing essential information for reproductive management, monitoring fetal development, and improving pregnancy outcomes. Accurate determination of fetal age is of great importance in the field of reproductive medicine, as it allows informed breeding and management decisions. Knowing gestational age allows for the prediction of calving dates, which in turn allows for better planning of subsequent breeding cycles. In addition, the ability to adjust nutritional plans according to the gestational stage ensures proper fetal development and maternal health (Jones and Reed 2017).

It is of utmost importance to continuously monitor fetal growth and development, as this helps to identify potential problems and ensure healthy pregnancies. Regular ultrasonographic measurements help to follow fetal growth patterns and detect any deviations from normal development (Wojtasiak et al. 2020). Early detection of abnormalities and identification of growth delays or anomalies at an early stage allows for timely interventions, which in turn improves pregnancy outcomes.

Ultrasonographic fetal aging has important applications in veterinary practice and research. Veterinarians use these measurements to diagnose and treat reproductive problems, thus improving overall herd health. Accurate estimation of fetal age is a fundamental aspect of reproductive research, contributing to advances in goat breeding and management practices (Abdelghafar et al. 2011).

Techniques such as transrectal and transabdominal ultrasonography, when combined with measurements of crown-rump length (CRL), trunk diameter (TD), biparietal diameter (BPD), orbit diameter (OD), and placentome size, among others, offer reliable estimates of gestational age throughout pregnancy (Erdogan 2012; Jones and Reed 2017; Wojtasiak et al. 2020). Despite the necessity for skilled operators and high-quality equipment, the advantages of precise fetal age determination demonstrate the significance of ultrasonography in contemporary veterinary reproductive care.

6.1   Crown-Rump Length

The crown-rump length (CRL) is a highly reliable indicator of fetal age, particularly during the early stages of gestation. The CRL is measured from the superior aspect of the fetal head (crown) to the inferior aspect of the tailbone (rump, end of sacrum) (Fig. 2a–d). This measurement is typically taken during the first trimester of pregnancy. The CRL provides a highly accurate estimation of fetal age in the early stages of pregnancy, with a low margin of error. In goats, transrectal CRL measurements have been observed to be 0.5 cm as early as day 19 (Martinez et al. 1998).

During the early stages of gestation, it has been observed that the CRL of singletons and multiples differs between days 32 and 42 in sheep (Jones et al. 2016). This indicates that offspring in multiple births are shorter than singletons. It is challenging to obtain accurate CRL measurements in goats after day 50 due to the length of the fetus exceeding the ultrasound screen (Karadaev et al. 2016). However, in Shiba and Saanen goats, CRL was measurable from 30 or 21 days of gestation up to day 70 (Kandiel et al. 2015; Del'Aguila-Silva et al. 2021), respectively.

In recent years, several researchers have conducted ultrasonographic studies using fetometry to determine the correlation between CRL and gestational age in

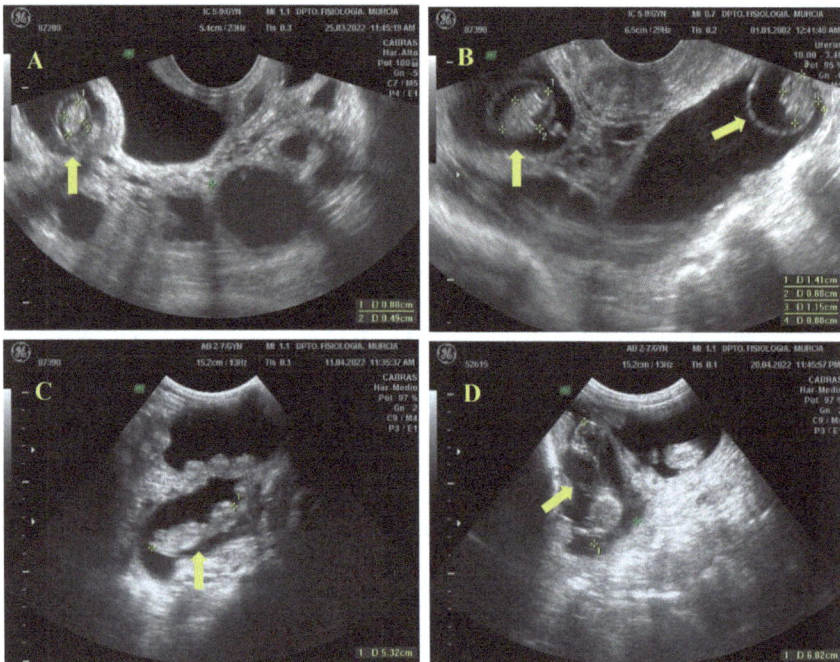

Fig. 2 Ultrasonographic images of measurements of CRL and TD (yellow arrows) in fetuses from Murciano-Granadina goats at different days ages of gestation: 25 (**a**), 31 (**b**), 45 (**c**), and 55 (**d**) days. Images (**a**) and (**b**) were obtained with an endovaginal probe. Images (**c**) and (**d**) were obtained with a transabdominal probe

various breeds of goats. Among the aforementioned studies, the following are worthy of particular mention.

A strong relationship between CRL and gestational age has been consistently observed during early gestation, regardless of the scanning approach ($R^2 = 0.94–0.95$) (Martinez et al. 1998; Karen et al. 2009). Abdelghafar et al. (2011) found a correlation ($R^2 = 0.90$) between CRL and gestational age in Saanen goats between the 5th and 10th weeks of gestation. In miniature Shiba goats, CRL was found to be highly correlated ($R^2 = 0.9848$) with gestational age (Kandiel et al. 2015).

Pati et al. (2016), by transabdominal ultrasonography, carried out 35–70 days of gestation, obtained results that clearly demonstrated that CRL was significantly correlated with gestational age ($R^2 = 0.988$) in Osmanabadi goats. Abubakar et al. (2016) observed a high correlation between CRL and gestational age in Jamnapari goats from day 28 to 100 of gestation, with a coefficient of determination $R^2 = 0.969$. Additionally, Rasheed (2017) reported that CRL was significantly and positively correlated with gestational age in goats during the first and second trimesters of pregnancy (5th–10th weeks) ($R^2 = 0.99$). Furthermore, Kuru et al. (2018) demonstrated that CRL is a highly effective predictor of gestational age in Abaza ($R^2 = 0.949$) and Gurcu goats ($R^2 = 0.942$).

In Beetal goats, Ejaz-ul-Haq et al. (2020) reported that CRL could be measured between days 34 and 65 postbreeding. They also demonstrated that there was a significantly high correlation between CRL and the gestational stage during this time ($R^2 = 0.93$). In Saanen goats, from 21 days up to day 70 of gestation, CRL was found to be highly correlated ($R^2 = 0.917$) with gestational age (Del'Aguila-Silva et al. 2021). Khand et al. (2021) conducted a study on Teddy goats from the third week to the 15th week of gestation, examining several fetal parameters. The results demonstrated a strong, positive correlation between CRL and gestational age ($R^2 = 0.98$). Muhammad and Aziz (2022) reported a strong correlation between gestational age and CRL ($R^2 = 0.917$) in Shami goats from 41 to 60 days. Recently, Qusay et al. (2023) obtained a very high correlation coefficient between embryo/fetal age and CRL ($R^2 = 0.97$) in different gestational ages (23–60 days).

6.2   Trunk Diameter

The diameter of the fetal trunk (TD) is also evaluated during ultrasonographic examination of pregnancy in goats. The diameter is determined in lateral imaging as the maximum length measured from the spine through the abomasum to the abdominal wall (Karadaev et al. 2016, 2018).

Kandiel et al. (2015) define the TD as the diameter measured at the height of the stomach and liver or the entry of the umbilical cord to the fetus. In the context of ultrasound imaging, the measurement of TD in goats is typically conducted in the transverse or sagittal plane (Gosselin et al. 2018). The initial TD measurement can be conducted from the 25th day of gestation (Ramírez-González et al. 2023) (Fig. 2a, b).

Suguna et al. (2008) reported a high correlation coefficient ($R^2 = 0.99$) between TD and gestational age. In Shiba goats, Kandiel et al. (2015) reported that TD demonstrated high predictive power for gestational age ($R^2 = 0.92$) and was detectable for a longer interval during pregnancy. Consequently, this parameter can be employed as a reliable estimator of the gestational stage. During the embryonic and fetal periods, there is a rapid growth and maturation of tissues and organs located in the abdominal and thoracic cavities, including the lungs, heart, liver, gastric chambers, intestines, and kidneys. This justifies the linear increase in the parameter throughout the gestational period.

Kuru et al. (2018) demonstrated that TD was one of the most effective predictors of gestational age for Abaza goats ($R^2 = 0.949$). Ejaz-ul-Haq et al. (2020) also reported that the diameter of the conceptus trunk increased linearly and demonstrated a significant correlation ($R^2 = 0.95$) with the gestational stage in Beetal goats. The latter results coincide with those obtained by Kumar et al. (2015c), who, in the same breed of goats, found that TD was highly correlated with gestational age.

In a similar study, Del'Aguila-Silva et al. (2021) observed a high correlation between TD and gestational age ($R^2 = 0.92$) in Saanen goats. The viewing interval was extended from day 42 until shortly before delivery on day 148. Also, a high correlation ($R^2 = 0.98$) was reported by Khand et al. (2021) from the third week to the 15th week of gestation in Teddy goats. A lower correlation ($R^2 = 0.812$) was recorded in a study between the measurements of TD and the gestational age in Shami goats (Muhammad and Aziz 2022). Finally, Qusay et al. (2023) reported that the correlation coefficient between embryo/fetal age and TD was recorded as very high ($R^2 = 0.98$) from 42 to 90 days of gestation.

6.3   Biparietal Diameter

The biparietal diameter (BPD) is a measurement that is routinely performed in the prenatal examination of pregnancy to estimate the birth date in several goat breeds

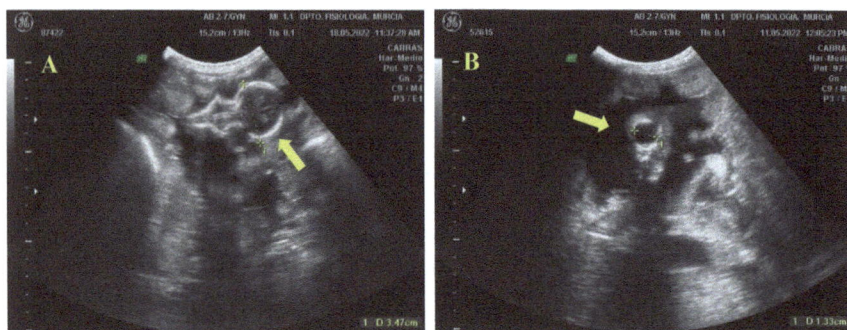

Fig. 3 Ultrasonographic images of measurements of biparietal diameter (BPD) (**a**) at day 82 of gestation and orbit diameter (OD) at day 75 of gestation (**b**) in fetuses from Murciano-Granadina goats (yellow arrows) obtained with a transabdominal probe

(Haibel et al. 1989). For its measurement, a symmetrical view of the fetal skull must be obtained This involves measuring the maximum oval-shaped size of the skull (looking at both eye sockets from the front), the visualized closed contour of the skull, and both well-visualized eye orbits. The BPD can be measured from one parietal end to the other (Lee et al. 2005; Ramírez-González et al. 2023) (Fig. 3a).

In a study by Nwaogu et al. (2010), which involved examining red Sokoto goat fetuses with known gestational ages ranging from 57 to 124 days, a high correlation coefficient ($R^2 = 0.98$) was found between BPD and gestational age. Abdelghafar et al. (2011) observed a strong positive correlation between BPD and gestational age ($R^2 = 0.91$) in Saanen goats between 6 and 23 weeks of gestation.

Abubakar et al. (2016) observed a highly correlated relationship between BPD and gestational age ($R^2 = 0.98$) in Jamnapari goats from day 28 to 100 of gestation. Rasheed (2017) reported that gestational age in goats can be accurately estimated by transabdominal ultrasonography, measuring fetal BPD during the first and second trimesters of pregnancy from 6 to 14 weeks ($R^2 = 0.95$). Finally, Ejaz-ul-Haq et al. (2020) demonstrated that BPD could be measured from days 34 to 125 post-breeding, with an increase from 2 to 4 months of gestation. Furthermore, they found a significant correlation between BPD and gestational stage ($R^2 = 0.83$).

6.4 Orbit Diameter

An estimation of the gestational stage, which depends on measuring the fetal eye by means of ultrasound, has been approved in animals and humans. The measurement of the eye orbit diameter (OD) must be taken once it is observed to have a rounded shape at its maximum size, thus measuring the diameter of the circle formed and should be assessed from the lateral view of the head when it is rounded and at maximal size (Ramírez-González et al. 2023) (Fig. 3b).

Fetal OD in goat fetuses can usually be determined at about the 49th day of gestation (Karadaev et al. 2016), although some authors suggest that earlier imaging is possible (Kandiel et al. 2015; Yazici et al. 2018). Variations in the time when OD measurements can be made may be due to differences in the characteristics of specific breeds.

Lee et al. (2005) reported a high correlation ($R^2 = 0.87$) between OD and gestational age in pregnant Korean black goats between days 60 and 135 of gestation. Also, Nwaogu et al. (2010) reported that OD is a reliable parameter for the estimation of gestational age in red Sokoto goat fetuses between 57 and 124 days with a high correlation ($R^2 = 0.92$). In miniature Shiba goats, the data recording of the fetal OD indicated a linear growth function with gestational stage with a high correlation ($R^2 = 0.9239$) between days 21 and 126 of gestation (Kandiel et al. 2015).

However, for Del'Aguila-Silva et al. (2021), OD showed a moderate coefficient of determination ($R^2 = 0.789$) in Saanen goats, a result similar to that obtained by Santos et al. (2018) in ovine fetuses for eye socket ($R^2 = 0.818$) between days 42 and 148 of gestation.

6.5   Placentome Size

Placentomes are vital structures in ruminant pregnancies, including goats, as they facilitate the exchange of nutrients and gases between the mother and the developing fetus. In the fetal period, the placentomes are perceived as concave surfaces directed towards the uterine lumen. As the pregnancy progresses, the concave shape of the placentomes results in isoechoic images compared to the uterine wall and presents as either a "C" (Fig. 4a) or "O" shape (Fig. 4b), depending on the angle. Furthermore, a visual reduction in the echogenicity of these structures is also observed throughout pregnancy (Del'Aguila-Silva et al. 2021).

The measurement of placentomes via ultrasonography could be a valuable tool for assessing fetal development and placental health during gestation. They can be detected and measured from approximately 30 days of gestation onwards and throughout the pregnancy. As the pregnancy progresses, the size and number of these structures increase. The changes in morphology result in an increase in diameter, which begins between days 32 and 40 and reaches its maximum by days 91–100 (Doize et al. 1997; Hussein 2017).

Transabdominal ultrasonography is the most widely used method for measuring placentomes in goats due to its noninvasive approach and comprehensive visualization capabilities. Transrectal ultrasonography is most effective from around 20 to 40 days of gestation when the placentomes are forming and can be closely examined. This allows for detailed imaging and measurement of the placentomes in the early stages of pregnancy.

The size and development stage of placentomes can help estimate the gestational age of the fetus, aiding in effective reproductive management. Doize et al. (1997) found in does a moderate correlation between gestational age and placentome diameter ($R^2 = 0.70$), and Nwaogu et al. (2010), in Red Sokoto goats, and Del'Aguila-Silva et al. (2021), in Saanen goats, reported a low correlation ($R^2 = 0.45$ and $R^2 = 0.543$, respectively). However, in Shiba goats, Kandiel et al. (2015) found a high correlation between gestational age and placentome diameter ($R^2 = 0.899$). In

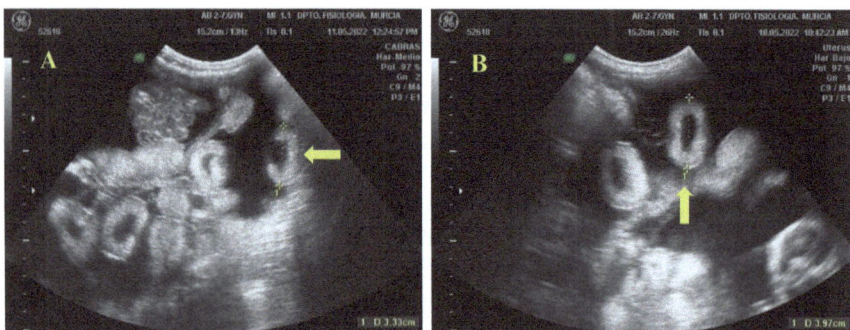

Fig. 4 Ultrasonographic images and sizes of placentomes at day 75 (**a**) and at day 88 of gestation (**b**) (yellow arrows) in fetuses from Murciano-Granadina goat obtained with a transabdominal probe

Egyptian native goats, placentome size was strongly correlated with gestational age between days 25–130 of gestation ($R^2 = 0.905$) (Karen et al. 2009) and also in Sahel Nigerian goats ($R^2 = 0.90$) (Waziri et al. 2017).

Suguna et al. (2008) reported that the diameter of the placentome was recorded from day 42 and demonstrated a significant increase, reaching a maximum diameter on day 130 of gestation. A significantly high correlation ($R^2 = 0.99$) was observed between the placentome diameter and gestational age. Rasheed (2016) found that the placentome reached its maximum diameter around day 126 and that the determination of gestational age according to placentome measurement is not reliable after day 90. Furthermore, the results indicated a significant difference in placentome diameter size between single and twin pregnancies.

Regular measurement of placentomes enables monitoring of fetal growth and development, allowing for early detection of potential issues and timely interventions. Individual variations in placentome development and size may affect the interpretation of measurements, requiring consideration of other fetal parameters for a comprehensive assessment.

6.6 Other Measures for Fetal Age Determination

In addition to the aforementioned measures for fetal age determination, some researchers have conducted trials and studies using other measures that may also be of interest and that sometimes have a high degree of correlation with gestational age. The following are some of the most interesting ones, although there are still some others to be analyzed.

- *Longitudinal and transverse axes of the heart.* The longitudinal (long) and transverse (short) axes of the fetal heart should be measured from a transverse section of the four-chambers view at the diastolic state during a period of absent fetal body movement. These axes can be measured from the 42nd day of gestation (Karen et al. 2009). Lee et al. (2005), in Korean black goats, found a high correlation between both long and short axes and gestational age ($R^2 = 0.9168$ and $R^2 = 0.8819$, respectively). Likewise, Kandiel et al. (2015), in Shiba goats, reported that the longitudinal and transverse axis of the goat fetal heart is an applicable estimate of fetal age as verified by its linear increment with gestational stage and its high correlation coefficients ($R^2 = 0.9382$ and 0.9589, respectively).
- *Femur length.* Some researchers have found good correlations between femur length and gestational age in goats. Thus, Kandiel et al. (2015) reported a high correlation with fetal age ($R^2 = 0.9278$). A similar correlation between femur length and gestation was reported in Egyptian Balady goats ($R^2 \geq 0.90$; Karen et al. 2009) between 55 and 125 days of gestation. A slightly higher relationship ($R^2 = 0.95$) between the femur length and gestational stage was obtained by Abdelghafar et al. (2012) in Saanen goats. Zongo et al. (2018) measured in Sahelian goats, the length of the tibia and femur from day 42 to 120 of gestation,

finding for both measures a good correlation with gestational age ($R^2 = 0.93$ and $R^2 = 0.92$, respectively). In Beetal goats, Ejaz-ul-Haq et al. (2020) reported that the femur length had a moderate value of the coefficient of determination ($R^2 = 0.83$) and was measurable between days 48 and 100 postbreeding. Finally, in Saanen goats, Del'Aguila-Silva et al. (2021) found a correlation between femur length and gestational age ($R^2 = 0.917$), extending the viewing interval from day 49 until day 119.

* *Fetal heart rate (FHR)*. The FHR during gestation is rapid before day 40 in the goat (200–250 beats/min; Karen et al. 2009) and decreases with the progression during caprine pregnancies (Serin et al. 2010; Ejaz-ul-Haq et al. 2020; Ramírez-González et al. 2023). Different correlations between FHR and fetal age were reported in Egyptian native goats ($R^2 = 0.55$) (Karen et al. 2009) and Saanen goats ($R^2 = 0.77$) (Serin et al. 2010). In this same breed, Del'Aguila-Silva et al. (2021) found a poor relationship between FHR and fetal age ($R^2 = 0.45$). Recently, Ejaz-ul-Haq et al. (2020) reported a medium value of the coefficient of determination between FHR and fetal age ($R^2 = 0.76$).

7   Doppler Ultrasound

Doppler ultrasonography is a highly effective imaging modality that has significantly advanced the field of veterinary reproductive medicine (Serin et al. 2010). The Doppler effect in ultrasound is based on the fact that when ultrasound waves collide with moving structures, the frequency of the reflected waves differs from that of the emitted waves, in contrast to when these waves collide with static surfaces, whose reflection is identical to that of the emitted waves by the transducer (Velasco and Ruiz 2021).

In the context of ultrasonography, the moving structures that are analyzed are the cells that circulate through the blood vessels. Consequently, there will be positive parameters if the frequency of the reflected waves is higher than those emitted when the blood cells move toward the transducer and negative if the cells move away from the transducer, so the frequency of the reflected waves is lower than those emitted (Bollwein et al. 2016) (Fig. 5).

Doppler can be continuous, without discriminating between depth field and pulsed wave, in which the depth of the studied field can be selected. Conversely, three methods have been described for representing the Doppler signal. Color Doppler enables the user to observe the change in Doppler frequency on a red-blue color scale, with red indicating an approach towards the transducer and blue an away movement (Fig. 5a–f). Power Doppler is more sensitive than color Doppler in indicating the presence of flow within a structure without providing information on the speed or direction of the flow. Finally, spectral Doppler analysis examines the velocity of vessel flow over a specific period, resulting in the generation of a waveform that represents the hemodynamics of the blood vessel (Rubio et al. 2014) (Fig. 5g–h).

Doppler ultrasonography is a fundamental tool for evaluating ovarian function and blood flow, which are key indicators of follicular activity and ovulation. The vascularization of the CL can be evaluated using Doppler techniques, which provide

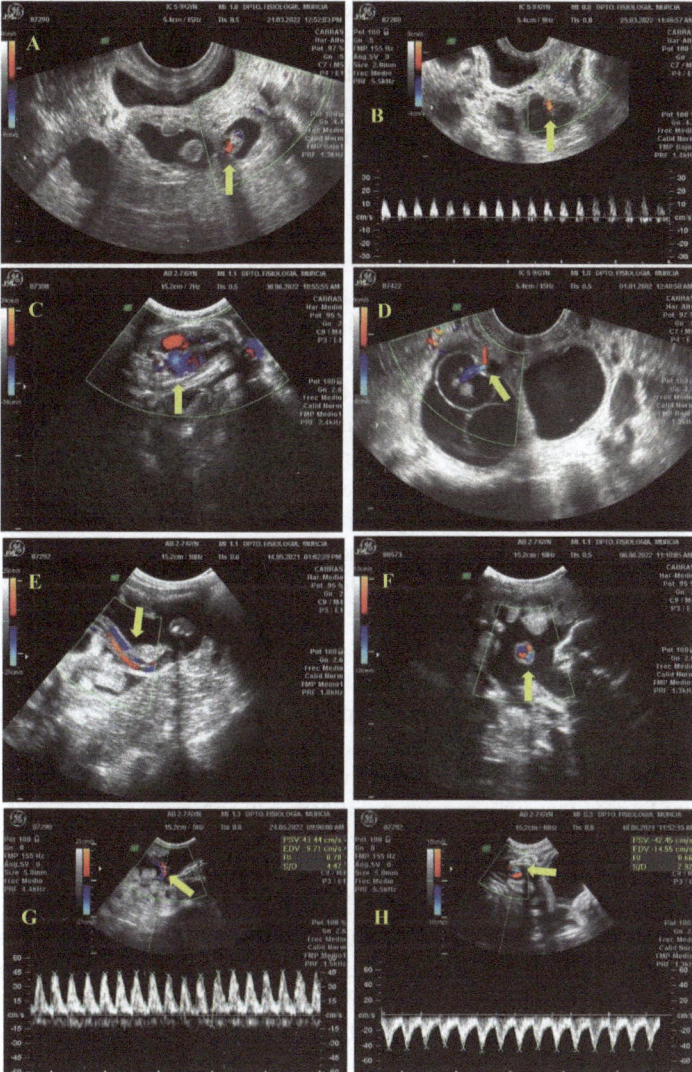

Fig. 5 Ultrasound images of color and spectral Doppler in fetuses from Murciana-Granadina goats at different days of gestation. (**a**) Ultrasonographic image obtained with an endovaginal probe at day 25 of gestation. The embryo is visible within the amniotic sac, and the fetal heartbeat is identifiable by the color Doppler (yellow arrow). (**b**) Determination of the fetal heart rate (FHR) at 25 days of gestation with an endovaginal probe by color and spectral Doppler. (**c**) Ultrasonographic image obtained with a transabdominal probe by color Doppler of the fetal heart at 125 days of gestation. The heart chambers with the blood flow inside can be clearly observed. (**d**) The umbilical cord (yellow arrow) of the embryo (32 days of gestation) was identified at an early stage using an endovaginal probe. (**e**) Longitudinal view obtained with a transabdominal probe of the umbilical cord by color Doppler ultrasound (yellow arrow). (**f**) Sectorial view obtained with a transabdominal probe of the umbilical cord by color Doppler ultrasound (yellow arrow). (**g**) Umbilical arterial (yellow arrow) flow waves by color and spectral Doppler, where blood flow is directed towards the transducer and are shown with positive values above baseline. (**h**) In this image, the direction of flow is away from the transducer, and the values are negative, below the baseline

insights into luteal function and progesterone production (Arashiro et al. 2018). Doppler ultrasonography facilitates the diagnosis and monitoring of pregnancy by providing detailed information about fetal and placental blood flow.

The presence of increased blood flow to the uterus and the detection of embryonic and fetal heartbeats in early pregnancy serve to confirm the existence of a gestational period. Doppler ultrasonography permits the assessment of FHR and blood flow in major vessels, thereby facilitating the monitoring of fetal well-being (Fig. 5a–c). The assessment of placental blood flow can provide insights into the health and function of the placenta, which are crucial for fetal development. Doppler ultrasonography is a valuable tool for evaluating uterine and placental blood flow, which are essential for maintaining a healthy pregnancy.

Doppler ultrasonography is a valuable diagnostic tool for the identification of various reproductive pathologies through the detection of abnormal blood flow patterns. An increase in blood flow to the uterine wall may indicate the presence of inflammation or infection, such as endometritis. Doppler ultrasound is useful in differentiating between active and inactive ovarian cysts based on their vascularization. Abnormal vascular patterns may indicate the presence of tumors or other pathological conditions in the reproductive tract (Bartlewski 2019). Doppler technology has been employed to investigate vessel flows, including those of the uteroplacental arteries, umbilical cord, aorta, caudal vena cava, and fetal venous ductus (Serin et al. 2010).

The application of ultrasound analysis using Doppler technology in goats has not been extensively studied, nor have Doppler parameters in abnormal pregnancies (Kumar et al. 2015a). In small ruminants, Doppler ultrasound studies of the waveforms in umbilical artery blood flow have been conducted with the objective of identifying fetuses that may require monitoring during pregnancy or planning for delivery. Conversely, umbilical artery pulsations have been extensively employed as a marker of cardiac functionality and fetal health and development (Ramírez-González et al. 2023) (Fig. 5c–f).

The umbilical cord in small ruminants is described as floating in the amniotic fluid that surrounds the fetus. It is formed by a bundle of blood vessels comprising two arteries and two umbilical veins, arranged in a characteristic spiral position (Elmetwally and Meinecke-Tillmann 2018) (Fig. 5e–f). The umbilical cord serves to connect the developing fetus with the maternal placenta. The umbilical cord serves to transport oxygenated blood and nutrients from the mother through the umbilical veins while simultaneously removing waste materials and deoxygenated blood for elimination. This process assists the fetal metabolism (Kellow and Feldstein 2011). Studies have demonstrated that the vascular system in the umbilical cord adapts to hemodynamic changes to ensure a blood supply for the developing placenta and fetus (Kumar et al. 2015a; Ramírez-González et al. 2023).

To study the correct functioning of the placenta and the well-being of the fetus, a series of parameters have been analyzed by Doppler ultrasonography. These parameters provide information on the optimal course of pregnancy (Schmidt et al. 1991). Initially, the study of blood flow in the umbilical vessels relied on invasive Doppler measurements of the cord. Nevertheless, the advancement and refinement of color Doppler ultrasonography techniques have permitted the noninvasive study of blood flow (Elmetwally and Meinecke-Tillmann 2018).

The principal Doppler parameters of blood flow typically investigated are peak systolic velocity (PSV), end-diastolic velocity (EDV), mean velocity (MV), the systolic velocity/diastolic velocity ratio (S/D) and the Doppler indices, resistance index (RI), and pulsatility index (PI) (Wojtasiak et al. 2022; Ramírez-González et al. 2023). PSV represents the maximum velocity of blood flow through the lumen of the examined vessel during one systolic phase. EDV represents the blood velocity at the end of a cardiac cycle, just before the onset of systole. Finally, the MV is the speed of the blood that passes through the interior of the vessels at a specific moment (Irion and Clark 1990; Fasulkov et al. 2021; Martínez-Díaz et al. 2022) (Fig. 5g–h).

Because of the inherent difficulties in analyzing blood flow velocity data, indices such as RI, PI, and S/D have been employed as more refined elements in this context. The RI is defined as the ratio between the systolic and diastolic flow velocities, and it is used to assess the resistance encountered by the blood flow as it traverses the lumen of the vessels. The maximum value of the blood flow is assumed in this calculation. The RI has been described as applicable to those vessels whose blood flow persists during diastole. The PI is a ratio that relates the PSV and the EDV to the MV during the cardiac cycle. It represents the speed of blood flow, which is suitable for vessels in which the flow is absent during diastole (Serin et al. 2010; Bollwein et al. 2016; Wojtasiak et al. 2022; Ramírez-González et al. 2023). Both the RI and the PI are influenced by the heart rate, with an increase observed when blood perfusion is reduced Finally, the S/D ratio is a well-studied measure of placental resistance in the blood flow circulation (Wright and Ridgway 1990) (Fig. 5g–h).

The analysis of umbilical artery blood flow waveforms, including vascular pulsatility and maternal–fetal vessel resistance, has become a standard procedure in the assessment of fetal well-being. This is due to the close relationship between these parameters and the potential for pregnancy-related alterations. Nevertheless, the accuracy of Doppler measurements can be limited due to the small size of the vessels or the distance between the transducer and the vessel (Schmidt et al. 1991).

8   Three-Dimensional (3D) Ultrasonography in Goat Reproduction

Three-dimensional (3D) ultrasonography represents a significant advancement in veterinary imaging, offering detailed and comprehensive visualization of reproductive structures. In the field of goat reproduction, 3D ultrasonography offers valuable insights into fetal development, placental health, and overall reproductive management (Karadaev et al. 2019) (Fig. 6).

3D ultrasonography entails the acquisition of multiple 2D images, which are then reconstructed into a 3D model using specialized software. This technique provides a volumetric representation of the reproductive structures, thereby enhancing the depth and detail of the imaging (Kumar et al. 2015b).

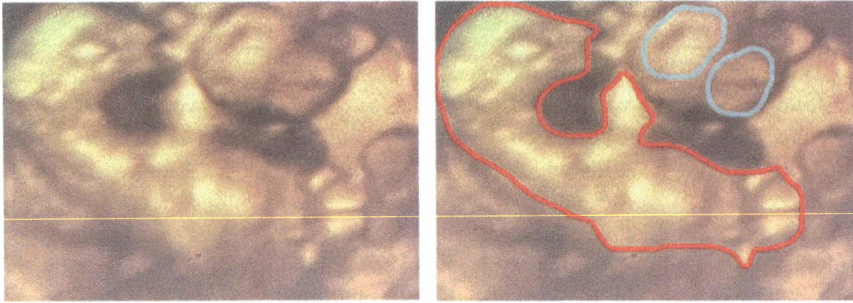

Fig. 6 Caprine fetus of Murciano Granadina goat in 3D (red). Placentomes in 3D (blue)

A conventional 2D ultrasound transducer is employed to capture a series of sequential images. The images are collected either manually or automatically by moving the transducer in a controlled manner across the area of interest. Specialized software processes the 2D images and constructs a 3D model, which allows for manipulation and viewing from various angles. This model provides a comprehensive overview of the reproductive anatomy and any pathological changes. Advanced systems offer real-time 3D imaging, also known as four-dimensional (4D) ultrasonography. This technology continuously updates the 3D model to reflect real-time movements and changes (Chandolia and George 2011).

It is necessary to utilize modern ultrasound machines that are equipped with 3D capabilities. The machines are equipped with advanced transducers and software for image reconstruction. The imaging software must be capable of handling large data-sets and rendering detailed 3D models. It is, therefore, essential that the ultrasound machine in question is equipped with the requisite software for volume rendering, multiplanar reconstruction, and surface rendering to facilitate a thorough analysis (Galián et al. 2021a).

The use of 3D ultrasound facilitates a comprehensive assessment of fetal development by providing detailed images of the fetus from a variety of perspectives, allowing a detailed assessment of fetal morphology, including the detection of structural abnormalities and congenital defects (Markov and Dimitrova 2006).

By measuring various fetal parameters in 3D, veterinarians can more accurately monitor growth patterns and ensure that the fetus is developing normally. 3D ultrasound can accurately measure the volume of the placenta, providing a complete understanding of placental health and function than 2D measurements. The technology can detect abnormalities in the shape and structure of the placenta, which can indicate potential problems with fetal nutrition and development.

3D ultrasound provides a comprehensive view of the entire reproductive tract, enabling better diagnosis and treatment of reproductive disorders. Detailed 3D images help to assess ovarian structures, follicles, and uterine health, which is vital for breeding management and diagnosis of reproductive problems. In the postpartum, 3D ultrasound helps assess the uterus for retained placenta or other complications. It allows early detection of fetal and placental abnormalities, facilitating timely intervention (Markov and Dimitrova 2006).

The introduction of 3D ultrasound represents a significant advance in veterinary practice, setting new standards in reproductive imaging and care. Training in 3D ultrasound allows veterinarians to stay at the forefront of technological advances, contributing to their professional development and expertise (Galián et al. 2021a).

Despite its benefits, the use of 3D ultrasound in goat reproduction presents some challenges. High-quality 3D ultrasound equipment is expensive, and not all veterinary practices may have access to this technology. In addition, accurate 3D imaging and interpretation require considerable training and expertise, and, finally, image acquisition and reconstruction can be time-consuming and require patience and precision from the operator.

9 Final Considerations

Currently, ultrasonography represents an indispensable component of reproductive management in small ruminants. Continued advances are expanding the scope of applications and enhancing the effectiveness of ultrasonography. The integration of AI in ultrasonography is a rapidly developing field of research. AI algorithms are being developed with the objective of assisting in image analysis, improving diagnostic accuracy, and reducing operator dependence. The utilization of telemedicine for remote ultrasound consultations is on the rise, facilitating access to expert veterinary care in remote or underserved areas. The latest innovations include highly portable and even handheld ultrasound devices that connect to smartphones or tablets. This facilitates reproductive monitoring, thereby making it more accessible and convenient for veterinarians and farmers alike.

Acknowledgments The author would like to express gratitude to all members of the staff at IMIDA Murcia and the Veterinary Teaching Farm of the Veterinary Faculty of Murcia University.

Competing Interests The author has no conflicts of interest to declare that are relevant to the content of this chapter. All figures in this chapter are the property of the author.

References

Abdelghafar RM, Bakhiet AO, Ahmed BH (2007) B-mode real-time ultrasonography for pregnancy diagnosis and fetal number in Saanen goats. J Anim Vet Adv 6(5):702–705
Abdelghafar RM, Ahmed BH, Ibrahim MT, Mantis P (2011) Prediction of gestational age by transabdominal real-time ultrasonographic measurements in Saanen goats (Capra hircus). Global Veterinaria 6(4):346–351
Abdelghafar RM, Ahmed BH, Abdelrahim MS, Ibrahim MT (2012) The accuracy of gestational age predicted from femur and humerus length in Saanen goat using ultrasonography. Acta Vet Brno 81:295–299. https://doi.org/10.2754/avb201281030295

Abubakar F, Kari A, Ismail Z, Rashid BA, Haruna UT (2016) Accuracy of transrectal ultrasonography: in estimating the gestational age of Jamnapari goats. Malays Appl Biol J 45:49–54

Amer HA (2008) Determination of first pregnancy and foetal measurements in Egyptian Baladi goats (Capra hircus). Vet Ital 44(2):429–437

Amer HA (2010) Ultrasonographic assessment of early pregnancy diagnosis, fetometry and sex determination in goats. Anim Reprod Sci 117:226–231. https://doi.org/10.1016/j.anireprosci.2009.05.015

Andrabi SEH, Gulavane SU (2015) Real time ultrasonography for determining viable foetal numbers in goats. J Anim Res 5(4):875–878. https://doi.org/10.5958/2277-940X.2015.00145.X

Anya KO, Ekere SO, Ogwu DO (2017) Early pregnancy diagnosis using trans-abdominal ultrasonography in West African dwarf goats. Niger Vet J 8(4):311–318. https://doi.org/10.4314/nvj.v38i4.6

Arashiro EKN, Ungerfeld R, Clariget RP, Pinto PHN, Balaro MFA, Bragança GM, Brandão FZ (2018) Early pregnancy diagnosis in ewes by subjective assessment of luteal vascularisation using colour Doppler ultrasonography. Theriogenology 106:247–252. https://doi.org/10.1016/j.theriogenology.2017.10.029

Azevedo EM, Dos Santos MH, Filho CR, Freitas Neto LM, Bezerra FQ, Neves JP, Lima PF, Oliveira MAL (2009) Migration time of the genital tubercle in caprine and ovine fetuses: comparison between breeds, sexes and species. Acta Vet Hung 57(1):147–154. https://doi.org/10.1556/avet.57.2009.1.15

Bartlewski P (2019) Applications of Doppler ultrasonography in reproductive health and physiology of small ruminants. Revista Brasileira de Reprodução. Animal 43(2):122–125

Bollwein H, Heppelmann M, Lüttgenau J (2016) Ultrasonographic Doppler use for female reproduction management. Vet Clin N Am Food Anim Pract 32:149–164. https://doi.org/10.1016/j.cvfa.2015.09.005

Chandolia RK, George J (2011) Application of 3D/4D ultrasonography in canine gyaecology and obstetrics. Indian J Canine Pract 3(2):99–102

Del'Aguila-Silva P, Cirino dos Santos F, Correia Santos VJ, Rodrigues Simões AP, Ramirez Uscategui RA, Padilha-Nakaghi LC, Amoroso L, Russiano Vicente WR, Rossi Feliciano MA (2021) B-mode ultrasound and ecobiometric parameters to assess embryonic and fetal development and estimate gestational age in goats. Theriogenology 75:123–133. https://doi.org/10.1016/j.theriogenology.2021.09.002

Doize F, Vaillancourt D, Carabin H, Belanger D (1997) Determination of gestational age in sheep and goats using transrectal ultrasonographic measurement of placentomes. Theriogenology 48(3):449–460. https://doi.org/10.1016/s0093-691x(97)00254-9

Ejaz-ul-Haq M, Hameed N, Khan MIR, Abbas Q, Sohail T, Rehman A, Moshin I (2020) Temporal changes in physical signs of estrus and validation of fetal parameters for estimation of gestational stage through B-mode ultrasonography in Beetal goats. Pak Vet J 40(4):425–430. http://pvj.com.pk/pdf-files/40_4/425-430.pdf

Elmetwally MA, Meinecke-Tillmann S (2018) Simultaneous umbilical blood flow during normal pregnancy in sheep and goat foetuses using non-invasive colour Doppler ultrasound. Anim Reprod 15(2):148–155. https://doi.org/10.21451/1984-3143-AR2017-976

Erdogan G (2012) Ultrasonic assessment during pregnancy in goats—a review. Reprod Domestic Anim 47(1):157–163. https://doi.org/10.1111/j.1439-0531.2011.01873.x

Fasulkov IR, Karadaev M, Vasilev N, Hristov K, Fedev I (2021) Doppler ultrasound measurements of the blood flow velocity in the fetal heart and aorta in Bulgarian White milk goats. Vet Med Sci 7(4):1297–1302. https://doi.org/10.1002/vms3.463

Galián S, Peinado B, Ruiz S, Poto A, Almela L (2021a) Uso de la ecografía para el diagnóstico y seguimiento de la gestación en la cabra Murciano-Granadina. Ed. Académica Española. ISBN: 978-620-3-87147-0

Galián S, Peinado B, Ruiz S, Poto A, Almela L, Castillo J, Lozano S (2021b) Ultrasound of gestation in Murciano Granadina goats. Archivos de Zootecnia 70(269):104–111. https://doi.org/10.21071/az.v70i269.5424

Ginther OJ (2014) How ultrasound technologies have expanded and revolutionized research in reproduction in large animals. Theriogenology 81:112–125. https://doi.org/10.1016/j.theriogenology.2013.09.007

Gosselin VB, Volkmann DH, Dufour S, Middleton JR (2018) Use of ultrasonographic fetometry for the estimation of days to kidding in dairy does. Theriogenology 118:22–26. https://doi.org/10.1016/j.theriogenology.2018.05.041

Haibel GK (1990) Use of ultrasonography in reproductive management of sheep and goat herds. Vet Clin N Am Food Anim Pract 6(3):597–613. https://doi.org/10.1016/S0749-0720(15)30835-5

Haibel GK, Perkins NR, Lidl GM (1989) Breed differences in biparietal diameters of second trimester Toggenbubrg, Nubian and Angora goat fetuses. Theriogenology 32:827–834. https://doi.org/10.1016/0093-691X(89)90471-8

Hussein K (2017) Detection of single and multiple pregnancy depending on placentomes measurement in Shami goats in Iraq by ultrasonography. Iraqi J Vet Med 41(2):118–123. https://doi.org/10.30539/iraqijvm.v41i2.60

Irion GL, Clark KE (1990) Direct determination of the ovine fetal umbilical artery blood flow waveform. Am J Obstet Gynecol 162:541–549. https://doi.org/10.1016/0002-9378(90)90426-8

Jones AK, Reed SA (2017) Benefits of ultrasound scanning during gestation in the small ruminant. Small Rumin Res 149:163–171. https://doi.org/10.1016/j.smallrumres.2017.02.008

Jones AK, Gately RE, McFadden KK, Zinn SA, Govoni KE, Reed SA (2016) Transabdominal ultrasound for detection of pregnancy, fetal and placental landmarks, and fetal age before day 45 of gestation in the sheep. Theriogenology 85(5):939–945, e931. https://doi.org/10.1016/j.theriogenology.2015.11.002

Kandiel MM, Watanabe G, Taya K (2015) Ultrasonographic assessment of fetal growth in miniature Shiba goats (Capra hircus). Anim Reprod Sci 162:1–10. https://doi.org/10.1016/j.anireprosci.2015.08.007

Karadaev M, Fasulkov I, Vassilev N, Petrova Y, Tumbev A, Petelov Y (2016) Ultrasound monitoring of the first trimester of pregnancy in local goats through visualisationand measurements of some biometric parameters. Bulgarian J Vet Med 19(3):209–217. https://doi.org/10.15547/bjvm.909

Karadaev M, Fasulkov I, Yotov S, Atanasova S, Vasilev N (2018) Determination of the gestational age through ultrasound measurements of some uterine and foetal parameters in Bulgarian local goats. Reprod Domestic Anim 53(6):1456–1465. https://doi.org/10.1111/rda.13305

Karadaev M, Fasulkov I, Vasilev N, Hristov K, Fedev I (2019) Three-dimensional (3D) ultrasound investigations for monitoring of the second and third pregnancy trimester in goats. Tradition and modernity. Vet Med 4(2):72–76

Karen AM, Fattouh ESM, Abu-Zeid SS (2009) Estimation of gestational age in Egyptian native goats by ultrasonographic fetometry. Anim Reprod Sci 114(1–3):167–174. https://doi.org/10.1016/j.anireprosci.2008.08.016

Karen AM, Samir H, Ashmawy T, El-Sayed M (2014) Accuracy of B-mode ultrasonography for diagnosing pregnancy and determination of fetal numbers in different breeds of goats. Anim Reprod Sci 147(1–2):25–31. https://doi.org/10.1016/j.anireprosci.2014.03.014

Kellow ZS, Feldstein VA (2011) Ultrasound of the placenta and umbilical cord: a review. Ultrasound Q 27(3):187–197. https://doi.org/10.1097/RUQ.0b013e318229ffb5

Khand FM, Kachiwal AB, Laghari ZA, Lakho SA, Khattri P, Soomro SA, Korejo NA, Leghari A (2021) Early pregnancy diagnosis and fetometry by real-time ultrasonography in teddy goat. Pak J Zool 53(3):853–858. https://doi.org/10.17582/journal.pjz/20190315060355

Koker A, Ince D, Sezik M (2012) The accuracy of transvaginal ultrasonography for early pregnancy diagnosis in Saanen goats: a pilot study. Small Rumin Res 105:277–281. https://doi.org/10.1016/j.smallrumres.2012.02.013

Kumar K, Chandolia RK, Kumar S, Jangir T, Luthra RA, Kumari S, Kumar S (2015a) Doppler sonography for evaluation of hemodynamic characteristics of fetal umbilicus in Beetal goats. Vet World 8(3):412–416. https://doi.org/10.14202/vetworld.2015.412-416

Kumar K, Chandolia RK, Kumar S, Pal M, Kumar S (2015b) Two-dimensional and three-dimensional ultrasonography for pregnancy diagnosis and antenatal fetal development in Beetal goats. Vet World 8(7):835–840. https://doi.org/10.14202/vetworld.2015.835-840

Kumar K, Chandolia RK, Kumar S, Pal M, Kumar S, Pandey AK (2015c) Prediction of gestational age in Beetal goats by ultrasonic fetometry. Indian J Small Ruminants 27:35–44. https://doi.org/10.5958/0973-9718.2015.00070.7

Kuru M, Oral H, Kulaksiz R (2018) Determination of gestational age by measuring defined embryonic and foetal indices with ultrasonography in Abaza and Gurcu goats. Acta Vet Brno 87:357–362. https://doi.org/10.2754/avb201887040357

Lee Y, Lee O, Cho J, Shin H, Choi Y, Shim Y, Choi W, Shin H, Lee D, Lee G, Shin S (2005) Ultrasonic measurement of fetal parameters for estimation of gestational age in Korean black goats. J Vet Med Sci 67(5):497–502. https://doi.org/10.1292/jvms.67.497

Mali AB, Amle MB, Markandeya NM, Kumawat BL (2019) Comparative efficacy of trans-rectal and trans-abdominal ultrasonography for early diagnosis of pregnancy and embryonic ageing in goats. Indian J Small Ruminants 25:171–175. https://doi.org/10.5958/0973-9718.2019.00046.1

Markov D, Dimitrova V (2006) Prenatal diagnosis of structural anomalies of the fetus by 3D/4D ultrasound. Akusherstvo i Ginecologiia 45(3):32–36

Martinez MF, Bosch P, Bosch RA (1998) Determination of early pregnancy and embryonic growth in goats by transrectal ultrasound scanning. Theriogenology 49:1555–1565. https://doi.org/10.1016/S0093-691X(98)00101-0

Martínez-Díaz S, García-Vázquez FA, Luongo C, Ruiz S (2022) Assessment of arterial flow of the umbilical cord in goat fetuses of the Murciano-Granadina race using spectral doppler ultrasonography. Anales de Veterinaria de Murcia 36:513941. https://doi.org/10.6018/analesvet.513941

Medan MS, Abd El-Aty A (2010) Advances in ultrasonography and its applications in domestic ruminants and other farm animal reproduction. J Adv Res 1(2):123–128. https://doi.org/10.1016/j.jare.2010.03.003

Medan M, Watanabe G, Absy G, Sasaki K, Sharawy S, Taya K (2004) Early pregnancy diagnosis by means of ultrasonography as a method of improving reproductive efficiency in goats. J Reprod Dev 50(4):391–397. https://doi.org/10.1262/jrd.50.391

Muhammad RS, Aziz DM (2022) Estimation of gestational age in Shami goats based on transabdominal ultrasonographic measurements of fetal parameters. Iraqi J Vet Sci 36(4):839–846. https://vetmedmosul.com/article_173308.html

Nwaogu IC, Anya KO, Agada PC (2010) Estimation of foetal age using ultrasonic meaurements of different foetal parameters in red Sokoto goats (Capra hircus). Veterinarski Arhiv 80:225–233

Padilla-Rivas GR, Sohnrey B, Holtz W (2005) Early pregnancy detection by real-time ultrasonography in Boer goats. Small Rumin Res 58:87–92. https://doi.org/10.1016/j.smallrumres.2004.09.004

Pati P, Sahatpure SK, Patil AD, Kumar U, Jena B, Nahak AK, Sharma LP (2016) Prediction of gestational age in Osmanabadi goats by ultrasonic measurement of crown-rump length. Indian J Vet Sci Biotechnol 12(2):75–78. https://doi.org/10.21887/ijvsbt

Philip LM, Abhilash RS, Francis BP (2017) Accuracy of transvaginal ultrasonography for early pregnancy diagnosis in Attappady black goats. Malays J Vet Res 8:35–41

Qusay A, Laith Y, Hayman A (2023) The effectiveness of ultrasonography in predicting pregnancy, litter size, viability and embryo/fetal age during first three months post-breeding in Iraqi Shami does. Univ Thi-Qar J Agric Res 12(2):237–228. https://doi.org/10.54174/utjagr.v12i2.286

Ramírez-González D, Poto A, Peinado B, Almela L, Navarro-Serna S, Ruiz S (2023) Ultrasonography of pregnancy in Murciano-Granadina goat breed: fetal growth indices and umbilical artery Doppler parameters. Animals 13(4):618. https://doi.org/10.3390/ani13040618

Rasheed YM (2016) Ultrasonic estimation of gestation age in goats via placentomes diameter. Iraqi J Vet Med 40(2):100–106. https://doi.org/10.30539/iraqijvm.v40i2.120

Rasheed YM (2017) Assessment of gestational age in goats by real-time ultrasound measuring the fetal crown-rump length, and bi-parietal diameter. Iraqi J Vet Med 41(2):106–112. https://doi.org/10.30539/iraqijvm.v41i2.58

Rubio I, Tirapu M, Gómez H, Zabalza J (2014) Ecografía Doppler: principios básicos y guía práctica para residentes. Eur Soc Radiol. https://doi.org/10.1594/seram2014/S-0379

Santos MHB, Rabelo MC, Filho CRR, Dezzoti CH, Reichenbach HD, Neves JP, Lima PF, Oliveira MA (2007a) Accuracy of early fetal sex determination by ultrasonic assessment in goats. Res Vet Sci 83:251–255. https://doi.org/10.1016/j.rvsc.2006.12.001

Santos MHB, Moraes EP, Bezerra FQ, Moura RT, Paula-Lopes F, Neves JP, Lima PF, Oliveira MAL (2007b) Early fetal sexing of Saanen goats by use of transrectal ultrasonography to identify the genital tubercle and external genitalia. Am J Vet Res 68:561–564. https://doi.org/10.2460/ajvr.68.5.561

Santos MHB, Moura RTD, Chaves RM, Soares AT, Neves JP, Reichenbach HD, Lima PF, Oliveira MAL (2018) B-mode ultrasonography and ecobiometric parameters for assessment of embryonic and fetal development in sheep. Anim Reprod Sci 197:193–202. https://doi.org/10.1016/j.anireprosci.2018.08.028

Schmidt KG, Di Tommaso M, Silverman NH, Rudolph AM (1991) Doppler echocardiographic assessment of fetal descending aortic and umbilical blood flows: validation studies in fetal lambs. Circulation 83:1731–1737. https://doi.org/10.1161/01.CIR.83.5.1731

Serin G, Gökdal Ö, Tarimcilar T, Atay O (2010) Umbilical artery Doppler sonography in Saanen goat fetuses during singleton and multiple pregnancies. Theriogenology 74(6):1082–1087. https://doi.org/10.1016/j.theriogenology.2010.05.005

Suguna K, Mehrotra S, Agarwal SK, Hoque M, Singh SK, Shanker U, Sarath T (2008) Early pregnancy diagnosis and embryonic and fetal development using real time B mode ultrasound in goats. Small Rumin Res 80:80–86. https://doi.org/10.1016/j.smallrumres.2008.10.002

Velasco A, Ruiz S (2021) New approaches to assess fertility in domestic animals: relationship between arterial blood flow to the testicles and seminal quality. Animals 11:12. https://doi.org/10.3390/ani11010012

Vinoles-Gil C, Gonzalez-Bulnes A, Martin GB, Zlatar FS, Sale S (2010) Sheep and goats. In: DesCoteaux L, Colloton J, Gnemmi G (eds) Ruminant and camelid reproductive ultrasonography. Willey-Blackwell, Hong Kong, pp 181–210

Waziri MA, Ikpe AB, Bukar MM, Ribadu AY (2017) Determination of gestational age through trans-abdominal scan of placentome diameter in Nigerian breed of sheep and goats. Sokoto J Vet Sci 15(2):49–53. https://doi.org/10.4314/sokjvs.v15i2.7

Wojtasiak N, Stankiewicz T, Udała J (2020) Ultrasound examination of pregnancy in the domestic goat (Capra hircus) - a review. Sci Ann Polish Soc Anim Prod 16(2):65–78. https://doi.org/10.5604/01.3001.0014.2019

Wojtasiak N, Stankiewicz T, Błaszczyk B, Udała J (2022) Ultrasound parameters of embryo-fetal morphometry and Doppler indices in the umbilical artery during the first trimester of pregnancy in goats. Pak Vet J

Wright JW, Ridgway LE (1990) Sources of variability in umbilical artery systolic/diastolic ratios: implications of the Poiseuille equation. Am J Obstetrics Gynecol 163(6):1788–1791. https://doi.org/10.1016/0002-9378(90)90750-2

Yazici E, Ozenc E, Celik HA, Ucar M (2018) Ultrasonographic foetometry and maternal serum progesterone concentrations during pregnancy in Turkish Saanen goats. Anim Reprod Sci 197:93–05. https://doi.org/10.1016/j.anireprosci.2018.08.017

Zongo M, Kimsé M, Kulo EA, Sanou D (2018) Fetal growth monitoring using ultrasonographic assessment of femur and tibia in Sahelian goats. J Anim Plant Sci 36(1):5763–5768

Ultrasonographic Pregnancy Diagnosis in the Mare

Duccio Panzani ⓘ, Juan Cuervo-Arango ⓘ, and Diana Fanelli ⓘ

Abstract

Pregnancy diagnosis is crucial in managing horse reproduction, influencing breeding strategies, mare health monitoring, and herd productivity. Accurate diagnosis ensures effective resource allocation and appropriate care for pregnant mares. Ultrasonography, a noninvasive imaging technique, uses sound waves to create body images. In equine reproductive management, it visualizes the conceptus and detects abnormalities during the embryonic and fetal stages. One key advantage is early pregnancy detection, enabling timely care and management. Ultrasonography also assesses embryo viability and monitors fetal development, identifying potential issues for prompt intervention. While ultrasonography is a powerful tool, its effectiveness depends on the operator's skill and the mare's condition. Practitioners must tailor their diagnostic approach to individual cases. Despite its strengths, understanding its limitations is crucial. By enabling early pregnancy detection, regular monitoring, and identifying potential issues, ultrasonography significantly contributes to reproductive success. It ensures the health and welfare of the mare and her offspring, ultimately enhancing overall equine well-being. In conclusion, ultrasonography is vital in equine reproductive management, playing a key role in pregnancy diagnosis, monitoring, and promoting equine welfare. Effective use of this tool can greatly enhance the success of equine reproduction efforts.

D. Panzani (✉) · D. Fanelli
Department of Clinical Sciences, Università di Pisa, Pisa, Italy
e-mail: duccio.panzani@unipi.it; diana.fanelli@unipi.it

J. Cuervo-Arango
Department of Animal Medicine and Surgery, Universidad CEU-Cardenal Herrera, CEU Universities, Alfara del Patriarca, Valencia, Spain
e-mail: juan.cuervo@uchceu.es

© The Author(s), under exclusive license to Springer Nature Switzerland AG 2024
J. C. Gardón, K. Satué Ambrojo (eds.), *Assisted Reproductive Technologies in Animals Volume 1*, https://doi.org/10.1007/978-3-031-73079-5_2

Keywords

Mare · Ultrasound · Pregnancy diagnosis · Embryo development · Fetal growth

1   Introduction

Ultrasonographic pregnancy diagnosis in mares is a cornerstone of modern equine reproductive management. The ability to accurately detect and monitor pregnancy is essential for effective breeding programs, ensuring the health and productivity of both individual mares and entire herds. The introduction of ultrasonography has revolutionized this field, providing veterinarians and breeders with a powerful tool to visualize the conceptus and assess the reproductive health of mares throughout gestation.

Ultrasonography, also known as ultrasound imaging, is a noninvasive diagnostic technique that uses high-frequency sound waves to produce images of the internal structures of the body. In the context of equine reproduction, it is primarily used to visualize the uterus and its contents, allowing for the detection of pregnancy as well as the assessment of embryonic and fetal development. The technique involves the use of a transducer, which emits sound waves that penetrate the body tissues and reflect back to the device, creating detailed images based on the echoes received.

One of the primary advantages of ultrasonography in equine pregnancy diagnosis is its ability to detect pregnancy at a very early stage. This early detection is crucial for several reasons. Firstly, it enables the timely implementation of appropriate care and management strategies, which can significantly enhance the likelihood of a successful pregnancy. For example, detecting a twin pregnancy early on can allow for the necessary intervention to reduce the risk of complications associated with multiple embryos. Additionally, early pregnancy diagnosis helps in making informed decisions about rebreeding strategies if a mare is found to be not pregnant.

Beyond early detection, ultrasonography plays a vital role in monitoring the ongoing health and development of the pregnancy. Regular ultrasonographic examinations can provide detailed information about the viability of the embryo and the growth of the fetus. This is particularly important for identifying any potential issues or abnormalities that may arise during gestation. Conditions such as placentitis, fetal growth retardation, or uterine infections can be detected through ultrasonographic monitoring, allowing for prompt intervention and treatment to address these problems.

The ability to visualize the conceptus and its associated structures in real time is another significant benefit of ultrasonography. This real-time imaging allows veterinarians to assess the immediate status of the pregnancy and make on-the-spot decisions regarding the care and management of the mare. For instance, observing the heartbeat of the embryo or fetus can confirm its viability, providing reassurance to both the veterinarian and the breeder.

Despite its numerous advantages, the effectiveness of ultrasonography in equine pregnancy diagnosis is influenced by several factors. The skill and experience of the operator play a critical role in obtaining accurate and meaningful images. Proper training and expertise are required to correctly interpret the images and make accurate diagnoses. Additionally, the condition and temperament of the mare can affect the quality of the images obtained. In some cases, sedation or special handling may be necessary to facilitate a thorough examination.

In conclusion, ultrasonographic pregnancy diagnosis is an indispensable tool in the field of equine reproductive management. Its ability to provide early and accurate detection of pregnancy, monitor embryonic and fetal development, and identify potential issues makes it a valuable asset for veterinarians and breeders. Understanding and effectively utilizing ultrasonography can significantly enhance the success of breeding programs, ensuring the health and welfare of mares and their offspring and ultimately contributing to the overall well-being and productivity of the equine population.

2 B-Mode Ultrasonography

B-mode ultrasonography, also known as brightness-mode ultrasonography, is a noninvasive research tool that has found extensive use in the field of equine reproduction. This technology leverages the principles of sound wave reflection and scattering to generate real-time, two-dimensional images of the body.

The process begins with a mobile transducer, which emits a pulsed high-frequency sound beam. This sound beam travels through the body, encountering various acoustic interfaces along its path. These interfaces are essentially changes in the density or elasticity of the medium through which the sound beam is passing. Each time the sound beam encounters such an interface, a fraction of the sound energy is reflected or scattered.

The reflected or scattered sound waves are then captured by the transducer, which converts them into electrical signals. These signals are processed to produce a two-dimensional image on a screen. The brightness of each point on this image corresponds to the intensity of the reflected sound wave, thus giving this technique its name—B-mode or brightness mode ultrasonography.

In essence, an ultrasound image generated by B-mode ultrasonography is a 2D map of the tissue's acoustic reflectivity. It provides a detailed view of the internal structures of the body, allowing for accurate diagnosis and monitoring.

This technology has revolutionized clinical diagnosis and reproductive outcomes in equine practice. It offers several advantages over traditional methods. For instance, it allows for the diagnosis of pregnancy at an earlier stage than with rectal palpation. This is particularly beneficial in managing equine pregnancies, as it enables the early detection of twins, which can be challenging to manage and carry significant risks for the mare and foals.

Moreover, B-mode ultrasonography can detect signs of impending early embryonic death, allowing for timely intervention and potentially improving the chances of a successful pregnancy. Thus, the advent of B-mode ultrasonography has greatly enhanced the ability of veterinarians and equine breeders to manage and optimize reproductive outcomes in horses. It continues to be an invaluable tool in equine reproduction, contributing significantly to advancements in the field.

3   Doppler Ultrasonography

Doppler ultrasonography is a sophisticated and powerful diagnostic tool that provides real-time, dynamic information about the structure of blood vessels and the hemodynamic aspects of blood flow. This method can determine the presence of blood flow, its direction, and its speed. It achieves this by subdividing into two main categories: color Doppler, which includes color flow and power flow, and pulsed Doppler.

The color Doppler technique visualizes blood flow in different colors depending on the direction of flow, which can be particularly useful in identifying areas of turbulence or abnormal flow patterns. The power flow technique, on the other hand, is more sensitive and can detect even low-velocity flow, making it ideal for studying small vessels and low-flow states.

Pulsed Doppler, meanwhile, allows for the selective examination of blood flow in a specific area of interest within the vessel. This is particularly useful when assessing vessels that contain multiple flow directions, such as in areas of stenosis or regurgitation.

This technique has been extensively used to assess blood flow in various parts of the mare's reproductive system, including the uterus and the ovary. It has proven to be invaluable in determining both physiological and pathological changes in the mare's reproductive tract.

In the context of mare reproduction, Doppler ultrasonography techniques have been employed to study a wide range of aspects. These include the dynamics of follicular development, the vascularization and functionality of corpora lutea, and the vascularization of the uterus in both pregnant and nonpregnant states.

The study of follicular dynamics involves tracking the growth and regression of ovarian follicles, which are the structures that contain and release eggs. Understanding these dynamics can provide important insights into a mare's fertility and reproductive cycle.

The corpora lutea, structures that form on the ovary after an egg has been released, play a crucial role in maintaining pregnancy by producing the hormone progesterone. Assessing their vascularization and functionality can, therefore, provide important information about a mare's reproductive health and pregnancy status.

Finally, studying the vascularization of the uterus in both pregnant and nonpregnant states can provide valuable insights into the health of the uterine environment, which is critical for successful conception and pregnancy.

In summary, Doppler ultrasonography is a versatile and powerful tool in the field of equine reproduction, providing real-time, dynamic insights into the vascular and hemodynamic aspects of the mare's reproductive system. Its applications range from assessing follicular dynamics to studying the vascularization and functionality of corpora lutea and from evaluating uterine health to monitoring pregnancy status.

With its ability to provide such comprehensive and detailed information, it is no wonder that this technique has become an indispensable part of equine reproductive medicine.

4   Embryonic Stage

The process of pregnancy in mares begins with the fertilization of the oocyte, marking the start of the embryonic stage. This stage extends up to approximately the 40th-day post-conception, a period during which organogenesis, or the formation of organs, is generally considered to be complete.

The fertilized oocyte, at the morula stage, transitions from the oviduct to the uterus around 6–6.5 days after ovulation (Betteridge et al. 1982; Battut et al. 1997). It is at this juncture that the capsule, a layer unique to equids, begins to form. This capsule develops between the inner surface of the zona pellucida coincides with the onset of the blastocyst stage (Betteridge 2010), and disappears around day 21 (Ginther 1998). The presence of the capsule prevents the embryonic vesicle (EV) from hatching, thereby maintaining an approximately spherical shape until about day 19. Upon entering the uterine lumen, the embryo commences a continuous movement facilitated by uterine contractions. This movement continues until the 15th day in ponies or the 16th day in mares, at which point the mechanisms of fixation at the base of one uterine horn and orientation of the embryonic pole on the opposite side of the mesometrium attachment typically occur (Ginther 1985). This transuterine migration is universally recognized as the embryo's signal for maternal recognition of pregnancy in equids. Unlike pigs and ruminants, where pregnancy recognition signals derived from the conceptus have been identified (Geisert et al. 1990; Bazer et al. 1997), horses lack such a signal. However, studies have shown that restricting the embryo's mobility before fixation in just one horn demonstrates the necessity for embryo to be free to move throughout the entire uterine lumen for the maintenance of pregnancy (McDowell et al. 1988). This mechanism is attributed to the absence of a countercurrent transfer system between the uterine vein and the ovarian artery in the equids (McCracken et al. 1999). Therefore, it is the movement of the embryo that provides a sufficient pregnancy signal to prevent the production of Prostaglandin F2 alpha (PGF2α) and the subsequent luteolysis, or degradation of the corpus luteum. This intricate process underscores the complex and unique nature of equine reproduction (McCracken et al. 1999).

The process of embryo fixation is complex and involves several factors. One of these factors is the attainment of a certain diameter by the embryo, which is typically around 25 mm. This growth in size is accompanied by an increase in uterine tone, which in turn increases the resistance within the uterine lumen to the embryo's mobility. This combined effect results in the fixation of the embryo (Ginther 1985).

Following the fixation, the pole of the embryo vesicle assumes a ventral position. This position is opposite to the attachment of the mesometrium, a term used to describe the orientation of the embryo within the uterus (Ginther 1985). The positioning of the embryo undergoes further changes between the 25th and 40th day of gestation. During this period, the embryo moves dorsally towards the "ceiling" of the vesicle. This movement is facilitated by the development of the allantoic sac and the reduction of the yolk sac.

The yolk sac, which initially plays a crucial role in the embryo's development, gradually diminishes in size. Concurrently, its blood vessels are progressively incorporated into the umbilical cord. This process marks the transition from the embryonic stage to the fetal stage, where the umbilical cord serves as the primary conduit for nutrients and waste products between the fetus and the mother.

In summary, the process of embryo fixation and subsequent development involves a series of complex and intricately coordinated events, each playing a crucial role in ensuring the successful growth and development of the fetus.

5   Embryonic Stage Ultrasonography

The science of ultrasonographic pregnancy diagnosis in mares has seen significant advancements since the first report was published by Palmer and Driancourt in 1980. This initial report indicated that pregnancy could be detected as early as 14 days post-ovulation, marking a significant milestone in equine reproductive science (Palmer and Driancourt 1980). Fast forward four decades, and the technology has evolved to allow for even earlier detection. Today, it is possible to detect pregnancy as early as 9 days post-ovulation when the EV measures approximately 3 mm in diameter. However, the majority of diagnoses are conducted between 12 and 14 days post-ovulation, when the EV typically measures between 14 and 16 mm in diameter (Ginther 1983; Bergfelt and Adams 2011). It is important to note that while earlier detection is possible, it may result in false-negative diagnoses. This is because the EV is still very small at this stage, and it may not be easily detectable (Ginther 1986; Abshenas et al. 2009). In recent years, the use of computerized analysis of power Doppler has allowed for even earlier pregnancy diagnosis. This technology enables the detection of pregnant mares 8 days after ovulation, even before the recovery of the embryo (Nieto-Olmedo et al. 2020). However, despite these advancements, there are still practical limitations to the use of this technology in the field. One significant challenge is the absence of an automatic image analyzer in the ultrasonographic machine software. This makes the technique less practical for field use, as it requires manual analysis of the images.

In conclusion, while the field of ultrasonographic pregnancy diagnosis in mares has seen significant advancements over the past four decades, there are still challenges to be overcome. The ongoing development of technology and techniques promises further improvements in the accuracy and timeliness of pregnancy detection in the future.

The process of diagnosing a pregnancy in horses, particularly before the 14th day, is often undertaken for a specific reason. This is typically in instances where there may be multiple ovulations. The early diagnosis allows for more time to potentially reduce a twin pregnancy before it becomes fixed.

As has been previously documented, the equine conceptus, or early-stage embryo, can be visualized by ultrasound before day 19 (Stolla et al. 2001). It appears as an anechoic, or non-reflective, spherical structure. This structure has two specular, or mirror-like, echoic poles, one at the top (dorsal) and one at the bottom (ventral).

During the initial stages of the pregnancy, a transrectal ultrasonographic examination is usually performed. This is even the case for miniature horses. However, when dealing with particularly small animals, certain precautions need to be taken. These include the use of a petroleum-based lubricant, administering low-dose sedation if necessary, and possibly a transrectal infusion of lidocaine (60 mL of a 2% solution).

In some cases, smooth muscle relaxants such as N-butylscopolammonium bromide (administered intravenously at a dose of 0.3 mg/kg) or propantheline bromide (also administered intravenously but at a dose of 0.014 mg/kg) may be used. In extreme cases, a rigid probe extension can be used.

Before embarking on the search for the conceptus, it is of utmost importance to thoroughly examine the ovaries. This is done to identify any instances of multiple ovulations that may have occurred but have not been diagnosed yet. Multiple ovulations can lead to twin pregnancies, which may not be the desired outcome in certain situations.

In scenarios where there is a suspicion of a twin pregnancy, it is advisable to bring forward the pregnancy diagnosis to 12 days post-ovulation. This allows for more time to intervene and reduce the pregnancy before it becomes firmly established, a process known as fixation.

A significant proportion of unexpected twin pregnancies are the result of undiagnosed asynchronous double ovulations. These are instances when two separate ovulations occur at different times within the same menstrual cycle. Often, ultrasonographic examinations cease after the detection of the first ovulation, potentially missing the second ovulation.

Therefore, it is strongly recommended that every single follicle that is larger than 30 mm be monitored for at least 2 days following the first detected ovulation. This practice increases the chances of identifying any subsequent ovulations, thereby providing a more accurate diagnosis.

In recent times, with the advancements in reproductive technology and the successful implementation of in-vitro-produced embryos, there have been rare instances (4/254, 1.6%; Dijkstra et al. 2020) of monozygotic twins' development. These are twins that develop from a single fertilized egg that splits into two or, exceptionally, three embryos. These embryos appear around the 25th day of pregnancy within the same EV. This is a relatively rare occurrence but is becoming more common with the increased use of in vitro fertilization techniques (Image 1).

The first step in the process involves a thorough evaluation of the ovaries for the presence of corpora lutea. Following this, a meticulous scan of the uterus is required to locate the EV. The growth rate of the EV is quite specific. It tends to increase by approximately 3.2 mm per day from the first day it becomes visible (around the 9th day of pregnancy when it is roughly 3 mm in size) until it becomes fixed in place.

Day 10   Day 14   Day 16

Day 19   Day 21   Day 25

Day 30   Day 35   Day 40

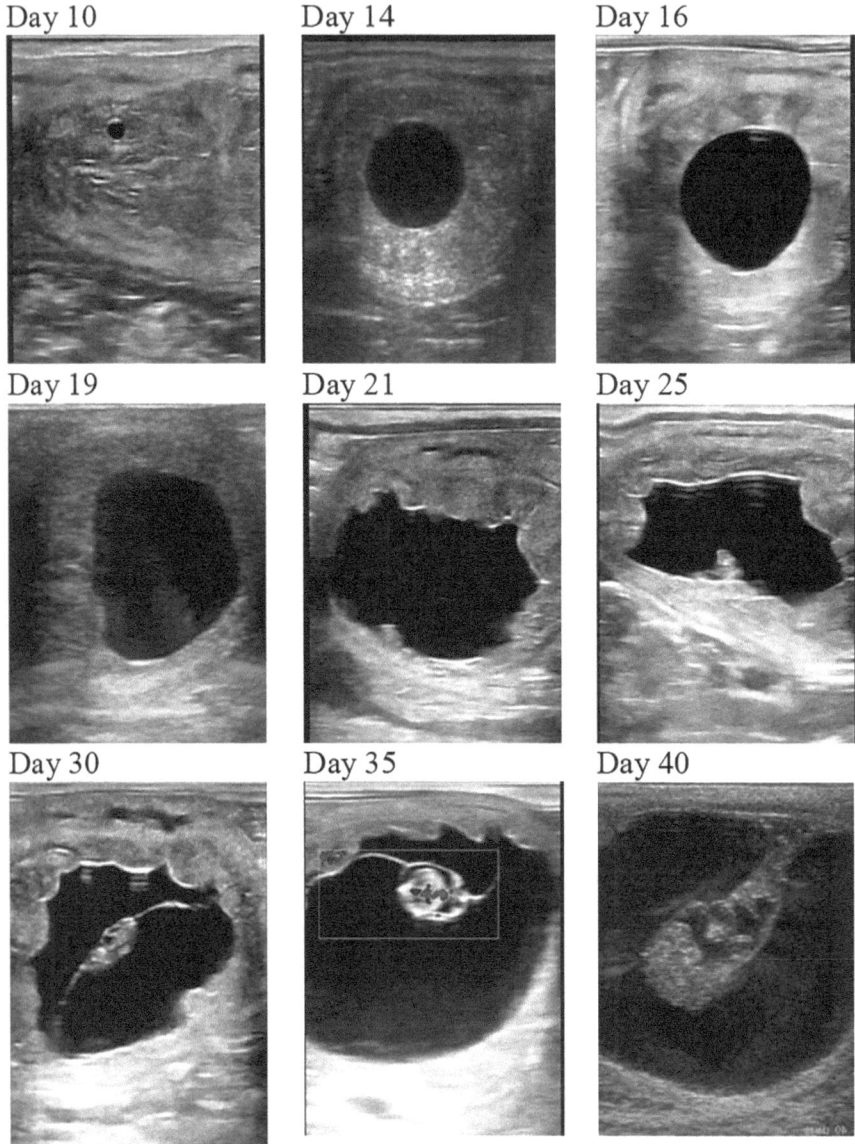

Image 1 Ultrasonographic image of a twin pregnancy at 27 days of gestation (image provided by Antony Claes, Utrecht University, The Netherlands)

From the day of fixation until the 28th day of pregnancy, the growth rate slows down to about 0.5 mm per day. Then, from the 28th day until the 46th day of pregnancy, the growth rate increases again to approximately 1.6 mm per day (Ginther 1983).

If a pregnancy diagnosis turns out to be negative, it is crucial to repeat the test 1–2 days later. This is especially important in the case of older mares. Mares that are between 16 and 24 years old tend to produce significantly smaller embryos compared to their younger counterparts, who are between 2 and 15 years old (Camevale and Ginther 1992; Panzani et al. 2016). Delayed embryo development in older mares has been linked to poor oocyte quality and a higher risk of embryo loss (Ball et al. 1986; Carnevale et al. 1993, 2000). Insemination after ovulation can also lead to a smaller embryo diameter. This is likely due to the time required for sperm capacitation and/or the aging of the oocyte (Huhtinen et al. 1996). A reduction in the production of Prostaglandin E (PGE) in older mares has been suggested as a reason for prolonged oviductal transit. This results in delayed embryo development, as the embryo requires more time to exit the oviduct (Weber et al. 1991; Kenney 1993).

Images and descriptions of ultrasonographic findings during the embryonic stage are described in Fig. 1.

In the process of confirming the health and viability of a pregnancy, there are several key indicators to consider. One of the primary pieces of evidence is the presence of the EV. However, there are also secondary signs that play a crucial role in this assessment.

One such secondary sign is uterine edema, which is characterized by the accumulation of fluid in the uterine wall. This condition can be a significant indicator of the health of the pregnancy. Another important factor is the maintenance of the corpus luteum, a temporary endocrine structure in female ovaries that is essential for the establishment and maintenance of pregnancy.

During the diagnosis of pregnancy, it is expected that there should be no signs of uterine fluid or endometrial edema. Moreover, the presence of a functional corpus

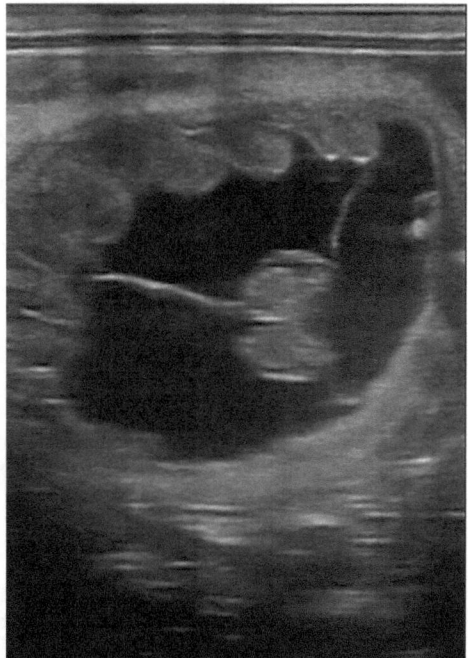

Fig. 1 Embryo stage ultrasonographic characteristics from day 10 to day 40

luteum (CL) is a must. The absence of these conditions can potentially indicate complications in the pregnancy.

The evaluation of uterine edema requires careful attention, especially when there is a change in the ultrasonographic machines or imaging parameters being used. Each machine has its unique image definition, and what may not appear edematous using one machine could appear so on another. This discrepancy can lead to potential misdiagnosis or confusion.

Therefore, when transitioning to a new machine, it is highly recommended to compare the uterine images produced by the new device with those from the previous one. This comparison can help ensure consistency in the interpretation of the images and avoid potential misinterpretations due to differences in image definitions between machines. This practice ultimately contributes to the accurate assessment of pregnancy health and viability (Image 2).

Physiologically, the equine Corpus Luteum (CL) can exhibit two fundamental morphologies: it can either have a central nonecogenic area or lack one. Interestingly, the production of progesterone (P4), a hormone crucial for pregnancy, does not differ between these two morphologies. Typically, the concentration of P4 increases until it reaches its peak on the 8th day after ovulation. Following this, there is a gradual decline of about 40% until the onset of luteolysis (on the 15th day. If a pregnancy is not detected, there is a drastic 60% decline in P4 concentration on the subsequent day (Ginther and Santos 2015). To evaluate the echogenicity (the ability to produce an echo) of the B-Mode CL, a seven-point grayscale was introduced, where 0 represents black and 7 represents white. The echogenicity of the parenchymatous portion of the CL (excluding the central nonecogenic area, if present) gradually shifts from 5 (on the day of ovulation) to approximately 3.25 on the 8th day, rising to around 4.25 by the 15th day (Pierson and Ginther 1985). Blood is

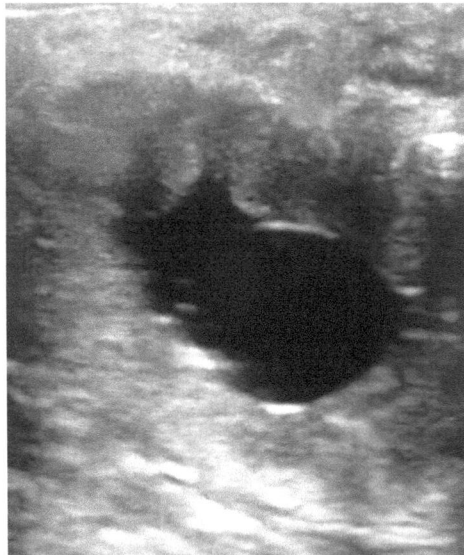

Image 2 Day 10 embryo vesicle surrounded by Grade 4 fluid (McKinnon and Squires 1988)

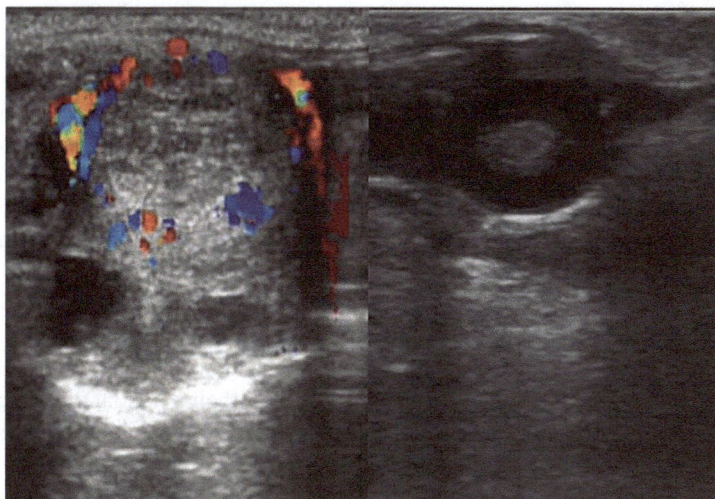

Image 3 Color Doppler of a functional corpus luteum and B-Mode ultrasonography of a dead CL (showing the "black ring of death", McCue personal communication

semi-echogenic, meaning it partially reflects ultrasound waves. Therefore, changes in CL vascularization, or blood supply, can also be assessed in B-Mode through corresponding decreases and increases in echogenicity (Pierson and Ginther 1985). More recently, the vascularization pattern of the CL has been demonstrated by analyzing luteal blood flow using color Doppler ultrasonography. Around 80% of the CL's area is vascularized by the 9th day, decreasing to approximately 55% by the 15th day (Ginther et al. 2007; Image 3). In summary, during a pregnancy diagnosis, the presence of a highly echogenic or poorly vascularized CL, possibly associated with slight uterine edema, in the presence of an EV should alert the practitioner to potential issues.

In equine gestation, a plateau in progesterone (P4) concentration is observed after the 15th day, which persists until the 35th day. A similar pattern is observed in the cross-sectional area of the primary CL, which remains around 5–6 cm^2 from day 15 to day 40 of pregnancy (Bergfelt et al. 1989; Šichtař et al. 2013; Ginther and Santos 2015). This resurgence in the dimensions of the primary CL and its P4 output coincides with the onset of equine chorionic gonadotropin (eCG) production by the endometrial cups (Bergfelt et al. 1989).

6 Embryonic Stage Pregnancy Abnormalities

A uterine cyst is a type of abnormal growth that can occur within the uterus. It is important to note that these cysts can sometimes be mistaken for pregnancy, particularly when a diagnosis is made between 9 and 21 days following ovulation. This is because the physical characteristics of a uterine cyst can mimic those of an early-stage pregnancy.

Endometrial cysts can either be lymphatic or glandular in nature. These cysts can be mistaken for the EV. If these cysts grow large enough to obstruct the migration of the embryo within the uterus, they can potentially lead to early pregnancy loss.

In terms of their appearance, endometrial cysts are often seen as hypoechoic areas, meaning they appear darker on an ultrasound image. These areas are enclosed by a hyperechoic membrane, which appears brighter on an ultrasound. This contrast in echogenicity helps to distinguish the cysts from the surrounding uterine tissue.

Furthermore, these cysts are sometimes perfectly spherical, which can make them resemble an EV. They can also, as EV, have hyperechoic specular echoic poles, which are bright areas seen at the top and bottom of the cyst on an ultrasound image. This characteristic further contributes to the potential for these cysts to be mistaken for an EV (Stanton et al. 2004; Image 4). In conclusion, uterine and endometrial cysts can present diagnostic challenges due to their potential to mimic early-stage pregnancy. Therefore, a careful and thorough examination is necessary to ensure an accurate diagnosis.

When comparing these two structures, one of the key differences lies in the growth and movement of the EV. Over time, the EV is expected to expand and shift to a different location within the lumen. This change in position is typically observed during subsequent examinations.

It is highly recommended that every mare that is affected by a cystic endometrium undergo a thorough evaluation. This evaluation should include noting down the number of uterine cysts present, their specific locations within the uterus, and the diameter of each cyst.

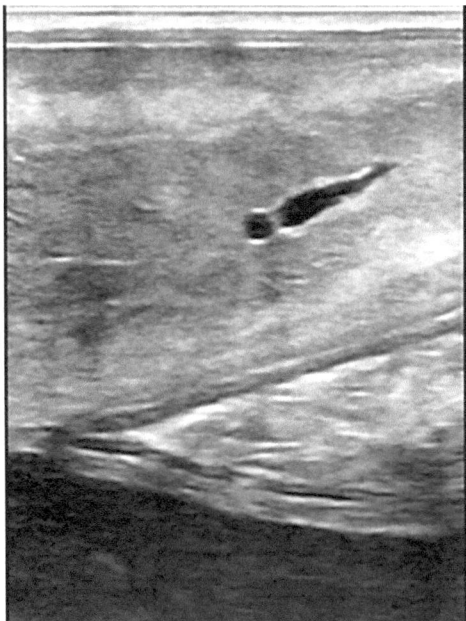

Image 4 Day 10 embryonic vesicle (left) close to an endometrial cyst (right)

When we delve into the comparison of these two distinct structures, we find that a significant difference is rooted in the growth pattern and movement of the EV. As time progresses, the EV undergoes a transformation. It is anticipated to not only expand in size but also shift its position within the lumen. This relocation is not a sudden occurrence but a gradual process that can be observed until days 15–16. Regular examinations provide an opportunity to track this change in position, offering valuable insights into the behavior of the EV.

In the case of mares affected by a cystic endometrium, it is of utmost importance to conduct a comprehensive evaluation. This is not a mere suggestion but a strong recommendation for pregnancy success. The evaluation process should be performed before insemination and must be thorough and meticulous, leaving no stone unturned.

The evaluation should document the number of these cysts present within the uterus. Each cyst should be accounted for, providing a clear picture of the extent of the condition. Furthermore, the specific locations of these cysts within the uterus should be identified and noted. This information is crucial as it can help in understanding the spread and pattern of the cysts.

In addition to the count and location, the evaluation should also measure the diameter of each cyst. This measurement can provide insights into the size, which can be critical in determining the embryo movement and, subsequently, the pregnancy health status (Table 1; Ginther 2022).

In the field of embryology, the detection of an embryonic heartbeat using B-Mode ultrasound technology can often be a challenging task. This is particularly true before the 24th day of embryonic development. However, as previously introduced, with the use of advanced imaging techniques, such as color or power Doppler, it is

Table 1 Fetal age estimation after 90 days of pregnancy in different breeds according to different parameters (eye diameter, trunk diameter, biparietal diameter) x: fetal age (days), y: mm, w = bpm, z: age group ($0 = \geq 15$ years old and $1 = <15$ years old)

Breed	Part of the fetus's body	Measurement interval	Equation	Authors
Thoroughbred and Standardbred	Eye ∅[a]	90–term	$y = 0.77 + 0.14x$	Kähn.and Leidl (1987)
	Trunk ∅	60–180	$y = 56.33 + 1.52x—0.0004x^2$	
	Skull biparietal ∅	90–240	$y = 10.08 + 0.0013x^2$	
Dutch warmblood	Eye ∅[b]	100–term	$y = -5.62 + 0.21x - 0.0003x^2$	Hendriks et al. (2009)
	Skull biparietal ∅		$y = -26.64 + 0.597x - 0.0008x$	
	Aorta ∅		$y = -3.67 + 0.077x - 0.088z$	
	Heart rate		$y = 167.97 - 0.249w$	
Quarter horses and thoroughbred	Skull biparietal ∅	75–term	$y = 1.26 + 0.29x$	Renaudin et al. (2000)
	Aorta ∅	60–term	$y = -6.74 + 0.09x$	

[a]Eye vitreous body length
[b](Eye orbit length + width)/2

usually possible to discern the embryonic heartbeat as early as the 18th day (Stolla et al. 2001).

There are instances where the embryo does not appear on the anti-mesometrial pole of the uterine horn base, which is often referred to as the 6 o'clock position (Ginther and Silva 2006). This misorientation, while not typical, does not seem to interfere with the normal progression of pregnancy unless it is associated with other signs such as an inadequate tone of the uterus (Image 5; Newcombe 2000).

A relatively rare occurrence in the realm of pathologic pregnancies is the uterine body pregnancy. Out of 3527 pathologic pregnancies studied, only 10 cases were identified as uterine body pregnancies. These pregnancies often do not develop properly and have a high loss rate. This is likely due to growth retardation, especially if the embryo becomes fixed in the caudal body portion of the uterus (Jobert et al. 2005).

In approximately 4% of pregnancies, it is not possible to detect a proper embryo within the EV. This event is known as a trophoblastic or anembryonic vesicle. Typically, these vesicles are identified early on as small-for-age pregnancies. However, despite careful monitoring, no embryo appears between days 21–30 of the pregnancy (Vanderwall et al. 2000; Image 6).

Image 5 Day 28
"upside-down" pregnancy

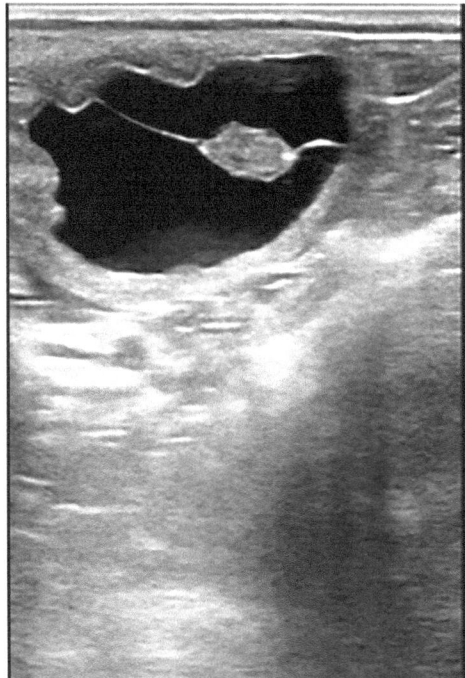

Image 6 Day 28
an-embryonic or
trophoblastic pregnancy

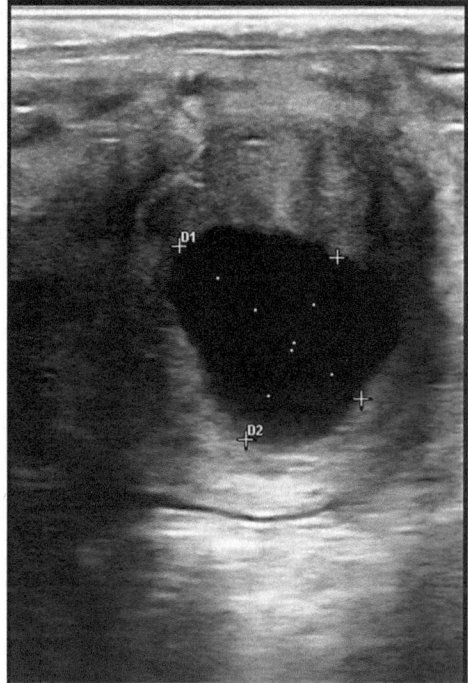

7 Fetal Stage

The fetal phase of a horse's development is a critical period that begins once organogenesis, or the formation of organs, is complete. This phase concludes at parturition or birth. The 40th day of pregnancy is typically chosen as the commencement of the fetal stage. This is due to several significant developments that occurred around this time, including the formation of the umbilical cord, the replacement of the yolk sac with the allantois sac, and the onset of fetal activity (Ginther 1998).

By the 48th day, the fetal amniotic vesicle, which is a fluid-filled sac that surrounds and protects the fetus, reaches the lower wall of the allantoic sac. This is a result of the lengthening of the umbilical cord. The fetus continues to develop on the floor of the uterus. However, its presentation, or position, may change up until the 180th to 270th day of pregnancy. At this point, mechanisms within the fetus orient it, usually in a cranial presentation and dorso-pubic position. This position is maintained until the first stage of parturition, when the position changes to dorso-sacral.

During the fetal stage, practitioners may need to predict the fetal age. Various measurements can be used to estimate this, including the eye diameter, biparietal diameter, aortic root diameter, femur length, thoracic diameter, and heart rate.

The size of the fetus can be influenced by several factors. These include the size of the breed, the age of the mare, and whether the mare has given birth before (parity). For instance, mares giving birth for the first time (primiparous mares) tend to have fetuses with larger eye diameters. Similarly, mares younger than 15 years old have been observed to have a larger aortic root diameter (Hendriks et al. 2009).

Equations were derived from data collected from different breeds of horses to predict the fetal age. These equations are described in Table 1. It is important to note that these are general guidelines, and individual variations may occur.

In a study involving a group of ponies similar to the Shetland breed, a regression equation was developed to predict the number of days remaining before birth. This equation is given by:

DPB = 265.16 − (0.21 × vitreous body length in mm), where DPB stands for "Days Before Birth" (Turner et al. 2006).

However, it is important to note that the heart rate alone is not a reliable indicator for dating a pregnancy. The variability of the heart rate depends on the activity level of the fetus, with lower frequencies observed when the fetus is at rest (Dascanio 2021).

It has been documented that measurements of the eye and skull can be taken transrectally from around 100 days of pregnancy up to term. On the other hand, the fetal aorta becomes visible through transabdominal ultrasound only around 200 days after gestation (Renaudin et al. 2000).

In addition to these, other parameters have been considered for estimating fetal age. These include the combined utero placental thickness (CTUP), femur length, intercostal distance, and the size of the gonads and kidneys (Renaudin et al. 2000).

In mares, CTUP is primarily used to diagnose placentitis rather than to date pregnancies. This is due to its large standard deviation, which makes it less reliable for determining gestational age (Troedsson et al. 1997; Renaudin et al. 1997, 2000, 2022; Bucca et al. 2005; Kimura et al. 2018; Renaudin and Conley 2023).

8 Conclusion

Ultrasonographic pregnancy diagnosis in mares has revolutionized equine reproductive management by providing a reliable, noninvasive method for early detection and monitoring of pregnancy. Through advancements in ultrasonographic technology and technique, veterinarians can accurately confirm pregnancy as early as 10–14 days post-ovulation, assess embryonic development, and identify potential complications that may affect the mare's reproductive success.

The ability to visualize and evaluate the embryo, fetal membranes, and uterine environment allows for timely intervention and management decisions that enhance pregnancy outcomes. Early detection of twins, for instance, facilitates necessary management strategies to reduce the risk of pregnancy loss or complications. Moreover, serial ultrasonographic examinations throughout gestation enable ongoing assessment of fetal health and development, providing invaluable information for predicting and managing parturition.

Overall, the integration of ultrasonographic pregnancy diagnosis into routine equine reproductive practice underscores its importance in achieving successful breeding programs. By allowing for early and accurate pregnancy detection and continuous monitoring, ultrasonography helps ensure the health and well-being of both the mare and the developing fetus, ultimately contributing to the advancement of equine reproductive medicine.

References

Abshenas J, Homayoon Babaei D, Mahdavi I, Alimolaei DM, Shafipour A, Babaei H (2009) Ultrasonographical measurement of caspian mare embryonic vesicle and embryo on days 8 to 44 after ovulation. Iranian J Vet Surg (IJVS) 2(1)

Ball BA, Little TV, Hillman RB, Woods GL (1986) Pregnancy rates at days 2 and 14 and estimated embryonic loss rates prior to day 14 in normal and subfertile mares. Theriogenology 26(5):611–619. http://linkinghub.elsevier.com/retrieve/pii/0093691X86901688

Battut I, Colchen S, Fieni F, Tainturier D, Bruyas JF (1997) Success rates when attempting to non-surgically collect equine embryos at 144, 156 or 168 hours after ovulation. Equine Vet J Suppl 25(25):60–62. https://doi.org/10.1111/j.2042-3306.1997.tb05102.x

Bazer FW, Spencer TE, Ott TL (1997) Interferon Tau: a novel pregnancy recognition signal. Am J Reprod Immunol 37(6):412–420. https://doi.org/10.1111/j.1600-0897.1997.tb00253.x

Bergfelt DR, Adams GP (2011) Pregnancy. In: McKinnon AO, Squires EL, Vaala WE, Varner D (eds) Equine reproduction, vol 2, 2nd edn. Blackwell Publishing, pp 2065–2079

Bergfelt DR, Pierson RA, Ginther OJ (1989) Resurgence of the primary corpus luteum during pregnancy in the mare. Anim Reprod Sci 21:261

Betteridge KJ (2010) The structure and function of the equine capsule in relation to embryo manipulation and transfer. Equine Vet J 21(S8):92–100. https://doi.org/10.1111/j.2042-3306.1989.tb04690.x

Betteridge KJ, Eaglesome M, Mitchell D (1982) Development of horse embryos up to twenty two days after ovulation: observations on fresh specimens. J Anat 135(1):191–209

Bucca S, Fogarty U, Collins A, Small V (2005) Assessment of feto-placental well-being in the mare from mid-gestation to term: Transrectal and transabdominal ultrasonographic features. Theriogenology 64(3):542–557. https://doi.org/10.1016/j.theriogenology.2005.05.011

Camevale EM, Ginther OJ (1992) Relationships of age to uterine function and reproductive efficiency in mares. Theriogenology 37

Carnevale EM, Bergfelt DR, Ginther OJ (1993) Aging effects on follicular activity and concentrations of FSH, LH, and progesterone in mares. Anim Reprod Sci 31:287

Carnevale EM, Ramirez RJ, Squires EL, Alvarenga MA, Vanderwall DK, McCue PM (2000) Factors affecting pregnancy rates and early embryonic death after equine embryo transfer. Theriogenology 54(6):965–979. https://doi.org/10.1016/S0093-691X(00)00405-2

Dascanio JJ (2021) Prediction of fetal age. Equine Reproductive Procedures:291–294

Dijkstra A, Cuervo-Arango J, Stout TAE, Claes A (2020) Monozygotic multiple pregnancies after transfer of single in vitro produced equine embryos. Equine Vet J 52(2):258–261. https://doi.org/10.1111/evj.13146

Geisert RD, Zavy MT, Moffatt RJ, Blair RM, Yellin T (1990) Embryonic steroids and the establishment of pregnancy in pigs. J Reprod Fertil Suppl 40:293–305. https://doi.org/10.1530/biosciprocs.13.0021

Ginther OJ (1983) Fixation and orientation of the early equine conceptus. Theriogenology 19(4):613–623. http://linkinghub.elsevier.com/retrieve/pii/0093691X83901814

Ginther OJ (1985) Dynamic physical interactions between the equine embryo and uterus. Equine Vet J Suppl 17(S3):41–47. https://doi.org/10.1111/j.2042-3306.1985.tb04592.x

Ginther OJ (1986) Ultrasonic imaging and reproductive events in the mare. Equiservices, 4343 Garfoot Road

Ginther OJ (1998) Equine pregnancy: physical interactions between the uterus and conceptus. 44:73–104

Ginther OJ (2022) The dynamic equine embryo from postfixation (day 17) to the end of the embryo stage (day 40). J Equine Vet Sci 108. https://doi.org/10.1016/j.jevs.2021.103808

Ginther OJ, Santos VG (2015) Natural rescue and resurgence of the equine corpus luteum. J Equine Vet Sci 35(1):1–6. https://doi.org/10.1016/j.jevs.2014.10.004

Ginther OJ, Silva LA (2006) Incidence and nature of disorientation of the embryo proper and spontaneous correction in mares. J Equine Vet Sci 26(6):249–256. https://doi.org/10.1016/j.jevs.2006.04.001

Ginther OJ, Gastal EL, Gastal MO, Utt MD, Beg MA (2007) Luteal blood flow and progesterone production in mares. Anim Reprod Sci 99(1–2):213–220. https://doi.org/10.1016/j.anireprosci.2006.05.018

Hendriks WK, Colenbrander B, van der Weijden GC, Stout TAE (2009) Maternal age and parity influence ultrasonographic measurements of fetal growth in Dutch warmblood mares. Anim Reprod Sci 115(1–4):110–123. https://doi.org/10.1016/j.anireprosci.2008.12.014

Huhtinen M, Koskinen E, Skidmore JA, Allen WR (1996) Recovery rate and quality of embryos from mares inseminated after ovulation. Theriogenology 45:719–726

Jobert ML, LeBlanc MM, Pierce SW (2005) Pregnancy loss rate in equine uterine body pregnancies. Equine Vet Educ 17(3):163–165. https://doi.org/10.1111/j.2042-3292.2005.tb00360.x

Kähn W, Leidl W (1987) Ultrasonic measurement of the equine fetus in utero and sonographic imaging of fetal organs. 94(9):509–515

Kenney RM (1993) A review of the pathology of the equine oviduct. Equine Vet J 25(S15):42–46. https://doi.org/10.1111/j.2042-3306.1993.tb04823.x

Kimura Y, Haneda S, Aoki T, Furuoka H, Miki W, Fukumoto N, Matsui M, Nambo Y (2018) Combined thickness of the uterus and placenta and ultrasonographic examinations of uteroplacental tissues in normal pregnancy, placentitis, and abnormal parturitions in heavy draft horses. J Equine Sci 29(1):1

McCracken JAMC, Custer EE, Lamsa JC (1999) Luteolysis : a neuroendocrine-mediated event. Physiol Rev 79(2):263–324

McDowell KJ, Sharp DC, Grubaugh W, Thatcher WW, Wilcox CJ (1988) Restricted conceptus mobility results in failure of pregnancy maintenance in mares. Biol Reprod 39(2):340–348. https://doi.org/10.1095/biolreprod39.2.340

McKinnon AO, Squires EL (1988) Equine embryo transfer. The veterinary clinics of North America. Equine Practice 4(2):305–333. https://doi.org/10.1016/S0749-0739(17)30643-0

Newcombe JR (2000) Embryonic loss and abnormalities of early pregnancy. Equine Vet Educ 12(2):88–101. https://doi.org/10.1111/j.2042-3292.2000.tb01771.x

Nieto-Olmedo P, Gaitskell-Phillips G, Martín-Cano FE, Ortiz-Rodríguez JM, Peña FJ, Ortega-Ferrusola C (2020) Endometrial blood flow area is a good marker for diagnosis of pregnant mares on day 8 post-ovulation before performing the uterine lavage for embryo recovery. J Equine Vet Sci 89:103075. https://doi.org/10.1016/j.jevs.2020.103075

Palmer E, Driancourt MA (1980) Use of ultrasonic echography in equine gynecology. Theriogenology 13(3):203–216. http://linkinghub.elsevier.com/retrieve/pii/0093691X80900825

Panzani D, Vannozzi I, Marmorini P, Rota A, Camillo F (2016) Factors affecting recipients' pregnancy, pregnancy loss, and foaling rates in a commercial equine embryo transfer program. J Equine Vet Sci 37:17–23

Pierson RA, Ginther UJ (1985) Ultrasonic evaluation of the corpus luteum of the mare. Theriogenology 23(5):795–806

Renaudin CD, Conley AJ (2023) Pregnancy monitoring in mares: Ultrasonographic and endocrine approaches. In: Reproduction in domestic animals. Wiley, vol 58, Issue S2, pp 34–48. https://doi.org/10.1111/rda.14392

Renaudin CD, Troedsson MH, Gillis CL, King VL, Bodena A (1997) Ultrasonographic evaluation of the equine placenta by transrectal and transabdominal approach in the normal pregnant mare. Theriogenology 47(2):559–573. http://eutils.ncbi.nlm.nih.gov/entrez/eutils/elink.fcgi?d bfrom=pubmed&id=16728008&retmode=ref&cmd=prlinks

Renaudin CD, Gillis CL, Tarantal AF, Coleman DA (2000) Evaluation of equine fetal growth from day 100 of gestation to parturition by ultrasonography. J Reprod Fertility Suppl 56:651–660

Renaudin CD, Kass PH, Bruyas JF (2022) Prediction of gestational age based on foetal ultrasonographic biometric measurements in light breed horses. Reprod Domest Anim 57(7):743–753. https://doi.org/10.1111/rda.14116

Šichtař J, Rajmon R, Hošková K, Řehák D, Vostrý L, Härtlová H (2013) The luteal blood flow, area and pixel intensity of corpus luteum, levels of progesterone in pregnant and nonpregnant mares in the period of 16 days after ovulation. Czeh J Anim Sci 58(11):512–519. https://cabi-digitallibrary.org

Stanton MB, Steiner JV, Pugh DG (2004) Endometrial cysts in the mare. J Equine Vet Sci 24(1):14–19. https://doi.org/10.1016/j.jevs.2004.12.003

Stolla R, Chen YH, Bollwein H (2001) Examination of embryonic death in mares using colour Doppler and B-mode sonography. Pferdeheilkunde 17(6):543–547. https://doi.org/10.21836/PEM20010601

Troedsson M, Renaudin C, Troedsson MH, Renaudin CD, Zent WW, Steiner JV (1997) Transrectal ultrasonography of the placenta in normal mares and mares with pending abortion: a field study. https://www.researchgate.net/publication/265206107

Turner RM, McDonnell SM, Feit EM, Grogan EH, Foglia R (2006) Real-time ultrasound measure of the fetal eye (vitreous body) for prediction of parturition date in small ponies. Theriogenology 66(2):331–337. https://doi.org/10.1016/j.theriogenology.2005.11.019

Vanderwall DK, Squires EL, Brinsko SP, McCue PM (2000) Diagnosis and management of abnormal embryonic development characterized by formation of an embryonic vesicle without an embryo in mares. JAVMA 217(1):58–63

Weber JA, Freeman DA, Vanderwall DK, Woods GL (1991) Prostaglandin E2 hastens oviductal transport of equine embryos1. Biol Reprod 45. https://academic.oup.com/biolreprod/article/45/4/544/2763008

Part II

Applying ART to Female Horses and Cattle

Commercial Equine Embryo Transfer Programs: Some Topics on How to Improve Efficiency

Luis Losinno ⓘ, Hernán Ramírez Castex ⓘ, and Melina Soledad Pietrani ⓘ

Abstract

Commercial equine embryo transfer programs are increasingly conditioned to achieve high productive efficiency, not only in the production of pregnancies and derived products from donor mares but also in doing so in the shortest time possible, respecting animal welfare procedures and at a low cost, pressured by increasingly efficient and commercially established in vitro embryo production programs. Multiple strategies regarding standard procedures have been and continue to be developed to increase the embryo recovery rate, post-transfer pregnancy rates, and decrease embryo loss rates, including the induction of multiple ovulations, which allows almost doubling the recovery rate per cycle, re-flushing techniques, massage during flushing, control of intervals between prostaglandin injection and ovulation (ITO), transfer techniques using cervical forceps, embryo vitrification, and many other simple, yet effective, procedures collectively aimed at achieving the goal of greater productive efficiency.

Keywords

Embryo transfer · Efficiency · Embryo vitrification · Cycle management · Transfer technique

L. Losinno (✉)
Laboratorio de Producción Equina, Universidad Nacional de Rio Cuarto, Río Cuarto, Argentina

Equine Academy, Rio Cuarto, Argentina

H. R. Castex
BIOTEQ, Paine, Chile

M. S. Pietrani
Universidad Nacional de Villa María, Villa María, Argentina
e-mail: mpietrani@unvm.edu.ar

J. C. Gardón, K. Satué Ambrojo (eds.), *Assisted Reproductive Technologies in Animals Volume 1*, https://doi.org/10.1007/978-3-031-73079-5_3

1   Introduction

Currently, accumulated experience in empirical and scientific knowledge sets certain standards and expected results in commercial embryo transfer (ET) programs. These standards are generally known by ET clients and, therefore, are the ones who put pressure on professionals. Moreover, from the Veterinarian's point of view, these standards and results set the levels of efficiency that their own system should have to keep up with the ET programs worldwide to be commercially competitive and economically viable.

In commercial ET programs, recording the activities and procedures (mare's treatments and medications, artificial insemination information, embryo flushings, embryo transfers, etc.) should be a common practice. It is difficult to perform a technical and "objective" analysis of a complex system such as an ET program if numerical data is not available. Impressions, opinions, and intuitions are not enough to carry out a system diagnosis, hindering the ability to draw conclusions from possible failures and critical points to improve. Currently, digital data technology facilitates real-time data processing, loading, reports, and online consultations, providing veterinarians with a friendly and productive tool. All this "processed" information is critical to ET systems helping to identify points to improve and optimize overall efficiency.

Despite there having been considerable improvements in ET results since the 1980s, there are still significant losses regarding system economy, time, genetics, donor mare's cycles, semen doses, client's expectations, etc., in other words, in efficiency.

Even though the expected post-transfer pregnancy rates depend on different factors such as type of ET program, breed, mares age, semen quality, nutrition management, recipient mares, staff training, etc., it is possible to establish expected values regarding embryo recovery rate (ERR), pregnancy rate (PR), and early embryonic loss (EEL) for commercial programs. An overall efficiency greater than 50% is considered acceptable. This translates into a 70% ERR per cycle and a PR post-transfer higher than 70% (Efficiency = ERR × PR = $0.7 \times 0.7 = 0.49$), which is equivalent to two flushings per pregnancy as the expected theoretical value. From that point, improvement can be a result, at least, of three simple things: (1) a systematic diagnosis of the program point by point, (2) control of small operational details, and (3) permanent professional update. There are also large-scale, stabilized commercial programs with efficiencies greater than 65% (approximately 1.2–1.5 embryo flushing per pregnancy) (Hartman 2011; Cuervo-Arango et al. 2018a, b).

Considering a 70% ERR, at least 30% of the estrous cycles in which donors have been inseminated and recipients prepared (semen, supplies, food, time, work) have been unproductive (negative flushing). These costs must be "absorbed" by the productive cycles of the same mare (or by something or someone else). In addition, considering an average of 70% PR, it can be assumed that 30% of the embryos recovered and transferred did not produce a pregnancy. Reported EEL rates in ET programs until day 60 are 8–15%. Considering an average of 10%, one in 10 pregnancies detected between days 12–14 and 60 will be lost.

The objective of this chapter is to analyze some critical points of equine ET and their impact on efficiency rates. In this analysis, we will not consider some critical factors that have a major impact on the efficiency of ET programs, such as semen quality, laboratory management, etc., and we will only focus on some specific points recently studied in scientific publications.

2   Donor Mare

2.1   Ovulation Induction

Donor mares' cycle is frequently manipulated to increase efficiency in ET programs. This can be achieved by increasing the number of ovulations/seasons by advancing the first ovulation of the year and using PGF after every flushing, increasing the number of ovulations/cycle by the use of FSH or GnRH analogs and inducing ovulation to inseminate close to ovulation and also to know the precise time of ovulation and predict the time of embryo recovery, especially the collection of small embryos for vitrification.

Spontaneous multiple ovulation is a relatively common event in certain breeds, such as Thoroughbred (22.4%) and Polo horses (29%). Induction of ovulation at the correct time and with the appropriate drug can condition ovulation rates (single or multiple) in a predictable time window and thereby can significantly modify the ERR (Tables 1 and 2).

Significantly higher ERR in donor mares with spontaneous or induced simple (106% vs. 53%) or double ovulations (160% vs. 87%) have been reported (Losinno 2009). Panarace et al. (2014) reported that double-ovulation cycles had higher and statistically different embryo recovery rates per cycle compared with single-ovulation cycles (Table 3).

Superovulation in mares has been attempted by many investigators over the past 40 years. Hormone regimens such as equine chorionic gonadotropin (eCG), gonadotropin-releasing hormone (GnRH), immunization against inhibin, equine pituitary extracts (EPE), and partially purified preparation of equine follicle stimulating hormone (eFSH) have been used to induce superovulation in cycling mares. In addition, recombinant eFSH treatment in cycling and anovulatory mares has been

Table 1 Ovulation rate in young Polo Argentino donor mares using hCG or Biorelease Deslorelin (LADes®)

Treatment	Cycles (n)	Ovulation rate (%)	
		Simple	Multiple
hCG	171	63.2[a]	36.8[d]
LAD	390	56.9[b]	43.1[e]
Control	849	76.1[c]	23.9[f]

[a-c]; [de-f] $p < 0.001$. hCG: Ovusyn®, Syntex SA; LAD: BET LADes®, USA
Adapted from Losinno et al. (2008)

Table 2 Ovulation rate in old Polo Argentino donor mares using hCG or Biorelease Deslorelin (LADes®)

Treatment	Cycles (n)	Ovulation rate (%)	
		Simple	Multiple
hCG	48	58.3[a]	41.7[d]
LAD	266	50.7[b]	49.2[e]
Control	327	68.8[c]	31.2[f]

[a-c]; [d-f] $p < 0.001$. hCG: Ovusyn®, Syntex SA; LAD: BET LADes®, USA
Adapted from Losinno et al. (2008)

Table 3 Embryo recovery rates according to the number of ovulations per cycle

Ovulations/cycle	Cycles (n)	Embryos/cycle (%)
1	698	348/698 (49.8)
2	137	144/137 (105.1)

Adapted from Panarace et al. (2014)

Table 4 Absolute or average data and standard deviations for the variables measured between the control group and the group treated with deslorelin acetate to obtain multiple ovulations

Variables	Control	Tx
Cycles (n)	56	56
Days from treatment to ovulation	8.5 ± 2.9	4.9 ± 1.1
Mares with ≥2 ovulations/cycle	0.0	46 (82%)
Ovulations (n)	56	102
Ovulations/cycle (average)	1.0	1.82
Embryos recovered	32	63
Embryos recovered/cycle (%)	57.0	112.5
Embryos recovered/ovulation (%)	57	61

Adapted from Nagao et al. (2012)

Table 5 Effect of multiple buserelin treatment on multiple ovulation rate in mares

Group	Cycles	Mean age	MO (%)	Double (%)	Triple (%)	Quadruple (%)	Mean ovulation per cycle	Interval Tx to ovulation (Days)
Control (historic)	459	13.6	24.4[a]	22.9[a]	1.3[a]	0.2[a]	1.21[a]	NA
Buserelin	13	13.6	92.3[b]	53.8[b]	30.7[b]	7.7[b]	2.38[b]	4.8

Mares in the treatment group were administered 3.75 mg buserelin every 8 h starting when more than 1 follicle reached 15 mm or larger. Different superscripts ([a,b]) indicate a significant difference ($P < 0.05$). Tx: treatment. MO: multiple ovulation. Buserelin: Suprefact, 1 mg/mL. NA: not applicable
Adapted from Newcombe and Cuervo-Arango (2018)

tested, but the success rate was limited or inconsistent (Roser and Meyers-Brown 2012).

Nagao et al. (2012) reported successful induction of double ovulation in 82% of mares using twice daily injections of 100 mg slow-release deslorelin acetate (Table 4). In a more recent study, Newcombe and Cuervo-Arango (2018) evaluated the effect of repeated treatments with buserelin on multiple ovulation (MO) rates in mares, obtaining an average of 2.38 ovulations per cycle (Table 5).

Table 6 Superovulation, embryo recovery, and pregnancy rates in seasonally anovulatory mares treated with recombinant eFSH (response to treatment in one cycle with rFSH versus PBS in anovulatory mares)

ANOV Group	n	Days of treatment (mean)	Positive response	>35 mm follicles (mean)	Ovulations per mare (mean)	Embryos recovered per mare (mean)	Transferred embryos	PR (%)
1 (rFSH)	10	5.9 ± 0.6^a	8/10	5.2 ± 0.9^a	5.1 ± 1.0	1.9 ± 0.5	14	57
2 (PBS)	10	10 ± 0^b	0/10	0 ± 0^b	0	0	0	–

Different superscripts ([a,b]) indicate a significant difference ($P < 0.05$)
Adapted from Roser et al. (2020)

Table 7 Superovulation, embryo recovery, and pregnancy rates in seasonally anovulatory mares treated with recombinant equine FSH (response to treatment in one cycle with rFSH in ovulatory mares)

Group OV	n	Days of treatment (mean)	Positive response	>35 mm follicles	Ovulations per mare (mean)	Embryos recovered per ovulated mare (mean)	Transferred embryos	PR (%)
3 (reFSH)	10	7	8/10	5.2	4.6	2.7	22	72

Adapted from Roser et al. (2020)

Another use of ovulation induction agents was tested in a recent study by Carmo et al. (2023), who administered low doses of hCG and GnRH agonists before induction of ovulation to promote oocyte maturation in aged mares, improving embryo recovery.

Although there is no eFSH commercially available at the time for inducing MO in mares, there are several reports on the use of this drug to improve ERR. Ross et al. (2012) evaluated the response to follicular stimulation treatment with recombinant equine FSH, using the same dose and protocol, in anovulatory and ovulatory mares and reported an 80% positive response in treated ovulatory and anovulatory mares (Tables 6 and 7).

2.2 Interval from PGF2α Treatment to Ovulation (ITO)

There is evidence that the time interval between treatment with PGF2α and ovulation (ITO) has a significant effect on ERR. Pietrani et al. (2019) determined that ERR increased along with the ITO, with the lowest ERR (30.7%) for mares with an ITO of ≤3 days, and the highest (78.3%) in mares with an ITO of 10 days (Fig. 1). This is consistent with the studies conducted by Pycock in 2007, Cuervo-Arango, and Newcombe in 2015 and 2017, in which a negative effect of short ITOs (<6 days)

1 or more embryos = positive flush
0 embryos = negative flush

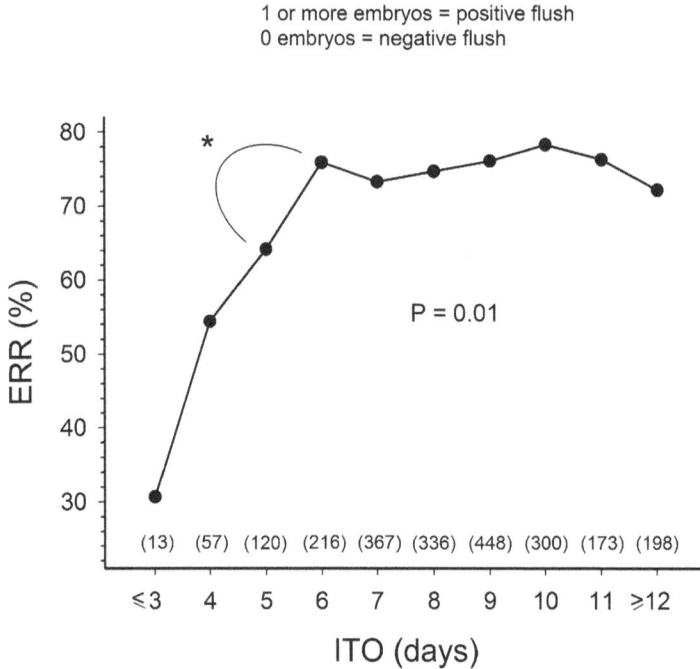

Fig. 1 Embryo recovery rate for different ITOs (adapted from Pietrani et al. 2019)

on PR and ERR was determined in different breeds and reproductive programs. Knowing that fertility is reduced in mares that respond with ovulation quickly after treatment with PGF2α, perhaps the correct indication would be waiting for the follicle to regress or for the mare to spontaneously undergo estrus. The advantage of shortening the cycle with PGF in these mares becomes a disadvantage, knowing that the ERR in mares with short ITO is lower. Mares likely to have a short ITO if treated with PGF (presence of a large diestrous follicle at the time of embryo flushing) would have a lower number of cycles per season if chosen to be left untreated but probably the same number of embryos per season, saving veterinary work, materials, and the cost of semen. Knowledge of the various responses of mares to PGF administration can help improve reproductive efficiency in intensive production systems.

2.3   Embryo Recovery

Embryo recovery rate varies between 26% and 160%, according to several reports presented over the last 30 years. This variability may be due, among other things, to semen quantity and quality, age and reproductive status of the donor, number of days post-ovulation in which the flushing is performed, number of ovulations/estrus cycle, flushing technique, and operator training. In a controlled study, ERR from old

Table 8 Relative diameter of the embryo (µm) in relation to flushing day

Collection day	Embryos (n)	Mean ± D.E. (µm)	Range (µm)
6.5	20	191.8 ± 13.2	150 a 325
7	183	354.0 ± 13.9	150 a 900
8	35	623.9 ± 72.9	150 a 2.500

subfertile mares was half of that for fertile young mares (31% and 71%, respectively). In addition, subfertile mares produced a high proportion of abnormal embryos compared with fertile mares (Marinone et al. 2015).

It is important to consider that the ERR can be expressed in the following ways: (a) the number of embryos recovered/cycle (EPC) and (b) the number of embryos/ovulation (EPO). The result of EPC and EPO varies considerably, especially in Polo Argentino mares, due to their high MO rates (30%) and high repeatability of the event on the same mare (70%). The second is the most accurate index and the best indicator of efficiency and/or skills of the operator but is the least used in practice. The interval between embryo flushings on a donor mare treated with PGF can be as short as 14–17 days during the breeding season, doubling the number of ovulations from 7 or 8 natural ovulations per season to 14–16.

2.4   Uterine Flushing Day

The majority of embryos enter the uterus after oviductal transport, approximately 5.5–6 days post-ovulation (PO), but an average of 144 to 156 h is the most accurate way to consider it (Robinson et al. 2000). Attempts to recover embryos on day 6 PO may result in slightly lower recovery rates compared to days 7, 8, or 9. In our opinion, this assumption may not accurately reflect reality since there are at least two factors that can contribute to considering the result of early flushing as negative: (1) veterinary training to search embryos smaller than 200 microns because most veterinarians are used to working with embryos 7–9 days PO (generally greater than 600–800 microns and many times visible to the naked eye) (Table 8), (2) the use of stereo microscopes with very low optical quality and resolution. Ten-day embryos are more difficult to manipulate and require training and appropriate materials. Otherwise, they can result in lower post-transfer success rates, although recent reports have shown similar ERR and PR for 10-day embryos as for 7– to 8-day PO embryos (Wilsher 2010; Garcia del Gaiso, 2018-personal communication-). The preferred day for embryo recovery is 7.5 days PO, but day 8 is a frequent second alternative. It must also be considered that an old mare's embryo flushing should be carried out on days 9, 10, or 11 PO, considering the prolonged oviductal transit and the lower growth rate of the embryo.

It has been reported, and it is also a perception among veterinarians, that the use of frozen semen in ET programs affects ERR (obtaining lower rates than with fresh or cooled semen) and that the size of the recovered embryos is smaller than the expected for the PO day (Table 9). A common practice is to perform uterine flushing

Table 9 Relative diameter of the embryo at days 7 or 8 post ovulation in mares inseminated with refrigerated or frozen semen (McCue and Squires 2015)

Semen	Day 7 (μm)	Day 8 (μm)
Cooled	401.9 ± 19.6[a]	716.9 ± 104.9[c]
Frozen	258.2 ± 33.3[b]	383.5 ± 54.9[d]

[a,b]$p < 0.05$; [c,d]$p = 0.0553$
Adapted from McCue and Squires (2015)

at days 8 or 9 in mares inseminated with frozen semen. The exact reason for the smaller size of embryos from frozen semen AI has not been determined yet, but it has been hypothesized that it could be due to a delay in fertilization and/or oviductal transport (McCue and Squires 2015). A recent study conducted by Cuervo-Arango et al. (2018c) reported that embryo diameter is influenced by embryonic age but not by the type of semen used (frozen or cooled). The discrepancy with other studies that reported smaller embryo diameter in mares inseminated with frozen semen, compared to those inseminated with fresh semen, appears to originate from differences in insemination timing relative to ovulation and the moment of ovulation.

Recently, Ortiz-Rodríguez et al. (2021) conducted a study involving equine embryos of 8, 10, and 12 days of age recovered from mares inseminated with frozen versus fresh semen. This study revealed a significantly different (down-regulated) transcriptional profile in gene expression involved in oxidative phosphorylation pathways, DNA binding, DNA replication, and immune response, many of them associated with embryonic death in mice. Although it is a preliminary study with a reduced number of embryos ($n = 12$), the results suggest that cryopreservation has an effect on the sperm epigenome and affects early embryonic development.

2.5 Embryo Flushing Technique and Re-flushing

Embryo flushing technique varies greatly among veterinarians. Embryo collection is performed by transcervical uterine lavage with sterile Lactated Ringer's solution (especially in South America) or "complete" flush media (non-PBS) using an 80 cm long silicone Foley catheter (28–34 FR caliper size). The uterus is filled with 0.5–1 l during each flush (total volume of 1–3 l) and effluent is recovered in a closed two-way system with an online filter. The uterus is washed completely, knowing that the equine embryo moves in this period due to myometrial contractions and can be found in any area within the lumen. Despite so many efficient and practical approaches based on operator experience, we prefer to collect the first liter without rectal assistance if possible and split the second liter on two 0.5 l flushes with very active guidance by the rectum during entry and exit. If the fluid output is delayed, which is common in some mares, especially old ones, we use an extra liter and 10 IU of oxytocin IV. A recent study carried out by Sala-Ayala et al. (2023) evaluated the effect of the embryo flushing technique, the number of flushing attempts, and the previous experience of the operator on embryo recovery, concluding that 77% of embryos were recovered in the first filter (23% in the third filter). The experienced operator obtained a higher ERR (10/14) than the beginner (3/14). Finally, even though ERR in the flushing techniques tested did not differ, more embryos were recovered in the first filter in the technique involving uterine massage compared with the control group.

It has been reported that 60–70% of the embryos are recovered in the first flushing ("first liter"). In case of a negative flushing (generally using 2 or 3 l) when the characteristics of the donor, its previous cycle, and semen quality were acceptable, it is recommendable to perform a new flushing (re-flush) immediately after the first one, using 20 IU of oxytocin IV. The recovery rates expected in the second attempt range between 5% and 15%, which is very positive economically and commercially. A second (and possibly complementary) option is to re-flush the mare 6–12 h after the first negative attempt. Although the effluents can be more turbid and with a greater amount of debris, positive flushings have been reported in 9.7% of the attempts (McCue and Squires 2015).

As discussed previously, most veterinarians administer PGF at the end of each embryo flushing (EF) to induce a rapid return to estrus. However, in certain circumstances, this may not be recommended, like in mares with large follicles at the EF. Martínez-Boví et al. (2024) determined that the fertility of mares after repeated EF without administration of PGF is not affected; however, there is a risk of unwanted pregnancy in donor mares following a negative EF (16.7%). Retrospective data indicate that between 10 and 30% of the mares with negative flushing and that were not treated with PGF2α after the flushing are diagnosed as pregnant, which correlates with previous expressed data (McCue and Squires 2015).

Multiple EF in donor mares has been associated with a decrease in the likelihood of recovering embryos (Squires et al. 1982). Martínez-Boví et al. (2024) reported that the likelihood of obtaining a pregnancy or an embryo in mares bred in the first cycles of the study was similar to that of the last cycles (5–6 breedings later).

3   Recipient Mare

Selection and management of the recipient mare is crucial for the success of ET. Recipient mares should ideally be young (3–12 years old), multiparous, or have had one uncomplicated parturition, without medical conditions, have adequate body condition score (3/5), optimal reproductive status and behavior; however, as most of the veterinarians that work on ET knows very well, that is an ideal scenario but very unusual to find in commercial ET programs.

In our opinion, an adequate recipient mare must have an endometrial biopsy score of 1 or 2A (maximum), according to Kenney's score (1978). Studies performed in our laboratory have shown significant EEL in recipient mares that were not subjected to selection by biopsy (Table 10) (Castañeira et al. 2008).

Morelli et al. (2023) tested a real-time method to select the most appropriate recipient mare. In this study, the corpus luteum (CL) characteristics were evaluated by B-mode ultrasonography to determine area, Color-Doppler ultrasonography to determine blood perfusion, and a blood sample was collected for plasma P4 assay.

Table 10 Pregnancy rates and embryonic loss in recipient mares

Mares	n	PR (%)		EEL (%)	
		Day 14	Days 15–130	Days 31–60	Total
With biopsy	1194	72.5	5.6[a]	1.6[c]	7.2[e]
Without biopsy	926	71.8	8.2[b]	5.4[d]	13.2[f]
Total	2129	72.2	6.8	3.2	9.8

[a,b]; [cd]; [ef] $p < 0.001$. PR: pregnancy rate; EL: embryonic loss
Adapted from Castañeira et al. (2008)

They determined that PR/ET was greater in the high luteal perfusion group, which implies that these mares would have higher chances of maintaining pregnancy in ET programs.

In countries where large numbers of embryos are produced per year (more than 20,000), it may be difficult to acquire mares to use as ideal recipients. Therefore, new alternatives have emerged to supply the equine ET industry. In this regard, mules ($2n = 63$) have been suggested for more than 40 years as recipient options for equine ET. Some mules cycle normally and develop pre-ovulatory follicles, which ovulate and form active CLs that produce enough progesterone to sustain the pregnancy. In non-cycling mules, progesterone supplementation is necessary to sustain pregnancy (Fig. 2).

3.1 Synchronization

Reproductive management and synchrony of recipient mares have been considered critical factors that affect pregnancy rates in ET programs. Optimal pregnancy rates are obtained when embryos are transferred to mares synchronized—1 (ovulated 1 day before the donor) to +3 (ovulated 3 days after the donor) (Stout 2003). However, Wilsher et al. (2010) reported that embryos recovered at day 10 PO can be transferred to ovulated recipients—2 to +5 days in relation to the donor without a significant decrease in pregnancy rates. Other studies (Jacob et al. 2012) conducted on a significant number of mares demonstrated an acceptable pregnancy rate (70%) in recipient mares that ovulated 4 or 5 days after the donor. Gibson et al. (2017) determined that embryos transferred to recipients with a significant asynchrony with the donor (+5) suffered a delay in development, but they did not seem unregulated. This is probably because equine embryos are more tolerant to asynchrony between donor and recipient than other species. Oliveira Neto et al. (2018) reported acceptable pregnancy rates (not significantly different) in recipient mares in different reproductive stages (deep anovulatory, transitional anovulatory, early estrus, late estrus, late diestrous, and early diestrous treated with PGF2α). The recipients were treated with long-acting injectable altrenogest (300 mg/ml, IM) 4 days before the transfer, the same day of the transfer, and then maintained until day 120 of pregnancy with long-acting injectable progesterone (Table 11). For any commercial program, this could be a practice to consider, reducing the time used to check the recipients and the total number of recipients required with conventional methods. Cuervo-Arango et al.

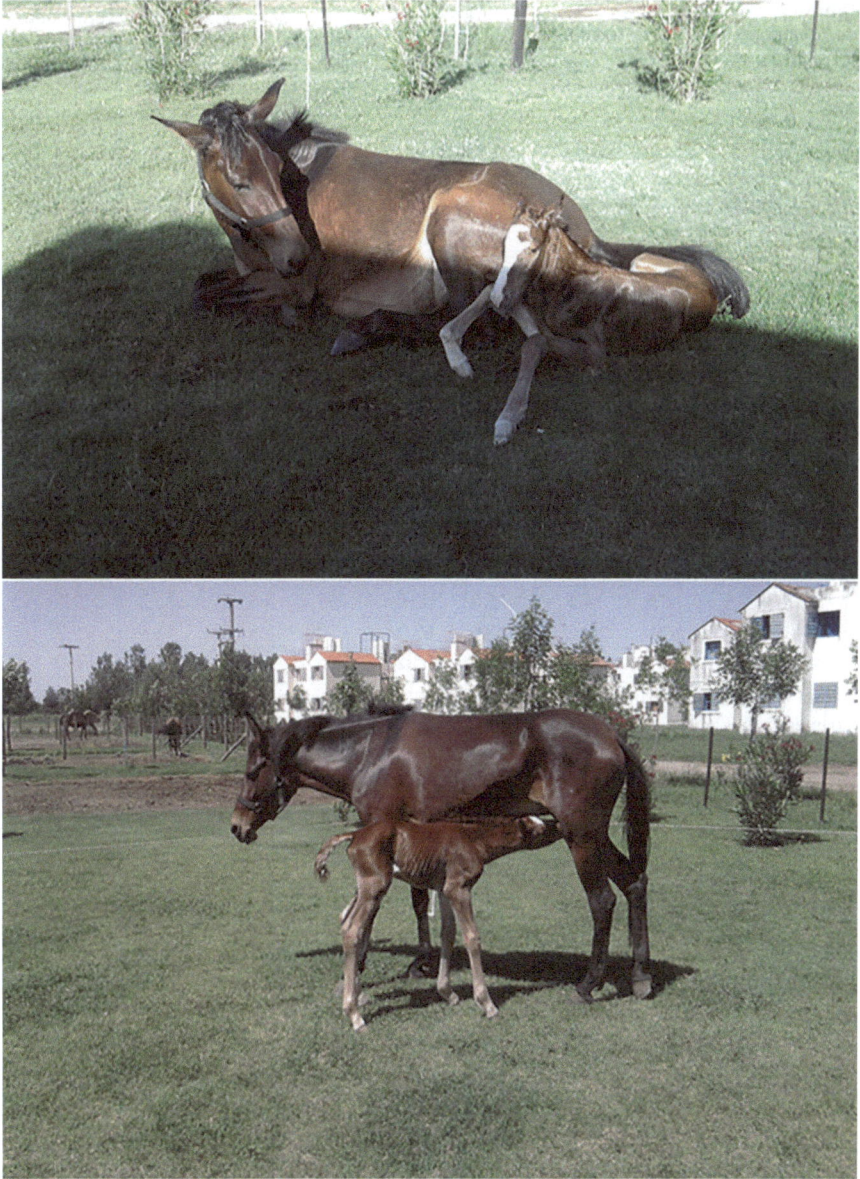

Fig. 2 Mule recipient foaling a horse

(2018a, b) determined the effect of the recipient's stage after ovulation on PR, embryonic vesicle growth, and EEL for flushed embryos classified according to embryonic diameter. The recipient's day of ovulation did not influence either PR or EEL ($P > 0.05$). Small in vivo embryos had similar PR and EEL following ET into recipients with less advanced or similar uterine stages. However, small embryos had

Table 11 Pregnancy rates post-embryo transfer in synchronized recipient mares treated with long-acting Altrenogest

Group	% Pregnancy (pregnancy/embryo transferred)
Control	75 (15/20)
Deep anovulatory	90 (18/20)
Anovulatory (spring)	90 (18/20)
Early estrus	75 (15/20)
Late estrus	75(15/20)
Late diestrus	65 (13/20)
Early diestrus	40 (8/20)
Early diestrus + PgF2alfa	75 (15/20)

Adapted from Oliveira Neto et al. (2018)

Table 12 Relationships between in vivo embryo diameter and recipients' day of ovulation on embryo survival

Day of ovulation	Embryo diameter							
	<300 μm				>300 μm			
	n	PR 12d	EEL	PR 45d	n	PR 12d	EEL	PR 45d
4–5	30	80.0	20.8	63.3	0	76.7	6.5	71.7
6–7	27	74.1	15.0	63.0	98	85.7	7.1	79.6
8–9	19	84.2	25.0	63.2	109	82.6	8.9	75.2
Overall	76	78.9	20.0[a]	63.2[x]	267	82.4	7.7[b]	76.0[y]

Within the overall row, different superscripts ([a,b]; [x,y]) indicate significant differences ($P < 0.05$)
Adapted from Cuervo-Arango et al. (2018a, b)

a higher EEL and corresponding lower PR at day 45 of pregnancy than larger embryos (Table 12).

3.2 Estrus Length

Recipient mares are usually monitored to detect ovulation and the presence of free uterine fluid before ovulation. Cuervo-Arango et al. (2017) showed that PR in recipient mares is positively correlated to the number of days of endometrial edema during the estrous cycle before ET. According to this study, the PR of a recipient with 3 or more days of endometrial edema during estrus, before embryo transfer, was significantly higher (83.1%; 157/189) than recipients with less than 3 days of edema before ET (63.6%; 77/121). Recipients with zero days of edema before ET had the lowest pregnancy rates (50%, 11/22) (Table 13). This suggests the important role of estrogens during estrus in uterine receptivity and embryonic survival post-ET. Based on the most recent studies, and probably the next ones that will come, it would be important to have closer and more detailed control of the recipient estrous cycle, in particular the number of days of endometrial edema, considering its impact on pregnancy rate. If possible, ovulation should not be induced in recipient mares to maximize estrus length. In the same way, PGF to induce estrus should only be used when recipients have a mature CL (>6 days) to avoid the risk of a partial regression of the

Table 13. Estrus characteristics of recipient mares and the likelihood of pregnancy and early embryonic death after embryo transfer

Type of cycle	ITO (days)	n	Median length of edema (days)	MO (%)	P (%)	EEL (%)
PGF induced	3–6	81	2	25.9	65.4[a]	20.8
	7–8	46	3	30.4	82.6[b]	10.5
	≥9	83	4	39.8	84.3[b]	17.1
Spontaneous		122	3	26.2	74.6[b]	14.3
1st Ov. of year		18	6	16.7	88.9[b]	6.3

Adapted from: Cuervo-Arango et al. (2018a, b)

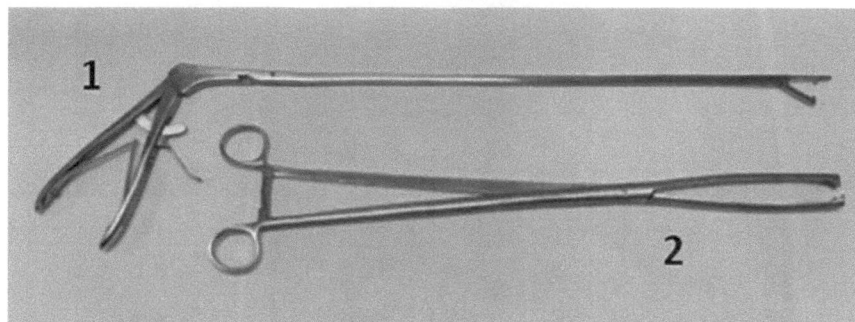

Fig. 3 Wilsher's equine cervical forceps (1) and bovine cervical forceps (2)

CL and ovulation without signs of estrus (Cuervo-Arango and Newcombe 2012). In addition, PGF should be administered only to recipient mares with follicles <25 mm to allow a longer estrus.

3.3   Embryo Transfer Method

Initially, in the late 1970s, equine embryos were largely transferred by surgical approach through ventral midline; later, in the 1980s, through flank laparotomy. Nonsurgical transcervical transfer technique was reported in those years but with lower pregnancy rates, probably due to excessive manipulation of the cervix leading to prostaglandin release and uterine contamination. Subsequently, in the early 1990s, high pregnancy rates were reported by nonsurgical transfer in ET centers in Argentina and Brazil, which indicated an important operator's experience effect and procedural quality on the results. This system was quickly adopted by commercial programs, and it is the most used nowadays.

Wilsher and Allen (2004) developed a technique involving the insertion of a duck-billed speculum into the vagina (to allow visualization and grasping of the external cervical os) and straightening of the cervix with the toothed Wilsher forceps. This facilitates the transfer of gun entry to the uterus, with minimal manipulation and fundamentally without contact with the vaginal wall (Figs. 3, 4, and 5).

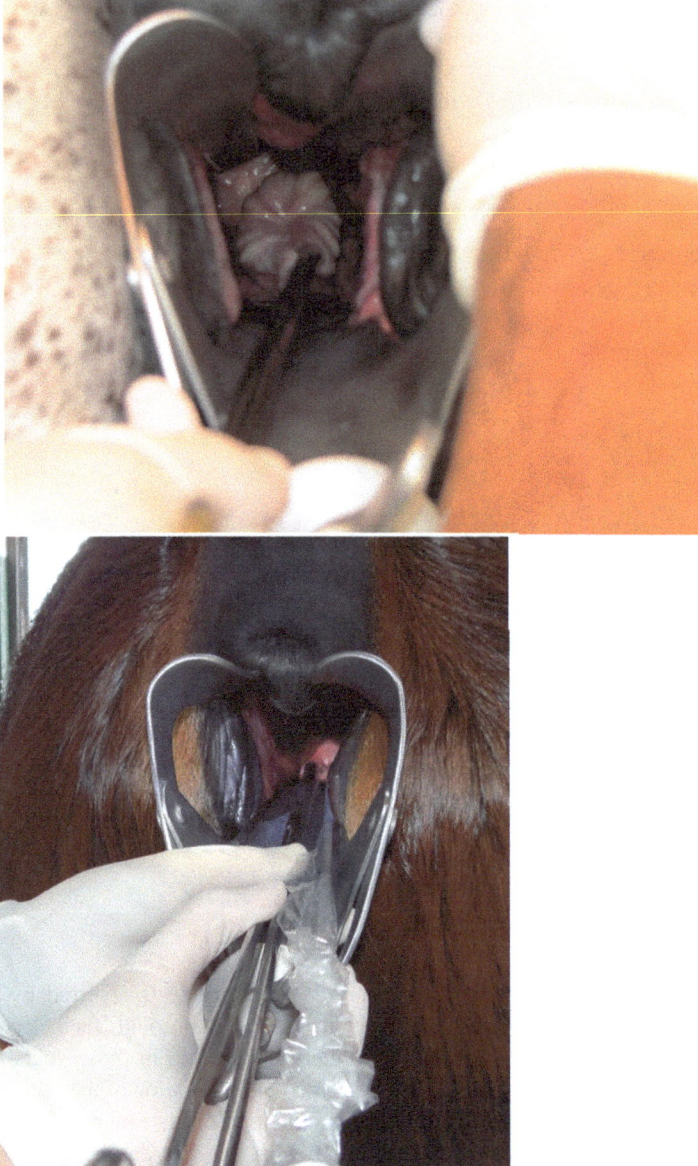

Fig. 4 Polansky speculum and cervical forceps during the embryo transfer procedure

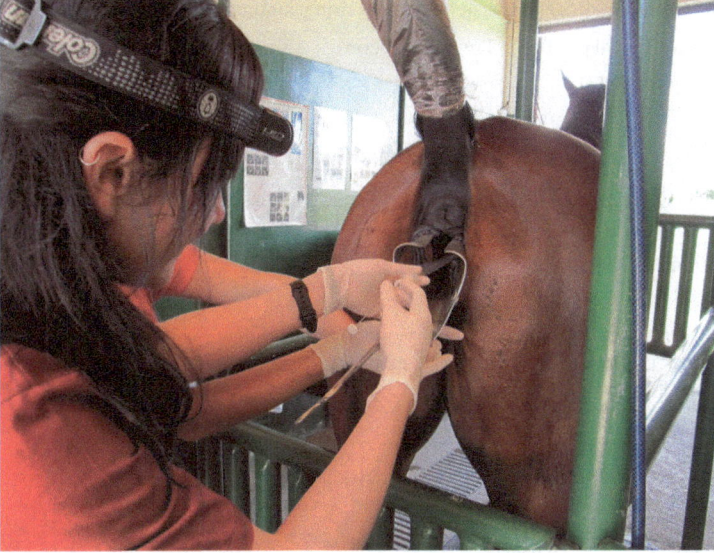

Fig. 5 Embryo transfer procedure using the Wilsher ET technique

Diverse technical and biological factors may influence the overall PR, but they are mostly related to operator skills. Recent studies by Cuervo-Arango et al. (2018a, b) and Ramirez et al. (2023) conducted in commercial programs compared pregnancy rates obtained through conventional transfer and Wilsher cervical forceps transfer involving different operators and expertise. The results of both studies demonstrate a significantly higher pregnancy rate on average using the Wilsher technique. Also, they showed that the effect of operator experience on PR can be overcome with the use of the Wilsher technique (Table 14). Ramirez et al. (2023) also determined that bovine cervical forceps are equivalent to the specifically designed equine forceps reported by Wilsher and Allen (Table 15).

4   Embryo Vitrification

Vitrification is a physical cryopreservation process where a liquid solution is transformed into a particular amorphous and stable solid, called vitreous, when it freezes rapidly at ultra-low temperatures (cryogenic) using high concentrations of cryoprotectants. As a result, intra and extracellular fluids become more viscous as the medium cools, preventing the binding of water molecules and, consequently, ice crystal formation. The cooling process must be very fast to obtain a vitreous state almost immediately (seconds). This state has the ionic and molecular distribution of a liquid, which is why the harmful effects (mechanical and chemical) of the ice crystals formed during conventional cryopreservation are avoided.

Table 14 Characteristics and transfer outcomes using conventional versus Wilsher techniques

Veterinarian	Clinical position	Previous experience	Technique	n	PR (%)	EEL (%)
1	Resident	Medium	Conventional	53	50.9	11.1
2	Specialist	High	Conventional	52	78.8	4.9
3	Specialist	High	Conventional	74	79.7	8.5
4	Specialist	Low	Wilsher	91	93.4	7.2
5	Specialist	Low	Wilsher	68	91.2	11.4
6–7	Intern	None	Wilsher	11	90.9	10.0

Adapted from Cuervo-Arango et al. (2018a, b)

Table 15 Pregnancy rate and early embryonic loss in recipient mares transferred with different embryo transfer techniques (Conventional: manual and transvaginal; Wilsher: cervical forceps and transvaginal) and operators experience

Operator	Technique	n	PR (%)	EEL (%)
Low experience	Conventional	86	56.98	6.12
High experience	Conventional	259	72.20	5.88
Low experience	Wilsher	81	83.95	4.41
High experience	Wilsher	147	86.39	4.72

Adapted from Ramirez et al. (2023)

Table 16 Success rate when attempting to nonsurgically collect equine embryos at 144, 156, or 168 h after ovulation

Post-ovulation time			
	144 h Day 6	156 h Day 6.5	168 h Day 7
Mares (n)	20	27	23
Flushings (+)	0	9	12
Recovery rate (%)	0	53	52
Embryo diameter average (μm)	–	188 ± 9	244 ± 15
Range of embryo diameter (μm)	–	162–245	159–365

Adapted from Battut et al. (1998)

Vitrification of equine embryos is a simple process carried out in a short time (15 min per embryo) and does not require complex equipment. Once the embryo has been vitrified, it can be kept indefinitely in liquid nitrogen. Vitrified embryos should be transferred to recipients between 5 and 6 days PO, ideally 156–160 h PO.

It is known that, to be vitrified successfully with commercial embryo transfer kits, embryo size should be ≤300 μm (morula or early blastocysts); recovering such embryos when flushing the mare at day 6 PO can result in low ERR as the embryo only enters the uterus on day 5.5–6. In our opinion and experience, to obtain embryos of that size, it is essential to flush mares (young ones) between 156–160 h PO and not to take generic time values such as 5.5 or 6 days PO (Table 16). Donor mares should be ultrasound frequently (less than 6 h intervals) to maximize the precision in the ovulation determination in a range no greater than 3 hours. The probability of obtaining embryos smaller than 250 μm in diameter is very high, and the ERR should not be significantly lower since the operator is looking for (and must be prepared for) very small embryos that often stick to detritus and are hard to find. The quality of the stereomicroscope for the search and manipulation of these embryos is also very important. A study by Couto et al. (2023) evaluated the effect of re-flushing donor mares that resulted in negative results at 6.5 days, obtaining 16.6% and 20.6% ERR on second- and third-day lavage, respectively.

Because of the presence of the embryonic capsule and a large amount of blasto-coelic fluid, to achieve acceptable pregnancy rates following vitrification, large embryos require micromanipulator-assisted puncture and blastocoele collapse. Until recently, larger than 300 μm embryos were punctured and collapsed before vitrification. Sanchez et al. (2017) and Wilsher et al. (2018) showed that >300 μm embryos can be vitrified and transferred with acceptable pregnancy rates (70%) using specific protocols for vitrification and embryo treatment (aspiration of >90% of the blastocoele or punction) depending on the size (Table 17). This is an important advantage for cases in which "large" embryos (300–600 μm) are recovered, but no recipients are available for ET. Wilsher et al. (2019) showed that vitrification of embryos >550 μm yielded a significantly higher PR when they were aspirated vs. not-aspirated [13/18 (72%) vs. 1/10 (10%), respectively, whereas there was no difference for smaller embryos [8/10 (80%) vs. 9/12 (75%), respectively] (Table 18).

Sanchez et al. (2017) reported that slow freezing or vitrification with a commercial kit was equally effective with embryos <300 μm. With embryos larger than 300 μm, high pregnancy rates were obtained only when they were collapsed and vitrified using a hemi-straw.

These results demonstrate the benefits of embryo aspiration and vitrification, although micromanipulator-assisted puncture and blastocoele collapse require sophisticated materials and skills, which can be inconvenient for veterinarians in the field or in facilities where the necessary equipment is not available. Wilsher et al. (2021) reported high pregnancy rates following vitrification of embryos 300–560 μm using manual puncture (without using a micromanipulator) but failed to achieve any pregnancies with non-punctured embryos. However, Kovacsy et al. (2023) reported excellent pregnancy rates (19/24 [79%]) in <500 μm non-punctured embryos

Table 17 Pregnancy rates after transfer of frozen-thawed equine embryos of small (<180 μm) or medium (180–300 μm) diameter using vitrification or slow freezing

Method	Vitrification commercial kit		Slow freezing	
Embryonic size	<180 μm	180–300 μm	<180 μm	180–300 μm
Number of embryos transferred	61	11	8	7
Pregnancy rate (16)	55.7[a]% (34/61)	0[b]% (0/11)	75[a]% (6/8)	14.3[b]% (1/7)
Pregnancies (45)	33	0	6	1
Number of births	33	0	6	1

Values with different superscripts differ significantly ($p < 0.05$)
Adapted from Sanchez et al. (2017)

Table 18 Mean (± s.e.m.) embryo diameter, range of diameters, and pregnancy rates 9 days post-transfer for horse embryos punctured or punctured and aspirated, prior to vitrification

Treatment group	Mean ± s.e.m. embryo diameter (range; μm)	Pregnancy rate (%)	P value[a] (z-test)
All embryos			
Punctured only (n = 22)	646.4 ± 61.7 (300–1300)	10/22 (45%)	0.061
Punctured and aspirated (n = 28)	754.8 ± 59.1 (400–1600)	21/28 (75%)	
Embryos ≤550 μm			
Punctured only (n = 12)	436.3 ± 26.4 (300–550)	9/12 (75%)	0.816
Punctured and aspirated (n = 10)	467.0 ± 14.5 (400–550)	8/10 (80%)	
Embryos >550 μm			
Punctured only (n = 10)	898.5 ± 74.5 (565–1300)	1/10 (10%)	0.006
Punctured and aspirated (n = 18)	914.7 ± 66.1 (600–1600)	13/18 (72%)	

[a]Statistical difference between pregnancy rates within the groups
Adapted from Wilsher et al. (2019)

Table 19 Pregnancy outcomes with fresh, cooled, and vitrified embryos in a large-scale commercial program in the USA (Center A)

Embryo	n	Size (μm)	Pregnancy rate (day 11) %	Pregnancy rate (day 25) %
Fresh	3060	>300	79.6	68.9
Cooled	1479	>300	76.8	70.2
Vitrified	296	150–174	69	63
Vitrified	108	175–200	64	59
Vitrified	17	>200	71	53

Adapted from McCue and Squires, 2016

following a 15-min exposure to an equilibration medium containing 7.5%EG and 7.5%DMSO. These results show that embryo puncture in <500 μm embryos may not be necessary, providing an easy and successful way to vitrify them.

Early results in experimental embryo vitrification programs were promising but could not be replicated in commercial programs, even with experienced operators (Eldridge-Panuska et al. 2005). The technique fell into professional discredit until recently when very good results were reported on a large-scale commercial ET program comparable with fresh or cooled embryos (Tables 19 and 20). However, data and anecdotal communications of field results from commercial programs (which, unfortunately, are rarely published) are still very variable and difficult to compare due to the great diversity of scenarios in which they are generated.

Vitrification is useful in mares whose "quota" of pregnancies has been fulfilled, but it is desired to continue producing embryos until the end of the season. Another advantage is the import–export business of vitrified embryos. For aged donor mares, whose fertility is reduced every year, it is important to obtain and vitrify embryos during their fertile cycles. Embryo cryopreservation is also an option for owners who want to keep their mares in competition. Cryopreservation allows the

Table 20 Pregnancy outcomes with vitrified embryos in a large-scale commercial program in the USA (Center B)

Embryo size (μm)	n	Pregnancy rate (day 14) %	Pregnancy rate (day 45) %
<300 μm	239	58	51

Adapted from McCue and Squires (2016)

production of embryos from mares without the pressure of recipient synchronization. Embryo transfers can be programmed at the most appropriate time of the year and when recipients are available.

5 Conclusions

There is still much to learn from the data generated in commercial and experimental ET programs. Although this technique has been inserted into production systems for more than 40 years, as we have seen, the information generated can potentially increase efficiency indicators in commercial programs. We should not fall into the trap of considering that our own experience, although important, is the only source of knowledge (something unfortunately widespread in "successful" professional circles). The biomedical sciences, in particular biotechnologies, constantly generate available and new knowledge, more and more easily, for all those who have the intent to take it.

References

Battut I, Colchen F, Fieni F, Tainturier D, Bruyas JF (1998) Success rate when attempting to nonsurgically collect equine embryos at 144, 156 or 168 hours after ovulation. Equine Vet J 25:60–62

Carmo MT, Nunes S, Freitas-Dell'Aqua C, Alvarenga MA, Dell'Aqua JA Jr, Canisso IF (2023) A combination of GnRH-hCG in aged mares improves embryo recovery. J Equine Vet Sci 125:104646

Castañeira C, Alonso C, Vollenwieder A, Miragaya MH, Losinno L (2008) Uterine biopsy score and pregnancy loss in embryo recipient mares. In: Havemeyer foundation monograph series. VIIth international symposium on equine embryo transfer, pp 96–97

Couto GR, Vigano DWA, dos Santos GdC, Allen WR, Wilsher S (2023) Embryo recovery on consecutive days from Day 6.5 to obtain small embryos for vitrification: is it effective? J Equine Vet Sci 125:104643

Cuervo-Arango J, Newcombe JR (2012) Relationship between dose of cloprostenol and age of corpus luteum on the luteolytic response of early dioestrous mares: a field study. Reprod Domest Anim 47:660–665

Cuervo-Arango J, Mateu-Sánchez S, Aguilar JJ, Nielsen JM, Etcharren V, Vettorazzi ML, Newcombe JR (2015) The effect of the interval from PGF treatment to ovulation on embryo recovery and pregnancy rate in the mare. Theriogenology 83:1–7

Cuervo-Arango J, Claes AN, Ruijter-Villani M, Stout TA (2017) Likelihood of pregnancy after embryo transfer is reduced in recipient mares with a short preceding oestrus. Equine Vet J 50:386–390

Cuervo-Arango J, Claes AN, Stout TA (2018a) The effect of embryo transfer technique on the likelihood of pregnancy in the mare: a comparison of conventional and Wilsher forceps assisted transfer. Vet Rec 183(10):323. https://doi.org/10.1136/vr.104808

Cuervo-Arango J, Claes AN, Stout TAE (2018b) Small day 8 equine embryos cannot be rescued by a less advanced recipient mare uterus. Theriogenology 126:36. https://doi.org/10.1016/j.theriogenology.2018.11.026

Cuervo-Arango J, Claes AN, Stout TAE (2018c) Horse embryo diameter is influenced by the embryonic age but not by the type of semen used to inseminate donor mares. Theriogenology 115:90–93. https://doi.org/10.1016/j.theriogenology.2018.04.023

Eldridge-Panuska WD, Caracciolo di Brienza V, Seidel GE Jr, Squires EL, Carnevale EM (2005) Establishment of pregnancies after serial dilution or direct transfer by vitrified equine embryos. Theriogenology 63:1308–1319

Gibson C, Ruijter-Villani M, Stout T (2017) Negative uterine asynchrony retards early equine conceptus development and upregulation of placental imprinted genes. Placenta 57:175–182

Hartman DL (2011) Embryo transfer. In: McKinnon AO (ed) Equine reproduction, 2nd edn. Wiley, Blackwell, Oxford

Jacob J, Haag K, Santos G, Oliveira J, Gastal M, Gastal E (2012) Effect of embryo age and recipient asynchrony on pregnancy rates in a commercial equine embryo transfer program. Theriogenology 77:1159–1166

Kenney RMC (1978) Cyclic and pathologic changes of the mare endometrium as detected by biopsy, with a note on early embryonic death. J Am Vet Med Assoc 172(3):241–262

Kovacsy S, Ismer A, Funes J, Hoogewijs M, Wilsher S (2023) Successful vitrification of embryos ≤500μm without puncture or aspiration of the blastocoele. J Equine Vet Sci 125:104656

Losinno L (2009) Factores críticos del manejo embrionario en programas de transferencia embrionaria en equinos. In: Proceedings I Congreso Argentino de Reproducción Equina, 89–94. http://www.congresoreproequina.com.ar/articulos.html

Losinno L, Alonso C, Rodriguez D (2008) Ovulation and embryo recovery rates in young and old mares treated with hCG or deslorelin. In: Havemeyer Foundation Monograph Series, VIIth International Symposium on Equine Embryo Transfer, 96–97

Losinno L, Aguilar JJ, Lisa H (2009) Impact of multiple ovulations in a commercial equine embryo transfer program. Havenmeyer Foundation Monograph Series 3:81–83

Marinone AI, Losinno L, Fumuso E, Rodriguez EM, Redolatti C, Cantatore S, Cuervo-Arango J (2015) The effect of mare's age on multiple ovulation rate, embryo recovery, post-transfer pregnancy rate, and interovulatory interval in a commercial embryo transfer program in Argentina. Anim Reprod Sci 158:53–59

Martínez-Boví R, Sala-Ayala L, Querol-Paajanen A, Plaza-Dávila M, Cuervo-Arango J (2024) Effects of repeated embryo flushing without PGF2α administration on luteal function, percentage of unwanted pregnancy and subsequent fertility in mares. Equine Vet J 56:1–10

McCue P, Squires EL (2015) Equine embryo transfer. Teton New Media, 172 pp

Morelli KG, Lourenço GG, Marangon VR, Feltrin IR, Oshiro TSI, Pugliesi G (2023) Doppler ultrasonography: an improvement in real-time recipient selection in equine ET programs. J Equine Vet Sci 125, 104665

Nagao et al (2012) Induction of double ovulation in mares using deslorelin acetate. Anim Reprod Sci 136:69–73

Newcombe JR, Cuervo-Arango J (2018) Induction of multiple ovulation in mares following multiple treatment with low dose Buserelin. J Equine Vet Sci 66:95

Oliveira Neto IV, Canisso I, Segabinazzi L, Dell'Aqua CPF, Alvarenga MA, Papa FO, Dell'Aqua JÁ (2018) Synchronization of cyclic and acyclic embryo recipient mares with donor mares. Anim Reprod Sci 190:1–9

Ortiz-Rodríguez JM, Martín-Cano FE, Gaitskell-Phillips G, Álvarez Barrientos A, Rodríguez-Martínez H, Gil M, Ortega-Ferrusola C, Peña FJ (2021) Sperm cryopreservation impacts the early development of equine embryos by downregulating specific transcription factors. https://doi.org/10.1101/2021.05.12.443855

Panarace M, Pellegrini RO, Basualdo MO, Belé M, Ursino DA, Cisterna R, Desimone G, Rodríguez E, Medina MJ (2014) First field results on the use of stallion sex-sorted semen in a large-scale embryo transfer program. Theriogenology 81:520–525

Pietrani MS, Cuervo Arango J, Losinno L (2019) Effect of the interval from prostaglandin F2α treatment to ovulation on embryo recovery rate and subsequent pregnancy rate in a commercial equine embryo transfer program. J Equine Vet Sci 41:58

Ramirez H, Belmar F, Cuervo-Arango J, Losinno L (2023) Embryo transfer technique affects pregnancy rates in recipient mares. J Equine Vet Sci 125:104672

Robinson SJ, Neal H, Allen WR (2000) Modulation of oviductal transport in mares by local application of prostaglandin E2. J Reprod Fertil Suppl 2000(56):587–592

Roser JF, Meyers-Brown G (2012) Superovulation in the Mare: a work in Progress. J Equine Vet Sci 32(7):376–386

Roser JF, Etcharren MV, Miragaya MH, Mutto A, Colgin M, Losinno L, Ross PJ (2020) Superovulation, embryo recovery, and pregnancy rates from seasonally anovulatory donor mares treated with recombinant equine FSH (reFSH). Theriogenology 142:291–295

Ross PJ, MacDonough J, Bereterbide S, Garzaron A (2012) Superovulation of cycling donor mares using recombinant equine gonadotropins and different ovulation induction agents. J Equine Vet Sci 32(7):403. https://doi.org/10.1016/j.jevs.2012.05.019

Sala-Ayala L, Martínez-Boví R, Arnal-Arnal G, Querol-Paajanen A, Plaza-Dávila M, Cuervo-Arango J (2023) Effect of embryo flushing technique, number of flushing attempts and previous experience of the operator on embryo recovery. J Equine Vet Sci 125:104676

Sanchez R, Blanco M, Weiss J, Rosati I, Herrera C, Bollwein H, Burger D, Sieme H (2017) Influence of embryonic size and manipulation on pregnancy rates of mares after transfer of cryopreserved equine embryos. J Equine Vet Sci 49:54–59

Squires EL, Imel KJ, Iuliano MF, Shideler RK (1982) Factors affecting reproductive efficiency in an equine embryo transfer programme. J Reprod Fertil Suppl 32:409–414

Stout TAE (2003) Selection and management of the embryo transfer donor. Pferdeheilkunde 6:685–688

Willsher S, Allen WR (2004) An improved method for nonsurgical embryo transfer in the mare. Equine Vet Ed 16:39–44

Wilsher S, Clutton-Brock, Allen WR (2010) Successful transfer of day 10 horse embryos: influence of donor-recipient asynchrony on embryo development. Reproduction 139(3): 575–85

Wilsher S, Rigali F, Couto G, Camargo S, Allen WR (2018) Is aspiration of blastocoelic fluid essential to vitrify large equine embryos? Equine Vet J 50 (Suppl 52)19 (Abstr)

Wilsher S, Rigali F, Couto G, Camargo S, Allen WR (2019) Vitrification of equine expanded blastocysts following puncture with or without aspiration of the blastocoele fluid. Equine Vet J. ISSN 0425-1644. https://doi.org/10.1111/evj.13039

Wilsher S, Rigali F, Kovacsy S, Allen WR (2021) Successful vitrification of manually punctured equine embryos. Equine Vet J 53:1227. https://doi.org/10.1111/evj.13400

Developmental Programming and Assisted Reproductive Technologies in Cattle

Eliab Estrada-Cortés ⓘ, Luiz Gustavo Siqueira ⓘ, and Jeremy Block ⓘ

Abstract

The objective was to analyze the nature of developmental programming, the characteristics of some assisted reproductive technologies, and the importance of their relationship on the expression of important productive traits in the resultant offspring in cattle. Developmental programming refers to the phenomenon by which organisms during gestation are prepared to have a phenotype that fits better to the environment present once they are born. However, cattle are commonly exposed to conditions like undernutrition or heat stress that might program a phenotype in the opposite direction according to common objectives of productive systems like greater beef or milk production per animal unit. Mechanisms for developmental programming involve changes in epigenetic marks like DNA methylation, which regulates chromatin accessibility, gene expression, and consequently, organisms' physical and physiological status. Embryos experience extensive changes in epigenetics marks and, therefore, are highly susceptible to developmental programming. Assisted reproductive technologies (ART) are valuable tools for genetic improvement programs in cattle and normally involve the manipulation of gametes and embryos under non-natural environmental conditions. Therefore, animals derived from ART might express undesirable pheno-

E. Estrada-Cortés (✉)
Campo Experimental Centro-Altos de Jalisco, Instituto Nacional de Investigaciones Forestales, Agrícolas y Pecuarias, Tepatitlán de Morelos, Jalisco, México
e-mail: estrada.eliab@inifap.gob.mx

L. G. Siqueira
Embrapa Gado de Leite, Av. Eugenio do Nascimento, Juiz de Fora, Minas Gerais, Brazil
e-mail: luiz.siqueira@embrapa.br

J. Block
Department of Animal Science, University of Wyoming, Laramie, WY, USA
e-mail: jeremy.block@uwyo.edu

© The Author(s), under exclusive license to Springer Nature Switzerland AG 2024
J. C. Gardón, K. Satué Ambrojo (eds.), *Assisted Reproductive Technologies in Animals Volume 1*, https://doi.org/10.1007/978-3-031-73079-5_4

types like a reduction in their productivity. However, recent studies suggest that productive traits might also be positively programmed in ART-derived embryos if appropriate conditions or signals are provided.

Keywords

Developmental programming · Dairy and beef cattle · Productive traits · Epigenetic mechanisms · Early embryo development · Assisted reproductive technologies

1   Introduction

It is estimated that billions of people throughout the world suffer from undernutrition or food insecurity (FAO 2021). This situation causes nutritional deficiencies in certain populations with inevitable consequences for their health; for example, in pregnant mothers such deficiencies during the perinatal period give rise to low body development of children, which generates poor cognitive performance, among other problems (FAO 2021). Additionally, recent projections suggest that the world human population will be increasing during the next two decades (FAO 2022). This standpoint highlights the relevance of increasing the availability of products from plant and animal sources to feed the growing human population. Therefore, it is justified to use any sustainable strategy to increase productivity (i.e., the amount of food per animal unit) in the cattle production systems, especially under the altered environmental scenarios that may arise due to the effects of climate change and progressive limited use of land for agricultural purposes due to the required preservation of natural resources.

In the animal kingdom, developmental programming is a phenomenon by which organisms during gestation are prepared to have a phenotype that fits better to the environment present once they are born (Bateson et al. 2014). Even though this phenomenon seems to provide a better possibility of survival (Gluckman et al. 2019), sometimes, in domestic animals, the resultant phenotypes follow an opposite direction to the common productive objectives like obtaining fewer beef kilograms per animal unit. As an example, when pregnant beef cows are subjected to undernutrition, their calves have a lower body development and muscle fiber size during postnatal life (Bohnert et al. 2013; Costa et al. 2021; McLean et al. 2018). Theoretically, this occurs because individuals in utero receive programming signals indicating a suboptimal environmental condition, so they are programmed to have metabolism in an economy mode (Sirard 2021).

The impact of potential positive or negative developmental programming on productivity is unknown. However, considering that every animal's productive system is commonly exposed to factors that generate negative developmental signals like undernutrition (Costa et al. 2021) or heat stress (Laporta et al. 2020), productivity loss due to developmental programming might be considerable. Most evidence indicates that epigenetic marks, including DNA methylation, histone posttranslational modifications, and noncoding RNAs, are involved in the mechanism regulating such phenomenon through the regulation of chromatin accessibility and gene expression (Greenberg and Bourc'his 2019). These epigenetic marks are characterized by their plasticity, especially during early embryo development (Hansen et al. 2016). Thus, it could be possible to manipulate signals to reverse the negative developmental programming or even to induce positive effects.

The use of ART in genetic improvement programs is one of the most common and effective strategies to increase animal productivity (Mueller and Van Eenennaam 2022). Cows and bulls with the highest genetic merit for traits of interest for a milk or beef production system are typically selected to be multiplied by ART. Genetic gain is increased due to the intensity of selection and reduction of generational intervals (Baruselli et al. 2023). In cattle (Hasler 2023), two of the most common assisted reproductive technologies include embryo transfer (ET) from multiple ovulation and in vivo derived embryos (IVD) and ultrasound-guided transvaginal ovum pick up (OPU) and in vitro produced embryos (IVP). The use of ART involves the manipulation of gametes and preimplantation embryos in non-natural conditions and during a sensitive period of development, which can lead to alterations in epigenetic marks (Jiang 2023). Associations between the use of ART and some abnormalities like Large Offspring Syndrome, as well as alterations in body growth, mortality, and milk production, have been reported (Siqueira et al. 2017a, 2019; Li et al. 2022). However, embryo plasticity may allow for the capacity to reverse the negative effects of ART and potentially promote improvement in cattle performance if appropriate conditions or signals are provided during such processes (Siqueira et al. 2017b; Estrada-Cortés et al. 2021a). This chapter analyzes the nature of the developmental programming phenomenon, the characteristics of some assisted reproductive technologies, and the importance of their relationship on the expression of important productive traits in the resultant offspring in cattle.

2   Developmental Programming

Developmental programming has been described in several species, including insects (Koshikawa 2020), fish (Lema 2020), rodents (Khurana et al. 2023), domestic animals (Wang and Ibeagha-Awemu 2021), and humans (Barker and Thornburg 2013). Nutrition and temperature have been the most common factors identified in programming postnatal animal physiology and phenotype. As an example, in *Drosophila melanogaster*, pigmentation and body size in adults are programmed during the larvae stage by nutritional and temperature conditions (Koshikawa 2020). In epidemiological studies in humans, it has been observed that individuals born from a mother suffering some level of undernutrition have a higher risk of presenting low birth weight and metabolic diseases during adulthood (Barker and Thornburg 2013). Theoretically, individuals have a higher risk of presenting metabolic diseases because they were programmed to have a metabolism to survive under nutritional deficiencies, but the resolution of such nutritional problems after birth seemed to be the reason for the observed metabolic problems. These epidemiological studies gave rise to the developmental origins of the health and disease hypothesis, which increased the interest in this topic for human health and animal production.

2.1 Effects of Developmental Programming on Productive Efficiency in Cattle

In cattle, the implementation of experiments to determine the existence of developmental programming has been challenging because of the variety of environmental and animal factors that should be controlled to ensure reliable results; however, growing pieces of evidence have been gathered to support its presence during the last two decades. Even though most studies have focused on nutritional interventions (Perry et al. 2019), other factors like heat stress, disease, and parity, appear to influence postnatal productivity in cattle (Bafandeh et al. 2023; Carvalho et al. 2020; Laporta et al. 2020). Here, we describe some results regarding the developmental programming effects on important productive traits in dairy and beef cattle (summary in Fig. 1).

In dairy cattle, supplementation with methyl donor molecules (like methionine or choline) has received increased attention from different research groups because of their potential role in regulating developmental programming. In a recent experiment (Zenobi et al. 2022), Holstein cows were supplemented with rumen-protected choline for 21 days before the expected calving date, and positive effects were observed in offspring immunity and growth. The authors observed that calves born from supplemented cows had better efficiency in the absorption of immunoglobulin G, greater numbers of red and white blood in serum, greater solid and liquid food intake, and a lower incidence of fever during their first 21 days of age (Zenobi et al. 2022). In another experiment, rumen-protected choline was supplemented 24 days

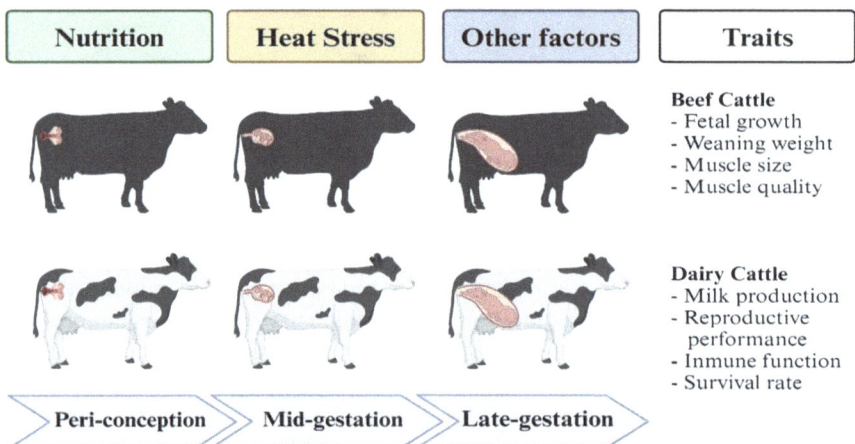

Fig. 1 Productive traits are influenced by environmental factors throughout different periods of gestation in dairy and beef cattle. The figure was created with BioRender.com

pre-calving, and the calf's immune response following bovine respiratory syncytial virus vaccination was evaluated. Calves born from cows supplemented with choline had a lesser haptoglobin concentration, but the apparent efficiency of immunoglobulin G absorption from colostrum and immune response to vaccination was not clear and depended on the body condition score and metabolic status of cows before calving (Swartz et al. 2022).

Some studies have highlighted the sensitivity of the Holstein fetus to the effects of developmental programming induced by exposure to heat stress during pregnancy. For example, when cows were exposed to heat stress during the last 45 days of pregnancy, their daughters produced less milk during their first and second lactations (Laporta et al. 2020). Moreover, daughters of cows exposed to heat stress had a lower survival rate during their life in the herd (Laporta et al. 2020). The reduction in productivity seems to be associated with heat stress impairments in ductal development and epithelial and stromal cell proliferation in the mammary gland (Dado-Senn et al. 2022). The lower animal survival seems to be associated with heat stress impairments in the immune function of resultant calves (Dahl et al. 2016).

In a recent observational study (Mozaffari-Makiabadi et al. 2023), cows were grouped by timing of exposure to heat stress, including those exposed during the first, the second, and third trimesters or only during the third trimester of pregnancy. The lowest milk production was observed in daughters born from cows exposed to heat stress in the first trimester (Mozaffari-Makiabadi et al. 2023). In contrast, reproductive performance (evaluated by conception rate, service per conception, and days open) was the poorest in daughters born from cows exposed to heat stress in the third trimester of pregnancy (Mozaffari-Makiabadi et al. 2023).

Other studies have supported the idea that dam parity can program productive traits, including milk production, reproductive performance, and health. In a big data set evaluated in Holstein cattle, it was observed that daughters from first parity animals produced around 445 kg more milk in their first two lactations as compared to daughters from multiparous animals (Hinde et al. 2014). In another study that included more than 11,000 observations, offspring born from multiparous dams had a greater first-service conception rate, as well as fewer services per conception and days open (Bafandeh et al. 2023). Additionally, it was observed that daughters from primiparous Holstein cows had a lesser probability of pregnancy loss as a heifer and clinical (uterine and non-uterine) disease during their first lactation than multiparous cows (Carvalho et al. 2020).

Few studies have been performed to explore the inheritance of developmental programming effects through multiple generations in cattle, but a piece of evidence was demonstrated in the experiment conducted by Laporta et al. (2020). Specifically, granddaughters of dams exposed to heat stress during the last 45 days of pregnancy had lesser milk production during their first lactation. Also, while not statistically significant, these granddaughters had a numerically lower survival rate.

In beef cattle, several studies have demonstrated the effects of nutrient restriction at different periods of pregnancy on postnatal growth and muscle development of offspring. For example, maternal protein restriction for 60 days pre and 90 days postconception reduced crown-rump length in 32-day-old embryos and reduced fetal body weight at 98 days of gestation (Copping et al. 2014). Offspring derived from cows exposed to protein restriction during the first (McLean et al. 2018), second (Underwood et al. 2010), or third trimester (Bohnert et al. 2013) of pregnancy also had reduced body weights at weaning. This type of nutritional restriction during pregnancy seems to influence offspring development because of its negative effects on the number of muscle fibers, as was described in the Longissimus dorsi of calves at 30 and 450 postnatal days of age (Costa et al. 2021). Such nutritional interventions also negatively affect measures of meat quality, including marbling score, back fat thickness, and Longissimus dorsi size area (McLean et al. 2018; Radunz et al. 2012; Underwood et al. 2010).

Although most studies have focused on nutritional restriction, a high nutritional plan or strategic supplementation at different stages of pregnancy has resulted in positive effects on body growth, muscle characteristics, and health. For example, when cows were maintained on a high nutritional plan under a pasture-based grazing system during the second and third pregnancy periods, the resultant offspring had greater birth and weaning weights (Cafe et al. 2006). Similar results have also been reported when cows are gaining body condition scores (Marques et al. 2016) or are receiving some type of supplementation (Barcelos et al. 2022) during the second or third trimester of pregnancy. Likewise, offspring from supplemented cows during the second or third gestation period had a greater ribeye area, subcutaneous fat thickness, muscle fiber number (Marquez et al. 2017), and a better feed intake pattern (Nascimento et al. 2022). Reduction of morbidity has also been observed in calves from supplemented cows (Larson et al. 2009), which could suggest that supplementation promotes immune function in the resulting offspring. Currently, several research groups are focused on the generation of nutritional strategies to manipulate developmental programming according to production objectives by using nutrients with important biological functions in cells like methionine (Silva et al. 2021), choline (Haimon et al. 2022), or polyunsaturated fatty acids (Brandão et al. 2020).

2.2   Epigenetic Mechanisms

Environmental factors like nutrition, heat stress, and production of embryos by ART seem to control developmental programming through alterations in epigenetic mechanisms (DNA methylation, histone posttranslational modifications, and noncoding RNAs), which regulate DNA function, cellular gene expression, and define the phenotype of the organisms without altering the DNA sequence (Sirard 2021). In cattle, mechanistic studies to evaluate direct epigenetic mechanisms are limited; however, some studies have established the association between epigenetic marks and such environmental effects. As an example, DNA methylation in some genomic regions was different in the muscle between Angus calves derived from mothers subjected to different nutritional restrictions (Devos et al. 2021). Holstein cows derived from mothers exposed to heat stress during the last 46 days of gestation had reduced milk production, which was associated with alterations in the DNA methylation profile in mammary gland tissue (Skibiel et al. 2018). Muscle tissue from 86-day fetuses derived from in vitro-produced embryos showed a different DNA methylation profile compared to fetuses from artificial insemination (Li et al. 2020). Likewise, different DNA methylation profiles have been observed in embryos and tissues of calves derived from IVP embryos compared with IVD embryos and calves (Canovas et al. 2021; Rabaglino et al. 2021).

In general, DNA methylation and histone posttranslational modifications regulate regional chromatin structures to deny or to allow, under certain circumstances, accessibility to protein machinery in charge of the transcription process (Larson et al. 2021). Methylation of the fifth carbon in the DNA cytosines (5-methylcytosine) is the most studied mark, and it has mostly been associated with gene silencing because of its role in X-chromosome inactivation and cell differentiation, among other processes (Greenberg and Bourc'his 2019). Several histone modifications have been described. Three methylations of lysine residue 27 in histone 3 (H3K27me3) have been associated with gene repression, but three methylations of lysine residue 4 in histone 3 (H3K4me3) have been associated with gene expression (Larson et al. 2021). DNA methylation and histone modifications are stable epigenetic marks, so they can be heritage to descendants through gametes (Lismer et al. 2020) and to perdure for years (Tobi et al. 2014). Growing information is also being gained in the epigenetic regulation by noncoding RNAs, which involves the modulation of transcription, posttranscription, and translation (Zhang et al. 2021).

2.3   Susceptibility of Embryos to Developmental Programming

In mammals, early embryo development is probably the period in which organisms are more sensitive for developmental programming since a process called epigenetic reprogramming occurs just after fertilization, during the preimplantation period (Velázquez et al. 2023). The objective of this process is to remove epigenetic marks established in paternal and maternal gametes during uterus development to achieve their biological function, fertilization, and the formation of a new organism (Greenberg and Bourc'his 2019). Once fertilization takes place, paternal and maternal chromosomes start suffering an extensive removal of epigenetic marks (global demethylation), and the newly formed embryo starts acquiring new ones (de novo methylation), which are required to regulate gene expression that ensures the success of the new organism's development (Duan et al. 2019).

In cattle, a global DNA demethylation takes place through the first embryo cell divisions and is completed around the 8-cell stage, and from this point on, embryos begin the remethylation (or de novo) process (Duan et al. 2019; Jiang et al. 2018). Chromatin remodeling of H3K4me3 and H3K27me3 also occurs in bovine preimplantation embryos (Org et al. 2019), but the complete image of this process is still under study. A recent experiment indicates that chromatin accessibility during the 2–4 cell stage in embryos is low, while the accessibility is high during the 8 to morula stage, but accessibility is low again during the blastocyst stage (Ming et al. 2021). These results are aligned with the pattern of gene expression observed in previous studies since the highest chromatin accessibility coincides with the major embryo genomic activation, the time when embryos are transcriptionally independent (Graf et al. 2014; Jiang et al. 2014). It is important to highlight that mammalian embryos rely on maternal transcripts heritage from oocytes during a short period after fertilization (Schulz and Harrison 2019).

Another important biological event occurring during early embryo development is the first cell differentiation, also referred to as the first cell fate decision. Some totipotent cells from the morula stage become trophectoderm to later form the placenta and extraembryonic tissues, whereas other cells differentiate into the inner cell mass to later form germinal layers that give rise to the different embryonic tissues (Gerri et al. 2020). The function of specific transcription factors and the presence of epigenetic marks are involved in the establishment and maintenance of cell differentiation (Greenberg and Bourc'his 2019). A recent study demonstrated that histone modifications, H3K4me3 and H3K27me3, are very active in the trophectoderm and inner cell mass (Org et al. 2019), and the allocation of trophectoderm and inner cell mass cells during bovine blastocyst formation can be altered due to environmental conditions (Hansen et al. 2014). Current evidence suggests that embryo environment influences epigenetic marks, gene expression, cell differentiation, and potentially the phenotype that individuals might express during pre and postnatal life (Estrada-Cortés et al. 2020, 2021a; Li et al. 2022; Siqueira and Hansen 2016).

3   Assisted Reproductive Technologies

Assisted reproductive technologies refer to procedures that involve the manipulation of reproductive cycles as well as gametes and embryos (Muller and Van Eenennaam 2022). Traditional ART includes hormonal control of the reproductive cycles of donor animals to obtain IVD or IVP embryos for embryo transfer (Fig. 2). In cattle, these technologies are typically used to implement genetic improvement programs. The main objective is focused on generating highly valuable offspring capable of increasing productivity traits (related to milk or beef) and, at the same time, contributing to a reduced carbon footprint per animal unit to achieve the sustainability of cattle production systems (Baruselli et al. 2023).

3.1   In Vivo Derived Embryos

Procedures to obtain IVD embryos are described in detail elsewhere (Bo and Mapletoft 2014; Canovas et al. 2021; Mapletoft and Bo 2022; Rabaglino et al. 2021), but such literature was used to highlight some biological events and procedures which are relevant to this section and chapter (see Fig. 2). Briefly, the production of IVD embryos involves hormonal manipulation of the estrous cycle, ovarian stimulation with exogenous FSH, artificial insemination, multiple ovulations, and nonsurgical recovery of embryos (uterine flushing), embryo evaluation, loading into straws, as well as the embryo transfer and/or embryo cryopreservation. At the

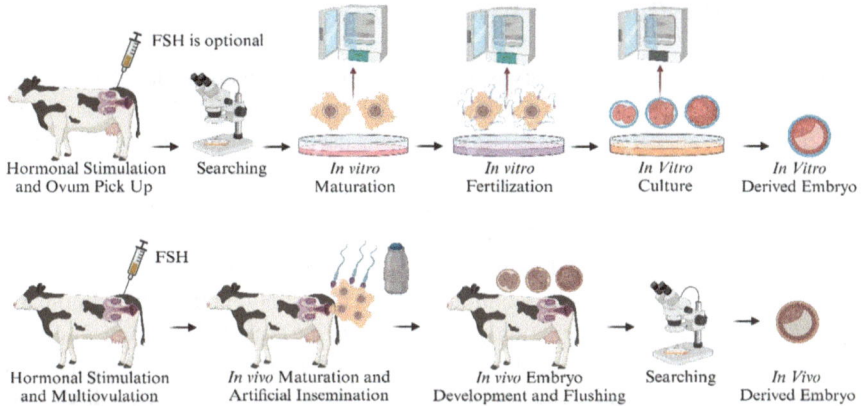

Fig. 2 Key biological events and managements in which embryos are exposed to non-natural environments throughout the assisted reproductive technologies. The figure was created with BioRender.com

beginning of the procedure, control of the estrous cycle typically includes the use of hormones like gonadotropin-releasing hormone (GnRH) to synchronize the emergence of a new follicular wave and insertion of an intravaginal device releasing progesterone to reduce the frequency of secretion of gonadotropins. Subsequently, the ovary is stimulated by administration of FSH to promote follicular growth. Practically, donors receive FSH applications twice a day over 4 days to achieve successful follicle stimulation since this hormone's half-life is only around 5–6 h. An application of prostaglandin F2α is included in the protocol either at day 3 or 4 after the beginning of the FSH stimulation to induce luteolysis, and 24 h after progesterone device removal, GnRH is administered to stimulate ovulation. Artificial insemination can be performed following detected estrus or at a fixed time.

Embryo recovery is performed 7 days after estrus or induced ovulation. This process involves the use of a Foley catheter, which is introduced into the uterus and fixed in place by insufflation of the cuff. The Foley catheter is attached to Y-junction tubing, which is connected at one end to a bag containing flush fluid and at the other end to an embryo collection filter. Flush fluid is introduced into the uterus and subsequently allowed to exit via gravity flow. This process is repeated 3–5 times using up to 0.5 L of medium per uterine horn. Recovered fluid is allowed to pass through the embryo collection filter. After the procedure is completed, the filter is rinsed into a petri dish, and embryo searching is carried out under a stereomicroscope. Embryos are subsequently evaluated to determine their stage of development and quality grade. If recipient cows are prepared, embryos are placed in a holding media and loaded in straws to carry out embryo transfer. However, if recipients are not prepared, embryos are exposed to a cryoprotectant solution (like ethylene glycol) and subjected to a slow freezing protocol; then they are stored in liquid nitrogen until embryo transfer.

3.2   In Vitro-Derived Embryos

Procedures to obtain IVP embryos from OPU are described in detail elsewhere (Demetrio et al. 2020; Estrada-Cortés et al. 2021b; Ortega et al. 2017; Viana et al. 2022; Zolini et al. 2019), but such literature was used to highlight some of the biological events and management strategies that are relevant to this chapter (Fig. 2). Briefly, IVP embryo production involves ultrasound-guided transvaginal OPU with or without stimulation with FSH, oocyte evaluation, and in vitro embryo production (maturation, fertilization, and culture), as well as embryo transfer and/or embryo cryopreservation. Follicular stimulation protocols typically include the administration of GnRH (or a follicle ablation) to synchronize the emergence of a new follicular wave and the insertion of an intravaginal progesterone device to reduce the frequency of secretion of gonadotropins. Subsequently (± 36 h later), follicles are stimulated to grow synchronously to achieve a size of around 8 mm using exogenous FSH, which is administered every 12 h for 3 days. The OPU procedure is then performed between 40 and 52 h after the last FSH application, depending on the age, breed, and physiological stage of the donor. Although FSH stimulation is not required, the proportion of oocytes becoming viable embryos following OPU-IVP is generally improved when follicle stimulation is performed. However, in certain cases, there may be little or no benefit to the use of FSH stimulation. For example, embryo production using OPU-IVP in cows with high antral follicle populations, such as *Bos indicus* breeds, may not be improved by follicle stimulation.

During the OPU procedure, donors are restrained in a squeeze chute, the perineal area is washed, and they receive epidural anesthesia. Once the donor is ready, an OPU plastic handle is introduced into the vagina and placed into the fornix vagina. The OPU handle contains a convex ultrasound probe, and a metal aspiration guide attached to a needle and a plastic tube attached externally to a 50-mL collection tube; this tube is connected by a flexible tubing to a vacuum pump to create negative vacuum pressure. Follicles identified by ultrasound are punctured with the needle, and follicular fluid is recovered into the 50 mL tube. The use of prewarmed collection media supplemented with heparin for the OPU procedure is necessary to avoid clotting. The recovered follicular fluid is filtered to remove the blood and clarify and then subsequently rinsed into a petri dish using a prewarmed oocyte handling medium. Oocytes are then identified and recovered from the petri dish using a stereomicroscope. Following collection, oocytes are washed in an oocyte handling medium, and the quality of each oocyte is assessed, and those of acceptable quality are placed into oocyte maturation medium.

Oocytes are typically maintained at 38.5 °C in maturation media for approximately 22 h. Subsequently, in vitro fertilization is carried out by using a frozen-thawed straw of conventional or sex-sorted semen. Co-incubation between gametes is performed in a specialized fertilization medium at 38.5 °C in a humidified atmosphere with 5% CO_2 (v/v) for 8–20 h. Afterwards, cumulus cells are removed, and presumptive zygotes are placed in a specialized culture medium at 38.5°C in a humidified atmosphere that contains 5% O_2 (v/v) and 5% CO_2 (v/v) for 7 days. When recipient cows are prepared, embryos are evaluated to determine the stage of development and quality before they are loaded into straws to carry out fresh embryo transfer. If recipient cows are not available, embryos can be cryopreserved using either vitrification or slow-rate cryopreservation and then in liquid nitrogen until embryo transfer.

3.3   Differences Between IVD and IVP Production

According to data collected by the International Embryo Technology Society, the number of bovine embryos produced and transferred worldwide has increased over the last decade, with a greater level of use of these technologies in North America, Europe, and South America (Viana 2022). In Europe, the IVD process is primarily used for embryo production, but in North and South America, IVP is now the predominant embryo production method utilized in cattle (Viana 2022). It is expected that the use of these technologies at the commercial level, particularly IVP, will continue to increase once the consistency of results, pregnancy outcomes, and costs of utilization are improved (Hansen 2020).

Since 2016, worldwide production and transfer of IVP embryos have been greater than for IVD (Viana 2022). This increased use of IVP is likely due to certain advantages of OPU-IVP as compared to IVD. As an example, OPU-IVP can be performed two (with hormonal stimulation) to four times a month (without hormonal stimulation) compared to once a month for IVD embryos. Thus, a greater number of embryos and offspring can potentially be produced by OPU-IVP (Ruíz-Lopéz 2022) in a given timeframe. When IVP is carried out without follicle stimulation, labor and donor management are simpler and less expensive as compared to IVD (Crowe et al. 2021). Oocytes collected by OPU-IVP can be obtained from first-trimester pregnant cows and from animals with some reproductive disorders, and limited semen can be used more efficiently at in vitro fertilization (Crowe et al. 2021; Galli et al. 2001).

3.4   Genetic Gain by Assisted Reproductive Technologies

In genetic improvement programs, the rate of genetic gain is accelerated with a higher selection intensity, selection accuracy, and heritability of the desirable traits, as well as a wider genetic variation and a reduced generational interval (Georges et al. 2019). Understanding the factors that affect the rate of genetic gain allows for a better appreciation of how reproductive technologies and molecular technologies, such as genomic testing (identification of single nucleotide polymorphisms and their association with productive traits), can be applied to enhance genetic gain (Muller and Van Eenennaam 2022).

Within a population, selection intensity refers to the proportion of cows chosen to be mothers of the next generation. As the proportion of selected cows is reduced, the selection intensity is increased (Hayes et al. 2013). Therefore, it is important to generate information for desirable characteristics and to identify animal performance so that the best animals can be selected as progenitors (Georges et al. 2019). Selection accuracy refers to the measure of how well the animal breeding values were estimated (Muller and Van Eenennaam 2022). These estimations depend on phenotypic evaluations of progeny performance (milk or beef-associated traits) in different populations; the more descendants evaluated in different environments, the better (Garrick 2011). The use of models for predictions for specific traits through genomic testing contributes to increased accuracy in the selection of progenitors (Garrick 2011).

Genetic variation refers to the diversity of genes present within a population, and heritability refers to the percent of the total variation within a population for a particular trait because of the genetics of animals (Kemper and Goddard 2012). Generation interval has been defined as the time between the birth of progenitors and their replacement, in other words, how quickly genes are propagated between generations (Muller and Van Eenennaam 2022). In general, the use of embryo production technologies allows for increased selection intensity by using only elite progenitors expressing valuable traits in a population for embryo production (Georges et al. 2019). On the other hand, generation interval can be reduced by using oocytes collected from elite heifers and semen from young elite bulls (Crowe et al. 2021). Commercially, embryos can be produced from oocytes collected from heifers around 10 months and semen from 12-month-old sires (Crowe et al. 2021).

The use of genomic testing has allowed for increased reliability in the calculation of estimated breeding values, and the application of this technology can allow for the identification of elite animals as newborn calves and even at the embryonic stage. Consequently, techniques for oocyte collection and IVP embryo production from young heifer calves between 2 and 6 months of age are being evaluated (Baldassarre et al. 2018). Such techniques could further reduce the generation interval. Additionally, improvements in stem cell culture, in combination with the use of ART and genomic testing, are opening more possibilities for genetic improvement in cattle. Recently, a process called "in vitro breeding" (Goszczynski et al. 2019) has been proposed. This technology consists of the derivation of embryonic stem cells from embryos produced using gametes from elite progenitors. The derived embryonic stem cell lines would then be selected following the application of genomic testing, and identified lines would be used to produce gametes in vitro. Subsequently, a new generation of embryos would be produced using the in vitro-derived gametes, and the process could be repeated for several generations. The ability to derive gametes from embryonic stem cells in vitro has not yet been reported in cattle, but it has been done in other species, including mice and humans (Goszczynski et al. 2023). If this technology does come to fruition, it could have a significant impact on the rate of genetic gain, primarily through reduction in the generation interval.

4 Assisted Reproductive Technologies and Developmental Programming

Embryos derived from assisted reproductive technologies are subjected to non-natural environmental conditions (Fig. 2). In the case of IVD, as an example, follicles are exposed to non-physiological FSH concentrations to stimulate follicle growth and multiple ovulations. For IVP, oocytes might or might not be exposed to non-physiological FSH concentrations, but in either case, oocytes are exposed to artificial conditions during the in vitro maturation, fertilization, and culture processes. For both IVD and IVP, procedures also involve the exposure of gametes and embryos to an artificial environment like holding media, microscope manipulation, and artificial insemination, either with the use of frozen-thawed conventional or sex-sorted semen.

The artificial conditions in which embryos are produced by ART differ considerably from what occurs during a normal estrous cycle, especially for IVP embryos, which spend more time outside of the reproductive tract (Hansen 2020). As mentioned above, early embryo development constitutes a window in which organisms are susceptible to being programmed due to the various important biological events that occur during this time (Velázquez et al. 2023). ART-derived embryos receive signals associated with an abnormal environment and are possibly programmed to have a phenotype that is contrary to the objectives of cattle production systems.

4.1 Effects of Assisted Reproductive Technologies on Cattle Productivity

In cattle, the execution of experiments to evaluate the effects of ART on the productive phenotype of the offspring is challenging since many factors at the animal, farm, and environment levels need to be controlled. In dairy cattle, characteristics of general management, including the availability of information on genetic merit and data collection practices, provide the opportunity to conduct studies with a certain level of control to evaluate the effects of ART. In fact, studies in dairy cattle have indicated that ARTs can affect the productivity of the resultant offspring. On the other hand, recent studies in beef cattle support the idea that the postnatal productivity of ART-derived offspring may be positively programmed. Such effects of ART on the postnatal performance of dairy and beef offspring will be described in this section (Fig. 3).

4.2 Evidence of Negative Effects

Siqueira et al. (2017b) conducted a retrospective cohort observational study to test whether the milk production of Holstein cows was altered due to ART. Data was collected from a large commercial dairy farm that routinely implemented ART and milk production from females derived from the transfer of either IVD or IVP

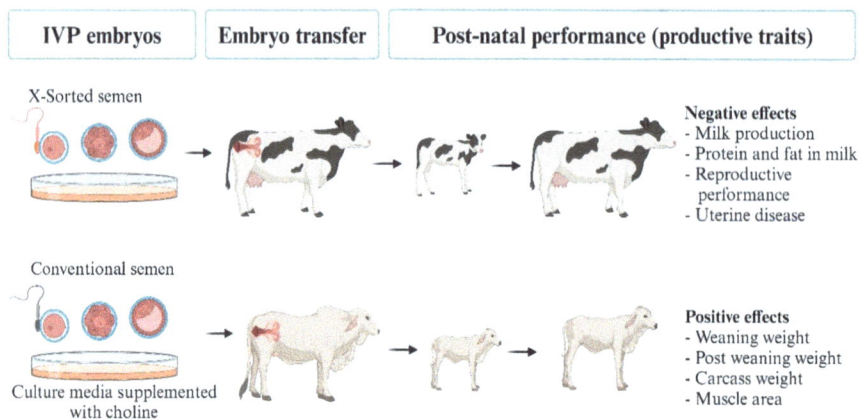

Fig. 3 Postnatal performance (positive and negative programming effects) of animals derived from in vitro-produced embryos in dairy and beef cattle. The figure was created with BioRender.com

embryos (IVD n = 183, IVP using conventional semen n = 218, and IVP using reverse sexed-sorted semen n = 430) was compared to that of females produced using AI (n = 2037). In the statistical analysis, the genetic merit of individual cows was included in the model to control for this confounding factor. It was observed that cows from IVP using reverse sexed-sorted semen had lower milk, fat, and protein yield adjusted to 305 days than AI-derived cows. Cows derived from IVP using conventional semen had a non-statistical reduction of milk and fat yield but a significant reduction of protein yield compared to AI-derived cows.

Recently, Lafontaine et al. (2023a) conducted another large-scale retrospective cohort study to test the same hypothesis as Siqueira et al. (2017a, b). The authors used data generated from 2012 to 2019 that was provided by Lactanet Inc. (Sainte-Anne-de-Bellevue, QC, Canada) and Holstein Canada. The dataset included cows derived from IVD (IVD n = 3585 animals and n = 24,192 lactations) and IVP (n = 732 animals and n = 1299 lactations) and was analyzed in comparison with AI-derived cows (n = 304,163 animals and n = 576,448 lactations). In this study, no statistical differences were found for milk production traits among cows derived from IVD, IVP, and AI technologies; this was true, even with and without the inclusion of genetic merit in the model. According to the authors, this outcome might be due in part to a lack of statistical power once data was adjusted for multiple comparisons among the ART methods.

Although no effects of the ART method were observed for milk production, Lafontaine et al. (2023a) observed that cows conceived from IVD and IVP had a longer interval from first service to conception than cows derived from AI. In this study, authors used a life performance index (LPI, which includes productive, durability, health, and fertility traits) and a daughter fertility index (DFI, which includes interval calving to first service, no return rate to 56 d for cows, interval first service to conception, among others) to calculate genetic progress. Considering the difference between parents and daughters for the LPI and DFI, it was estimated that

IVP-derived cows had a slower rate of genetic progress compared to AI-derived cows. Authors suggest these outcomes might be partially attributed to detrimental epigenetic effects from IVP. In another large-scale retrospective cohort study, Lafontaine et al. (2023b) also evaluated the intergenerational effects of ART on gestation and health characteristics of resultant adult Holstein cows. It was observed that cows derived from IVD and IVP embryos had a longer gestation length than AI-derived cows. Likewise, it was observed that cows from IVP embryos had a greater probability of presenting ovarian cysts and uterine disease compared to AI-derived cows.

4.3   Evidence of Positive Effects

In contrast to previous results, some reports suggest that appropriate signals provided to embryos produced by ART during early development might result in an improvement in the postnatal productivity of offspring. It can be hypothesized that if embryos are susceptible to being programmed in a way that reduces productivity, the plasticity of the epigenetic programming phenomenon can also provide the opportunity to enhance postnatal productivity. This idea has been tested in several studies evaluating the effects of certain components included in embryo culture medium for IVP on the postnatal phenotype of the resulting offspring (Hansen 2023). The identification of molecules present in the reproductive tract and their biological functions at the cellular level have been key for the basis of such studies (Denicol et al. 2014; Estrada-Cortés et al. 2020; Sang et al. 2020; Tríbulo et al. 2018). In this research, colony-stimulating factor 2 (CSF2) and choline have been evaluated due to their potential effects as developmental programming molecules (Estrada-Cortés et al. 2020; Hansen et al. 2014).

In an early study (Kannampuzha-Francis et al. 2015), Holstein embryos were exposed to CSF2 in the culture media and transferred into recipients. The authors observed that the resultant CSF2-derived heifers showed a higher accelerated body growth until 13 months of age than heifers derived from a control culture medium. According to parallel studies that used the same experimental design, this change in offspring phenotype seems to be associated with alterations in gene expression and DNA methylation during embryonic and fetal life (Li et al. 2020; Siqueira and Hansen 2016).

In another study (Estrada-Cortés et al. 2021a), Brahman embryos were exposed to choline in the culture medium and transferred to recipients. The authors observed that the resultant offspring exposed to choline had a greater birth weight and weaning weight than calves derived from culture medium containing vehicles only. This increase in body weight at weaning represented an average of 20 kg more and occurred in male and female calves. The same experimental design was used to determine whether such a phenotypic response could be repeated (Haimon et al. 2022, 2024). It was observed that calves derived from choline-treated embryos had higher weaning weight, post-weaning growth, hot carcass weight, and area of the longissimus   dorsi   muscle   compared   to   calves   derived   from   control treatment-embryos. Such differences in body weight at weaning were associated with DNA methylation alterations located in genomic regions linked with tissue growth and cellular proliferation (Estrada-Cortés et al. 2021a). Collectively, these results support the idea that some signals received by the embryo during preimplantation development might program the offspring in a way that postnatal productivity is enhanced.

5 Conclusions

Evidence from a variety of species supports the concept of developmental programming in mammals. Environmental factors like nutrition, heat stress, and assisted reproductive technologies constitute important modulators of this phenomenon in dairy and beef cattle and, consequently, in the expression of productive traits in the postnatal period. Early embryonic development represents a period that is sensitive to the impacts of developmental programming since essential biological events like epigenetic reprogramming occur during this time. The exposure of embryos derived from ART to artificial conditions can induce negative postnatal phenotypes, such as reduced milk production and reproductive performance in dairy cattle. However, recent evidence suggests that positive postnatal production outcomes might be induced if appropriate signals are provided to the embryos during the IVP culture period. Further research is necessary to identify the mechanisms that regulate this phenomenon. Such information could be important for the development of strategies to either improve postnatal productivity or to ameliorate the potential negative effects of environmental factors, which will be more drastic due to the effects of climate change.

References

Bafandeh M, Mozaffari Makiabadi MJ, Gharagozlou F et al (2023) Developmental programming of production and reproduction in dairy cows: I. Association of maternal parity with offspring's birth weight, milk yield, reproductive performance and AMH concentration during the first lactation period. Theriogenology 210:34–41. https://doi.org/10.1016/j.theriogenology.2023.07.012

Baldassarre H, Currin L, Michalovic L et al (2018) Interval of gonadotropin administration for *in vitro* embryo production from oocytes collected from Holstein calves between 2 and 6 months of age by repeated laparoscopy. Theriogenology 116:64–70. https://doi.org/10.1016/j.theriogenology.2018.05.005

Barcelos SS, Nascimento KB, Silva TE et al (2022) The effects of prenatal diet on calf performance and perspectives for fetal programming studies: a meta-analytical investigation. Animals 12:2145. https://doi.org/10.3390/ani12162145

Barker DJ, Thornburg KL (2013) The obstetric origins of health for a lifetime. Clin Obstet Gynecol 56(3):511–519. https://doi.org/10.1097/GRF.0b013e31829cb9ca

Baruselli PS, de Abreu LÂ, de Paula VR et al (2023) Applying assisted reproductive technology and reproductive management to reduce CO2-equivalent emission in dairy and beef cattle: a review. Anim Reprod 20(2):e20230060. https://doi.org/10.1590/1984-3143-AR2023-0060

Bateson P, Gluckman P, Hanson ML (2014) The biology of developmental plasticity and the predictive adaptive response hypothesis. J Physiol 592(11):2357–2368. https://doi.org/10.1113/jphysiol.2014.271460

Bó GA, Mapletoft RJ (2014) Historical perspectives and recent research on superovulation in cattle. Theriogenology 81(1):38–48. https://doi.org/10.1016/j.theriogenology.2013.09.020

Bohnert DW, Stalker LA, Mills RR et al (2013) Late gestation supplementation of beef cows differing in body condition score: effects on cow and calf performance. J Anim Sci 91(11):5485–5491. https://doi.org/10.2527/jas.2013-6301

Brandão AP, Cooke RF, Schubach KM et al (2020) Supplementing Ca salts of soybean oil to late-gestating beef cows: impacts on performance and physiological responses of the offspring. J Anim Sci 98(8):skaa247. https://doi.org/10.1093/jas/skaa247

Cafe LM, Hennessy DW, Hearnshaw H et al (2006) Influences of nutrition during pregnancy a lactation on birth weights and growth to weaning of calves sired by Piedmontese or Wagyu bulls. Austr J Exp Agric 46:245–255. https://doi.org/10.1071/EA05225

Canovas S, Ivanova E, Hamdi M et al (2021) Culture medium and sex drive epigenetic reprogramming in preimplantation bovine embryos. Int J Mol Sci 22(12):6426. https://doi.org/10.3390/ijms22126426

Carvalho MR, Aboujaoude C, Peñagaricano F et al (2020) Associations between maternal characteristics and health, survival, and performance of dairy heifers from birth through first lactation. J Dairy Sci 103(1):823–839. https://doi.org/10.3168/jds.2019-17083

Copping KJ, Hoare A, Callaghan M et al (2014) Fetal programming in 2-year-old calving heifers: peri-conception and first trimester protein restriction alters fetal growth in a gender-specific manner. Anim Prod Sci 54(9):1333–1337. https://doi.org/10.1071/AN14278

Costa TC, Du M, Nascimento KB et al (2021) Skeletal muscle development in postnatal beef cattle resulting from maternal protein restriction during mid-gestation. Anim (Basel) 11(3):860. https://doi.org/10.3390/ani11030860

Crowe AD, Lonergan P, Butler ST (2021) Invited review: use of assisted reproduction techniques to accelerate genetic gain and increase value of beef production in dairy herds. J Dairy Sci 104(12):12189–12206. https://doi.org/10.3168/jds.2021-20281

Dado-Senn BM, Field SL, Davidson BD et al (2022) In utero hyperthermia in late gestation derails dairy calf early-life mammary development. J Anim Sci 100(10):skac186. https://doi.org/10.1093/jas/skac186

Dahl GE, Tao S, Monteiro APA (2016) Effects of late-gestation heat stress on immunity and performance of calves. J Dairy Sci 99(4):3193–3198. https://doi.org/10.3168/jds.2015-9990

Demetrio DGB, Benedetti E, Demetrio CGB et al (2020) How can we improve embryo production and pregnancy outcomes of Holstein embryos produced in vitro? (12 years of practical results at a California dairy farm). Anim Reprod 17(3):e20200053. https://doi.org/10.1590/1984-3143-AR2020-0053

Denicol AC, Block J, Kelley DE et al (2014) The WNT signaling antagonist Dickkopf-1 directs lineage commitment and promotes survival of the preimplantation embryo. FASEB J 28(9):3975–3986. https://doi.org/10.1096/fj.14-253112

Devos J, Behrouzi A, Paradis F, Straathof C et al (2021) Genetic potential for residual feed intake and diet fed during early- to mid-gestation influences post-natal DNA methylation of imprinted genes in muscle and liver tissues in beef cattle. J Anim Sci 99(5):skab140. https://doi.org/10.1093/jas/skab140

Duan JE, Jiang ZC, Alqahtani F et al (2019) Methylome dynamics of bovine gametes and in vivo early embryos. Front Genet 28(10):512. https://doi.org/10.3389/fgene.2019.00512

Estrada-Cortés E, Negrón-Peréz VM, Tríbulo P et al (2020) Effects of choline on the phenotype of the cultured bovine preimplantation embryo. J Dairy Sci 103(11):10784–10796. https://doi.org/10.3168/jds.2020-18598

Estrada-Cortés E, Ortiz W, Rabaglino MB et al (2021a) Choline acts during preimplantation development of the bovine embryo to program postnatal growth and alter muscle DNA methylation. FASEB J 35(10):e21926. https://doi.org/10.1096/fj.202100991R

Estrada-Cortés E, Jannaman EA, Block J et al (2021b) Programming of postnatal phenotype caused by exposure of cultured embryos from Brahman cattle to colony-stimulating factor 2 and serum. J Anim Sci 99(8):skab180. https://doi.org/10.1093/jas/skab180

FAO (2022) Contribution of livestock to food security, sustainable agri-food systems, nutrition, and healthy diets First session (16–18 March 2022). Sub-committee on livestock https://www.fao.org/3/ni005en/ni005en.pdf

FAO, IFAD, UNICEF, WFP and WHO (2021) The state of food security and nutrition in the world. Transforming food systems for food security, improved nutrition, and affordable healthy diets for all. Rome. https://doi.org/10.4060/cb4474en

Galli C, Crotti G, Notari C et al (2001) Embryo production by ovum pick up from live donors. Theriogenology 55(6):1341–1357. https://doi.org/10.1016/s0093-691x(01)00486-1

Garrick DJ (2011) The nature, scope and impact of genomic prediction in beef cattle in the United States. Genet Sel Evol 43(1):17. https://doi.org/10.1186/1297-9686-43-17

Georges M, Charlier C, Hayes B (2019) Harnessing genomic information for livestock improvement. Nat Rev Genet 20(3):135–156. https://doi.org/10.1038/s41576-018-0082-2

Gerri C, McCarthy A, Alanis-Lobato G (2020) Initiation of a conserved trophectoderm program in human, cow and mouse embryos. Nature 587(7834):443–447. https://doi.org/10.1038/s41586-020-2759-x

Gluckman PD, Hanson MA, Low FM (2019) Evolutionary and developmental mismatches are consequences of adaptive developmental plasticity in humans and have implications for later disease risk. Philos Trans R Soc Lond Ser B Biol Sci 374(1770):20180109. https://doi.org/10.1098/rstb.2018.0109

Goszczynski DE, Cheng H, Demyda-Peyrás S et al (2019) Review: in vitro breeding: application of embryonic stem cells to animal production. Biol Reprod 100(4):885–895. https://doi.org/10.1093/biolre/ioy256

Goszczynski DE, Navarro M, Mutto AA et al (2023) Review: embryonic stem cells as tools for in vitro gamete production in livestock. Animal 17(Suppl 1):100828. https://doi.org/10.1016/j.animal.2023.100828

Graf A, Krebs S, Zakhartchenko V et al (2014) Fine mapping of genome activation in bovine embryos by RNA sequencing. Proc Natl Acad Sci USA 111(11):4139–4144. https://doi.org/10.1073/pnas.1321569111

Greenberg MVC, Bourc'his D (2019) The diverse roles of DNA methylation in mammalian development and disease. Nat Rev Mol Cell Biol 20(10):590–607. https://doi.org/10.1038/s41580-019-0159-6

Haimon MLJ, Estrada-Cortés E, Amaral TF et al (2022) Culture with choline chloride programs development of the in vitro-produced bovine embryo to increase postnatal bodyweight, growth rate, and testes size. Reprod Fert Dev 35(2):125–125. https://doi.org/10.1071/RDv35n2Ab1

Haimon MLJ, Estrada-Cortés E, Amaral TF et al (2024) Provision of choline chloride to the bovine preimplantation embryo alters postnatal body size and DNA methylation. Biol Reprod 111(3):567–579. https://doi.org/10.1093/biolre/ioae092

Hansen PJ (2020) The incompletely fulfilled promise of embryo transfer in cattle-why aren't pregnancy rates greater and what can we do about it? J Anim Sci 98(11):skaa288. https://doi.org/10.1093/jas/skaa288

Hansen PJ (2023) Review: some challenges and unrealized opportunities toward widespread use of the in vitro-produced embryo in cattle production. Animal 17(Suppl 1):100745. https://doi.org/10.1016/j.animal.2023.100745

Hansen PJ, Dobbs KB, Denicol AC (2014) Programming of the preimplantation embryo by the embryokine colony stimulating factor 2. Anim Reprod Sci 149(1–2):59–66. https://doi.org/10.1016/j.anireprosci.2014.05.017

Hansen PJ, Dobbs KB, Denicol AC et al (2016) Sex and the preimplantation embryo: implications of sexual dimorphism in the preimplantation period for maternal programming of embryonic development. Cell Tissue Res 363(1):237–247. https://doi.org/10.1007/s00441-015-2287-4

Hasler JF (2023) Looking back at five decades of embryo technology in practice. Reprod Fertil Dev 36(2):1–15. https://doi.org/10.1071/RD23120

Hayes BJ, Lewin HA, Goddard ME (2013) The future of livestock breeding: genomic selection for efficiency, reduced emissions intensity, and adaptation. Trends Genet 29(4):206–214. https://doi.org/10.1016/j.tig.2012.11.009

Hinde K, Carpenter AJ, Clay JS et al (2014) Holsteins favor heifers, not bulls: biased milk production programmed during pregnancy as a function of fetal sex. PLoS One 9(2):e86169. https://doi.org/10.1371/journal.pone.0086169

Jiang Z (2023) Molecular and cellular programs underlying the development of bovine pre-implantation embryos. Reprod Fertil Dev 36(2):34–42. https://doi.org/10.1071/RD23146

Jiang Z, Sun J, Dong H et al (2014) Transcriptional profiles of bovine *in vivo* pre-implantation development. BMC Genomics 15(1):756. https://doi.org/10.1186/1471-2164-15-756

Jiang Z, Lin J, Dong H et al (2018) DNA methylomes of bovine gametes and *in vivo* produced preimplantation embryos. Biol Reprod 99(5):949–959. https://doi.org/10.1093/biolre/ioy138

Kannampuzha-Francis J, Denicol AC, Loureiro B et al (2015) Exposure to colony stimulating factor 2 during preimplantation development increases postnatal growth in cattle. Mol Reprod Dev 82(11):892–897. https://doi.org/10.1002/mrd.22533

Kemper KE, Goddard ME (2012) Understanding and predicting complex traits: knowledge from cattle. Hum Mol Genet 21(R1):R45–R51. https://doi.org/10.1093/hmg/dds332

Khurana P, Cox A, Islam B, et al, (Apr 2023) Maternal undernutrition induces cell signalling and metabolic dysfunction in undifferentiated mouse embryonic stem cells. Stem Cell Rev Rep; 19(3):767–783. doi: https://doi.org/10.1007/s12015-022-10490-1

Koshikawa S (2020) Evolution of wing pigmentation in Drosophila: diversity, physiological regulation, and cis-regulatory evolution. Develop Growth Differ 62(5):269–278. https://doi.org/10.1111/dgd.12661

Lafontaine S, Labrecque R, Blondin P et al (2023a) Comparison of cattle derived from *in vitro* fertilization, multiple ovulation embryo transfer, and artificial insemination for milk production and fertility traits. J Dairy Sci 106(6):4380–4396. https://doi.org/10.3168/jds.2022-22736

Lafontaine S, Cue RI, Sirard MA (2023b) Gestational and health outcomes of dairy cows conceived by assisted reproductive technologies compared to artificial insemination. Theriogenology 198:282–291. https://doi.org/10.1016/j.theriogenology.2023.01.002

Laporta J, Ferreira FC, Ouellet V et al (2020) Late-gestation heat stress impairs daughter and granddaughter lifetime performance. J Dairy Sci 103(8):7555–7568. https://doi.org/10.3168/jds.2020-18154

Larson DM, Martin JL, Adams DC et al (2009) Winter grazing system and supplementation during late gestation influence performance of beef cows and steer progeny. J Anim Sci 87(3):1147–1155. https://doi.org/10.2527/jas.2008-1323

Larson ED, Marsh AJ, Harrison MM (2021) Pioneering the developmental frontier. Mol Cell 81(8):1640–1650. https://doi.org/10.1016/j.molcel.2021.02.020

Lema SC (2020) Hormones, developmental plasticity, and adaptive evolution: endocrine flexibility as a catalyst for 'plasticity-first' phenotypic divergence. Mol Cell Endocrinol 15(502):110678. https://doi.org/10.1016/j.mce.2019.110678

Li Y, Tríbulo P, Bakhtiarizadeh MR et al (2020) Conditions of embryo culture from days 5 to 7 of development alter the DNA methylome of the bovine fetus at day 86 of gestation. J Assist Reprod Genet 37(2):417–426. https://doi.org/10.1007/s10815-019-01652-1

Li Y, Sena Lopes J, Coy-Fuster P et al (2022) Spontaneous and ART-induced large offspring syndrome: similarities and differences in DNA methylome. Epigenetics 17(11):1477–1496. https://doi.org/10.1080/15592294.2022.2067938

Lismer A, Siklenka K, Lafleur C et al (2020) Sperm histone H3 lysine 4 trimethylation is altered in a genetic mouse model of transgenerational epigenetic inheritance. Nucleic Acids Res 48(20):11380–11393. https://doi.org/10.1093/nar/gkaa712

Mapletoft RJ, Bó GA (2022) Volumen 2: Chapter 1a General sanitary procedures and considerations associated with *in vivo*-derived bovine embryos. International Embryo Technology Society Manual, 5th edn

Marques RS, Cooke RF, Rodrigues MC et al (2016) Impacts of cow body condition score during gestation on weaning performance of the offspring. Livest Sci 191:174–178. https://doi.org/10.1016/j.livsci.2016.08.007

Marquez DC, Paulino MF, Rennó LN et al (2017) Supplementation of grazing beef cows during gestation as a strategy to improve skeletal muscle development of the offspring. Animal 11(12):2184–2192. https://doi.org/10.1017/S1751731117000982

McLean KJ, Boehmer BH, Spicer LJ et al (2018) The effects of protein supplementation of fall calving beef cows on pre- and postpartum plasma insulin, glucose and IGF-I, and postnatal growth and plasma insulin and IGF-I of calves. J Anim Sci 96(7):2629–2639. https://doi.org/10.1093/jas/sky173

Ming H, Sun J, Pasquariello R et al (2021) The landscape of accessible chromatin in bovine oocytes and early embryos. Epigenetics 16(3):300–312. https://doi.org/10.1080/1559229 4.2020.1795602

Mozaffari Makiabadi MJ, Bafandeh M, Gharagozlou F et al (2023) Developmental programming of production and reproduction in dairy cows: II. Association of gestational stage of maternal exposure to heat stress with offspring's birth weight, milk yield, reproductive performance and AMH concentration during the first lactation period. Theriogenology 212:41–49. https://doi.org/10.1016/j.theriogenology.2023.09.002

Mueller ML, Van Eenennaam AL (2022) Awardee talk: synergistic power of genomic selection, assisted reproductive technologies, and gene editing to drive genetic improvement of cattle. J Anim Sci 100(Suppl 3):10–11. https://doi.org/10.1093/jas/skac247.018

Nascimento KB, Galvão MC, Meneses JAM et al (2022) Effects of maternal protein supplementation at mid-gestation of cows on intake, digestibility, and feeding behavior of the offspring. Animals (Basel) 12(20):2865. https://doi.org/10.3390/ani12202865

Org T, Hensen K, Kreevan R et al (2019) Genome-wide histone modification profiling of inner cell mass and trophectoderm of bovine blastocysts by RAT-ChIP. PLoS One 14(11):e0225801. https://doi.org/10.1371/journal.pone.0225801

Ortega MS, Wohlgemuth S, Tribulo P et al (2017) A single nucleotide polymorphism in COQ9 affects mitochondrial and ovarian function and fertility in Holstein cows. Biol Reprod 96(3):652–663. https://doi.org/10.1093/biolre/iox004

Perry VEA, Copping KJ, Miguel-Pacheco G et al (2019) The effects of developmental programming upon neonatal mortality. Vet Clin North Am Food Anim Pract 35(2):289–302. https://doi.org/10.1016/j.cvfa.2019.02.002

Rabaglino MB, Bojsen-Møller Secher J, Sirard MA et al (2021) Epigenomic and transcriptomic analyses reveal early activation of the HPG axis in in vitro-produced male dairy calves. FASEB J 35(10):e21882. https://doi.org/10.1096/fj.202101067R

Radunz AE, Fluharty FL, Relling AE et al (2012) Prepartum dietary energy source fed to beef cows: II. Effects on progeny postnatal growth, glucose tolerance, and carcass composition. J Anim Sci 90(13):4962–4974. https://doi.org/10.2527/jas.2012-5098

Ruiz LS (2022) Ovum pick-up (OPU) in cattle: an update. In: Biotechnologies applied to animal reproduction. Current trends and practical applications for reproductive management. CRC Press, Abingdon, pp 139–183

Sang L, Ortiz W, Xiao Y et al (2020) Actions of putative embryokines on development of the preimplantation bovine embryo to the blastocyst stage. J Dairy Sci 103(12):11930–11944. https://doi.org/10.3168/jds.2020-19068

Schulz KN, Harrison MM (2019) Mechanisms regulating zygotic genome activation. Nat Rev Genet 20(4):221–234. https://doi.org/10.1038/s41576-018-0087-x

Silva GM, Chalk CD, Ranches J et al (2021) Effect of rumen-protected methionine supplementation to beef cows during the periconception period on performance of cows, calves, and subsequent offspring. Animal 15(1):100055. https://doi.org/10.1016/j.animal.2020.100055

Siqueira LG, Hansen PJ (2016) Sex differences in response of the bovine embryo to colony-stimulating factor 2. Reproduction 152(6):645–654. https://doi.org/10.1530/REP-16-0336

Siqueira LGB, Dikmen S, Ortega MS et al (2017a) Postnatal phenotype of dairy cows is altered by *in vitro* embryo production using reverse X-sorted semen. J Dairy Sci 100(7):5899–5908. https://doi.org/10.3168/jds.2016-12539

Siqueira LG, Tribulo P, Chen Z et al (2017b) Colony-stimulating factor 2 acts from days 5 to 7 of development to modify programming of the bovine conceptus at day 86 of gestation. Biol Reprod 96(4):743–757. https://doi.org/10.1093/biolre/iox018

Siqueira LG, Silva MVG, Panetto JC et al (2019) Consequences of assisted reproductive technologies for offspring function in cattle. Reprod Fertil Dev 32(2):82–97. https://doi.org/10.1071/RD19278

Sirard MA (2021) How the environment affects early embryonic development. Reprod Fertil Dev 34(2):203–213. https://doi.org/10.1071/RD21266

Skibiel AL, Peñagaricano F, Amorín R et al (2018) In utero heat stress alters the offspring epigenome. Sci Rep 8(1):14609. https://doi.org/10.1038/s41598-018-32975-1

Swartz TH, Bradford BJ, Lemke M et al (2022) Effects of prenatal dietary rumen-protected choline supplementation during late gestation on calf growth, metabolism, and vaccine response. J Dairy Sci 105(12):9639–9651. https://doi.org/10.3168/jds.2022-22239

Tobi EW, Goeman JJ, Monajemi R et al (2014) DNA methylation signatures link prenatal famine exposure to growth and metabolism. Nat Commun 26(5):5592. https://doi.org/10.1038/ncomms6592

Tríbulo P, Siqueira LGB, Oliveira LJ et al (2018) Identification of potential embryokines in the bovine reproductive tract. J Dairy Sci 101(1):690–704. https://doi.org/10.3168/jds.2017-13221

Underwood KR, Tong JF, Price PL et al (2010) Nutrition during mid to late gestation affects growth, adipose tissue deposition, and tenderness in cross-bred beef steers. Meat Sci 86(3):588–593. https://doi.org/10.1016/j.meatsci.2010.04.008

Velazquez MA, Idriss A, Chavatte-Palmer P et al (2023) The mammalian preimplantation embryo: its role in the environmental programming of postnatal health and performance. Anim Reprod Sci 256:107321. https://doi.org/10.1016/j.anireprosci.2023

Viana J (2022) Statistics of embryo production and transfer in domestic farm animals. Embryo Technol Newslett 41(4). https://www.iets.org/Portals/0/Documents/Public/Committees/DRC/IETS_Data_Retrieval_Report_2022.pdf. Cited 13 May 2024

Viana JHM Dode MAN, Basso AC (2022) General sanitary procedures associated with *in vitro* production of embryos. In: International Embryo Technology Society Manual. 5th ed.

Wang M, Ibeagha-Awemu EM (2021) Impacts of epigenetic processes on the health and productivity of livestock. Front Genet 11:613636. https://doi.org/10.3389/fgene.2020.613636

Zenobi MG, Bollatti JM, Lopez AM et al (2022) Effects of maternal choline supplementation on performance and immunity of progeny from birth to weaning. J Dairy Sci 105(12):9896–9916. https://doi.org/10.3168/jds.2021-21689

Zhang Z, Zhang J, Diao L et al (2021) Small non-coding RNAs in human cancer: function, clinical utility, and characterization. Oncogene 40(9):1570–1577. https://doi.org/10.1038/s41388-020-01630-3

Zolini AM, Carrascal-Triana E, Ruiz de King A et al (2019) Effect of addition of l-carnitine to media for oocyte maturation and embryo culture on development and cryotolerance of bovine embryos produced *in vitro*. Theriogenology 133:135–143. https://doi.org/10.1016/j.theriogenology.2019.05.005

Part III

The Impact of Assisted Reproduction on Production Systems

Reproductive Management of Donkeys in Milk Production Programs

Luis Losinno ⓘ, Ana Flores Bragulat ⓘ, Luisina Chapero ⓘ, Liliana Rosetto ⓘ, and Melina Pietrani ⓘ

Abstract

Domestic donkeys have been related to ancient human civilizations for thousands of years as pack and transportation animals, and indeed still are, particularly in the driest and poorest countries of the world. Nonetheless, the beneficial effects for human health of donkey milk have been known and documented for over 2000 years and have been the basis for the current growth of commercial production systems, primarily aimed at producing natural but industrialized milk for children allergic to cow milk proteins (CMPA) and elderly adults. Additionally, it is used in the human cosmetic industry. Donkey milk production systems require organized and controlled programs where reproduction is one of the main axes, given that without foals, there is no milk production. These systems are mostly semi-extensive and based on pasture natural breeding. These conditions are more suitable for donkeys' behavioral patterns, reducing intensive human handling and, therefore, reducing stress, with positive, low-cost outcomes and negative impacts on the animals. However, considering future genetic improve-

L. Losinno (✉)
Laboratorio de Producción Equina, Universidad Nacional de Río Cuarto, Río Cuarto, Argentina

Equine Academy, Río Cuarto, Argentina

A. F. Bragulat
Laboratorio de Producción Equina, Universidad Nacional de Río Cuarto, Río Cuarto, Argentina

L. Chapero · L. Rosetto
Universidad Nacional de La Pampa, Santa Rosa, Argentina

M. Pietrani
Universidad Nacional de Villa María, Villa María, Argentina
e-mail: mpietrani@unvm.edu.ar

© The Author(s), under exclusive license to Springer Nature Switzerland AG 2024
J. C. Gardón, K. Satué Ambrojo (eds.), *Assisted Reproductive Technologies in Animals Volume 1*, https://doi.org/10.1007/978-3-031-73079-5_5

97

ment programs and the preservation of lines and breeds, assisted reproduction techniques such as embryo transfer, embryo cryopreservation, semen cryopreservation, and in vitro embryo production are being developed, although still in experimental phases.

Keywords

Dairy donkey · Milk production · Farm management · Reproduction · CMA

1   Introduction

Domestic equids derive from a small common ancestor from 50–60 million years ago and, more recently, the evolutionary lineage that gave rise to current species in the genus *Equus* (horses, donkeys, and zebras) approximately 4 million years ago. All of them evolved as wild animals, meaning their lives were exclusively directed by natural and sexual selection until the beginning of their domestication, around 6000 years before the present (BP). From then, the genus *Equus*, with a wide variation in the number of chromosomes, includes two domestic species whose lives are directed by artificial selection managed by humans (horses, donkeys, and their hybrids: mules and hinnies) and seven wild species, meaning undomesticated: zebras; wild asses from Asia and Africa; and the Przewalski's "horses" from Mongolia. There is also a third post-domestication category: feral animals (from the Latin *fera*: wild beast), which are domestic animals in freedom conditions, outside human control, where natural selection strongly operates in almost all the planet (goats, cats, rabbits, camels, pigs, dogs, horses, and donkeys), generally closed population groups with a tendency to high inbreeding levels.

As far as fossil records are concerned, the domestication of *Equus africanus africanus* (Nubian wild ass) and E*quus africanus somaliensis* (Somali wild ass) in Northeast Africa (one extinct in the wild and the other critically endangered today) was the species that generated what would later become a subspecies (*Equus africanus asinus*) that we now know as the domestic donkey worldwide. All donkey breeds belong to this same species. Over 6000 years ago, the donkey was initially hunted by humans as a meat provider and then domesticated in northern Africa. It was used mostly as a pack and transport animal. This animal has played a crucial role in facilitating transportation and agriculture in many historical contexts, thereby greatly facilitating commerce among regions. Over the years, donkeys were slowly replaced by machines in developed countries, which led to a decrease in their use for carrying loads, except in the poorest countries, where they are still involved in daily work and transportation. According to Tibary, donkeys and mules represent 60–80% of the equid population in arid and semi-arid countries in the world (Tibary and Bakkoury 1994). In economically developed regions like Europe, donkey populations have decreased dramatically and are more often bred as ornamental pets, with some exceptions like Italy, where the donkey milk industry is a productive economic niche. In Eastern countries like China, donkey production is intended for meat, pharmaceuticals, and milk. In recent years, globally, donkey milk has begun to gain importance due to its

Table 1 Estimated world distribution of donkey population (Adapted from Polidori and Vincenzetti 2019)

Continent/countries	Donkey population
Asia	15,000,000
Africa	10,000,000
Middle East	10,000,000
South America	9,000,000
Europe	1,500,000
North America	50,000
Australia	5,000,000
Total	50,550,000

qualities, leading to an interest in donkey production worldwide, which leads us to make donkey production more efficient.

Currently, the global donkey population (about 50 million animals) is continuously and progressively declining (FAO 2020) (Table 1). Particularly, most of the 189 donkey breeds developed by human selection over thousands of years on almost all continents are categorized as threatened, protected, or directly extinct in the wild. Many of the breeds described are not very accurate in morphological descriptions, registrations, and updated status. For example, from the 28 breeds described in Europe, 7 are in critical status (less than 100 jennies and 5 jacks according to FAO criteria), 20 are endangered (less than 1000 jennies and 20 jacks), and just 1 is not at risk (more than 1000 jennies and 20 jacks) (Camillo et al. 2018).

Nevertheless, this diversity of breeds, like in other domestic mammals, has generated enormous phenotypic variability in conformational and functional traits. This allows the adaptation to different environments and purposes for which the breeds have been developed. An example of this is the average body weight of adults, ranging between 80 and 480 kg, and the height, between 64 cm in miniature donkey breeds and 170 cm in the Mammoth breed. Donkeys' longevity extends between 30 and 35 years, during which they maintain their capacity for active labor even in advanced age.

Historically, donkeys have been associated with negative terms of insult and mockery, portraying these animals as stupid, inept, and bad-tempered (Person 2022). However, they have been used since ancient Egyptian times, as pack and transport animals by the poorest people on the planet and are still used today. This is due to some of their most evident temperamental and physiological characteristics, such as low maintenance requirements, robust health, docility, empathy, resistance to fatigue, adaptability, learning capacity (despite the unfounded human stereotype of stupidity), and impressive physical endurance. Unfortunately, their marked resistance to carry heavy loads, a tenacious approach to toil, and stoicism have been used against them by humans, being subjected to mistreatment, violence, and abuse for thousands of years, something that can still be verified today (Miragaya et al. 2018). Stoicism is typical predator-avoidance behavior in a prey species such as the donkey; appearing strong and normal reduces the chances of a predator attack. However, this stolidity does not lessen the donkey's ability to experience pain and distress (The Donkey Sanctuary).

2   Donkey Milk Production Programs

Donkeys have not only been overlooked by science until very recently (there is currently a significant flow of scientific information, especially in the last decade), but they have also not been considered a "productive" animal in classical terms, understanding this concept as one that generates some kind of "product," many of them post mortem such as meat and skin. The donkey has been a service animal rather than a product for humans. Perhaps that's why, when mentioning donkey milk and especially the formal production, generally, the first responses are astonishment or disbelief. However, for many years, it has been a common productive practice in smallholder systems or family farms in Mediterranean countries, the Balkans, North Africa, Latin America, and mainly China. Currently, there are specific commercial programs on low, medium, and large scales for milk production destined for direct human consumption, dairy by-products, or the human cosmetic industry.

Donkey production programs for commercial purposes have gained special relevance in recent years, mainly due to clinical evidence proving that donkey milk is a natural alternative for children with cow's milk allergy (CMA). Also, it has been proven to have a positive clinical impact on human health based on its bioactive peptide profile and the cosmetic industry (soaps and beauty creams). Donkey's milk composition is similar to human milk, has a proven tolerability, palatability, and an adequate concentration of nutrients, as well as a large number of bioactive compounds (Sarti et al. 2019; Martini et al. 2021; Di Salvo et al. 2023). This led to a significant increase in scientific information flow regarding this species, historically relegated by science.

Donkey milk production in the world is currently limited to a few countries, most of them in the Mediterranean area (Italy, Portugal, Spain, France, Serbia, Croatia, and Greece), Asia (China and India) and very few in America (Chile, Mexico, Brazil, and Argentina). Some aspects specific to the species deserve to be highlighted to quickly dispel any kind of comparison or free association with horses, much less with a dairy cow.

Donkeys evolved in arid and mountainous environments, with scarce water sources and low-quality forage, so generally, "native" donkey populations, i.e., not belonging to a particular registered breed, tend to be of a medium biotype (110–130 cm tall) and 100–200 kg weight. Like with most other domestic animals (such as horses, dogs, and cattle), "extreme" phenotypes and rarities in both biotype and coat are a product of human domestication and controlled breeding toward "functional" (larger size, for example) or purely aesthetic (coat color, dwarfs) phenotypes generically called "breeds."

The four classic categories in donkeys related to size are: (1) Large (over 130 cm and in some cases up to over 160 cm), the Mammoth breed being one of the largest in the world, for a male to be registered, it must measure at least 147 cm and a female 142 cm; (2) Medium (110–130 cm); (3) Small (90–110 cm); (4) Dwarf or Miniature (less than 90 cm). The latter two, the smallest ones, have been the most popular for export in recent years, especially to the United States, but they are also in vogue in Europe, especially in France, where the breeding of small donkeys has been developed in a highly selective manner. A poorly studied aspect is the relationship between coat color (black, brown, grey, piebald, skewbald, and white) and the aptitudes for adaptation to specific environments or agricultural activities. For example, black donkeys typical from Europe could have a better resistance to solar radiation and high humidity, while grey or skewbald donkeys may have a better adaptation to temperate mountain areas with large temperature variations between day and night.

Ideally, considering animal welfare regulations and characteristics of donkey biology patterns, it has been proposed as a basis for donkey milk production, a pasture-based system (i.e., where animals can graze all day or with certain hourly restrictions). They are gregarious migratory herbivores, meaning it is not very difficult to infer what they need for their homeostasis and fitness: to choose and eat grass, preferably of medium to low quality and lignified; freely walk and social interaction. As scientists, technicians, and teachers of animal science and production, we must try to promote the best possible life for the animals that serve us and from which we derive economic benefits for thousands of years and not consider them mere money-making objects.

Production scales can be categorized according to investment and potential profitability in terms of the number of milking jennies. Low scale (10–30), medium scale (40–100), and high scale (more than 100). Regarding investment, the objectives of the production system for categorization must be considered, from primary production of chilled fluid milk for delivery to industry to a complete production cycle with on-site industrialization. It is considered that the majority of the production systems are in the medium scale category, between 30 and 60 milking jennies, under mixed pasture-based systems, with daily milking and an average production that varies between 0.5 and 1.2 L/jenny/day with average lactations of 6 months. Quality standards and procedures are determined by each country, but as a general rule, milking must be carried out mechanically, in approved environments under biosafety standards, pasteurized, applying strict cleaning protocols and standardized animal welfare conditions since it involves the production of food with nutraceutical properties for babies. The product can be marketed as refrigerated, frozen, or lyophilized.

Donkey milk production in European countries has been estimated at 300 tons per year (Valle et al. 2018) and 270,000 tons/year in China, the world's leading producer, related to a long tradition and its high donkey population (2.7 million) (Seyiti and Kelimu 2021). There are no official reports in Latin America, but there is evidence of commercial production programs on different scales in Chile and Argentina, according to unofficial and recent data. Donkey milk production is an emerging and innovative productive phenomenon with a strong potential impact on public health, generation of industries, and specific jobs, transforming an excluded animal into a productive one. It is possible to create artificial systems that are sustainable and environmentally friendly, with animals and people working in them in harmony, creating responsible systems as a model for future generations.

3 Donkey Milk and Human Health

There is fossil evidence and written records of the cosmetic and biomedical use of donkey milk dating back 3000 years. More recently, documents in ancient medical treatises such as *Corpus Hippocraticum* (2480 BP); *Naturalis Historia* by Gaius Plinius Secundus (2000 BP); Valuable Prescriptions for Emergencies by Sun Simiao (Tang Dynasty 1800 BP); Compendium of Materia Medica by Li Shizhen (Ming Dynasty, 1000 BP) for the treatment of arthritis, cough, surgical wounds, ulcers, and dysentery. Moreover, one of the most renowned and impactful documents in the medical community for hundreds of years, the famous Al-Qanum, known as "The Canon of Avicenna," the monumental 14 volumes of medical work written around 1100 BP by the Persian physician Ibn Sina (Avicenna) where he advised the use of donkey milk for the treatment of cough, hemoptysis, ulcers, wounds, ascites, fever, fatigue, and asthma.

There are also graphic records of a widespread practice in Europe in the nineteenth century where donkeys were bred adjacent to human hospitals, especially those with areas for orphans. Nurses breastfed the babies directly from the jennies, which allows us to assume that donkey milk was considered the most similar to human milk (something currently experimentally demonstrated) and that it had clinically proven beneficial effects. The sale of donkey milk, through manual milking on demand, especially for children with respiratory problems or elderly people, can be observed today in many Mediterranean and Latin American countries as an ancestral cultural practice based on strong empirical evidence of its beneficial effects (Losinno 2022).

Scientific production relating to donkeys and donkey milk began in the mid-twentieth century and has grown significantly in the past few years (Figs. 1 and 2). Comparative compositional and microbiological studies with the milk of other domestic mammals emerged at the beginning of this century. The first randomized clinical trials in children diagnosed with cow's milk protein allergy (CMPA) were approved and conducted in Europe using hydrolyzed bovine milk formulas as control. These hydrolyzed bovine formulas are the standard recommendation in most countries that do not promote or that are unaware of the demonstrated effects of

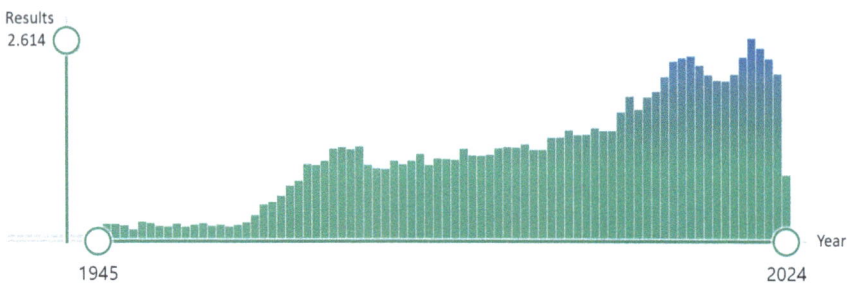

Fig. 1 Scientific publications per year using the DONKEY keyword in the PubMed database between 1945 (45) and 2023 (2138)

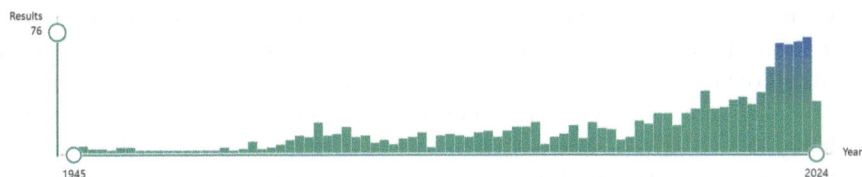

Fig. 2 Scientific publications per year using the DONKEY MILK keyword in the PubMed database between 1945 (0) and 2023 (45)

Table 2 Chemical composition (g/100 g) of the donkey, cow, and human milk

Milk	Dry matter	Fat	Total Protein	Whey Protein (g/L)	Casein (g/L)	Lactose	Energy (kJ/kg)
Human	12.4	3.3	1.6	7.0–8.3	3.2–4.2	6.7	2855
Jennie	9.61	1.2	1.7	5.0–8.0	6.4	6.3	1939
Cow	12.3	3.4	3.4	4.5	27.2	4.7	2983

Adapted from Polidori et al. (2015), Salimei et al. (2004) and Vincenzetti et al. (2017)

donkey milk. Recently (2021), the results of clinical trials using fortified donkey milk in premature babies (with birth weights of less than 1250 g) with remarkable positive effects have been published. In the last 5 years, studies have been more focused on metabolomics, proteomics, molecular compositional analysis (protein and fatty acid profiles), exosomes, and in vitro effects assays on human cell cultures evaluating gene expression patterns.

The biomedical use of donkey milk is partly due to its compositional similarity to human milk and its marked differences with ruminant milk. Donkey milk has a low protein concentration, especially caseins, considered the main cause of food allergies in human infants (CMPA), a high concentration of lactose, which makes it very sweet and palatable for babies, and a low-fat content (Table 2).

Donkey milk production systems are mainly based on "semi-intensive" reproductive management. In order to obtain optimal milk production and maintain animal health and welfare, donkey farm management has changed from traditional to more organized systems. This involves production programs that consider nutrition for the different categories, amount of milking per day, separation time from the jennies (number of hours and time of the day), specific milking facilities, and also reproductive design. Dairy farm management is highly correlated with welfare. In poorly managed systems, animals will be predisposed to chronic stress, which can impact their behavior, health, and milk production (Salari et al. 2022).

4 Donkey Reproduction Programs

Donkeys have a territorial, non-harem type of sociosexual behavior. In the classical pattern, jacks spend most of their time away from females. Jennies move freely, usually in groups, and those in estrus may form a sexually active group, similar to cows, displaying estrus behavior, mounting each other to attract the male and approaching him frequently. Females are gregarious and migratory animals, meaning they live in relatively stable social groups with their offspring and prefer to stay in proximity, something very similar to what happens with zebras. On the contrary, isolation and loneliness cause them anxiety and severe stress that can even lead to depression and anorexia, resulting in illness, so these situations should be avoided. There's an old saying that goes: "If you're going to buy a donkey, it's better to buy two," referring to avoiding isolation (unnatural) and providing a better quality of life in captivity. This situation can even be reversed by placing the donkey with another animal, such as a sheep, goat, or even a horse. We have successfully experimented by placing two breeding males, a stallion (equine) and a donkey (jack), together in a pen for years, and their behavior and libido greatly improved, decreasing the typical aggressiveness of the isolated animal (Fig. 3).

Unlike females, males are generally territorial, meaning they live relatively solitary lives in territories they mark as "their own" through feces and urine. Females are sexually active, and they approach males' territories during estrus, guided by the typical braying of donkeys that can be heard from at least 3 km away due to an anatomical modification in the larynx. Additionally, during the heat or copulatory receptivity period, female donkeys adopt a typical position in front of the male by lowering their heads and opening their mouths wide as if they were chewing (Figs. 4 and 5).

Gestation is longer than in mares (360–370 days) and a very important characteristic, especially in milk production programs, is that jennies only produce milk if they remain with their offspring, which means that if for any reason she loses it permanently, she will stop producing milk (Fig. 6). This is why in organized milk production programs; scheduled separations of the jennies and their offspring are carried out for a maximum of 3–6 h to achieve efficient milking.

Feeding of donkeys deserves separate comments, especially when considering production systems in which, due to ignorance, improvisation, and analogy, practices applied to other domestic species, such as horses or dairy cows, are transferred. Donkeys are highly efficient in utilizing dietary fiber for energy production, better than most other herbivores, and have a low tolerance to excess starch and other

Fig. 3 Stallion and Jack live together in a paddock

Fig. 4 Freely and natural copulative behavior in domestic donkeys in a pasture-based donkey milk production program

carbohydrates (CHO). A low-fiber and/or high-CHO diet is contraindicated as it endangers not only productivity but also the animals' lives through clinical conditions such as colic, laminitis, hyperlipidemia, metabolic syndrome, and dysbiosis.

4.1 Natural Breeding

Jennies in natural breeding programs may be pasture or hand-bred (Fig. 7). Hand breeding is less common due to the particularities of jack sexual behavior (Fig. 8). Jacks, when hand-bred, usually take a long time to achieve erection and mount, they also tend to perform several mounts without erection and appear to require space and liberty to approach and retreat from jennies before successful mount and ejaculation. Donkeys have long held the reputation of being slow and difficult to hand-breed. As a result, most are bred at pasture or in paddocks where they are allowed to interact freely with one or more jennies. By comparison to horse stallions, jacks typically take a long time to achieve erection and mount when handled for breeding. They also tend to proceed with erection only after mounting without erection, as is the normal sequence for free-ranging jacks. For hand-breeding, estrus jennies require minimal to no physical restraint, and they tend to be less aggressive than mares; hence, breeding accidents are less common. When a hand-breeding program is employed, jennies should be, if possible, teased and evaluated by manual palpation and ultrasound to enhance success. This implies more intensive management that may cause stress, but on the other hand, it allows better control of the estrus cycle, reduced number of services per cycle, and accurate conception date determination, among other things.

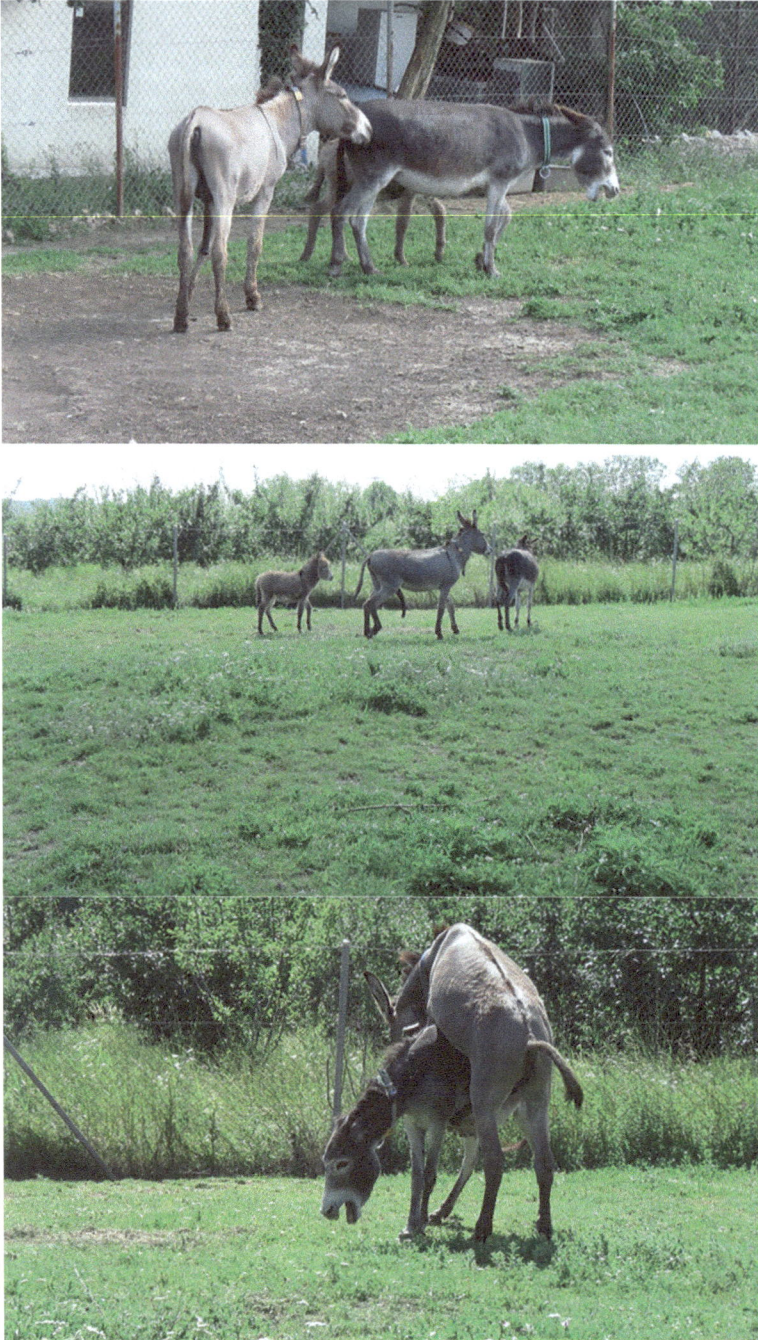

Fig. 5 Freely precopulatory sequence in a pasture-based donkey milk production program

Fig. 6 Prepartum jennies in a pasture-based donkey milk production program

Fig. 7 Milking jennies and jack in a pasture-based breeding program

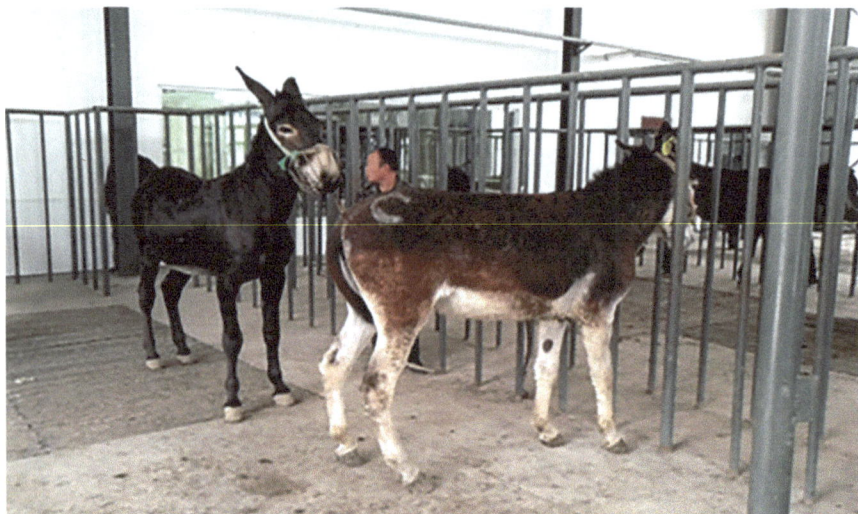

Fig. 8 Precopulatory behavior in a hand-bred donkey reproduction program

In pasture breeding, the jack is turned into pasture with the jennies. Usually, up to 20–25 jennies per jack for a prolonged period of time, approximately 90 days. Pregnancy exams can be performed at 15-day intervals; pregnant jennies may be replaced with barren ones. In dairy farms, jennies are grouped according to whether they are being milked or not. It is advisable to group jennies that are not being milked by age, leaving those younger than 30 months out of breeding. In pasture breeding, the jack is responsible for estrus detection and may mate with a jenny several times per day. An experienced pasture breeding jack will rarely attempt to breed a jenny that is not in estrus and, in general, is very tolerant with foals, although it can become aggressive with male donkeys reaching puberty. The advantages of pasture breeding are decreased personnel and breeding expenses, as well as decreased handling; this latter is of particular importance to reduce stress in donkeys. This minimal handling may become a disadvantage given that we may not be able to detect reproductive problems soon enough in jacks or jennies, leading to decreased pregnancy rates. This can be overcome by examination of jacks and jennies before pasture turnout and pregnancy exams every 15 days to determine conception day accurately and obtain a potential foaling date (Figs. 9, 10 and 11). Also, this allows early detection and reduction of twin pregnancies, which are rare but can occur. Per-season pregnancy rates in pasture breeding programs can range between 85 and 90% (personal data, Equslac™, Argentina).

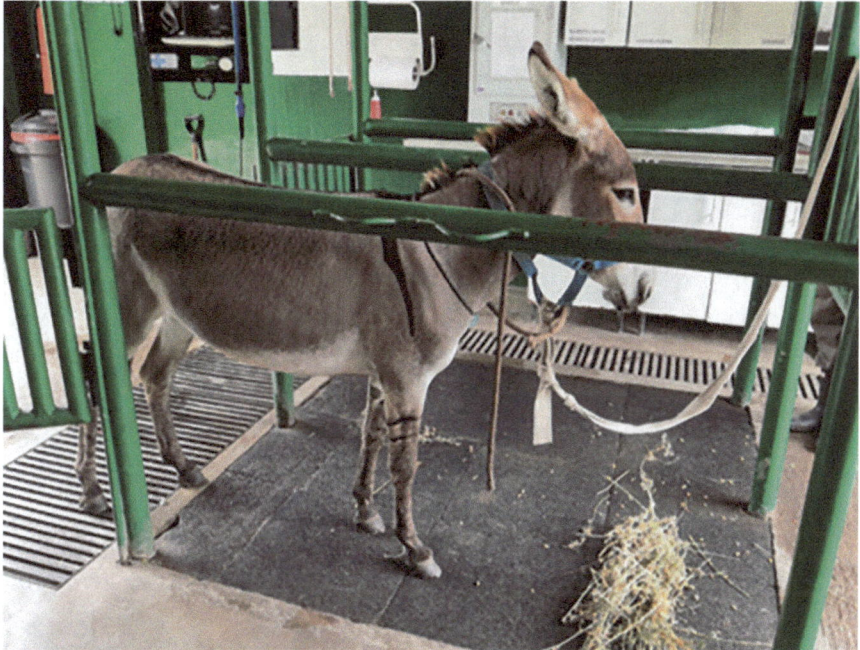

Fig. 9 Jenny getting used to stocks. Note the inappropriate size of the horse-designed stocks

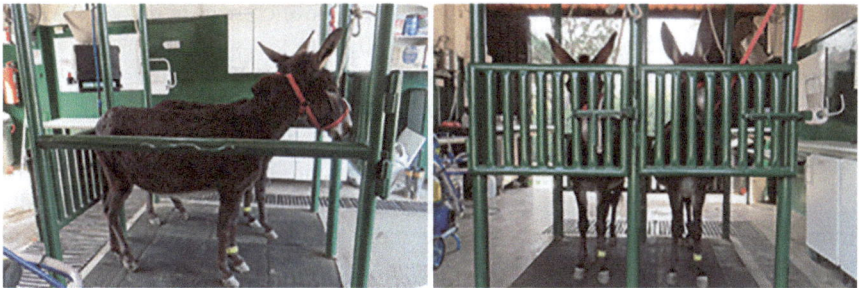

Fig. 10 Jennies in horse-designed stocks

4.2 Assisted Reproductive Technologies

The use of assisted reproductive technologies (ARTs) has become increasingly prevalent in equine species over the past few decades. Most ARTs applied to donkeys are adapted from horses (Oliveira et al. 2006), extrapolating information and knowledge and creating new information specific to donkeys. Within ARTs, the most widely used in donkeys are artificial insemination with fresh or chilled semen; however, there are others much more complex, such as semen freezing, embryo transfer (ET), intracytoplasmic sperm injection (ICSI), and somatic cell nuclear

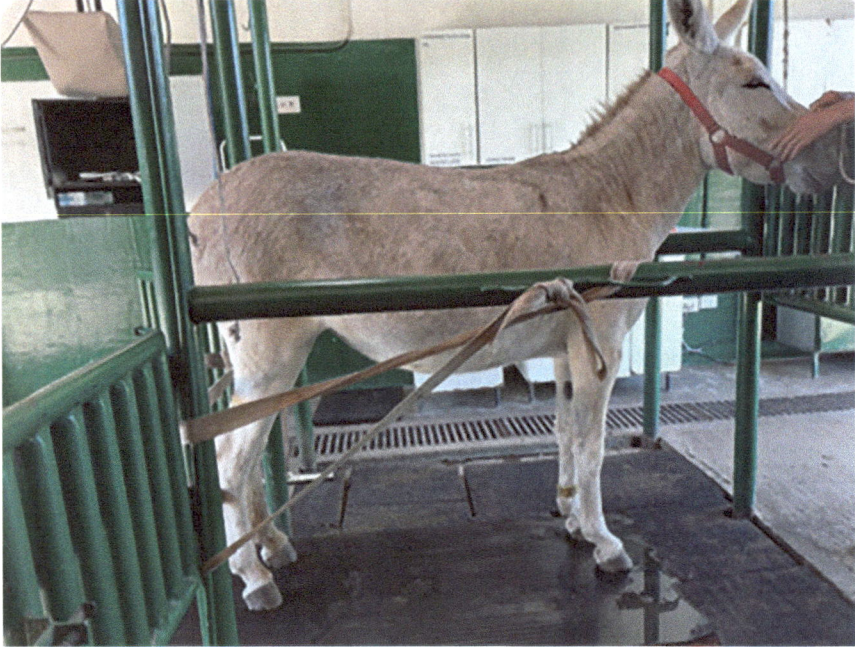

Fig. 11 "Adapted" mare's stock for jennies' reproductive procedures

transfer (SCNT) which are in very early stages of development in donkeys. While AI with fresh and cooled semen has given satisfactory results, the use of frozen semen presents distinct challenges. Despite advancements, the pregnancy rates achieved with frozen semen are significantly lower compared to fresh or chilled semen.

Although pregnancy rates in semi-intensive pasture breeding programs can be high, and this type of program has more advantages than disadvantages, especially from the animal welfare point of view, in certain situations, ARTs may be necessary to implement. ARTs can be applied for genetic improvement programs, inbreeding risk avoidance, preserving biodiversity, and preserving certain endangered donkey breeds.

These technologies offer a strategic approach to managing reproductive challenges and advancing the long-term sustainability of donkey populations. More studies and improvements in ARTs are necessary for them to be applied successfully in productive systems.

4.2.1 Ovulation Induction

Ovulation induction agents are routinely used in mares' reproductive cycles to optimize breeding management. Different ovulation induction and estrus synchronization protocols have been studied in jennies to optimize reproductive management, especially in semi-intensive reproduction programs (Table 3).

Table 3 Ovulation induction agents in jennies

Drug	Jennies (n)	Dose	Follicle diameter at treatment (mm)	Interval to ovulation (h)	Ovulation within 48 h (%)	References
Lecirelin	34	100 μg (IV)	36–40	42.8 ± 14.0	80	Carluccio et al. (2007)
hCG	27	2500 UI (IV)	36–40	42.4 ± 13.0	100	
Buserelin	14	40 μg (SC)	33 ± 1	24–48	71.4	Camillo et al. (2014)
LH	25	400 UI (IM)	35	24–48	92	Chang et al. (2019)
hCG	25	3000 UI (IM)	35	24–48	76	
Histrelin acetate	10	250 μg (IM)	29–32	36 to 48	100	Oliveira et al. (2020)
hCG	10	2500 UI (IV)	29–32	36 to 48	90	
Deslorelin acetate	12	0.75 mg (IM)	36.34 ± 0.71	48.79 ± 2.69	78.7	Bottrel et al. (2022)
hCG	13	1500 UI (IM)	37.85 ± 1.27	62.61 ± 7.20	60.9	

Estrus duration in jennies varies based on factors such as time of year, body condition score, and age. Ovulation induction can aid in narrowing the ovulation window and facilitating breeding management or ET. Human chorionic gonadotropin (hCG) and gonadotropin-releasing hormone (GnRH) agonists have been utilized to induce ovulation. As in mares, the recommended follicular size to apply the ovulation induction agent correlates with body size. Small-framed jennies ovulate small follicles (28–32 mm), and larger-framed jennies, such as the Catalan breed, ovulate 40–44 mm follicles. Jennies induced with hCG or lecirelin with periovulatory follicles (36–40 mm) may also ovulate additional smaller follicles (30–35 mm) (Carluccio et al. 2007).

4.2.2 Artificial Insemination and Semen Evaluation

Artificial insemination is a viable technique for donkeys and can be applied with fresh, cooled, and frozen semen. Jacks, similar to stallions, can undergo collection procedures; however, it is advisable to allow longer teasing periods for them. A collection session with a jack may last between 30 and 60 min, with younger jacks exhibiting slower collection times compared to mature ones. Various methods, including using an estrous jenny or mare, a dummy, or ground collection, can be utilized for jack semen collection (Fig. 12). Standard equine equipment can be employed for semen collection, evaluation, transportation, and artificial insemination in donkeys (Canisso et al. 2019).

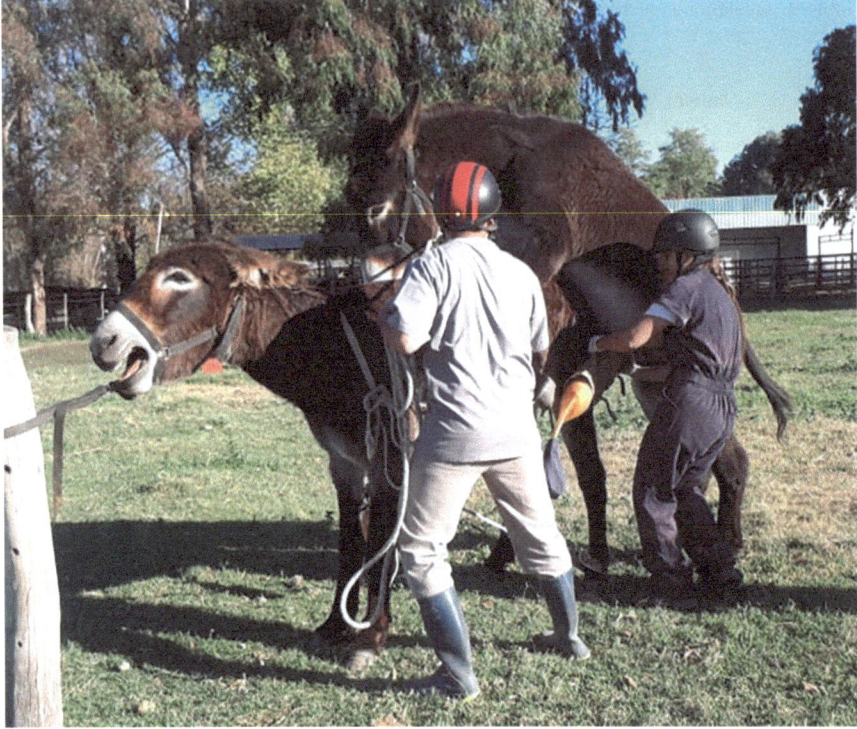

Fig. 12 Jack's semen collection with a Missouri artificial vagina using an estrus jenny

Compared with stallions, donkeys exhibit superior motility parameters, with higher velocities and progressive motility rates. Typically, sperm concentration ranges from 300 to 400 million sperm/mL, with 80–90% total and progressive motility. Younger donkeys tend to yield gel-free ejaculates ranging from approximately 30 to 50 mL, while older ones produce larger volumes, typically ranging from 60 to 90 ml (Miró et al. 2005). Younger donkeys tend to produce ejaculates with lower volume but higher concentrations than older jacks.

Artificial insemination with fresh semen in donkeys has been used for more than 50 years with per-season pregnancy rates between 60 and 80% (Svendsen 2008), and different breeding doses and extenders have been tested. Conception rates from artificial insemination of jennies with fresh jack semen typically vary from 40% to 86% and from 45% to 78% with cooled semen (Table 4).

Cooled semen can be stored at 4 °C for up to 48 h with satisfactory fertility rates post-insemination (Rota et al. 2008). High per-cycle pregnancy rates (50%) have been achieved in mares inseminated with frozen donkey semen (Oliveira et al. 2006). However, when it is used in jennies, pregnancy rates were 0–36% (Trimeche et al. 1998; Oliveira et al. 2006). Frozen semen induces a more severe inflammatory reaction in the uterus than fresh semen (Kotilainen et al. 1994), and it was

Table 4 Pregnancy rates in jennies after artificial insemination with fresh and cooled donkey semen (Adapted from Canisso et al. 2019)

Type	Breeding dose (million sperm)	Extender	Pregnancy rate (%)	References
Fresh	400	Skimmed milk	86 (6/7)	Vidament et al. (2009)
	500	Botusemen	40 (6/15)	Oliveira et al. (2016)
	1000	Botusemen	73 (11/15)	
	800	INRA 96	81 (25/31)	Carluccio et al. (2017)
Cooled	200	Skimmed milk	45 (9/20)	Vidament et al. (2009)
	460	Skimmed milk	78 (7/9)	
	460	INRA 82 + 2% egg yolk	64% (5/8)	
	500	Equiplus +1% egg yolk	50% (4/8)	Alonso et al. (2023)

hypothesized (Canisso et al. 2019), but not proven, that this may be the cause of the low pregnancy rates obtained with frozen donkey semen in this species.

4.2.3   Artificial Insemination Techniques

The artificial insemination technique is similar in all equine species. The pregnancy rate after AI with fresh semen does not differ significantly between jennies and mares. Jenny's reproductive anatomy is similar to that of mares, although there are some differences. The cervix in jennies points upwards or to the sides and is longer and smaller in diameter than in mares (Fig. 13). Protrudes 2–3 cm into the vagina, which may preclude intrauterine ejaculation, and also represents a challenge for intrauterine procedures like AI or ET.

Due to this, some studies have been carried out to evaluate pregnancy rates in jennies using different AI techniques. In the conventional artificial insemination technique, Jennie's tail is wrapped and tied to clean the perineal area. The pipette is introduced transvaginally and guided through the cervix; one finger is inserted into the cervix, and the pipette is passed into the uterine body. The semen-filled syringe is attached to the pipette to deliver semen with a gentle push. If the jenny is too small, the technique becomes more difficult to perform (Figs. 14 and 15).

An alternate technique for insemination could be the use of cervical forceps to hold the cervix in position using the Wilsher technique. A Polansky speculum is introduced in the vagina, and the cervix is held by the ventral segment with the cervical forceps and pulled backward to straighten it. Finally, an AI catheter is inserted through the cervix into the uterus. Flores Bragulat et al. (2022) reported pregnancy rates per cycle not statistically different using Wilsher and conventional AI techniques (54.5% and 75.0% conventional and Wilsher forceps techniques, respectively). Despite these results, the application of the Wilsher technique can be an especially useful tool in the AI of small donkey breeds (Figs. 16 and 17).

Also, the AI technique can vary depending on the place of semen deposition. The breeding dose can be deposited in the uterine body (used for fresh and cooled semen) or deep in the uterine horn. In the last case, semen is released as deep as

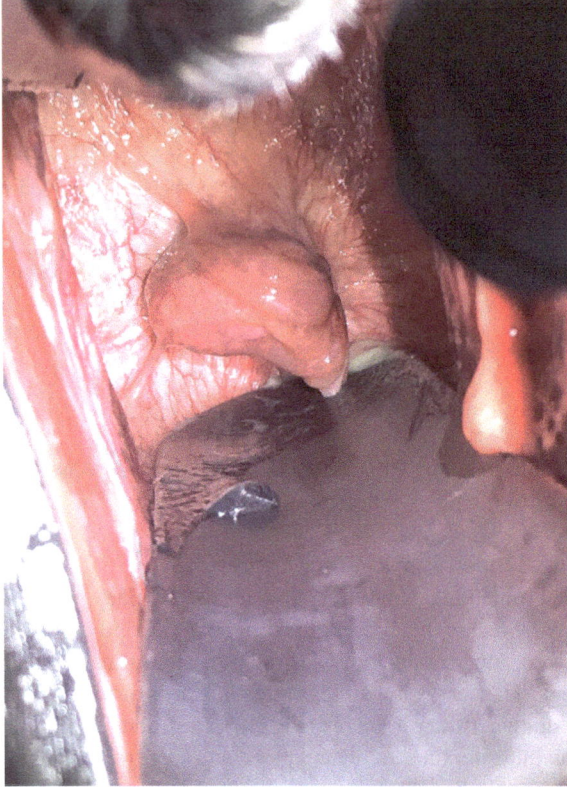

Fig. 13 Normal jennies' cervix observed with the use of a Polansky speculum. Note the cervical os pointing to the side

Fig. 14 Jennies in stocks ready for artificial insemination

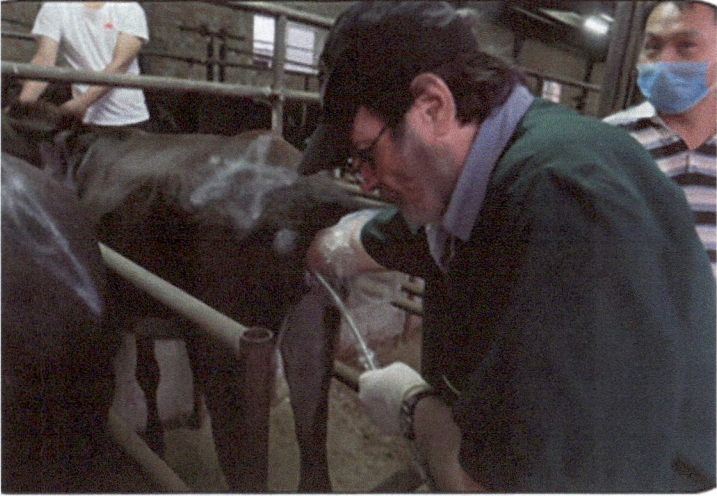

Fig. 15 Artificial insemination in a jenny using a conventional transvaginal technique

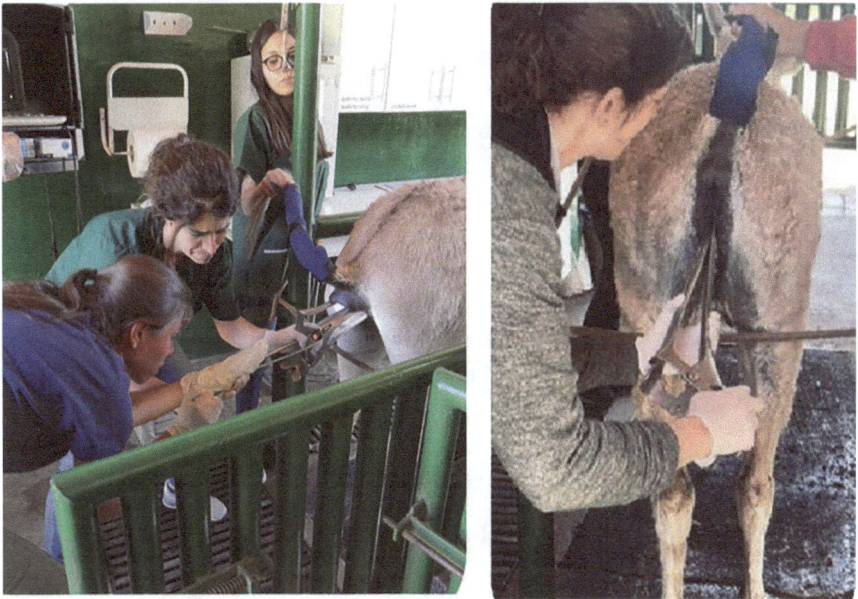

Fig. 16 Artificial insemination in jennies using the Wilsher technique and Polansky speculum

possible into the uterine horn ipsilateral to the ovary with the dominant follicle and/ or ovulation. Oliveira et al. (2016) found differences in pregnancy rates in jennies inseminated with the same breeding dose of frozen semen in the uterine body (0%) and deep horn (28%) (Fig. 18).

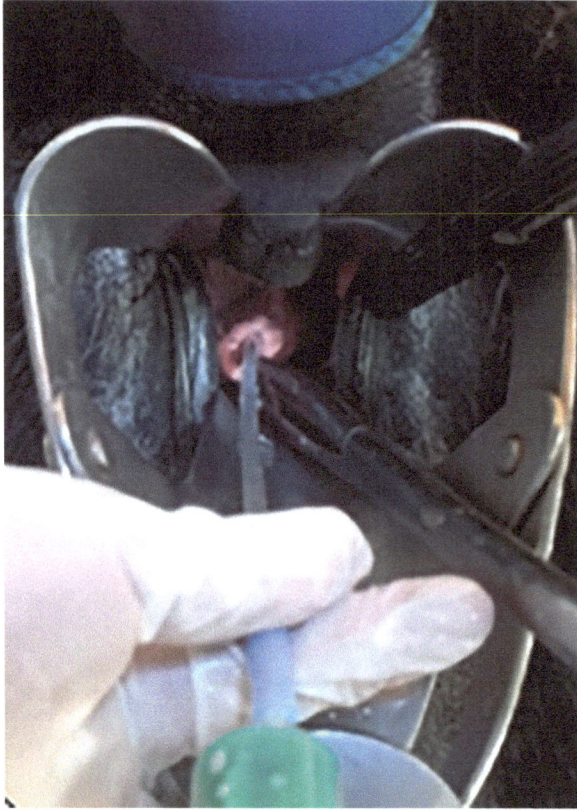

Fig. 17 Artificial insemination in a jenny using the Wilsher cervical forceps technique and standard AI pipette

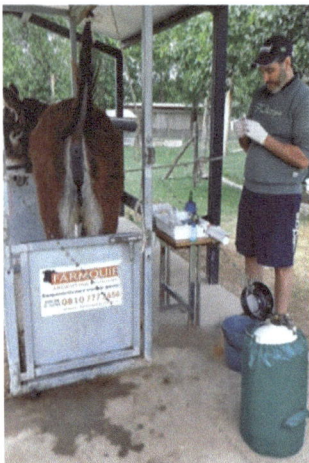

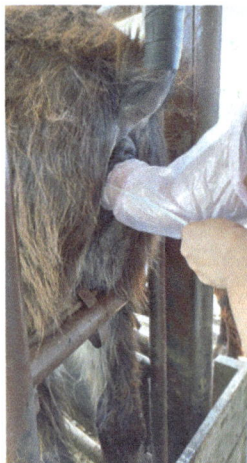

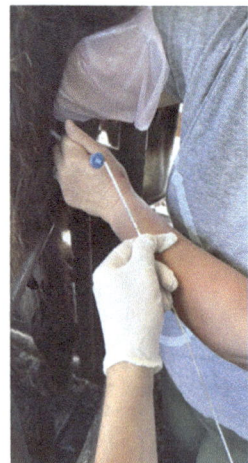

Fig. 18 Deep horn AI with frozen semen in a jenny

4.2.4   In Vivo Embryo Production and Transfer

In vivo embryo production by AI and subsequent uterine lavage for embryo recovery in donkeys has been carried out, mainly experimentally, for the past 40 years. It shares numerous similarities with ET in horses, although several challenges persist in the ET technique in donkeys, largely attributed to differences in reproductive anatomy and physiology between jennies and mares, the species from which the donkey ET technique was adopted. The diverse body sizes among jenny breeds and the unique conformation of the cervix (long and narrow) have posed additional challenges in catheterization. For the ET procedure, uterine lavage is performed using a non-surgical transcervical technique between 7 and 9 days post-ovulation, and embryos are processed and transferred to synchronized recipient donkeys between 5 and 8 days after ovulation. Embryo recovery rates following artificial insemination with fresh or chilled semen vary from 30% to 75%, showing no significant impact from photoperiod changes. Pregnancy rates in donkey-to-donkey ET can be up to 45% (Panzani et al. 2020). However, further ET studies in donkeys are necessary to confirm these results and, therefore, for this biotechnology to be applied in commercial programs.

To implement a donkey ET program, the system must have an adequate herd of recipients, according to the number of donor jennies, as is done in equine ET. Estrous cycles of donor and recipient jennies must be synchronized; hence, transrectal palpation and ultrasound examinations should be performed on a daily basis. Even though this technique is unlikely to be widely used in semi-extensive productive systems, it may be a useful tool in particular cases, such as old jennies, or to produce offspring from jennies with high milk yields.

4.2.5   In Vitro Embryo Production and Transfer

Regarding in vitro embryo production, despite the fact that highly complex ARTs such as cloning, Intracytoplasmic Sperm Injection (ICSI), follicular aspiration, and in vitro maturation have gained relevance and experience in horses in recent years, there are very few reports on this subject in donkeys and they are still in initial phases of development (Panzani et al. 2020; Flores Bragulat et al. 2023). Although these complex biotechnologies may not be applied in milk production systems, they could be extremely important for the preservation of certain breeds of donkeys that are considered endangered.

Ovum pick-up (OPU) and in vitro maturation (IVM) of cumulus-oocyte complexes (COCs) are essential steps for in vitro embryo production. Donkey oocytes have been successfully retrieved from ovaries obtained from abattoirs, utilizing methods such as scraping or follicle aspiration, as well as from live jennies through transvaginal ultrasound-guided follicle aspiration (Goudet et al. 2015). OPU techniques have been adapted for jennies in a manner similar to that for mares (Fig. 19), resulting in COCs recovery rates ranging between 34 and 76% (Flores Bragulat et al. 2023).

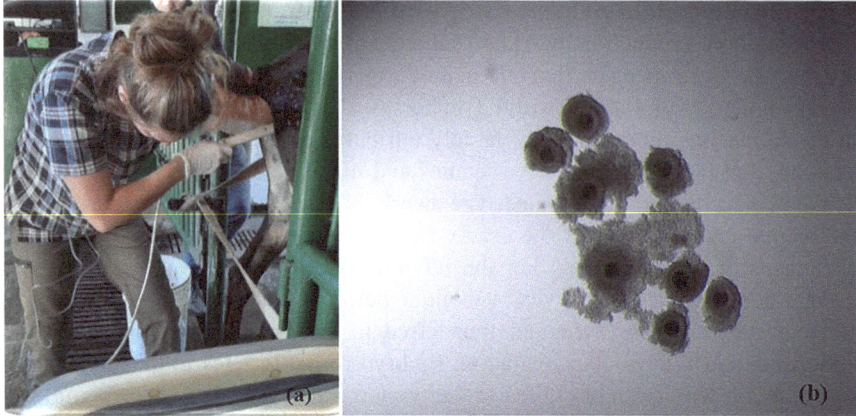

Fig. 19 Transvaginal procedure of follicle aspiration (OPU) in a jennie for ICSI program (**a**); aspirated oocytes from a jennie (**b**)

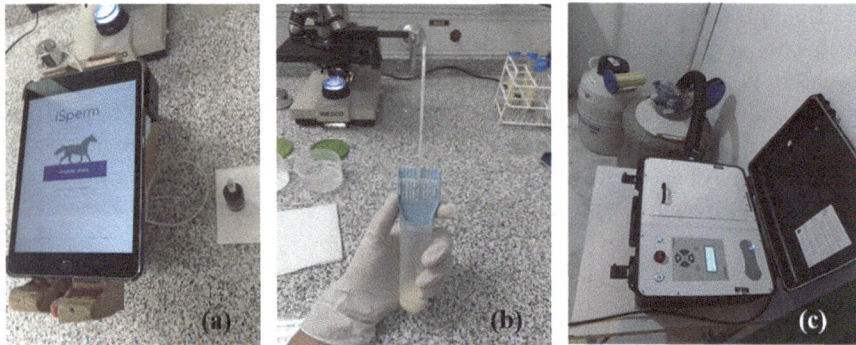

Fig. 20 Donkey semen processing for freezing. Sperm concentration, total, and progressive motility were evaluated using iSperm software (**a**). Semen being packaged in 0.5 ml straws (**b**). Automatic freezing machine (**c**)

4.2.6 Sperm Cryopreservation

Equine semen freezing protocols have been adapted for donkeys and post-thawed semen exhibits good viability and motility (Fig. 20).

In mares, per cycle pregnancy rates after insemination with jack frozen-thawed semen range from 38% to 53%; however, these rates significantly decrease (0–21%) when inseminations are performed in jennies (Miragaya et al. 2018). Several cryo-protectants such as glycerol, dimethyl sulfoxide, dimethyl formamide, and dimethyl acetamide have been assessed, yet pregnancy rates in jennies have remained low, ranging from 0 to 11% (Oliveira et al. 2006; Vidament et al. 2009).

Better post-thaw motility parameters were achieved by replacing glycerol with ethylene glycol or dimethylformamide (Oliveira et al. 2006; Acha et al. 2016; Kumar et al. 2019). Additionally, membrane integrity parameters improved with the

Table 5 Pregnancy rates in jennies after artificial insemination with frozen and vitrified donkey semen. EG: egg yolk; SP: seminal plasma

Type	Breeding dose (million sperm)	Extender	Pregnancy rate	References
Frozen	600	Skimmed milk +4% glycerol +10% EY	0% (0/17)	Trimeche et al. (1998)
		Skimmed milk +4% glycerol +10% EY (glycerol removal after thaw)	38.9% (8/21)	
Frozen	400	INRA 82 + 2.2% glycerol +2% EY	11% (4/38)	Vidament et al. (2009)
Frozen	500	INRA 96 + 2,2% glycerol +2% EY	20% (1/5)	Rota et al. (2012)
		INRA 96 + 2,2% glycerol +2% EY + SP (SP addition after thaw)	61.5% (8/13)	
Frozen	1000 (body)	Botucrio	0% (0/12)	Oliveira et al. (2016)
Frozen	1000 (deep horn)		28.26% (13/46)	
Vitrified	500	Gent	22% (2/9)	Diaz-Jimenez et al. (2020)
Frozen		INRA 96 + 2.2% glycerol +2% EY	10% (1/10)	
Frozen	500	INRA 96 + 2.2% glycerol +2% EY	22% (2/9)	Fanelli et al. (2022)
		INRA 96 + 2.2% glycerol +2% EY + SP (SP addition after thaw)	60% (6/10)	

use of non-permeable cryoprotectants (Bottrel et al. 2018). These findings suggest that donkey semen may not be suited to the same cryopreservation processes used for stallion semen. Extender trials should be conducted on an individual basis, as each donkey responds differently.

Trimeche et al. (1998) obtained a variation between 0% and 39% in pregnancy rates with frozen semen, while other studies achieved rates of 0%, 11%, 20%, and 28% (Oliveira et al. 2006; Vidament et al. 2009; Rota et al. 2012; Oliveira et al. 2016). The highest pregnancy rate reported so far was 61.5% by Rota et al. (2012) and 60% by Fanelli et al. (2022) using semen cryopreserved in INRA-96 with glycerol and re-diluted in homologous seminal plasma after thawing (Table 5).

Due to the lack of commercial repeatability in PR published in studies mentioned previously, assisted reproduction in donkeys faces a critical problem when designing and executing genetic improvement programs. In milk production programs, the movement of genetic material (such as cryopreserved semen) between countries and continents is highly complex, making cross-breeding trials between different productive lines very difficult.

Pregnancy rates per cycle in jennies inseminated with frozen semen are lower compared to those in mares, and the underlying reasons for this remain unclear. The understanding of jack sperm transport within the jenny's reproductive tract is limited, and the viable lifespan of frozen-thawed donkey semen in vivo has not been well-established. Therefore, employing higher doses, conducting deep-horn artificial insemination near ovulation, or resuspending frozen-thawed semen in seminal plasma and reducing post-mating induced endometritis (PMIE) may contribute to enhanced results when using frozen-thawed donkey semen (Canisso et al. 2019). To date, pregnancy rates similar to those obtained in mares and repeatable when frozen donkey semen is used for insemination in jennies have not been reported.

4.2.7 Embryo Cryopreservation

Efforts have been undertaken to cryopreserve donkey embryos with minimal damage, aiming to establish donkey germplasm banks. Cryopreservation serves to extend embryo viability before ET and can be achieved at −196 °C through freezing or vitrification methods. The presence of the glycoprotein capsule surrounding mare and jenny embryos has been proposed as a factor contributing to poor results following conventional freezing and vitrification of embryos larger than 300 mm in diameter (Panzani et al. 2020). Building on this premise, a few researchers have investigated the outcomes of vitrifying donkey embryos larger than 300 mm. Evaluation of vitrified/thawed embryos in vitro revealed a slight increase in cell death compared to fresh embryos. Notably, an in vivo study resulted in the successful birth of two healthy donkey foals following the transfer of 11 grade-one embryos (Panzani et al. 2012).

5 Practical Reproductive Management

Donkey milk production systems range from semi-intensive to semi-extensive, and this directly affects reproductive management. Semi-intensive systems will allow greater veterinary intervention and, therefore, the application of more complex reproductive biotechnologies. Semi-extensive systems usually use natural breeding and less complex reproductive biotechnologies such as periodic ultrasound examinations and artificial insemination in special situations. To ensure optimal reproductive management, adequate facilities such as stocks specially designed for donkeys are necessary. This allows a safe environment for the operator and donkey, allowing them to feel calm and without generating stressful situations. Donkeys are very intelligent and have great memory; therefore, subtle and progressive handling is recommended to generate a gradual habituation in different situations, such as entering the stocks or the dairy facilities. Donkeys' habituation to these facilities can be accompanied by stimulation with food, entry of other animals at the same time, correct lighting of spaces, and giving them time to observe. In case of fearful animals or reluctance to handle, sedation or covering the eyes with a mask or some fabric may be helpful.

Pasture breeding, in a program that has just begun, allows animals gradual habituation to handling, with check-ups every 15–20 days. Replacement of males every 2 years must be taken into account in this type of program to avoid inbreeding; another option could be to create different groups of jennies to be bred with non-related jacks. Subsequently, more complex reproductive biotechnologies could be applied in special cases with the aim of improving the genetic potential for milk production in future generations using animals with a high genetic potential for milk production as parents.

Even though all aspects of a production system are important, such as genetics, reproduction, and animal health, animal welfare and respect for the biological and behavioral aspects of this species are essential to work properly and obtain optimum results (Figs. 21, 22, 23, 24 and 25).

Fig. 21 Post-partum jennies in a pasture-based program

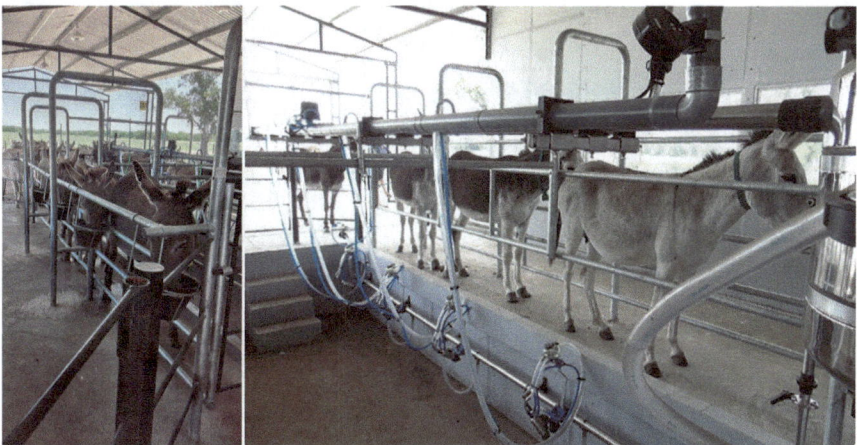

Fig. 22 Jennies during the mechanical milking process

Fig. 23 Jennies immediately after the milking process (right) with their offspring's waiting (left)

Fig. 24 Post-milking meeting and suckling

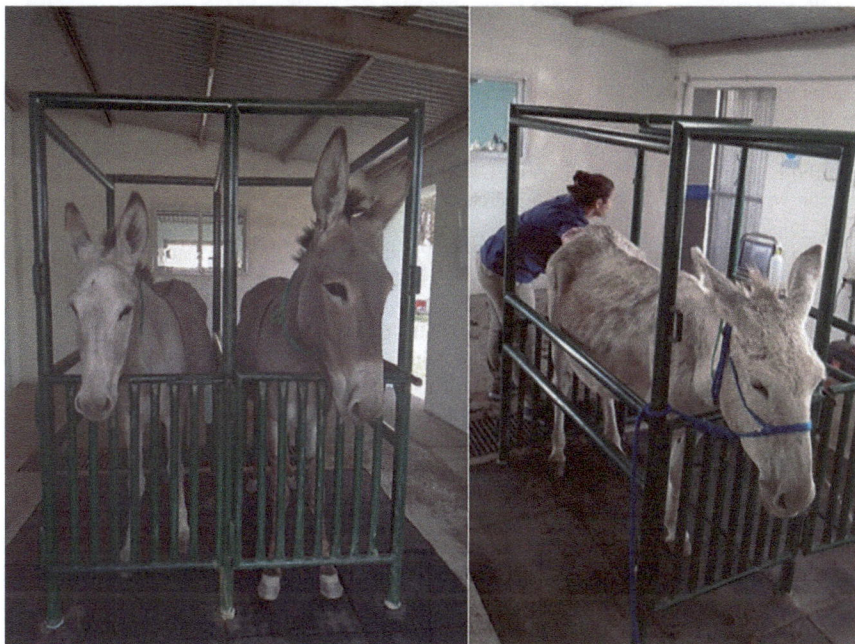

Fig. 25 Milking jennies in double examination stocks with a specific size for donkeys

6 Conclusions

Donkeys have been much less genetically manipulated by humans than other species, so intensive management that entails changes in social groups, isolation, aggressiveness, and sudden changes in diets generate significant stress that negatively impacts reproduction and their productive health status, especially in milk production programs. Hence, it is advisable to consider all these aspects when designing a reproductive program for donkeys and that changes or management guidelines such as serial ultrasounds, insemination, flushings, etc., are introduced sequentially and in a friendly manner.

References

Acha M, Acha D, Hidalgo M, Ortiz I, Gálvez MJ, Carrasco JJ, Gómez-Arrones V et al (2016) Freezability of Andalusian donkey (Equus asinus) spermatozoa: effect of extenders and permeating cryoprotectants. Reprod Fertil Dev 28(12):1990

Alonso CN, Castañeira C, Bragulat APF, Losinno L (2023) Effects of egg yolk extender on fertility of cooled donkey semen. J Equine Vet Sci 125:104575

Bottrel M, Acha D, Ortiz I, Hidalgo M, Gósalvez J, Camisão J et al (2018) Cryoprotective effect of glutamine, taurine, and proline on post-thaw semen quality and DNA integrity of donkey spermatozoa. Anim Reprod Sci 189:128–135

Bottrel M, Ortiz I, Hidalgo M, Díaz-Jiménez M, Pereira B, Consuegra C et al (2022) Hormonal management for the induction of luteolysis and ovulation in Andalusian Jennies: effect on reproductive performance, embryo quality and recovery rate. Animals 12(2):143

Camillo F, Vannozzi I, Tesi M, Sabatini C, Rota A, Paciolla E et al (2014) Induction of ovulation with buserelin in jennies: in search of the minimum effective dose. Anim Reprod Sci 151(1–2):56–60

Camillo F, Rota A, Biagini L, Tesi M, Fanelli D, Panzani D (2018) The current situation and trend of donkey industry in Europe. J Equine Vet Sci 65:44–49

Canisso IF, Panzani D, Miró J, Ellerbrock RE (2019) Key aspects of donkey and mule reproduction. Vet Clin North Am: Equine Pract 35(3):607–642

Carluccio A, Panzani S, Tosi U, Faustini M, De Amicis I, Veronesi MC (2007) Efficacy of hCG and GnRH for inducing ovulation in the jenny. Theriogenology 68(6):914–919

Carluccio A, Gloria A, Robbe D et al (2017) Reproductive characteristics of foal heat in female donkeys. Animal 11:461–465

Chang ZL, Li BX, Liu B, Yao L, Yu J, Jiang GM et al (2019) Effects of FSH and the weather during induced ovulation and timed artificial insemination to increase jenny conception rates. Sci Rep 9(1):3220

Di Salvo ED, Conte F, Casciaro M, Gangemi S, Cicero N (2023) Bioactive natural products in donkey and camel milk: a perspective review. Nat Prod Res 37(12):2098–2112. https://doi.org/10.1080/14786419.2022.2116706

Diaz-Jimenez M, Rota A, Dorado J, Panzani D, Fanelli D, Tesi M et al (2020) Comparison of uterine inflammatory response of jennies after artificial insemination with vitrified or frozen-thawed donkey sperm. J Equine Vet Sci 89:103036

FAO S. FAOSTAT database. Food Agric Organ UN Rome Italy 2020;1. http://www.fao.org/faostat/en/

Fanelli D, Tesi M, Monaco D, Diaz-Jimenez M, Camillo F, Rota A et al (2022) Deep-horn artificial insemination with frozen thawed semen after re-extension in autologous seminal plasma may improve pregnancy rates in jennies. J Equine Vet Sci 112:103932

Flores Bragulat AP, Alonso C, Castañeira C, Losinno L (2022) Wilsher cervical forceps for artificial insemination technique in jennies. J Equine Vet Sci 113:103993

Flores Bragulat AP, Ortiz I, Catalán J, Dorado J, Hidalgo M, Losinno L et al (2023) Time-lapse imaging and developmental competence of donkey eggs after ICSI: effect of preovulatory follicular fluid during oocyte in vitro maturation. Theriogenology 195:199–208

Goudet G, Douet C, Kaabouba-Escurier A, Couty I, Moros-Nicolas C, Barriere P et al (2015) Establishment of conditions for ovum pick up and IVM of jennies oocytes toward the setting up of efficient IVF and in vitro embryos culture procedures in donkey (Equus asinus). Theriogenology 86:528e35. https://doi.org/10.1016/j.theriogenology.2016.02.004

Kotilainen T, Huhtinen M, Katila T (1994) Sperm-induced leukocytosis in the equine uterus. Theriogenology 41(3):629–636. https://doi.org/10.1016/0093-691x(94)90173-g

Kumar P, Kumar R, Mehta JS, Chaudhary AK, Ravi SK, Chandra Mehta S et al (2019) Ameliorative effect of ascorbic acid and glutathione in combating the cryoinjuries during cryopreservation of exotic Jack Semen. J Equine Vet Sci 81:102796

Losinno L (2022) Producción de Burros I. Ediciones del Puente. Argentina. ISBN: 978-987-48610-92

Martini M, Altomonte I, Tricò D, Lapenta R, Salari F (2021) Current knowledge on functionality and potential therapeutic uses of donkey milk. Animals 11(5):1382

Miragaya MH, Neild DM, Alonso AE (2018) A review of reproductive biology and biotechnologies in donkeys. J Equine Vet Sci 65:55–61

Miro J, Lobo V, Quintero-Moreno A et al (2005) Sperm motility patterns and metabolism in Catalonian donkey semen. Theriogenology 63:1706–1716

Oliveira JV, Alvarenga MA, Melo CM, Macedo LM, Dell'aqua JAJ, Papa FO (2006) Effect of cryoprotectant on donkey semen freezability and fertility. Anim Reprod Sci:82–84

Oliveira JVD, Oliveira PVDLF, Melo E Oña CM, Guasti PN, Monteiro GA, Sancler Da Silva YFR et al (2016) Strategies to improve the fertility of fresh and frozen donkey semen. Theriogenology 85(7):1267–1273

Oliveira SN, Segabinazzi LGTM, Canuto L, Lisboa FP, Medrado FE, Dell'Aqua JA et al (2020) Comparative efficacy of Histrelin acetate and hCG for inducing ovulation in Brazilian northeastern jennies (Equus africanus asinus). J Equine Vet Sci 92:103146

Panzani D, Rota A, Romano C et al (2012) Birth of the first donkey foals after transfer of vitrified embryos. J Equine Vet Sci 32:419. (abstract)

Panzani D, Fanelli D, Camillo F, Rota A (2020) Embryo technologies in donkeys (Equus Asinus). Theriogenology 156:130–137

Person J (2022) Burros. Un retrato por Jutta Person. Ed. Adriana Hidalgo, España, 2022. ISBN: 978-84-19208-25-5

Polidori P, Ariani A, Vincenzetti S (2015) Use of donkey milk in cases of cow's milk protein allergies. Int J Child Health Nutr 4:174–179

Polidori P, Vincenzetti S (2019) The therapeutical, nutritional and cosmetic properties of donkey milk. Cambridge Scholars Publishing, Cambridge, pp 45–68

Rota A, Magelli C, Panzani D, Camillo F (2008) Effect of extender, centrifugation and removal of seminal 595 plasma on cooled-preserved Amiata donkey spermatozoa. Theriogenology 69:176–185

Rota A, Panzani D, Sabatini C, Camillo F (2012) Donkey jack (Equus asinus) semen cryopreservation: studies of seminal parameters, post breeding inflammatory response, and fertility in donkey jennies. Theriogenology 78(8):1846–1854

Salari F, Mariti C, Altomonte I, Gazzano A, Martini M (2022) Impact of variability factors on hair cortisol, blood count and milk production of donkeys. Animals 12:3009. https://doi.org/10.3390/ani12213009

Salimei E, Fantuz F, Coppolac R, Chiofalod B, Polidori P, Variscoe G (2004) Composition and characteristics of ass's milk. Anim Res 53(1):67–78. https://doi.org/10.1051/animres:2003049

Sarti L, Martini M, Brajon G, Barni S, Salari F, Altomonte I et al (2019) Donkey's Milk in the Management of Children with Cow's Milk protein allergy: nutritional and hygienic aspects. Ital J Pediatr 45(1):102

Seyiti S, Kelimu A (2021) Donkey industry in China: current aspects, suggestions and future challenges. J Equine Vet Sci 102:103642. https://doi.org/10.1016/j.jevs.2021.103642

Svendsen ED. (2008) The professional handbook of the donkey. Yatesbury: Whittet Books Limited. The Donkey Sanctuary, p 331. https://www.thedonkeysanctuary.org.uk/research/book-chapter/2597#:~:text=Stoicism%20is%20typical%20predator%2Davoidance,to%20experience%20pain%20and%20distress

Tibary A, Bakkoury M (1994) Particularités de la reproduction chez les autres espèces équines. Reproduction équine, Tome I: La jument. Actes Editions, pp 385–400

Trimeche A, Renard P, Tainturier D (1998) A procedure for Poitou jackass sperm cryopreservation. Theriogenology 50(5):793–806

Valle E, Pozzo L, Giribaldi M, Bergero D, Gennero MS, Dezzutto D, et al (2018) Effect of farming system on donkey milk composition. J Sci Food Agric 98(7):2801–2808

Vidament M, Vincent P, Martin FX, Magistrini M, Blesbois E.(2009) Differences in ability of jennies and mares to conceive with cooled and frozen semen containing glycerol or not. Anim Reprod Sci 112(1–2):22–35

Vincenzetti S, Pucciarelli S, Polzonetti V, Polidori P (2017) Role of proteins and some bioactive peptides on the nutritional quality of donkey milk and their impact on human health. Beverages 3:34. https://doi.org/10.3390/beverages3030034

Effect of Stress on Reproduction and Reproductive Technologies in Male and Female, Beef and Dairy Cattle

Sonia S. Pérez-Garnelo ⓘ, María José Utrilla ⓘ,
Aitor Fernández-Novo ⓘ, Ángel Revilla-Ruiz ⓘ,
Arantxa Villagrá ⓘ, and Susana Astiz ⓘ

Abstract

Stress represents a significant factor influencing reproductive outcomes in beef and dairy cattle, impacting both natural reproductive efficiency and the outcomes of assisted reproductive technologies. This chapter reviews various stressors and their physiological impacts on cattle, with a focus on the hypothalamic–pituitary–gonadal axis disruption, which can lead to irregularities in both the oestrus cycle and the semen quality, and, consequently, reduced fertility. The role of stress in altering the ovarian reserve, semen quality, and the timing of ovulation is examined. We also explore how stress compromises the efficacy of reproductive technologies such as artificial insemination, seminal evaluation, in vitro fertilization, and embryo transfer. Key interventions for mitigating stress effects, including environmental management, nutritional supplementation, and stress-reducing handling practices, are discussed. The integration of these management strategies has been shown to improve reproductive performance and the success rates of reproductive technologies in both beef and dairy herds. Our review high-

S. S. Pérez-Garnelo · S. Astiz (✉)
Dpto. Reproducción Animal, INIA-CSIC, Madrid, Spain
e-mail: sgarnelo@inia.csic.es; astiz.susana@inia.csic.es

M. J. Utrilla · A. Fernández-Novo
Dpto. Veterinaria, Facultad de Ciencias Biomédicas y de la Salud, Universidad Europea de Madrid, Villaviciosa de Odón, Spain
e-mail: mariajose.utrilla@universidadeuropea.es; aitor.fernandez@universidadeuropea.es

Á. Revilla-Ruiz
Facultad de Veterinaria, Universidad Complutense de Madrid, Madrid, Spain

A. Villagrá
Centro de Tecnología Animal-Institut Valencià d'Investigacions Agràries (CITA-IVIA), Segorbe, Spain
e-mail: villagra_ara@gva.es

J. C. Gardón, K. Satué Ambrojo (eds.), *Assisted Reproductive Technologies in Animals Volume 1*, https://doi.org/10.1007/978-3-031-73079-5_6

lights the need for a holistic approach to managing stress to enhance reproductive efficiency in cattle, providing a guide for producers to optimize both animal welfare and productivity.

Keywords

Heat stress · Cow · Bull · Hierarchy · Nutrition

1 Physiological Stress Response and Its Impact on the Reproductive Function

1.1 Introduction

The concept of stress was first defined by Hans Seyle in 1936 as a non-specific response of the body to any demand. Since then, the definition has evolved to include the perception of a threat to homeostasis because it increases the maintenance requirements of domestic animals (Collier et al. 2017). Collier et al. (2006) proposed a widely accepted definition of stress as the pressure on a biological system by an external event or condition. Stress is defined in numerous ways, such as the inability of animals to cope with their environment or the unfitness to adapt and reproduce effectively.

The threat that causes stress is referred to as a stressor, and it is the component of the environment that places a strain on a biological system (Moberg 2000). Stressors, which may arise from endogenous and exogenous sources, induce behavioral, metabolic, and physiological changes in the animals. Cattle face a variety of stressors throughout the production cycle that potentially constrain general productivity and well-being because of neuroendocrine disruption and stress-induced immunosuppression (Carroll and Forsberg 2007).

Depending on the origin (exogenous or endogenous stressors), perception, and length the stress suffered by the animals is classified into three categories: psychological stress, physiological stress, and physical stress. These three categories emerge from a range of psychological, physiological, and physical factors inherent in farming practices as a result of management decisions and interactions between humans and cattle (Brown and Vosloo, 2017; Lynch 2010). Psychological stress, also known as fear stress, is characterized by unpredictable fear responses in animals, as they are influenced by their perception of various factors such as restraint, human contact, exposure to new environments, and social dynamics (such as competition, aggression, leadership, and dominance; Grandin 1997). Physiological stress results from disruptions in endocrine or neuroendocrine function. The physiological stress depends on breed, sex, temperament, behavior, or weight (Carroll and Forsberg 2007). Physical stress is a consequence of injuries and it is associated with housing, stocking density, spatial allowance, thermal conditions (temperature, humidity, and wind), hunger, thirst, fatigue, and disease.

Stress can also be defined depending on the time interval, in which the animal suffers it. The response to stress can be categorized into acute and chronic phases. The acute phase is the immediate response to a stressor and it extends from hours to a few days. If the stressor persists or if the body perceives a prolonged threat, the response transitions into the chronic phase lasting several days to weeks (Hughes et al. 2014). The acute response is developed by homeostatic regulators of the endocrine and nervous systems and the chronic phase by homeorhetic regulators of the endocrine system. Both responses generate modifications in energy equilibrium and metabolic processes (Collier et al. 2017).

During the acute phase, the body undergoes rapid changes to prepare for a "fight or flight" response. This includes increased heart rate, heightened alertness, and mobilization of energy resources (glucose and fatty acids) to provide the necessary energy for dealing with the stressor (Hughes et al. 2014). Instead of the immediate energy mobilization seen in the acute phase, the chronic phase involves more sustained adjustments to maintain energy balance and cope with ongoing stress. During the chronic phase, the body may experience prolonged elevation of stress hormones, which can have negative effects on various organs and systems, including metabolism, immune response, and even changes in behavior. In fact, chronic stress has been associated with health issues such as hypertension, impaired immune function, and metabolic disturbances (Brown and Vosloo 2017).

This chronic response is associated with a reduction in productive efficiency attributed to several mechanisms, including a reduction in appetite and feed intake, an increase in energy and nutrient expenditure, and a higher vulnerability to infectious diseases (Chantziaras et al. 2018). Heat stress, for example, has been estimated to cause annual economic losses of $900 million in the US dairy industry (Carroll and Forsberg 2007). Over the last 25 years, the understanding of how environmental stress affects domestic animal productivity has improved (Collier et al. 2006). The ongoing debate among animal scientists regarding the definition and quantification of stress in relation to animal productivity and well-being has led to a growing appreciation and understanding of its effects. Producers and the scientific community increasingly recognize the importance of stress in livestock production (Carroll and Forsberg 2007). It is especially relevant to prevent stressors, reduce the level of stress, and, simultaneously, increase animal welfare. Otherwise, the impact of stress on animals prevents them from achieving their genetic potential (Von Borell et al. 2007).

Some kinds of stressors can be mitigated through management practices and nutritional strategies, while others, like thermal stress, pose significant challenges and economic burdens to the cattle industry (Carroll and Forsberg 2007).

1.2 Physiological Response to Stress

Endocrine, neuroendocrine, and immunologic pathways are affected because of stress (Carroll and Forsberg 2007). The response to stressors involves a series of events that begin with the detection and signaling of potential threats in the animal's biological mechanisms. These events induce the activation of neurophysiological mechanisms aimed at resisting and preventing considerable damage. Acute stress occurs when an animal is exposed to a stressor for a brief period, triggering the "fight or flight" response and immune system priming, which aids in adapting to short-term stressors (Hughes et al. 2014). During acute stress, the activation of the hypothalamic–pituitary–adrenocortical (HPA) axis increases the circulating cortisol levels in the blood. This fight and flight response generally allows a quick and complete re-adaptation to recover physiological balance (Grelet et al. 2022).

When a stressor continues persistently and the HPA axis is unable to effectively regulate its effects, the restoration of homeostasis becomes challenging, leading to allostatic overload. If this overload becomes chronic, it has detrimental effects on the immune system, reproductive system, and overall animal welfare (Fernandez-Novo et al. 2020). In fact, repeated or prolonged elevation of stress hormones can be detrimental to the overall health and well-being of animals.

The physiological response to any stressor is the same, independently of its origin: the stressor is detected by the specific receptor (olfactive, visual, tactile, etc.) which sends the first message to the central nervous system: encephalon (Von Borell et al. 2007). In the brain, areas responsible for processing sensory information, such as the hypothalamus and amygdala, identify and evaluate the stressor's threat. The rapid response to stressors is facilitated by the sympathetic adrenal medullary system (SAM) and the hypothalamic–pituitary–adrenocortical axis (HPA) aids in the process of linking the initial perception of the stress to an adequate response (Brown and Vosloo 2017). Neural pathways involved in the stress response are activated. The hypothalamus, specifically the corticotropin-releasing hormone (CRH)-producing neurons, detect the stress signal and begin to release CRH in response to the stressor stimulus (Mormède et al. 2011b). The release of CRH is the first step in the activation of the HPA axis. The released CRH then binds to CRH 1 receptors on the surface of hypothalamic cells. This binding triggers a cascade of intracellular signaling events. In brief, binding of CRH to the CRH 1 receptor activates the receptor-associated G protein, which in turn activates adenylate cyclase. This leads to an increase in the production of cyclic adenosine monophosphate (cAMP), which acts as an intracellular second messenger. The increased cAMP levels activate protein kinase A (PKA), an enzyme that phosphorylates other proteins within the cell. This activation of the intracellular signaling pathway culminates in the release of adrenocorticotropic hormone (ACTH) in the pituitary gland. ACTH travels to the adrenal glands and stimulates the production and release of glucocorticoids, such as cortisol. These glucocorticoids are the main stress hormones that help the body to cope with stressful situations, preparing the animal to fight against or to fight the stressor.

Chronic stress has the potential to impact the sensitivity of the HPA axis (Mormède et al. 2007). In situations of chronic stress, this system can become overtaxed. Chronic stress induces a cascade of physiological responses in cattle, notably placing significant strain on the HPA axis. This intricate system, crucial for maintaining homeostasis, becomes overburdened, leading to dysregulation and adverse health outcomes (Turner et al. 2002). When exposed to chronic stressors such as heat, transportation, or social disruption, cattle's hypothalamus perceives the threat and releases CRH (Lay et al. 1997). This CRH release, which follows in the same way as acute stress, stimulates the anterior pituitary gland to secrete ACTH (Rivier and Rivest 1991). This prompts the adrenal cortex to produce cortisol, the primary stress hormone in cattle (Wathes et al. 2007). Prolonged exposure to stress maintains elevated cortisol levels in circulation as the HPA axis remains chronically activated (Moberg and Mench 2000). This sustained cortisol release is associated with various physiological alterations, including immunosuppression and metabolic dysregulation. In a healthy system, cortisol exerts negative feedback on the hypothalamus and pituitary gland, suppressing further CRH and ACTH release. However, chronic stress disrupts this feedback loop, leading to prolonged activation of the HPA axis (Turner et al. 2002).

Moreover, chronic stress can manifest in altered behavior patterns, such as decreased feeding and social interactions, indicative of the animal's attempt to cope with the stressor (Hemsworth et al. 2011). Additionally, the sustained elevation of cortisol under chronic stress conditions suppresses the immune system, making cattle more susceptible to infections and diseases (Barnett et al. 1999). It also disrupts metabolic processes in cattle, leading to alterations in nutrient utilization and energy expenditure (Lay et al. 1997).

1.3 Specific Response to Stress by the Reproductive System

Stress affects the reproductive system through endocrine regulatory mechanisms (Fig. 1) (Dobson and Smith 2000). Responses to short- and long-term stressors differ as short-term stressors often fail to affect reproduction or even may have stimulatory effects (von Borell et al. 2007). However, a stressor inducing a chronic stimulation of the HPA axis has detrimental effects on the carefully orchestrated events in the reproductive system by disrupting the normal release of hormones from the hypothalamic–pituitary–ovarian (HPO) axis (Dobson et al. 2003). Chronic stressors have an impact on the reproductive axis at the hypothalamus [disrupting the Gonadotropin-Releasing Hormone (GnRH) secretion] and at the pituitary gland (disrupting the gonadotrophins secretion) mainly via corticotrophin-releasing hormone (CRH), arginine vasopressin (AVP), and opiates, with the direct effects of ACTH and adrenal hormones on the gonads being less important. High concentrations of cortisol suppress luteinizing hormone (LH) pulse frequency by acting centrally reducing the pulsatile GnRH secretion during the follicular phase. The reduction in endogenous GnRH/LH secretion ultimately deprives ovarian follicles of adequate gonadotrophin support leading to reduced estradiol production by the growing follicles (Dobson and Smith 2000). The coincidental occurrence of oestrus and the preovulatory LH surge is due to the fact that both phenomena depend, at least in part, on similar regulatory mechanisms.

Multiple research (Koenneker et al. 2023; Burnett et al. 2014; Dobson and Smith 2000) has discussed the interactions between the hypothalamic–pituitary–adrenal axis and the hypothalamus–pituitary–ovarian (HPO) axis. Cortisol plays an

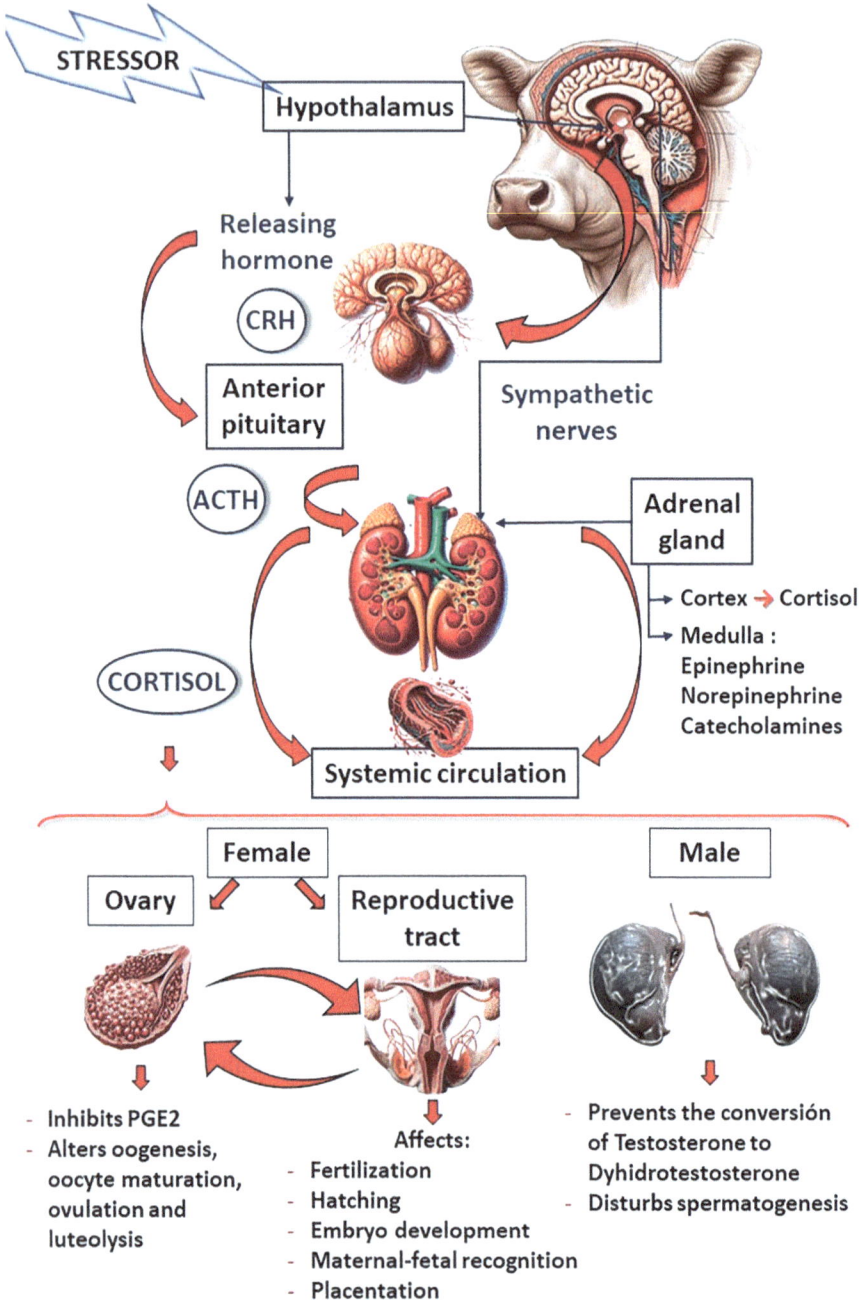

Fig. 1 Schematic representation of stress physiology and its link with reproductive function in males and females

important role in the stress response of animals, generated through the activation of the HPA axis when encountering stressors. High cortisol levels can suppress functions such as eating, growth, and reproduction as part of an adaptive response to stress (Burnett et al. 2014). LH is an important pituitary hormone in maintaining regular estrous cycles in cows. It stimulates the maturation of preovulatory follicles, leading to the secretion of estradiol. This prompts the hypothalamus to release a surge of GnRH, initiating the LH surge necessary for ovulation and corpus luteum formation. However, stress disrupts reproduction by reducing pulsatile LH release, lowering follicular estradiol levels, and/or inhibiting the LH surge (Lucy 2019).

These molecular events triggered by the chronic stress in cattle dysregulate the reproductive processes. Elevated cortisol levels negatively impact gonadal function, leading to reduced fertility and reproductive failure (Hansen 2009). In females, stress-induced alterations in hormone levels can lead to irregular estrous cycles, delayed ovulation, and reduced fertility (Dobson et al. 2001). Studies by Cooke et al. (2019) and Watters et al. (2013) have demonstrated that stress-induced cortisol secretion can inhibit GnRH pulsatility, leading to decreased LH release and consequently disrupted estrous cycles in cows. Estradiol low plasma levels are associated with stress exposure such as heat stress which hinders luteolysis, and eventually alters follicle dominance and disrupts ovulation (Collier et al. 2017). Moreover, chronic stressors such as heat stress have been shown to impair oocyte quality and embryonic development, contributing to suboptimal fertility outcomes (Roth 2020; Wolfenson et al. 2000). Exposure to stressors such as transportation, mastitis shortly after calving or lameness leads to lower estradiol concentrations, potentially weakening oestrus behavior and reproductive functions (Dobson et al. 2001). In general, cows experiencing health and nutritional issues seemed to undergo chronic stress, indicated by elevated cortisol levels, potentially impacting their fertility (Endo 2022).

Some stressors contribute to the large variations in the duration of oestrus (Von Borell et al. 2007). For example, social dominance reduces both the number of cows in oestrus at any one time, and the duration of oestrus of submissive cows. Dominant cows exhibit earlier signs of oestrus and display longer durations of oestrus (Orihuela 2000). Another example is the stress caused by the hot weather. It reduces oestrus duration and intensity with smaller follicles and lower estradiol, progesterone, and LH concentrations (Roth 2018; De Rensis and Scaramuzzi 2003). Also, undernutrition can be considered a metabolic stressor. It influences lower oestrus signs and ovulation rate, animals with better body condition scores subsequently greater, come into oestrus (Von Borell et al. 2007). As far as clinical conditions are concerned, lameness is associated with worse long-term reproductive performance after once the lameness has been treated and solved (Hughes et al. 2014). Luteal activity and hence the onset of oestrus commences later in postpartum cows treated for mastitis (Von Borell et al. 2007).

Oestrus/ovulation synchronization is widely used to enhance breeding efficiency in livestock before fixed-time artificial insemination. In fact, it has been proposed as a strategy to overcome fertility problems linked to the incapacity to observe oestrus. Nonetheless, stress can hinder the efficacy of these hormonal synchronization protocols. Research indicates that stressed animals may exhibit altered responses to hormonal treatments used for synchronization, leading to poor synchronization outcomes and decreased conception rates, when the level of stress is high (Brandão and Cooke 2021).

Moreover, stressors can negatively affect fertilization, embryo implantation, and embryo survival, with negative effects beyond ovulation. Chronic stressors like heat stress have been shown to reduce oocyte quality and impair fertilization rates (Roth et al. 2001a, b). High ambient temperatures during the breeding season can decrease the viability of sperm and prolong the time of pregnancy (Shabankareh et al. 2018). Artificial insemination (AI) is a common practice used in livestock breeding programs to introduce superior genetics, prevent disease transmission, and enhance reproductive efficiency. However, as previously described, stress can negatively impact the success of AI procedures (Abdallah et al. 2022). Studies have demonstrated that stressed females may have reduced conception rates following AI, possibly attributable to alterations in the uterine environment or impaired oocyte quality (Abdallah et al. 2022; Hansen 2014).

Moreover, when obtaining oocytes for subsequent in vitro fertilization and embryo transfer, it is essential to consider the stress factors that may affect both the oocyte collection process and the subsequent transfer procedure. Oocytes collected during the warm season exhibit lower quality than those collected in the cold season, expressed by reduced oocyte maturation and developmental competence (Roth 2018; Dobson et al. 2001). During periods of heat stress, embryo transfer with embryos produced during the cool season can improve fertility in the recipient cows (Hansen 2009; Baruselli et al. 2020; Dobson et al. 2001). The primary cause of infertility due to heat stress is attributed to damage to the oocyte and early embryo, which would be overcome by transferring an embryo produced without heat stress. Consequently, an enhancement in pregnancy rates can be observed with embryo transfer during warmer seasons, when compared to conception rates after insemination (Hansen 2020).

However, beyond fertilization, stress-induced alterations in hormonal profiles can impede embryo implantation and development in stressed cows. Chronic stress disrupts the balance of progesterone, a crucial hormone for maintaining pregnancy, triggering implantation failure (Minton and Parsons 2012). Moreover, heat stress reduces, particularly, uterine blood flow and alters endometrial receptivity, thereby impairing embryo attachment and its subsequent development (Hansen 2009). Additionally, stress-related immune suppression enhances the risk of uterine infections, further compromising embryo implantation (De Rensis and Scaramuzzi 2003). Various stressors can also impact directly on the embryo or fetus. Heat stress during early gestation stages increases embryonic mortality rates due to thermal-induced embryonic cell death (Edwards et al. 2013). Moreover, stress-induced alterations in maternal cortisol levels can affect placental function, nutrient transport, and fetal programming, leading to intrauterine growth restriction and higher rates of pregnancy loss (Sakatani et al. 2012).

Additionally, the frequent handling of the animals required in the artificial insemination or embryo transfer programs including repeated transrectal exams and administration of medications, induces stress that reduces fertility through higher rates of early embryonic loss (Ideta et al. 2009). This stress during pregnancy diagnosis (generally between days 28 and 40 after ovulation) and early gestation can have detrimental effects on pregnancy maintenance. It has also been demonstrated that handling stress or environmental disturbances during this critical period increases the risk of embryo loss or early embryonic death (Hansen 2020; Moggy et al. 2017). Additionally, stress-induced hormonal changes may interfere with maternal recognition of pregnancy and placental development, further compromising pregnancy maintenance (Arnott et al. 2012).

Regarding male fertility, stressors apparently affect testicular function through different mechanisms. As in females, there is evidence that stress causes a hormonal imbalance and contributes to infertility by suppressing the HPA axis. However, there are important differences between males and females in their response to stress and in the effect of stress on reproduction. Females are known to respond to stress more pronouncedly than males, producing higher levels of ACTH and cortisol, but in both sexes' products of the HPA axis inhibit reproductive function. However, sex differences in HPA function are due in part to differences in circulating gonadal steroid hormones. It appears that while testosterone may inhibit HPA function, estrogen may potentiate it. Androgens and estrogens modulate stress responses by binding to their cognate receptors in the central nervous system. It appears that in the case of androgens, control of the hypothalamic paraventricular nucleus is mediated transynaptically, whereas, in estrogens, modulation of the HPA axis may be due to changes in negative feedback mechanisms mediated by glucocorticoid receptors. Several studies suggest that gonadal steroid hormones, in particular, testosterone, moderate HPA activity in an attempt to prevent the detrimental effects of HPA activation on reproductive function (Handa et al. 1994). All this together suggests that females and males face stress situations and their effects differently, although both suffer its negative influence. In the case of males, their natural role forces them to compete more aggressively, so they need a mechanism to mitigate the negative effects caused by stress and its hormonal response.

Testosterone (T), released by the Leydig cells of the testis, acts both centrally and peripherally and is essential for sexual behavior, development, and maintenance of secondary sexual characteristics and libido, muscle mass increase, secretory activity of accessory glands, and optimal spermatogenesis (Mahmood et al. 2013). There is a positive correlation between testosterone concentration and several characteristics such as body weight, scrotal circumference, testicular measurements, seminal parameters, sexual behavior, and fertility both in *Bos indicus* and *Bos taurus* breeding bulls (Mahmood, et al. 2013).

High local levels of T are necessary for the maintenance of normal testicular (Steinberger 1971; Amann 1983) and epididymal function (Goyal 1983); LH stimulates T secretion by the Leydig cells, and follicle-stimulating hormone (FSH) optimizes T concentration in the seminiferous tubules by increasing the production of androgen-binding protein in the Sertoli cells (Steinberger 1971). Studies have shown that high cortisol levels profoundly affect LH and T production (Barth and Bowman 1994). In resting bulls, endogenous adrenal corticosteroid concentrations in peripheral blood were negatively associated with T and LH concentrations, suggesting that adrenal dysfunction could significantly influence testicular function in sires. A positive regulatory role of LH in testicular T production was also demonstrated. Thus, elevated cortisol concentrations are associated with LH and T basal concentrations (Welsh et al. 1979). Cortisol slows down the action of T produced by the Leydig cells, which may interfere with the development of genital organs, sperm production, and fertility of breeding bulls.

The mechanism of inhibition is through the inhibition of GnRH secretion, thereby suppressing LH release from the pituitary (Ren et al. 2010). In the same study on the effects of acute restraint stress on sperm motility and hormonal profiles in adult male rats, significant elevations in corticosterone, plasma adrenocorticotropic hormone, prolactin, and progesterone were observed, as well as alterations in sexual behavior, decreased FSH, and immunoreactive inhibin. Psychological stress in men has been reported to primarily reduce the serum level of total T and thus to affect seminal quality by decreasing sperm count, sperm motility, and morphologically normal sperm count. The increase in serum LH and FSH is secondary to the decrease in serum testosterone level (Bhongade et al. 2015).

In bulls, cortisol concentration peaks 10–20 min after the start of the stressor action (Chernenko et al. 2023). It has been proposed that the effects of stress on bulls and cows should be assessed when blood cortisol concentrations reach 70 ng/mL (Grandin 1997). However, in Holstein bulls subjected to transport, dehorning, or castration stress, cortisol peak was 21.77 ± 1.78 ng/mL with a minimum of 4.1 ± 0.17 ng/mL (Johnston and Buckland 1976) and another study showed that the peak of cortisol and T was achieved 30 min after the start of the stressor and remained at this level for 2–3 h (Welsh and Johnson 1981). On the other side, long-term stress decreases T production and concentration under the depressant effect of cortisol (Welsh and Johnson 1981), which interferes with spermatogenesis (Barth and Bowman 1994). In summary, stress most likely affects spermatogenesis through an endocrine mechanism (Barth and Bowman 1994), although the response to stress may also be due to the adrenal-hypothalamic hypophyseal gonadal axis (Morrell 2020).

Regarding the different stressors, the most important factor affecting male reproduction is heat, altering testicular thermoregulation mechanisms, disrupting and impairing spermatogenesis, leading to testicular degeneration that causes declining sperm quality, resulting in subfertile or temporary infertile animals with poor semen quality (Garcia-Oliveros et al. 2020). Therefore, we will deal more deeply with the pathophysiology of thermoregulation of the testicular tissue in the next sections of this chapter, as well as with the effects of heat stress on the male reproductive efficiency.

Beyond the heat, several factors can stress our bulls, but scarce information is available on how other stressors may affect semen quality and on the stress resistance of breeding bulls in order to be able to select stress-resistant animals (Barth 2012; Chernenko et al. 2023). Illness, malnutrition, extreme and prolonged climatic conditions, transportation, environmental changes, and pain are stressful conditions for bulls, and stress interferes with spermatogenesis through an endocrine mechanism (Barth and Bowman 1994; Napolitano et al. 2020; Chernenko et al. 2023).

Semen quality can change quite rapidly for the better or worse depending on whether a bull is suffering from, or just recovering from, a process detrimental to spermatogenesis. Some sperm defects appear to have a genetic basis and knowledge of genetic predispositions, together with the bull's history, can help in the interpretation of abnormal spermiograms. It appears that heat and stress affect testicular function through different mechanisms, which may result in the presence of different types of sperm abnormalities in ejaculates. Thus, a bull's spermiogram might indicate the cause of abnormal spermatogenesis (Barth and Bowman 1994; Cojkic and Morrell 2023).

It has been confirmed that an individual animal specificity of the stress response is characterized by changes in blood parameters, especially cortisol, but also in the development of testes and testicular-related structures, histological structure of testes, and sperm production. This was confirmed by an increase in serum cortisol concentrations 1 h after stress and a higher SC2/SC1 ratio (serum cortisol concentration 1 h before stress/serum cortisol concentration 1 h after stress), with lower morphological and histological parameters of the testes, sperm production, and sperm quality. For this reason, it has been suggested that it would be useful to select replacement breeding bulls on the basis of their resistance to stress for certain handling conditions in order to optimize their reproductive performance (Chernenko et al. 2023). Some testicular abnormalities are associated with the presence of excessive fat in the scrotum due to an increase in testicular temperature, which makes the bull less reslient to heat stress (Palmer 2016). Stress response in animals is characterized by changes in cortisol, T, creatinine-phosphate-kinase concentrations, and aminotransferase activity (ALT and AST). Since T affects testes development any disruption of its action by stress can lead to morphological and histological changes, which can affect spermatogenesis and semen quality (Kastelic et al. 2018). In addition, an alteration in T synthesis in young bulls leads to an extension of puberty (Byrne et al. 2017). In animals, individuality in stress resistance is observed both in the peak blood concentration reached by these compounds as a result of stressor exposure and in the changes in their concentrations and activity compared to reference values (Johnston and Buckland 1976; Chernenko et al. 2023).

In the end, bulls affect the reproductive efficiency of breeding herds, irrespective of whether they are used for natural breeding or AI. Thus, impairment of bull fertility results in great economic losses, particularly in extensive cattle production systems. Sperm quality evaluation could be a good indirect parameter to assess bull welfare and even in the identification of health and stress problems (Barth and Bowman 1994; Cojkic and Morrell 2023). Taking into account the different stressors and situations that may affect the fertility and mating ability of our bulls is of outstanding relevance in beef and dairy herds, where natural mating is part of the reproductive programs.

1.4   Stress and Epigenetics

Furthermore, the prenatal period plays a crucial role in shaping individuals' responses to their environment over their lifetime. Recent literature supports the concept that prenatal and early life stressors (such as high environmental temperatures) have a major impact on the growth and development of cattle, creating a suboptimal environment for the gestating fetus, and altering the phenotype of the newborn calf. This refers to the so-called Developmental Origins of Health and Disease (DOHaD) theory. The mechanism behind this is a dynamic process of DNA methylations and demethylations, which play an important role in cell functionality (Steger et al. 2011). DNA methylation is essential in genomic imprinting, gene expression regulation, X chromosome inactivation, and embryonic development (Surani 1998), among others. These metabolic alterations through epigenetic changes, prepare the newborn calf for similar conditions ("matching") after birth, improving its survival during periods of suboptimal conditions. However, when conditions in postnatal life are not according to the conditions during the fetal life, the "mismatch" between the prenatal and postnatal environment results in metabolic features in the offspring, with long-term implications for health and production (Van Eetvelde and Opsomer 2017).

Some evidence are found related to stressed mothers during pregnancy, affecting the offspring. Indeed, stress experienced by cattle during pregnancy has significant implications for the welfare and performance of their offspring (Arnott et al. 2012). During very early pregnancy, nutritional stress can have long-lasting effects on embryo/fetal viability and subsequent fetal growth. Inadequate maternal nutrition induces a suboptimal uterine environment, affecting embryo implantation, placentation, and nutrient supply (Van Eetvelde et al. 2017), due to, in part, hormonal imbalance and uterine dysfunction (Stevenson and Britt 2017). This triggers epigenetic modifications in developing embryos, altering gene expression patterns and metabolic programming (Sinclair and Watkins 2013).

Exposure to heat stress has been observed to trigger short- and long-term negative productive consequences in dairy heifers through in utero programming (Dado-Senn et al. 2020). Heat stress suffered early in the development of the embryos, programs structural and functional changes in the fetus that will persist and alter physiological function and health into adulthood (Barker 1998). In fact, it has been demonstrated that maternal hyperthermia early during pregnancy alters the fertility of the offspring during their adulthood (Kipp et al. 2021).

However, there has been limited examination of the relationship between paternal experiences and offspring development, although, since spermatogenesis is a continuous process, experiences that have the ability to alter epigenetic regulation in parents can actually change the developmental pathways of the offspring (Mychasiuk et al. 2013).

Paternal stress exposure in rats is known to influence the development of stress-related behaviors in offspring, which highlights the importance of the factors conferring disease risk and/or resistance. Although the effect of parental stress on offspring is described, the mechanisms that explain it are the subject of speculation in the scientific literature and probably involve a complex interaction of molecular, physiological, and behavioral factors (Mashoodh et al. 2023).

In mammals, the intergenerational influence of mothers is likely to be due to both pre- and postnatal maternal interactions that have developmental programming effects on offspring. However, for males, paternal effects, particularly in species where there is little or no post-conception parent–offspring interaction, suggest the existence of a germline inheritance mechanism. Much attention has been focused on sperm-mediated mechanisms (e.g., sperm RNA and DNA methylation) that influence the effects of paternal stress (Mychasiuk et al. 2013; Wang et al. 2021).

A strong paternal epigenetic contribution to embryogenesis depends on sperm DNA containing multiple regulatory elements, including methyl groups that promote the activation or silencing of genes on contact with the egg. It has been demonstrated that stress alters methylation patterns in the germline of F2 male mice (Franklin et al. 2010). Thus, paternal stress for 27 consecutive days before mating, altered the behavior of all offspring, reduced stress reactivity in male offspring, and altered DNA methylation patterns in offspring at 21 days after birth. Global methylation increased in the hippocampus of both male and female offspring and showed a reduction in the frontal cortex of female offspring, so paternal stress during spermatogenesis affects in a sex-dependent manner (Mychasiuk et al. 2013). Studies using embryo transfer in mice have also shown the role of sperm-mediated factors and how paternal pre-conceptional stress can predict the neurobiological and behavioral phenotypes of the offspring (Mashoodh et al. 2018). However, this work also highlighted how pre- and postnatal maternal interactions may influence these findings, over any epigenetic effects in sperm. This maternal influence could originate prenatally (e.g., increased feeding) or postnatally, from maternal care (licking/feeding and lactation), or even from the microbiomes of the mothers. In a subsequent study, females mated with stressed males have less successful pregnancies, gain less weight during gestation, and nurse offspring at reduced frequencies compared to females mated with control males; however, the male condition did not affect the female licking and grooming of pups. Paternal stress resulted in sex-specific behavioral outcomes in offspring and gene expression of corticotropin-releasing hormone was increased in the developing hypothalamus of female offspring of stressed fathers, whereas mRNA levels of brain-derived neurotrophic factor were decreased in male offspring.

By understanding the intricate interactions between stress and reproduction as elucidated by these studies, strategies can be developed to mitigate the impact of stress, thus improving both animal welfare and productivity in cattle populations (Thatcher et al. 2006; Bartolome et al. 2009).

2 Hierarchy, Herd Handling, and Nutritional Management Stress: Interference on Reproductive Outcomes

The previously described general and specific mechanisms included in the response to stress by the animals translate to different impacts on the reproductive efficiency of dairy and beef heifers, cows, and bulls, depending on which the stressors were. This section delves into the intricate relationship between hierarchy within the herd, handling practices, and nutritional management, which can turn into stressors and their collective impact on reproductive success.

2.1   Hierarchy in Cattle Herds

2.1.1   Cows and Heifers' Dominance

The social group constitutes a component of an intricate and dynamic environment for individuals, wherein various strategies have evolved to ensure survival and sustain group viability (Miranda-de la Lama et al. 2013). In intensive production systems, the social behavior of cattle notably differs from their natural habitat due to constraints on natural substrate, space, and resources, including resting and feeding places. Additionally, alterations occur in the composition and size of the group (Galindo and Broom 2000). The biological cost of intensive conditions on individual animals can be considerable, particularly concerning social competition for resources and heightened aggression when resources are limited (Estevez et al. 2007). Social competition among animals is often governed by dominance relationships, where the concept of social dominance is defined as "a priority of access to an approach situation or away from an avoidance situation that one animal has over another" (Francis 1988). Social dominance serves two main functions: providing dominant individuals with priority access to limited resources and reducing aggression levels in a group, benefiting individual animals by avoiding potentially costly interactions with conspecifics, whether dominant or subordinate (Lahn 2020). Dominance facilitates successful coexistence in social communities, however, social interactions with rank-influencing individuals often involve conflicts and prolonged exposure to high levels of social competition can lead to chronic stress, negatively impacting reproductive outcomes (Fiol et al. 2019). Lower-ranking individuals may face reduced access to resources, such as food, resting places, shade, mating opportunities, and overall activity inhibition, while dominant animals generally enjoy priority access to limited resources (Barroso et al. 2000). Therefore, access to feeding areas and resting spaces may be monopolized by dominant individuals, limiting the energy intake and fitness of subordinate cows, both factors affecting indirectly reproductive efficiency (Palmer et al. 2005). In fact, dominant cows typically experience less intensive stress and have preferential access to high-quality nutrition, contributing to enhanced reproductive efficiency (Mee 2008). Conversely, subordinate cows show disruptions in hormonal balance and reduced fertility, probably due to a higher level of stress (Le Neindre et al. 1995). Furthermore, subordinate individuals are more prone to reproductive disorders such as anoestrus or irregular oestrus cycles (Schütz et al. 2006). Once cows are pregnant, dominant ones, cope better with the physiological demands of pregnancy, leading to higher rates of embryo survival and successful calving (O'Driscoll et al. 2017). In contrast, subordinate cows experience higher rates of embryonic loss or abortion (Cooke et al. 2012). Finally, social stressors influence also maternal behavior and milk production, affecting the health and development of newborn calves (Færevik et al. 2006).

All this may be ameliorated through enough space/animals in the installations, guaranteeing access to resources by the low-ranking animals in our herds.

The social hierarchy in the herd plays a pivotal role directly in behavior and physiology, including also the reproductive function, with immediate implications for both productive and reproductive outcomes (Drews 1993). Studies have shown that dominant heifers often exhibit earlier puberty and enhanced reproductive development compared to subordinate counterparts (Rodríguez-Sánchez et al. 2015). In dairy cows, fertility and milk yield were greater in cows with higher social rank (Dobson and Smith 2000), and high-ranked beef cows were rebred earlier during the postpartum period in comparison with low-ranked cows (Landaeta-Hernández et al. 2013).

Social hierarchy profoundly influences the expression of oestrus behavior in cattle. Dominant individuals often exhibit more pronounced oestrus signs, including increased vocalization, mounting behavior, and receptivity to mating advances. Subordinate cows, on the other hand, experience suppressed or delayed oestrus due to social stress associated with their lower rank (Savoini et al. 2014). Individuals occupying higher-rank positions experience an increase in cortisol levels compared to those in lower ranks when confronted with specific situations of restriction and handling. This is likely attributed to the need for heightened stimulation to effectively cope with these challenges (Solano et al. 2004). This knowledge is important, in order to minimize the risk of inducing stress.

Therefore, understanding how social rank influences reproduction can help us to optimize fertility in cattle farming systems.

2.1.2 Male Dominance

It is certain that among almost all animals there is a struggle between the males for the possession of the female. This fact is so notorious that it would be superfluous to give instances (Darwin 1871, p. 259).

In bovid species, social interactions and fights between males result in the establishment of a social hierarchy in which dominant males have preferential access to females. Most of the studies of animal behavior have focused on wild animals, but relatively little research concerning the behavior of domestic animals and the effects of male hierarchy on stress, reproductive performance, and semen quality have been conducted.

Like cows, bulls kept in groups develop social orders, fighting for dominance (Lamb 1976; Costa et al. 2016). This order of dominance within a herd is observed while drinking and feeding, where certain animals are always the first to drink or feed and subordinated bulls stay behind until the others have finished. Social interactions between animals often involve some degree of conflict, and rank has a pronounced effect on the individual. The hierarchy can cause stress especially, similar to that described in female cattle, when the animals are confined in small spaces such as yards, and feedlots. Some studies have reported an aggression increase when resources (food, water, space) are limited, with a substantial decrease in average daily gain following increased aggression (Partida et al. 2007). In a study conducted on young beef bulls of the Gasconne breed during the fattening period, social rank influenced physiological indicators of welfare (Miranda-de la Lama et al. 2013). Therefore, minimizing aggressive interactions could improve performance and productivity, and for this reason, it is important to maintain social stability throughout the fattening period (Partida et al. 2007; Miranda-de la Lama et al. 2013). In addition, higher-ranking bulls show higher cortisol levels, which suggests that they are more sensitive to handling than subordinate bulls, perhaps because of their need to quickly, face immediate competitive challenges (Miranda-de la Lama et al. 2013).

Bulls used in natural service in a single sire herd are dominant in the herd and exert control over the cows, which is natural and desirable (Lamb 1976). However, hierarchy is problematic in multiple-sire herds, because the dominant sire keeps the other bulls away from the cows, which makes almost all cows mated by one bull. In this situation, the calving percentage may be lower because the dominant bull cannot mate all cows and may spend time protecting his rank rather than acting as a herd sire. Consequently, in farms with natural mating, the herd should be divided into single-sire-sized groups (Lamb 1976). Other considerations for managing social dominance include: (1) separating bulls by age when possible, because mature bulls are more dominant and the division by age groups reduces the chance of injury from fighting, and allow younger bulls with genetic potential to mate females; (2) altering bull-to-female ratio as the breeding season progresses, removing some bulls after the first 21-day cycle to reduce chances of fighting; (3) rotation of bulls during the breeding season, initially using the most experienced bulls, with higher mating ability, and replacing them with younger bulls during the second 30 days of the breeding season; (4) implementing breeding groups by reducing the number of cows, so that only three to four bulls or fewer are needed per pasture; and (5) establishing social order earlier by introducing the new bulls into an already established bull group prior to the breeding season, thus avoiding the establishment of the hierarchy once they are with the cows, thus avoiding stress and possible injuries due to competition between bulls (Cojkic and Morrell 2023). However, there are few scientific studies on how breeding groups of bulls of the same age and size influence animal welfare and, consequently, fertility results.

Although cortisol blocks specific T target cells, when aggressive behavior is provoked under acute stress conditions, there is a significant increase in both cortisol and T, especially at the beginning of the immobilization phase (Welsh and Johnson 1981). However, the struggle for dominance, accompanied by an increase in plasma T concentration as an aggressiveness factor, is observed mainly in dominant bulls, while in the rest of the bulls, there is a decrease in this hormone (Knol 1991). In subordinate males, this is considered an adaptive reaction, as it reduces aggressive motivation, to avoid conflict and stress. Thus, aggressiveness can be considered as a reaction to stress, although temporary, protective, and adaptive, and it is not necessarily associated with a low resistance of the animal to stress (Knol 1991; Chernenko et al. 2023).

Variation in male reproductive success may depend on both the ability of males to compete for mating and the ability of their sperm to compete for fertilization if females are mated with more than one male (Parker 1998). Thus, subordinate males could counteract their initial disadvantage in precopulatory competition, because some studies have reported a negative relationship between social status and sperm quality, explaining the mismatch between copulation rate and siring success by the sperm depletion of dominant males (Preston et al. 2001).

Regarding bulls entering a sire center for semen dose production, AI centers house their bulls under optimal conditions, in clean and ventilated buildings with facilities that protect them from inclement conditions, fed with nutritionally correct rations, cared for by trained employees and health status is continually monitored. In order to maximize the potential of superior bulls, husbandry management is conducted to reduce physical, social, and environmental stress. Therefore, AI center bulls of different ages are kept in relatively static conditions and without access to females, to avoid hierarchy fights, which prevents the sires from expressing their natural behavior, and indirectly induces negative emotional states limiting the expression of their natural behavior. Despite its relevance, there is a lack of information about bull welfare assessment in AI centers or about how a reduced level of welfare can affect their productivity (Napolitano et al. 2020; Cojkic and Morrell 2023).

This isolation from females and the fact that semen collection is not a normal reproductive behavior, requiring a close human–animal interaction, there are many critical points in the semen production process that may induce inappropriate behaviors (Cojkic and Morrell 2023). Keeping animals isolated will prevent them from expressing natural social behaviors (Napolitano et al. 2020). On the other hand, bulls that are raised and housed in all-male groups develop monosexual tendencies (Lamb 1976). In this situation, hierarchy problems occur in which the dominant bulls will try to force the weaker animals, which in their constant attempts to move away, lose masculinity and become undesirable as breeding animals (Viljoen 1968; Lamb 1976). It is therefore important the division by age groups (Lamb 1976).

In brief, regarding the effect of hierarchy and dominance in male and female cattle, understanding its impact on reproductive outcomes is essential for implementing an effective management strategy in cattle farming systems. Providing adequate space, resources, and environmental enrichment helps to minimize social stress and promote the well-being of all herd members, regardless of their rank within the hierarchy. Additionally, optimizing nutrition and health management practices mitigates the negative effects of social stress on fertility and pregnancy success. Finally, concrete management tips for bulls, such as clustering them and establishing groups of similar age is a valuable approach.

2.2   Herd Handling Practices: Effects on Reproduction

During the execution of routine husbandry procedures involving human–animal interactions, there is a potential for the induction of elevated levels of stress (Galindo and Broom 2000). Common practices confining the animals in a handling race are rectal palpation, drug administration, artificial insemination, semen collection, or medical interventions. These can trigger behavioral reactions in livestock that increase the risk of injury to both the cattle and workers (Barrozo et al. 2012). Moreover, the stress response to these circumstances inhibits physiological systems such as reproduction and immunity (Grandin 1997). As previously described, these stressors activate physiological responses in cattle, including the release of stress hormones such as cortisol, which can have detrimental effects on reproductive function (Cook et al. 1997; Crookshank et al. 1979).

Factors enhancing the stress associated with handling include rough handling, loud noises, unfamiliar environments, and overcrowding (Hemsworth et al. 2015). Rough treatment or excessive restraint, can trigger the release of stress hormones, compromising fertility (Grandin 1997). Conversely, gentle, and skilled handling promotes a calm environment, facilitating normal reproductive processes.

Farming practices affect similarly the reproductive performance of bulls that are often handled and restrained in squeeze chutes, branded with a hot iron, castrated, dehorned, changed of housings, transported, and semen collected. Therefore, these practices should be conducted in a way that minimizes stress and decreases excitement, restlessness, and panic, providing calm to the animals. Just the simple removal of the animal from its herd and the consequent social isolation affects animal welfare and causes stress.

The Bull Breeding Season Evaluation (BBSE) is an essential on-farm practice to detect animals with fertility problems in the herd that allows the strategic use of sires, thus having a direct economic impact on the farm. BBSE includes the evaluation of size, musculature, body condition, and the absence of diseases and physical defects that may compromise sires copulatory ability, but also necessarily involves qualitative evaluation of the ejaculates. BBSE is necessary when acquiring animals (i.e., in animal auctions), and yearly for all mating bulls, in order to detect subfertile sires. Bulls should be assessed 60–75 days before the breeding season begins, allowing time to replace unsatisfactory bulls or treat and retest deferred bulls (Ball and Furman 1972; Palmer et al. 2005; Kastelic et al. 2012; Napolitano et al. 2020).

There are two main methods to obtain the ejaculates of the bulls. Artificial vagina (AV) is the method of choice for semen collection in AI centers which is designed to effectively simulate natural mating and requires the bull's active participation and animal training. The protocol for ejaculate collection includes the sexual stimulation, preparation, and collection of ejaculates. Sexual stimulation starts by exposing the bull to a mounted animal and is caused by the presence of other animals and various visual, olfactory, and auditory stimuli that sexually arouse the bull. When the bull becomes sexually aroused, will have a penile erection and will try to mount the "dummy animal." Another bull or bullock is used as a mounted animal to provide sexual stimulation for the bull. Cows can also be used as mount animals ("dummy" cows), however, their use is not advocated, since animal welfare concerns for females arise (Stafford 1995). The ejaculation results from tactile stimuli to the bull's glans penis, the temperature (internal water temperatures between 45 and 60 °C), and the turgidity of the AV.

Sires training for semen collection in AV is time consuming and not all the animals can be trained. Thus, semen collection with AV in the field is not possible in most of the cases. On the contrary, electroejaculation (EE) is the most contrasted technique and the recommended one by the reference guides for BBSE (Ball and Furman 1972; Stafford 1995; Kastelic et al. 2012; Barth 2013; Beggs 2013; Koziol and Armstrong, 2018). EE involves the stimulation of an electrical current of emission, erection, and ejaculation. The use of EE for semen collection arose in the 1930s when an American pathologist observed that prisoners executed in the electric chair ejaculated during execution. Electrical stimulation of pelvic nerves to cause ejaculation in animals was pioneered by Gunn in 1936 using rams as experimental animals.

Current electroejaculators utilize a sine-wave pulse at a frequency of 20–30 cycles and it has been demonstrated that increased electrical intensity tends to cause unnecessary muscle contraction (Marden 1954; Stafford 1995). The maximum voltage of commercial electroejaculators is 16 V, with a maximum current of <900 mA, with most bulls, ejaculating with electrical impulses <8 or 9 V (Palmer et al. 2005).

The technique execution is very simple (Fig. 2). Bulls should be suitably restrained for EE ensuring the safety of both, the animal and those handling it. Afterward, a thorough cleaning and sanitizing of the preputial zone is performed. Briefly, the preputial fur is cut, and the prepuce is washed internally with a physiological saline solution (0.9% sodium chloride, warmed at 37 °C), using a 60 mL syringe and dried with sterile swabs. Feces are removed from the rectum and once transrectal examination of the bull's ampullae, seminal vesicles, prostate, pelvic urethra, and inguinal rings has been concluded, it is recommended to massage these areas for 10–60 s. The massage is intended to sexually excite the bull, thus reducing the number of electrical impulses needed to obtain ejaculation, and causing relaxation of the anal sphincter prior to probe insertion (Palmer et al. 2005; Sarsaifi et al. 2013). The electroejaculator probe is then inserted into the rectum, with the electrodes in the ventral position. Commercial devices usually have two modes of operation; manual in which electric stimuli are applied according to animal response, or automatic, in which electric stimuli are applied with a pre-programmed cycle (Whitlock et al., 2012). The manual mode is recommended (Palmer et al. 2005). In both cases, the procedure simply consists of applying alternating electrical stimuli with increasing intensity (voltage ranging from 8 to 16 V), in a sequence pattern with intermittent electrical stimulation.

Based on the behavioral responses of the animals during EE, several studies have indicated that the procedure can cause pain, discomfort, and stress to the animals (Stafford 1995; Barth and Waldner 2002; Mosure et al. 1998; Palmer et al. 2005; Orihuela et al. 2009a, b; Damián and Ungerfeld 2011; Whitlock et al. 2012; Abril-Sánchez et al. 2017; Costa et al. 2018). The physical reaction of conscious animals to EE has led to growing concern about the welfare of bulls, rams, and bucks

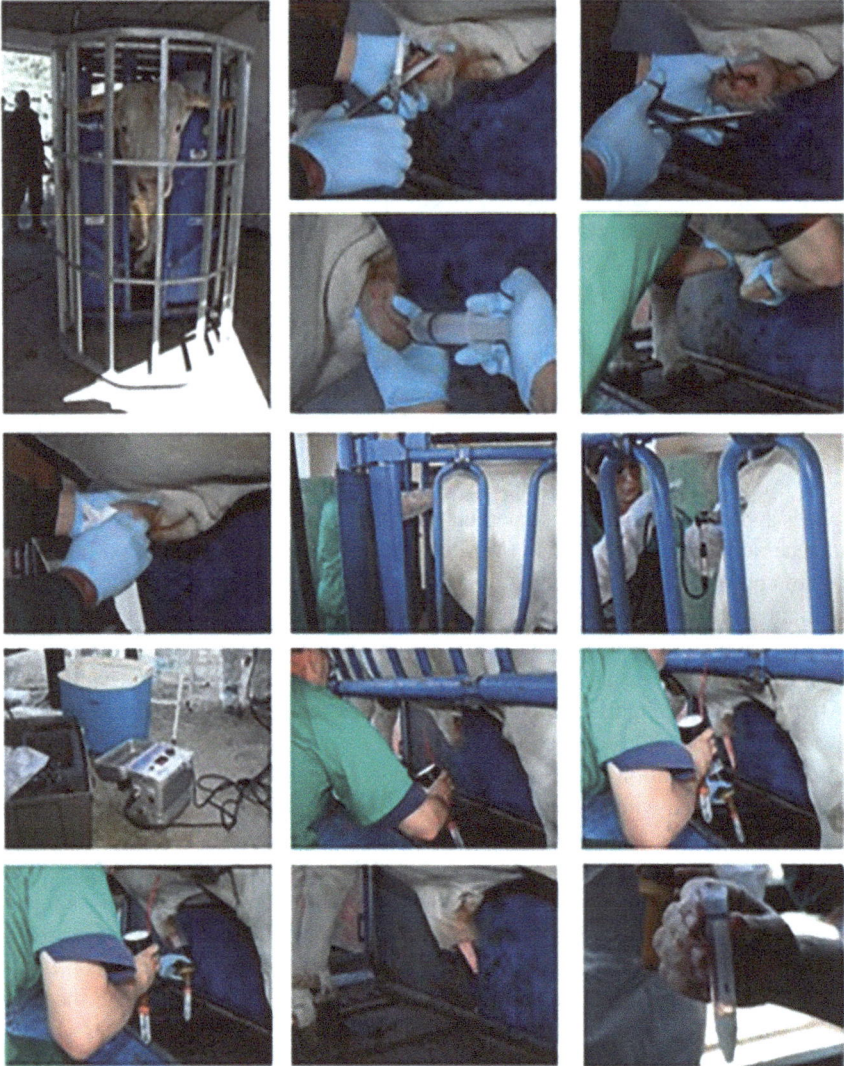

Fig. 2 Picture sequence of the protocol used for semen collection in bulls by EE

subjected to this procedure (Carter et al. 1990; Stafford 1995; Palmer et al. 2005; Mujitaba et al. 2022). By contrast, it has also been suggested that the technique only provokes a stress response when used frequently, causing adverse effects on animal welfare in this case (Pugliese et al. 2022). Moreover, it is not clear the intensity of the pain linked to EE. In bulls, plasma cortisol levels increased markedly within 15′ of EE and remained elevated for 2–4 h (Welsh and Johnson 1981). Maximal cortisol levels were similar between bulls when only a rectal probe was introduced without electrical stimulation and when the rectal probe was introduced and electrical stimulation was applied (Welsh and Johnson 1981). These results indicate that the activation of the HPA axis may be due to the manipulation of the animal and not that the EE involves a painful stimulus. However, early studies in which plasma cortisol measurements were performed to assess the stress level of the animals during the procedure are not very reliable (Mosure et al. 1998).

In recent years, European welfare organizations are increasingly pressing member States and producers to stop its use (Falk et al. 2001; Scientific Advisory Committee on Animal Health and Welfare 2022; Mujitaba et al. 2022). Between 1991 and 1995, the European Union banned the importation of frozen semen doses obtained by EE (Mosure et al. 1998; Scientific Advisory Committee on Animal Health and Welfare 2022). Although controversial, today it is still considered an acceptable procedure by most animal welfare committees in the world (Palmer 2005). However, its use has been banned in some European countries, such as Denmark and the Netherlands (Palmer et al. 2005; Scientific Advisory Committee on Animal Health and Welfare 2022). In the United Kingdom, legislation requires that the procedure is performed by a trained veterinarian (Scientific Advisory Committee on Animal Health and Welfare 2022) and is encouraged to be executed under anesthesia (Palmer 2005).

In female cows, stressful handling stress has been associated with reduced fertility due to disruptions in ovarian function and hormonal balance, with the molecular and enzymatic mechanisms previously described. Handling stress can disrupt the expression of oestrus behavior in cattle, leading to irregularities in the oestrus cycle and reduced fertility. Indeed, studies have shown that transportation or handling in unfamiliar environments can delay or suppress oestrus behavior in cattle (Cook et al. 1997). This delay in oestrus expression can result in missed breeding opportunities and decreased conception rates, ultimately impacting herd reproductive performance (Cooke et al. 2012). Handling stress may affect the synchronization of oestrus cycles in artificially inseminated cattle, leading to reduced success rates in reproductive programs (Alvarez et al. 2000). This stress-induced alterations in cortisol levels can adversely affect oocyte quality and fertilization rates, further compromising fertility (Alvarez et al. 2000). High-stress handling practices during artificial insemination procedures have also been linked to decreased pregnancy rates and increased embryonic mortality in cattle (Fisher et al. 2001).

Once the animals are pregnant are further sensible to handling stress, which can significantly impair pregnancy success and calf survival. Stress-induced elevations in cortisol levels during pregnancy can lead to placental dysfunction, reduced uterine blood flow, and impaired fetal development, increasing the risk of pregnancy loss or abortion (Cooke et al. 2012). Stress during late gestation can trigger premature parturition, low birth weights, and compromised calf health (Alvarez et al. 2000). Furthermore, handling stress on pregnant cows and heifers can affect maternal behavior and milk production, influencing indirectly the calf growth and survival in its postnatal period (Hemsworth et al. 2015). In addition, prolonged exposure to stressors such as handling during pregnancy or parturition can impair ovarian activity and follicular development in the dam after calving, leading to irregular

oestrus cycles and decreased conception rates after calving, in the subsequent reproductive cycle (Grandin 1997).

In order to reduce the adverse effects of these situations, it has been evidenced
that cattle show reduced cortisol levels upon repeated exposure to a stressor, such as
traversing through a handling race, setting the principle of acclimation (Solano et al.
2004), which consists in familiarizing cattle with chutes and facilities. This strategy
is called acclimation and is an effective management practice to minimize handling
stress and optimize reproductive outcomes in cattle (Barrozo et al. 2012), that need
to be frequently handled. Strategies to reduce stress during handling include providing proper training for handlers, designing facilities to minimize noise and conges-
tion, and implementing low-stress handling techniques (Grandin 1997). Additionally,
adopting gradual acclimation protocols for new environments or procedures can
help mitigate stress responses in cattle (Hemsworth et al. 2015). Furthermore,
ensuring adequate rest periods and access to comfortable resting areas after and
before intensive handling can promote overall well-being and reproductive health in
cattle (Cook et al. 1997).

2.3   Nutritional Management: Effects on Reproduction

Nutritional stress is a significant concern in cattle production, affecting animal
health, productivity, and welfare (National Research Council 2000; Suttle 2010).
Cattle are susceptible to nutritional stress when their dietary intake does not meet
their nutritional requirements, leading to imbalances in essential nutrients, energy
deficits, and metabolic disorders (Russell et al. 1992). Nutritional stress in cattle can
arise from various factors, including inadequate feed quality or quantity, poor
dietary balance, changes in feeding management practices, environmental conditions, and competition for resources within the herd (Van Soest 1994; Aboagye et al.
2015). Insufficient forage availability, drought, poor pasture management, and feed
shortages are common causes of nutritional stress, particularly in extensive grazing
systems (Stewart et al. 2013). Additionally, suboptimal feeding practices, such as
underfeeding, overfeeding, or improper diet formulation, can result in nutrient deficiencies or excesses, contributing to nutritional stress in cattle. Nutritional stress
adversely affects cattle physiology and performance (Weiss 2006; Reynolds et al.
2003). Inadequate nutrition compromises immune function, making cattle more
susceptible to diseases and infections (Loor et al. 2005). It also impairs reproductive
efficiency, leading to reduced conception rates, delayed puberty, and extended calving intervals (Diskin and Kenny 2016). Furthermore, nutritional stress can impact
growth rates, body condition scores, and milk production in dairy cows, resulting in
decreased profitability and economic losses for producers (Bach et al. 2005).

As commented above, social dominance and competition among cattle have a
direct impact on access to food, thereby influencing heifer growth. This raises welfare concerns and emphasizes the need to explore alternative management strategies
(de Vries et al. 2014). Moreover, the dominance hierarchy in heifers subjected to
continuous competitive environments directly influences their reproductive

development, body growth, and certain metabolic parameters through its influence over the metabolic status of prepubertal heifers and their reproductive development (Abeni et al. 2019).

In fact, follicular development and puberty onset in heifers are directly associated with nutritional status and consequently, changes in metabolic profile (Funston et al. 2012). Dominant heifers demonstrate higher average daily gains (ADG) compared to subordinate heifers, in agreement with the presence of larger dominant follicles at a younger age (Perry 2012). Dominant heifers also exhibit a "more positive" energy status, leading to greater follicular development compared to subordinate heifers. Moreover, higher glucose concentrations in dominant heifers may be linked to a lower age at first breeding (Brickell et al. 2009). By contrast, insufficient energy or nutrient intake can result in poor body condition scores, delayed puberty, and increased embryonic loss in heifers (Chagas et al. 2007). Finally, heifers, even those subordinates, provided with adequate nutrition exhibit improved reproductive performance, highlighting the importance of proper nutritional management and assuring access to resources to all animals of the herd (Diskin and Kenny 2016).

As social stress does, nutritional stress can also disrupt the expression of oestrus behavior in cattle leading to irregularities in the oestrus cycle and reduced fertility. Added to the specific effect of stress, insufficient energy or protein intake delays or suppresses oestrus by itself, and induces extended intervals between oestrus periods (Wiltbank et al. 2016) and irregular oestrus cycles (Stevenson and Britt 2017). Undernourished cows may exhibit decreased sexual receptivity and mounting behavior, further compromising reproductive efficiency (Rhoads et al. 2009).

Beyond behavior expression, nutritional stress impairs ovarian functionality in cattle, affecting follicular development, ovulation, and hormone production. Inadequate energy or protein intake can lead to the formation of small, poorly developed follicles and reduced ovulation rates (Lamb 1976). Moreover, nutritional deficiencies disrupt the pulsatile secretion of gonadotropin-releasing hormone (GnRH) and luteinizing hormone (LH), essential for follicular growth and ovulation (Martinez et al. 2015).

All these negative effects translate to prolonged intervals of calving to conception, increased embryonic mortality rates (Diskin and Morris 2008), decreased conception rates, and longer calving intervals (Stevenson and Britt 2017).

During very early pregnancy, nutritional stress can have long-lasting effects on embryo/fetal viability and subsequent fetal growth (Van Eetvelde et al. 2017), due to, in part, hormonal imbalance and uterine dysfunction (Stevenson and Britt 2017). This triggers embryonal mortality directly, and increases the risk of pregnancy loss, stillbirths, and low birth weights in cattle due to the inefficient nutrient transfer through the placenta (Rhoads et al. 2009).

Nutrition plays a significant role in bull fertility, influencing body vigor, physical soundness, structural conformation, masculinity, and general health (Byrne et al. 2023). Bulls are often managed nutritionally as a single group, even though there may be large differences in age, size, and body condition (Geary et al. 2021). Although the effect of underfeeding may be obvious, overfeeding also has negative effects on the reproductive performance of bulls. Excess fat increases scrotal fat and consequently, the temperature in the scrotum, which reduces both sperm production and semen quality. Excess fat also increases stress on the bull, limiting his ability to move and search for females in oestrus. Managing bull nutritional needs during the breeding season is difficult, and often the bulls are kept on the same nutritional plane as cows. However, it is necessary to assess the body condition of the bulls as well as observe their ability to serve the cows, because bulls often lose from 100 to 200 pounds during the breeding season. A bull that becomes extremely thin during the breeding season must be replaced because its ability to serve could be reduced (Walker et al. 2008a, b). It is necessary to consider that since the duration of sper-matogenesis in the bull is 60 days, the nutritional effects of overfeeding or under-feeding on sperm quantity and quality will have a subsequent effect. Similarly to that observed in the female, testicular size and hence sperm production can be influ-enced by nutrition. Supplementation with protected proteins in Braham crossbred bulls for 60 days maintained live weight and over a short period, changes in scrotal circumference were directly proportional to changes in live weight. According to testicular weight, the non-supplemented group showed a decrease of 21.3 g of tes-ticular tissue per every 10 kg loss in live weight, which on a daily sperm production basis, could represent a decline of 268×10^6 sperm/10 kg live weight loss. These marked decreases in testicular size and, therefore, in sperm production capacity may affect fertility at the next mating. Epididymal and seminal vesicle weights were slightly lower in the non-supplemented bulls, so a lower storage capacity was observed in these bulls, which could be related to a lower sperm concentration of the ejaculates and perhaps to fertility. However, neither pituitary function nor androgen function, as measured by LH-induced GnRH release or LH or T-release was signifi-cantly influenced by the nutritional treatment (Ndama et al. 1983).

Barth (2012) published a case study on bull semen quality and infertility, detected after BBSE, in which the effect of nutrition, body condition, testicular degeneration, and various stress situations are reflected in sperm morphology, acrosome integrity, and DNA condensation. Thus, 87% of bulls passed their BBSE after bringing hay bales into pasture for 1 month at a location where pastures were still in winter dor-mancy and dead forage was very scarce. One month before supplementation, the main characteristics of the abnormal spermiograms were high percentages of distal midpiece reflexes, detached heads, and proximal cytoplasmic droplets. These poor semen quality results obtained in the first BBSE attempt were interpreted as a typi-cal picture of stress related to weight loss. Similarly, poor motility as a consequence of a high incidence of distal midpiece reflexes was found in bulls that were at pas-ture during and after a storm and evaluated for breeding soundness a few days after. These sperm defects may be caused by stress due to insufficient availability of pas-ture due to deep snow caused by the storm, and its sudden occurrence only 3–4 days after the storm could implicate the cauda epididymis as the site of development of this defect. After 10 days, bulls were re-evaluated and ejaculates showed good mass motility and a great reduction in the incidence of distal midpiece reflexes.

Consequently, ensuring adequate nutrition for both, females and males, is vital for successful reproduction in cattle. Effective management strategies are essential for mitigating nutritional stress and optimizing cattle health and productivity

(Overton and Waldron 2004; Bewley et al. 2008; Winton et al. 2024). Proper nutritional management involves formulating balanced diets tailored to the specific requirements of different production stages and the physiological needs of cattle. Regular monitoring of body condition score, weight gain, and production parameters enables early detection of nutritional deficiencies or imbalances, allowing timely intervention and adjustment of feeding practices. Providing access to high-quality forage, supplemental feed, and clean water sources for all individuals of our herds is critical for assuring a non-stressing environment, the optimal nutritional requirements, and sustaining optimal performance, especially during periods of increased demand or environmental challenges. Understanding that the effects of malnutrition are beyond the strictly nutritional deficiencies and that it includes the negative consequences of a situation of chronic stress is crucial. By optimizing feeding practices and nutritional management producers can minimize the risk of nutritional stress and promote optimal performance and profitability in their cattle operations.

In conclusion, the complex interplay between hierarchy, herd handling practices, and nutritional management significantly influences reproductive outcomes in female and male, dairy and beef cattle. Recognizing these interconnected factors and implementing informed management practices are pivotal for sustaining healthy and productive herds.

3 Thermal Stress

3.1 Introduction

Global warming and climate change are now a universally accepted fact of life. Climate change is causing global temperature rise, warming oceans, shrinking ice sheets, shrinking snow cover, rising sea levels, shrinking Arctic Sea ice, and is causing extreme weather events and ocean acidification. Furthermore, rising temperatures and extreme weather events negatively affect animal physiology, welfare, health, and reproduction. According to Hansen (2009), Heat Stress is defined as "an environment that acts to drive body temperature above set-point temperature," i.e., the increase in body temperature above its regulated set point, which can compromise cellular function, including that of germ cells. Thermal stress appears when an animal's ability to maintain its thermal balance is disrupted. It occurs when the environmental temperature is out of range of the thermoneutral zone. This zone is defined as the temperature range in which an animal can thermoregulate. The thermoneutral zone ranges from 5 to 25 °C for lactating dairy cows (Constable et al. 2017) and from 15 to 25 °C for young calves (Davis and Drackley 1998). However, there is scarce scientific information for adult beef cattle with the range 9–19 °C being proposed (Morignat et al. 2015).

The thermoregulation is the process used by an animal to maintain its body temperature. It implies a balance between body heat gain and body heat loss. The body heat gain is due to the metabolic heat that includes the maintenance heat, heat increases due to exercise, growth, lactation, gestation, and feeding. Therefore, heat gain increase is triggered by increasing metabolic activity (increasing feed intake and decreasing feed conversion), shivering and sustaining muscular contraction, and peripheral vasoconstriction (Constable et al. 2017; Abbas et al. 2020). Animals lose heat, exchanging it with the environment through radiation, convection, conduction, and latent heat loss by evaporation (Kadzere et al. 2002). In addition to the heat generated by metabolism, body heat may be increased due to the environmental temperature (Fuquay 1981) and radiant energy (Dahl et al. 2020). These factors not only increase heat but also impair heat loss. Hence, the lack of shading strongly aggravates heat stress (Dahl et al. 2020).

Climatic factors that may affect the magnitude of heat stress include temperature, as well as amounts of humidity, solar radiation, and wind (Gwazdauskas 1985), but others, such as photoperiod and thinning of the ozone layer, will also affect the amount of solar radiation reaching the animal, while wind speed and rainfall may mitigate its effects (Dikmen and Hansen 2009).

Cattle tolerate high temperatures in the absence of high relative humidity because they are efficient at releasing heat through evapotranspiration (Johnson and Vanjonack 1976). Thus, the Temperature-Humidity Index (THI) is a measure for assessing the risk of heat stress that combines the joint effects of temperature and humidity and is usually used to evaluate the impact of elevated temperatures in cattle production (Armstrong 1994). It has been reported that a THI value of 68 is the threshold of the thermal comfort zone and that values higher than 75 are indicative of severe heat stress (De Rensis et al. 2015). When the mean THI is above 68, dairy cows show signs of heat stress, such as hyperpnoea, increased rectal temperature, and reduced milk yield (Collier et al. 2011). Panting is increased in beef cattle when THI is above 74 (Mader et al. 2006). It has also been stated that *Bos taurus* species enter heat stress at THI 70 and that thermoregulatory mechanisms cease to be effective above THI 80 (McDowell et al. 1976). However, the *Bos indicus* species is more heat resistant, initiating heat-related tachypnoea and transpiration at temperatures 8 °C higher than those of *Bos taurus* (Allen et al. 1963).

Heat stress can cause alterations in the reproductive function via two basic mechanisms. First of all, because homeokinetic changes regulating body temperature and re-stabilizing the balance (redistribution of the body's blood flow; reduction of food intake to reduce metabolic heat production, etc.), may compromise reproductive function and, secondly, the homeokinetic inability to regulate reproduction that compromise germ cells function itself (Hansen 2009).

Finally, regarding the effect of heat stress on the male reproductive function, it is one of the most important factors affecting the reproductive capacity of bulls. For all these reasons, global warming represents a serious threat to livestock production (Hansen 2009; Rahman et al. 2018; Llamas-Luceño et al. 2020; Capela et al. 2022; Yadav et al. 2022), in male and female cattle through the impairment of their reproductive efficiency (Llamas-Luceño et al. 2020; Kastelic et al. 1997).

Cold stress is also possible, and it is observed when THI is below 38 (Abbas et al. 2020). The clinical signs related to cold stress are weakness, decreased activity, cold extremities, and varying degrees of shock are common; bradycardia, weak arterial pulse, and collapse of the major veins are characteristic; the mucous membranes of the oral cavity are cool and there is a lack of saliva (Constable et al. 2017). Therefore, increased or decreased THI interferes with reproduction, as we are describing more in detail, in the next sections.

3.2   Heat Stress: Effect on Reproduction

As seen before, there are several, relevant stressors that affect animal productivity and immune function. However, heat stress is considered the most undesirable factor for animal health (Dahl et al. 2020). Moreover, heat stress, both directly (mediated by hyperthermia) and indirectly (through reduced nutrient intake and altered behavior), reduces the synthesis of valuable products (milk and meat) and leaves animals more vulnerable to disease (Bernabucci et al. 2010).

In a hot environment, heat-stressed cattle try to regulate their body temperature reducing the metabolic heat production by decreasing feed intake and yield (milk production and growth) (Kadzere et al. 2002). On the other hand, cutaneous blood flow increases, diverting heat from the body's core to the periphery (Choshniak et al. 1982). However, as ambient temperature increases, heat exchange drops due to the reduction of environment-body surface temperature difference, and, therefore, evaporative heat loss becomes predominant (Maia et al. 2005).

Direct exposure to heat in lactating dairy cows deleteriously affects health, fertility, and milk production. Heat stress impairs both productivity parameters and the health of cows and bulls, but the most dramatic effect for dairy farmers is the decrease in fertility (De Rensis et al. 2015; López-Gatius 2012; Gernand et al. 2019).

Depending on the physiological phase at the moment of suffering heat stress, different consequences are observed in females and bulls (Dahl et al. 2020), which are reviewed in the following lines of this text.

3.2.1   Effect of Heat Stress on Female Reproductive Performance

Heat stress suffered by a non-pregnant cow once the transition period is over affects its reproductive efficiency (De Rensis and Scaramuzzi 2003). Follicular dynamics constitutes the recruitment of primordial follicles to the pool of growing follicles, with the final development of a preovulatory follicle. Early alterations of inhibin, estradiol, and progesterone have been shown to impair the development of primordial follicles, perhaps due to reduced folliculogenesis since a significant proportion of plasma inhibin originates from small and medium-sized follicles. (Sammad et al. 2020). Summer (mean THI = 79.9) plasma concentrations of inhibin are lower in dairy animals, meaning a decrease in the production of pituitary FSH, indicating less growth and development of follicles (Wolfenson et al. 1997). LH levels are decreased by heat stress, and the dominant follicle develops in a low LH environment which lowers estradiol secretion by the dominant follicle, leading to poor estrous expression and thus decreased fertility (De Rensis and Scaramuzzi 2003). Heat stress on cows demonstrates a delayed effect on follicular development, both in medium-sized and preovulatory follicles; this effect could be related to the low fertility of cattle in autumn (Roth et al. 2001a, b). Heat stress may induce a decrease in follicular dominance (Guzeloglu et al. 2001), when individual follicular dominance is reduced, more than one dominant follicle may develop, which may explain the increased twinning rates that can be observed under these adverse climate conditions (Ryan and Boland 1991).

Once the ovulation occurs, heat stress still disturbs the next steps required to acquire a successful pregnancy. Heat-stressed cows show decreased blood perfusion to the uterus and increased uterine temperature (Gwazdauskas et al. 1973; Roman-Ponce et al. 1978). This implies an inadequate genital tract environment for both, oocyte and sperm survival, reducing the reproductive performance of the cow. Hence, heat impairs reproduction performance, increasing inseminations/pregnancy with THI and with a drastic decrease in pregnancy per artificial insemination (P/AI) when THI reaches 80 ($\leq$16% of conception rate; Gernand et al. 2019).

Moreover, even if the gametes survive, the zygote formed will suffer the too-hot endometrial environment, showing a high rate of early embryonic death (Rivera and Hansen 2001). Indeed, ambient temperature affects embryos in the pre-attachment phase, although the magnitude of this effect decreases with embryos' age (Ealy et al. 1993). Heat stress has similarly been related to an increased rate of early pregnancy loss [late embryo-early fetal death (Cartmill et al. 2001)]. Some results indicate that heat stress can affect gestational success during the peri-implantation period, such that elevated values of the temperature-humidity index during the 21–30 days of gestation are a risk factor for subsequent early fetal loss (García-Ispierto et al. 2006).

Once the gestation is achieved, the exposure to heat stress through the in utero environment causes short- and long-term negative productive consequences in dairy heifers through in utero programming (Dado-Senn et al. 2020), as previously described in a former section in this chapter.

The response to heat stress in bovine embryos is associated with increased chaperone protein expression, oxidative stress-related genes, protein signaling, cell-to-cell interaction, and cell survival. This response affects embryonic development by decreasing the rate of division through altered cell morphology, increased ROS production, and a number of apoptotic cells. These effects are notably greater in heat-susceptible cattle such as *Bos taurus* breeds compared to *Bos indicus* breeds (Naranjo-Gómez et al. 2021). Therefore, in herds of temperate parts of the world, the introduction of animals of breeds derived from *Bos indicus* subspecies and native breeds from tropical countries is an alternative to control the negative effect of high-temperature conditions.

Within utero heat stress suffered later on, during gestation, an important period for fetal growth and development, is linked in Holstein to lighter calves at birth, when compared to cooled dams, with a mean ($\pm$SD) difference of 4.2 $\pm$ 2.7 kg. Compared to cooled dams, calves from heat-stressed cows had lower total plasma protein (6.3 vs. 5.9 g/dL), total serum IgG (1577.3 vs. 1057.8 mg/dL), and apparent colostrum absorption efficiency (33.6 vs. 19.2%); compromising fetal growth and immune function of offspring (Tao et al. 2012).

The negative effects of heat stress during late gestation are likewise observed. This period concurs with the dry-off period, i.e., the non-lactating state of the dairy cows, covering, usually, the last 6–8 weeks of pregnancy up to calving (Tao and Dahl 2013). Considering lactation is the last phase of the reproductive period, crucial to having a healthy calf, the effect of heat stress on udder development should also be contemplated. Although, heat stress has not been linked to a reduction of colostrum production, reducing heat stress during the last 3 weeks of gestation improved colostrum quality with a higher density (1.056 vs. 1.065 g/cm^3) being observed in cooled cows than that from heat-stressed cows (Karimi et al. 2015). Heat stress in late gestation reduces mammary gland involution in the first half of the dry period and affects cell proliferation as calving approaches. Cows exposed to heat stress before calving show reduced adipose tissue mobilization and a lower degree of insulin resistance during early lactation. Prepartum heat exposure suppresses likewise the immune capacity, altering prolactin signaling (Ouellet et al. 2020), being an essential lactogenic hormone in ruminant mammals (Lacasse et al. 2016). The mammary gland microstructure is affected by heat stress during the dry-off period, with heat-stressed cows showing fewer alveoli and a higher ratio of connective tissue in the mammary gland than cooled cows (Dado-Senn et al. 2019).

Heat stress suffered during the dry-off period and over the transition period has been linked to an increased incidence and persistence of uterine diseases of the dam, with an association between THI and the incidence of different postpartum diseases (mastitis, puerperal disorders, and retained placenta) being almost linear (Gernand et al. 2019). However, it is unclear whether the effects observed on whole blood cells (white, red, and platelets) could be directly related to the uterine environment (Molinari et al. 2023). In fact, the percentage of cows with retained placenta and metritis doubles in summer compared to winter (24.05% vs. 12.24%), worsening their reproductive efficiency, enlarging the interval calving–next pregnancy in 24 days on average (DuBois and Williams 1980). A later study links retained placenta and metritis to increased THI, with data from more than 22,000 cows, and postpartum disorders and retained placenta increasing linearly with THI values postpartum (Gernand et al. 2019).

Although much focus has been given to understanding the direct impact of heat stress on lactating cows, and the development of strategies to mitigate it, preweaned calves are often ignored in the application of heat stress reduction measures. The idea that calves are less susceptible to heat, combined with the absence of immediate milk production losses, may have led to the perception that cooling calves is not economically beneficial in the short term (Dahl et al. 2020). Like adult cows, when calves overcome their ability to dissipate heat and maintain homeostasis, heat stress occurs; however, the thermal neutral zone of young calves is narrower. Calves less than 3 weeks of age have a thermal neutral zone between 15 and 25.6 °C (Stull and Reynolds 2008). Above these temperatures, the calf will trade energy for growth to maintain body temperature. Physiological and behavioral effects (e.g., lower feed intake, maintenance energy needs, and decreased immunity) can lead to impaired growth, increased susceptibility to disease, and, in extreme cases, death (Dahl et al. 2020). Higher temperatures and humidity experienced by calves resulted in higher heifer age at first calving (Heinrichs et al. 2005). Later on, heat stress on young heifers depresses follicular development by impairing follicular selection, delaying the follicular wave, decreasing follicular dominance, and jeopardizing follicular steroidogenesis (Bilby et al. 2008).

While most data are studied in dairy cattle, many of the effects of heat stress can be extrapolated to beef cattle. In fact, reduced dry matter intake, decreased weight gain rate and reduced fertility of males and females have been described, with these consequences having been linked to a total annual economic loss of $370 million in the beef American industry (St-Pierre et al. 2003).

3.2.2   Testicular Thermoregulation

As in other mammals, bull testes are suspended in the scrotum and the intratesticular temperature is slightly below the core body temperature, in order to ensure spermatogenesis, sperm storage or to minimize mutations in gamete DNA (Bedford 2004). Hence, the testicular temperature in bulls must be 2–6 °C below body temperature to produce fertile sperm (Barth and Bowman 1994; Kastelic et al. 1997; Paul et al. 2009; Rahman et al. 2011; Morrell 2020; Capela et al. 2022). The optimal environmental temperature for sperm production has been reported to be 15–18 °C throughout the period of spermatogenesis, i.e., for 65–70 days before collection (Parkinson 1987). Similarly, the maintenance of testicular temperature around 32 °C is vital for normal spermatogenesis (Morrell 2020). Testicular temperature is regulated by an intricate and complex thermoregulatory system. First, by the vascular cone formed by the pampiniform plexus (a complex venous network surrounding the very coiled testicular artery) that originates a countercurrent exchange of temperature from warm blood entering the testis and cool blood draining from the testis, so that heat is transferred from the artery to the vein. In bulls, the efficiency of the testicular vascular cone for heat transfer averaged 91% (Sørensen et al. 1991). In addition, the skin overlying the cone is usually the warmest area of the scrotum, which makes it an important location for heat loss. Ultimately, the testicular thermoregulation mechanism and cooling are completed by two muscles that are involved in controlling the position of the testicles in relation to the abdomen: the tunica dartos and the cremated. The tunica dartos is a thin layer of smooth muscle underlying the scrotal skin. The tunica dartos is controlled by the sympathetic nervous system in response to ambient temperature changes and regulates scrotal surface area, allowing heat loss, especially at the level of the scrotal neck, scrotal sweating, and relaxation of scrotal muscles. When an increase in scrotal temperature occurs, vasodilatation of the scrotal arterioles is produced by the direct action of heat and by the reflex elimination of the sympathetic vasoconstrictor tone.

Sweating is another important contribution to testicular cooling. It has been shown that in the scrotum of the bull, the volume of sweat glands per unit of scrotal skin surface area is higher than in other body regions (Blazquez et al. 1988). Finally, the cremaster muscle controls the relative position of the scrotum from de body depending on environmental temperature. The cremaster is a striated muscle and, therefore, probably cannot maintain its contraction for long periods of time (Kastelic et al. 1997; Hansen 2009). A long and defined scrotal neck provides a significant area for heat loss and keeps the testes away from the abdomen, but may also predispose the testicles to trauma (Kastelic et al. 1997). In this sense, there are breed differences in scrotal morphology and thus, for example, when compared to bulls of other breeds, Belgian Blue bulls show detrimental characteristics in terms of reproductive capacity, which seem to be associated with their double-muscled properties (Residiwati et al. 2020). Bulls of this breed usually have a small scrotum with a poorly differentiated scrotal neck (Hoflack et al. 2006), which makes them more susceptible to heat stress and low fertility results when bulls are used in natural mating (Brito et al. 2004). In addition, the Belgian Blue is a *Bos taurus* breed, which is more sensitive to heat stress than the *Bos indicus* subspecies and then cattle crossed with *Bos indicus* (Gaughan et al. 2010a, b). Thus, different aspects of testicular thermoregulation, scrotum, and testicular vascular cone were analyzed in both species and crossbred bulls, showing great differences, particularly between *B. indicus* and *B. taurus* bulls, in several aspects of scrotal/testicular thermoregulation (Brito et al. 2004). It appears that *B. indicus* bulls have a better blood supply to the testes and consequently, perform a more effective heat transfer between the testicular artery and vein due to the special characteristics of the testicular vascular cone in this species, which makes them less susceptible to heat stress. Thus, *Bos indicus* breeds have developed adaptive mechanisms of testicular thermoregulation, by reducing testicular artery wall thickness and blood arteriovenous distance, whereas the adaptation mechanism of *B. taurus* breeds to restore homeostasis is based on adjustments of feed intake and T3/T4-based metabolism, higher heat loss capacity, less intense tachypnea and a more efficient sweating rate (Capela et al. 2022).

Different factors and circumstances, such as environmental heat, fever, fat deposition in the neck of the scrotum, scrotal frostbite, and trauma may all interfere with thermoregulation of the testicles (Barth and Bowman 1994).

Irrespective of the genotype differences described in relation to heat stress, thermoregulatory mechanisms work well to keep functional testes cool until an animal is exposed to an excessive heat load from the environment or fever-associated internal heat production. Therefore, breeding bulls with a normal scrotum and adequate scrotal circumference can tolerate heat stress up to certain limits. In crossbred beef bulls, testicular temperature determined at three different positions of the testis (top, middle, and bottom) was respectively 30.4, 29.8, and 28.8 °C at the scrotal surface; 33.3, 33.0, and 32.9 °C at the subcutaneous lining; and 34.3, 34.3, and 34.5 °C at the intratesticular plane. Thus, the temperature gradients from top to bottom were 1.6 °C for the scrotal surface, 0.4 °C at the subcutaneous level, and –0.2 °C inside the testis. Therefore, the temperature gradient was marked on the scrotal surface, minor in the scrotal subcutaneous tissues and nonexistent in the testicular parenchyma. These gradients may be due to the arrangement of the vasculature; the scrotum is apparently vascularized from the dorsal pole of the testis to the ventral pole, whereas the testicular artery branches dorsally from the inferior part of the testis to the superior part. The intraepididymal temperatures of the epididymal head, body, and tail were, on average, 35.6, 34.6, and 33.1 °C, and the gradient between the caput and the cauda averaged 2.5 °C. The epididymal head was warmer than the testicular parenchyma, but the tail, an important sperm storage place, was cooler

(Kastelic et al. 1995). In a subsequent study, it was reported that the scrotal surface and testes have opposite, complementary temperature gradients, resulting in a relatively uniform intratesticular temperature (Kastelic et al. 1996). Thus, it is well established that various preset functions contribute to the regulation of testicular temperature. Moreover, the testis displays a variety of mechanisms that are activated when exposed to heat stress, such as DNA repair, heat shock response, oxidative stress response, apoptosis, and cell death, and any disruption of these mechanisms can cause serious consequences for spermatogenesis.

3.2.3 Study Models to Evaluate the Effect of Heat Stress on the Testicular Function

Given the relevance of the effect of heat on the testicular function, several rodent models have been used to study this issue, such as temporary exposure of the testes to high temperatures (usually 40 °C), surgical induction of cryptorchidism resulting in prolonged exposure of the testes to core body temperature (37 °C), or housing males at elevated environmental temperatures during many hours (Zhu and Setchell 2004; Zhu et al. 2004; Cammack et al. 2006). However, scrotal insulation is the common model to study the effects of increased testicular temperature in bulls (Barth and Bowman 1994; Brito et al. 2003; Rahman et al. 2011), being the method of choice in an attempt to mimic the effects of environmental heat stress on spermatogenic cells under controlled conditions (Brito et al. 2003; Rahman et al. 2011). Scrotal insulation disrupts testicular thermoregulation by increasing the testicular temperature and also by interfering with scrotal sweating (Rahman et al. 2018). It has frequently been used to evaluate the nature and magnitude of sperm morphoanomalies in bulls (Vogler et al. 1991; Walters et al. 2006; Fernandes et al. 2008), because provides insights regarding the pathogenesis of specific sperm defects and is useful to determine the effects of increased testicular temperature on sperm production and semen quality (Brito et al. 2003). Isolation of the scrotal neck causes an increase in scrotal and testicular temperatures with adverse effects on semen quality that mimic the effects of whole scrotal isolation (Kastelic et al. 1996).

Scrotums are insulated by wrapping them with a sack made of insulating material (nylon cloth with polyester batting), held in place by VELCRO® and medical tape. The pouch closes tightly enough over the neck of the scrotum to prevent the bull from pulling a testicle out of the top, but not so tight as to pinch the neck of the scrotum. The scrotum is inserted into the sac and held loosely for 30 min to promote complete descent of the testicles. The scrotum and scrotal neck are then readjusted to the body wall. During isolation, the necessary precautions are taken to avoid interfering with testicular irrigation, but at the same time, enough is secured to avoid premature drop. After the final placement of the sac, a routine check-up is performed, and the scrotal temperature on the caudal surface of the scrotum is measured during isolation with a glass rectal thermometer (Rahman et al. 2011).

The time required to detect alterations in sperm production and semen quality after scrotal insulation varies depending on the stage of germ cell development at the time of insulation, and the time required for the damaged cells to be released into the seminiferous tubules and transported through the epididymis (Barth and Bowman 1994; Brito et al. 2003).

It has been described that insulation of the scrotal neck of bulls caused profound changes in the proportion of morphologically normal spermatozoa, observable from 8 days after insulation (Kastelic et al. 1997). In this study, it was observed that the spermatozoa stored in the epididymis at the time of scrotal insulation were the most affected. A similar study evaluated the effects of whole-scrotum or scrotal neck insulation on sperm production, semen quality, and testicular echotexture in *B. indicus* and crossbred bulls (Brito et al. 2003). The results showed that insulation of the scrotal neck in crossbred bulls did not significantly affect sperm production and semen quality. However, whole-scrotum insulation resulted in decreased sperm production, with a reduction in sperm concentration, and semen quality in *B. indicus* and crossbred bulls. The decrease in semen quality was characterized by a reduction in sperm motility and by a time-specific appearance of sperm defects, although with great individual variation among sires. Lower sperm concentration and motility were not associated with changes in testicular echotexture (Brito et al. 2003).

The scrotal insulation model has been widely used due to the fact that several studies reported that, since it prevents normal temperature regulation in the scrotum, could explain the profound changes in sperm concentration, motility, and morphological abnormalities responsible for the decrease in fertility observed in bulls subjected to natural heat stress. However, this method has also been questioned, as it severely perturbs the animal's ability to control temperature within the scrotum. Therefore, scrotal insulation is not an accurate representation of what may occur during natural heat stress, unless injury is prolonged or severe. The individual response to heat stress depends on factors such as the age of the animal, body composition, e.g., fat and its distribution, breed (genotype), animal resistance, as well as whether or not the animal can perform behavioral adaptations, such as lying on a cool surface or wallowing (Morrell 2020). Moreover, models used to study these systems usually involve intense and/or prolonged heat exposure, which are not the most typical heat stress conditions encountered by animals, either in nature or in production systems (Rizzoto et al. 2020).

3.2.4 Effect of Heat Stress on Spermiogenesis and Male Cattle Reproduction

Regarding the effect of heat stress on the male reproductive function, it is one of the most important factors affecting the reproductive capacity of bulls. Bull fertility is of vital importance in livestock production since one bull can cover 20 females under natural service conditions or hundreds of thousands by means of artificial insemination (Kastelic et al. 1997). Therefore, heat stress can notably impair the reproductive efficiency of the whole herd, affecting just the mating bulls.

Spermatogenesis is a complicated biological process that occurs in the testis' seminiferous tubules, followed by spermiogenesis in the epididymis. This process requires a temperature between 2 and 6 degrees below core body temperature to produce fertile sperm (Barth and Bowman. 1994; Kastelic et al. 1997; Paul et al. 2009; Rahman et al. 2011; Morrell 2020; Yadav et al. 2022; Capela et al. 2022). It has been stated that the optimal environmental temperature for sperm production ranges between 15 and 18 °C for the entire spermatogenesis period (Parkinson 1987). An increase in scrotal temperature arrests spermatogenesis and has a negative impact on testicular function, leading to degeneration of spermatozoa, reducing their fertilizing capacity and ultimately causing infertility (Paul et al. 2009). The sequential cellular events of the spermatogenic process are initiated at the basal compartment and conclude at the apical compartment of the seminiferous tubules (Durairajanayagam et al. 2015). In bulls, spermatogenesis takes place in three phases: (1) spermatocytogenesis, (2) meiosis, and (3) spermiogenesis lasting approximately 21, 23, and 17 days, respectively (Johnson et al. 2000; Fig. 3).

During the spermatocytogenesis phase, primordial germ cells divide by mitosis to form type A1 spermatogonia. Some type A1 spermatogonia revert back to stem cells and the process continues, from which new spermatogonia can be derived (Hochereau-de Reviers 1976). The remaining A1-type spermatogonia progressively divide by mitosis to form A2-, A3- and A4-type spermatogonia (Johnson 1995). A4-type spermatogonia divides again to form intermediate spermatogonia (Int-type) and then to form B-type spermatogonia. The B-type spermatogonia divides by mitosis at least once or twice to form primary spermatocytes. Meiosis comprises two stages. During meiosis I, primary spermatocytes duplicate their DNA and undergo progressive meiotic prophase nuclear changes known as preleptotene, leptotene, zygotene, pachytene, and diplotene before dividing to form secondary spermatocytes. Secondary spermatocytes undergo a second meiotic division resulting in round haploid cells called spermatids. Finally, during spermiogenesis, the spermatids gradually develop into spermatozoa, and nuclear chromatin condensation, flagellum formation, and acrosome development occur. The elongated spermatids approach the lumen of the seminiferous tubules and are subsequently released into the lumen of the seminiferous tubules while undergoing a further transformation process known as spermiation before moving to the epididymis for final maturation and storage (Yoshinaga and Toshimori 2003). Spermatocytes and/or spermatids are extensively remodeled during the meiotic and spermiogenic phases and thus, nucleoprotein histones are initially replaced by non-histone transition proteins, and finally, by protamines (Gaucher et al. 2010). Protamines participate in the packaging of sperm chromatin in such a way that the paternal genome remains protected and functionally inert, and a marked reduction in nuclear volume is reached (Ward 2010).

The immotile spermatozoa formed are released into the seminiferous tubular fluid and transported to the epididymis, where they acquire the motion and fertilizing capacity (Yanagimachi 1994). The epididymis is generally divided into three parts: caput, corpus, and cauda. In bulls, the transit of sperm through the epididymis lasts between 8 and 11 days (Vogler et al. 1993). To reach fertilizing capacity, sperm undergo maturational changes during their transit through the epididymal duct (Yanagimachi 1994). These include changes in plasma membrane lipids, proteins, and glycosylation, alterations in the outer acrosomal membrane, gross morphological changes in the acrosome, and cross-linking of nuclear protamines, outer dense fiber proteins, and fibrous sheath. Mature sperm are stored in the tail of the

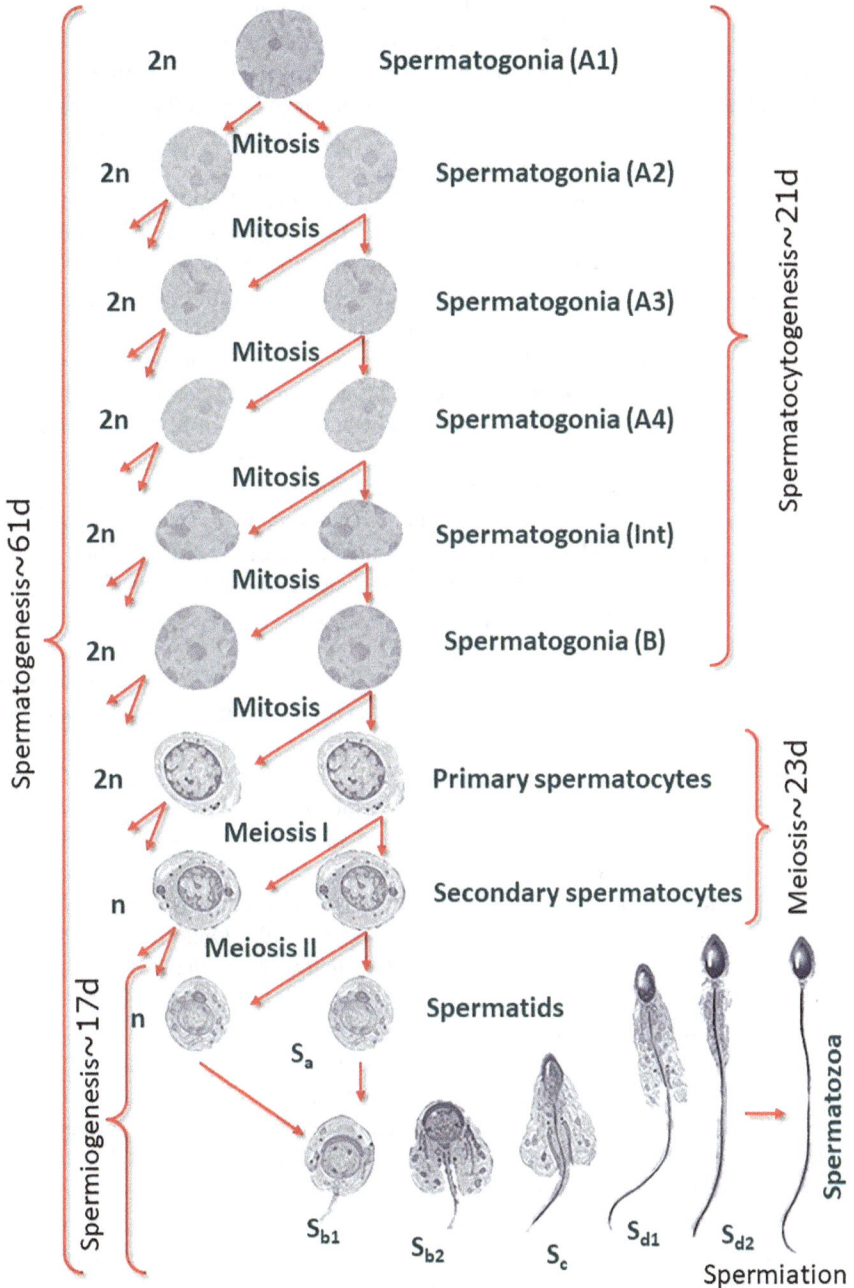

Fig. 3 Schematic representation of spermatogenesis and the time of each one of its phases

epididymis and proximal portion of the vas deferens just before ejaculation (Turner

1995). During ejaculation, the stored spermatozoa and epididymal fluid mix with the alkaline secretions of the male's accessory sex glands, forming semen.

A published study indicated that between 1973 and 2008, the number of days with a THI above 68 increased by 4.1% in certain areas of Central Europe (Solymosi et al. 2010), which may affect testicular function since the maintenance of testicular temperature around 32 °C is of great importance for normal spermatogenesis. High ambient temperatures, alone or in combination with high humidity, interfere with evaporative heat loss from the scrotal surface. This, together with elevated body temperature (due to interference with heat transfer out of the body) and high metabolic activity of the tissue itself, leads to elevated temperature within the testis (Morrell 2020). Although it appears that mammalian testes operate normally at the edge of hypoxia (Setchell et al. 1964), increased metabolism in the testis consequent to heat stress may not be attended by an increase in blood flow and the corresponding increase in oxygen demand (Galil and Setchell 1988), which causes testicular hypoxia (Barth and Bowman 1994; Setchell 1998), leading to tissue oxidative stress (Paul et al. 2009; Capela et al. 2022), by increasing production of reactive oxygen species (ROS). Hypoxia occurs when oxygen tension falls below that required for normal cellular function in a particular tissue (Paul et al. 2009). This can result when blood flow to the tissue is inadequate or from a reduced oxygen-carrying capacity (Höckel and Vaupel 2001). Hypoxia causes cell cycle arrest and apoptosis (Iida et al. 2002).

Increased testicular temperature elevates cellular metabolism, which results in higher production of reactive oxygen species (ROS), and consequently causes oxidative stress (Setchell 1998). The production of free radicals and reactive oxygen species (ROS), including superoxide anion and hydrogen peroxide, can positively affect certain sperm functions, including hyperactivation, capacitation, acrosome reaction (de Lamirande et al. 1997), and also in binding to the zona pellucida (Morrell 2020). However, elevated levels of these reactive species can be harmful and cause oxidative damage to DNA, and several studies relate male fertility problems to oxidative stress (Aitken 1995; Aitken et al. 1998; Ong et al. 2002). The testes possess an antioxidant enzymatic system that protects the germ cells from oxidative damage. These include superoxide dismutase (SOD), glutathione reductase (GSR), glutathione peroxidase (GPX), glutathione S-transferase (GST), and heme oxygenases (inducible HMOX1 and constitutive HMOX2; Bauché et al. 1994; Maines 1997; Paul et al. 2009). Spermatozoa possess high amounts of polyunsaturated fatty acids in the plasma membrane if compared to the low concentrations of antioxidant enzymes present in the scarce cytoplasm, making them highly susceptible to ROS damage (Nichi et al. 2017). In animal models, poor semen quality resulting from heat stress has been linked to increased ROS production and consequent lipid peroxidation (Aitken et al. 2016), which affects plasma membrane integrity, mitochondrial function, chromatin integrity (Paul et al. 2008), and the appearance of sperm abnormalities in ejaculates (Barth and Bowman 1994). Mitochondrial regulation of redox balance is crucial for the maintenance of cellular homeostasis. Oxidative stress activates apoptotic mechanisms that are initiated by an imbalance between ROS production and the antioxidant action to neutralize them

(Aitken et al. 2016). In addition, mitochondrial membrane dysfunction affects energy production and thus, sperm motility, and bull sperm motility is very susceptible to oxidative stress induced by hydrogen peroxide which affects sperm tail contractility (de Castro et al. 2016). If the lipid peroxidation cascade is not regulated, it can affect the integrity of sperm chromatin and increase the frequency of DNA breaks (Aitken and Krausz 2001).

However, although it was long believed that mammalian testes are close to hypoxia, that increased testicular temperature does not change blood flow and that hypoxia is responsible for decreased sperm quality and fertility, recent studies indicate that hyperthermia and not secondary hypoxia cause heat-induced changes in spermatozoa (Kastelic et al. 2019; Rizzoto et al. 2020).

Heat stress increases germ cell apoptosis, probably caused by DNA damage, leading to decreased testicular weight and infertility (Setchell 1998; Pérez-Crespo et al. 2008). Shortly after heat exposure, a reduction in testicular weight is observed, which is attributed to the loss of germ cells, mainly due to apoptosis. (Durairajanayagam et al. 2015). During spermatogenesis, physiological apoptosis of germ cells occurs to maintain germ cell quality and DNA damage is a precursor to apoptosis (Shaha et al. 2010). Apoptosis, or type I programmed cell death, involves identifiable cellular changes, such as DNA fragmentation, reduction of cell volume, and plasma membrane blebbing (Liu 2010). Significant apoptotic loss of germ cells following testicular heat stress can occur through intrinsic or extrinsic pathways, but both, the intrinsic and extrinsic pathways converge on caspase-3 and other executioner caspases and nucleases that trigger the events of programmed cell death (Durairajanayagam et al. 2015). The germ cell apoptosis response following heat stress takes place at specific developmental stages, with spermatocytes (diplotene and pachytene) and spermatids being the most susceptible to heat-induced changes (Setchell 1998), although the reason for this higher vulnerability has not been elucidated. Finally, autophagy, commonly referred to as type II programmed cell death, may also be involved in causing germ cell death. Autophagy is a process in which cells are phagocytosed by vesicles, degraded by lysosomes, and the resulting cellular components are recycled for energy generation (Zhang et al. 2012).

Despite the high rate of apoptosis, some cells may survive, and it has been shown that they eventually develop into mature spermatozoa containing damaged DNA in mice (Banks et al. 2005). This loss of integrity is generally attributed to oxidative stress, but could also be due to defective repair mechanisms or destruction of the testis in mice, such that spermatocytes are unable to develop properly (Banks et al. 2005; Setchell 1998). Germ cells have been shown to be sensitive to even short-duration heat stress, and thus exposure to temperatures of 43 °C for 5–45 min induces DNA fragmentation consistent with an apoptotic pattern in rats (Kaushik et al. 2019). During chromatin remodeling in spermatogenesis, DNA breaks are common; however, spermatozoa are not transcriptionally active and cannot repair DNA damage due to DNA condensation. Thus, spermatozoa can accumulate DNA damage without being repaired. Changes in area, perimeter, width, and length have been induced in sperm heads by heat stress and have been associated with a detrimental effect on IVF, while alterations in shape, lateral symmetry, anteroposterior symmetry, chromatin decondensation, and chromatin heterogeneity were associated with poor blastocyst development (Lucio et al. 2016). Spermatocyte and spermatid stages are the most temperature-sensitive cells in the spermatogenic cycle of the seminiferous tubules (Paul et al. 2008; Pérez-Crespo et al. 2008), because spermatocytes II, as postmeiotic cells, and spermatids, after chromatin condensation, lose the apoptotic response and the ability to repair DNA damage (Pérez-Crespo et al. 2008). In DNA, regions of low sperm chromatin compaction, in particular, where protamines are replaced by histones, have been identified as the most vulnerable to oxidation (Xavier et al. 2019). In mice, it has been shown that germ cell loss resulting from heat stress results in the formation of degenerated giant germ cells and "holes" in the seminiferous epithelium and that, despite the activation of the apoptosis mechanism, many of these damaged cells give rise to spermatozoa with damaged DNA (Paul et al. 2008). Furthermore, this study showed that in early embryos produced, heat stress conditions can induce mutations or deletions in sperm DNA, leading to the alteration of regulatory networks. In addition, early embryos produced under heat stress conditions can induce mutation or deletion of sperm germ cell DNA, leading to alteration of regulatory networks, so this study, in addition to providing evidence of temperature-dependent effects on germ cell DNA integrity, highlighted the importance of an intact paternal genome for normal embryo development.

Therefore, heat stress apparently affects testicular function through different mechanisms. The effects of mild, moderate, and severe natural heat stress differ and the consequences depend on the duration of the injury, among others (Morrell 2020). Manifestation of heat stress represents a complex interaction of factors, with the fertilizing ability of spermatozoa affected by several pathways (Garcia-Oliveros et al. 2020; Morrell 2020; Capela et al. 2022). Ejaculates from animals subjected to heat stress are characterized by low sperm concentration and motility; increased sperm abnormalities, such as proximal cytoplasmatic droplets; and by diadem, vacuoles, and other nuclear defects (Barth and Bowman 1994; Kastelic et al. 1996; Brito et al. 2003; Krishnan et al. 2017; Llamas-Luceño et al. 2020; Capela et al. 2022). Sperm are sensitive to the effects of heat stress at various time points of spermatogenesis. Minor defects have been observed on sperm cells that are located in the epididymis at the time of the insult; however, most detrimental effects are observed during meiotic stages, before DNA compaction has occurred. Subsequent chromatin remodeling also represents a stage where DNA is susceptible to attack by ROS, indicating that heat stress is more prejudicial for developing germ cells than for mature sperm. Thus, negative effects of heat exposure have been observed in semen collected between 14 and 42 days after artificial scrotal insulation, which indicates heat stress has more detrimental effects on developing germ cells than on mature spermatozoa (Rahman et al. 2011).

Transient mild testicular hyperthermia causes temperature-dependent germ cell death as well as a complicated stress response (Pérez-Crespo et al. 2008; Paul et al. 2009). The differentiation and maturation processes of spermatocytes and spermatids are particularly sensitive to thermal stress (Durairajanayagam et al. 2015). An elevated scrotal temperature results in testicular germinal atrophy and spermatogenetic arrest (Munkelwitz and Gilbert 1998). Due to their high mitotic activity, germ cells are more vulnerable to heat (Shiraishi et al. 2012). Heat stress has different effects on different cell types in the testis, such as oxidative stress, hypoxia, apoptosis, DNA damage (Pérez-Crespo et al. 2008; Paul et al. 2009), and autophagy (Zhang et al. 2012). In addition, it causes a disruption of the blood–testis barrier (BST) and alterations in hormone secretion (Morrell 2020; Yadav et al. 2022).

Moreover, since spermatogenesis in bulls lasts about 60 days, changes in ejaculate characteristics caused by increased testicular temperature may not be observed immediately. However, 2–4 weeks after heat stress, potential damage to spermatogenic cells can be found in the ejaculate (Vogler et al. 1993) and it may take 6–12 weeks before sperm quality returns to normal (Barth and Bowman 1994; Kastelic et al. 1996). Thus, sperm morphological abnormalities (mostly detached heads, cytoplasmatic droplets, and tail defects) appear in the ejaculates in chronological order depending on the stages of sperm maturation during thermal insult (Rahman et al. 2018).

In fact, despite the number of published studies on the effect of heat stress on testicular function, it should be noted that the chronology of events following heat stress is not yet fully understood and that this knowledge could help to better define the pathophysiology of testicular degeneration (Garcia-Oliveros et al. 2020).

The effect of heat stress on the level of blood testosterone concentration depends basically on how stressful the environmental conditions have been for the animal. Testosterone is produced by Leydig cells and results in fundamental for spermatogenesis and sperm function. Bulls subjected to scrotal insulation showed a profound (tenfold) reduction in testicular testosterone concentrations after 48 h (Rizzoto et al. 2020).

Although it is assumed that the adverse effect of heat stress on fertility is due, mainly, to its effects on gametogenesis, recent studies indicate that it can also alter seminal plasma proteins (Rahman et al. 2018), affecting indeed sperm functionality.

3.2.5   Effect of Heat Stress on Ejaculate Quality and Fertilizing Ability

The effects of heat stress exposure on the testis affect testicular physiology and may affect the quality of the seminal samples, or the fertilizing capacity of those spermatozoa (even when they present normal morphology and motility), and on the development of subsequent embryos produced both in vivo or in vitro (Setchell 1998). Bulls exposed to heat stress are more likely to suffer reduced testicular weight (with subsequent lower sperm production), decreased sperm motility and viability, altered hormone levels, and reduced fertility (Setchell 1998; Morrell 2020).

Most studies have focused on the effects of THI (Gantner et al. 2011) on semen production and quality, but the temperature difference between seasons has the same short-term importance as prolonged heat stress on quantitative and qualitative semen parameters. Thus, in bulls, more sperm abnormalities and reduced fertilization rates have been observed during seasons with increased environmental temperature (Parkinson 1987; Barth and Bowman 1994). Studies in temperate climates in Sweden (Valeanu et al. 2015) or northern Spain (Sabés-Alsina et al. 2017) have shown differences in semen quality according to the season of the year in which ejaculate collection was performed.

Several studies in bulls have reported the adverse effect of heat stress on sperm concentration, motility, and morphology, although the exact phases of spermatogenesis at which such defects occur have not yet been fully documented. (Vogler et al. 1993; Barth and Bowman 1994; Kastelic et al. 1997; Fernandes et al. 2008).

In two *Bos taurus* (Guernsey) bulls exposed to 37 °C with 81% relative humidity for 12 h/day during 17 consecutive days, 30–40% morphologically abnormal spermatozoa were observed in the ejaculates (mainly coiled tails and detached heads), as well as a marked decrease in sperm concentration and motility (Casady et al. 1953). In another study, in which the exposure of *Bos taurus* (Friesland) and *Bos indicus* (Afrikaner) bulls to heat stress (40 °C with a relative humidity of 35–45%) was shorter (12 h), there was a significant decrease in motility and percentage of live sperm in both breeds and a significant increase in the percentage of morphologically abnormal sperm (Skinner and Louw 1966).

Increased abnormal morphology, membrane damage, and DNA fragmentation have been observed in spermatozoa after 21 days of testicular insulation in rams (Alves et al. 2015). Rahman et al. (2011) similarly observed an increase in head morphoanomalies in bull spermatozoa between 2 and 6 weeks after scrotal insulation, when the germ cells were probably in the spermiogenesis and meiosis stages. The increase in morphological abnormalities of spermatozoa, especially sperm heads, after scrotal isolation is related to defective chromatin condensation during the acrosomal and Golgi stages of spermiogenesis, as spermatozoa are known to be vulnerable to heat stress when these cells are not fully protamine (O'Donnell 2015). Since the effects on DNA are not readily detectable when performing routine sperm quality checks, heat stress can cause decreases in fertility even when sperm is apparently normal (Setchell 1998).

The increase in testicular temperature particularly affected spermatozoa in the spermiogenic and meiotic phases of development at the time of scrotal insulation (Rahman et al. 2011). According to the authors, this increased susceptibility is due to defective remodeling of the sperm chromatin, which, in part, could be related to the lack of protamination in the DNA conformation and, in turn, could lead to alterations in the nuclear shape of the spermatozoa. This study also showed a significant decrease in progressive sperm motility as assessed by computer-assisted sperm analysis (CASA), which according to the authors is caused by high numbers of dead and morphologically abnormal spermatozoa. In this study, the predominant types of abnormalities were abnormal heads (acrosome defects, pyriform heads, microcephalic and macrocephalic heads), nuclear vacuoles, and tail defects. The increased presence of head anomalies and nuclear vacuoles are comparable to those described in other studies (Vogler et al. 1991; Barth and Bowman 1994; Fernandes et al. 2008). Heat stress can alter sperm motility 2 weeks after heat exposure, decreasing progressive motility up to 40%, which can recover first, 8 weeks after the injury (Rizzoto et al. 2020). Oxidative stress induced by heat stress promotes premature capacitation and an increase in the proportion of capacitated and acrosome-reacted spermatozoa (Zhu and Setchell 2004). This premature capacitation state reduces sperm lifespan, fertilization rates, and embryo development. In addition, sperm may be a carrier of metabolites that induce damage to the oocyte, through lipid peroxidation and antioxidant depletion, which can impair embryonic development (de Castro et al. 2016).

The effect of THI on in vitro fertility has also been evaluated. In a study with Holstein bull sperm collected after periods of high or low THI, no differences in sperm quality were observed, except for an increase in the proportion of dead cells in the high THI group compared to the low THI group (Llamas-Luceño et al. 2020). In this study, there were also no differences in the cleavage rate of fertilized oocytes, although there was a decrease in the total blastocysts produced in the high THI group. The study showed no differences in the total number of embryonic cells, the number of inner cell mass, or the number of apoptotic cells in the embryos of both groups. However, another study showed that the proportion of advanced blastocysts was higher in the group using semen collected in spring for IVF compared to winter or summer. In addition, the number of sperm attached to oocytes was lower in summer than in winter or spring (Sabés-Alsina et al. 2017). Oocyte DNA repair mechanisms can partially repair sperm DNA fragmentation but may be overwhelmed in the presence of extensive sperm DNA damage, leading to embryonic death. Oocyte quality exacerbates the effect of ROS-induced sperm DNA damage, negatively affecting the development of the resulting embryos (Rahman et al., 2011). Sperm with damaged DNA are able to fertilize the oocyte without affecting cleavage. However, after genome activation at the 8–16 cell stage; either the paternal DNA damage is repaired or the apoptotic machinery of the early embryo is activated, arresting further development (Fatehi et al. 2006).

3.2.6 Effect of Heat Stress of Male and Female on the Offspring

Regarding the influence of the heat suffered by the mothers during pregnancy on the embryos or fetuses, we find also evidence of a direct effect on the viability of the pregnancy or the future health of the offspring. For example, compared to cooled dams, calves born to heat-stressed cows have lower total plasma protein (6.3 vs. 5.9 g/dL), total serum IgG (1577.3 vs. 1057.8 mg/dL), and apparent colostrum absorption efficiency (33.6 vs. 19.2%); compromising fetal growth and immune function of offspring (Tao et al. 2012). Additionally, reduced weaning weights have been observed in in utero heat-stressed calves (difference of weight: 6.3 ± 3.7 kg; (Ouellet et al. 2021). This weight difference persists until puberty (in males and females), with heat-stressed calves weighing 2.4 kg less at birth and 6.6 kg less at weaning with a 10% decrease in ADG compared to chilled calves (Dado-Senn et al. 2020). Laporta et al. (2020) reported that in utero fetal heat stress affects long-term milk production of the resulting offspring up to three lactations. Considering a 305-day lactation, milk production of cows that suffered in utero heat stress was reduced on average, 671, 702, and 1983 kg/lactation in their first, second, and third lactations, respectively, compared to in utero cooled animals (Dahl et al. 2019). Moreover, heat stress affects the survival of offspring, with data showing that only 65.9% of heifers reach first lactation when they suffer in utero heat stress versus 85.4% of heifers from cooled cows (Monteiro et al. 2016).

As indicated above, exposure of bulls to heat stress affects the fertility of their semen. However, damage to sperm may not be enough to prevent fertilization totally and can produce offspring (Setchell 1998). The paternal contribution to the zygote involves not only DNA but also the centriole is required for subsequent embryonic divisions after fertilization. Paternal epigenetics (DNA methylation marks and several RNA types) plays an important role in successful embryonic development (Steger et al. 2011). Therefore, an alteration in the paternal epigenome caused by any stressor, including heat stress, on spermatozoa can have consequences and effects on subsequent generations.

The replacement of histone-bound chromatin by protamine-bound chromatin is responsible for the high sperm DNA compaction necessary for the safe delivery of sperm DNA to the oocyte by protecting it from oxidative stress in the female reproductive tract. The highly compacted sperm nucleus blocks sperm DNA transcriptional activity. On the other hand, the lack of protamines in the sperm nucleus results in DNA damage and can cause subfertility or infertility in bulls (Dogan et al. 2015). As noted above, following exposure to heat stress, the resulting bull spermatozoa are associated with altered chromatin conformation, possibly due to aberrant DNA contamination (Rahman et al. 2011). This study showed that heat stress affects sperm remodeling procedures (protamine-DNA compaction), which resulted in a higher incidence of protamine deficiency or poorly protaminated sperm in ejaculates from sires subjected to heat stress. The most susceptible population were spermatozoa that were at the spermiogenic and meiosis stages at the time of heat stress, with the ejaculates showing the greatest increase in protamine deficiency, causing defective protamination of chromatin that mainly affects during the histone to protamine transition period.

Furthermore, it has been indicated that environmental stress produces changes in the DNA methylation pattern of spermatozoa (Wei et al. 2014). DNA methylation is essential in genomic imprinting, gene expression regulation, X chromosome inactivation, and embryonic development (Surani 1998). Aberrant methylations in genes at the promoter and imprinted loci are associated with infertility and the appearance of sperm defects in men (Pacheco et al. 2011). Several studies have demonstrated that the use of DNA methylation inhibitors may reduce global germ cell methylation patterns that may, at least in part, be responsible for abnormal embryogenesis (Rahman et al. 2018).

The erasure of epigenetic marks acquired immediately after fertilization is key to the acquisition of nuclear totipotency in mammalian zygotes. Cytosine in the paternal pronucleus is actively and almost completely demethylated, followed by de novo methylation prior to the two-cell stage, while cytosine methylation levels gradually decrease in the maternal pronucleus in bovine zygotes (Rahman et al. 2014), being this dynamic reprogramming of DNA methylation of the pronuclear stage essential for normal embryogenesis.

The effect of heat stress during spermatogenesis causes alterations in sperm chromatin, which can affect DNA methylation dynamics in the zygote, not only in the spermatozoa self (Rahman et al. 2014). Reduced fertilization rates and perturbed DNA methylation reprogramming were observed in the paternal pronucleus following in vitro fertilization of oocytes with heat-stressed sperm cells (Rahman et al. 2014). This aberrant methylation pattern might persist through subsequent generations resulting in transgenerational epigenetic inheritance. However, no studies have focused on epigenetic transgenerational inheritance in cattle the impact of heat stress on sperm quality, and the consequences for the offspring have not yet been determined on bulls. Furthermore, spermatozoa RNA involved in zygote cleavage may be targeted by oxidative stress, which can be deleterious even before embryo genome activation (de Castro et al. 2016).

3.2.7   Strategies to Mitigate the Effects of Heat Stress on Reproduction

Fortunately, proven strategies exist to mitigate at least partially the heat stress suffered by the livestock. Heat stress can be reduced to some extent by different strategies, which include management, husbandry, and facilities that minimize heat stress (Krishnan et al. 2017). In the case of bulls, the choice of breed, the age of the animals, the optimal time for obtaining ejaculates, as well as adequate management that allows adopting behavioral mechanisms that aid physiological homeostasis, are factors that can mitigate the effects of heat stress on bull fertility (Morrell 2020).

The most immediate way to avoid the consequences of heat stress on sperm quality is to choose breeds that are well adapted to the local environment, including climatic conditions, and the genetic selection of heat-tolerant breeds (Krishnan et al. 2017). Thus, in very hot and humid climates it would be appropriate to avoid the use of European breeds and focus on *B. indicus* breeds (Morrell 2020), which, as indicated above, are less susceptible to thermal stress. The identification of heat-tolerant animals within high-producing breeds, able to maintain high productivity and survivability when exposed to heat stress conditions, is the key. Depending on individual variations between bulls in terms of their resistance to heat stress, sires can be classified as either "heat-tolerant" (produce good quality spermatozoa all year regardless of temperature) or "heat-sensitive" (produce good quality sperm only during the cool season; Netherton et al. 2022). In this study, the purposeful heating of bulls to 40 °C for 12 h confirmed that individual animals had different degrees of heat sensitivity.

Although several studies have evaluated seasonal effects on semen quality, few have used a metric that can be used across different climatic zones or with a representative number of bulls. Moreover, some studies focused on weather conditions at the time of semen collection, ignoring the effect of high temperatures during the spermatogenesis of these ejaculates. Finally, the factor age of the bull introduces confusion. Thus, taking into account THI 30 days before ejaculation, total motility and normal morphology of fresh semen from young bulls was higher when spermatogenesis occurred in summer compared with other seasons, whereas with frozen semen, the lowest percentage was observed when spermatogenesis occurred in winter. In older bulls, minor differences in total motility of fresh sperm were observed between the seasons when spermatogenesis occurred, and total motility after thawing was highest when spermatogenesis occurred in summer (Llamas-Luceño et al. 2020). Thus, according to a previous study conducted in Ireland (Murphy et al. 2018), the differences between seasons were relatively small and, although statistically significant, might not be biologically meaningful or have a commercial impact. Moreover, the response to elevated THI varied according to the age group, with young bulls tending to decrease with THI in summer. In contrast, all of the semen quality variables in older bulls tend to decrease with THI in autumn (Llamas-Luceño et al. 2020).

Therefore, breeding or semen collection should be limited to the most favorable season of the year for good sperm quality, keeping in mind that spermatogenesis takes 2 months to complete.

Another important factor is to provide adequate feed to cattle to optimize production under adverse climatic conditions, since the energy balance is closely related to their fertility (Krishnan et al. 2017). It has been stated that animals subjected to mild to severe heat stress should receive an additional supplement to their maintenance requirements. Thus, using the type and intensity of panting as an index, increases the estimated maintenance energy requirements by up to 7% for rapid shallow breathing and 11–25% for deep open-mouth panting (National Research Council (US) Subcommittee on Environmental Stress 1981). Also, dietary supplements of vitamins, trace elements, and minerals can ameliorate the adverse effects of heat stress (Krishnan et al. 2017). Among nutrients, lipids have a crucial role in male fertility since they can be consumed as a source of energy and additionally, they are critical components of spermatozoa membranes (Santos et al. 2008). Dietary omega-3 supplementation or its precursors improved in vitro quality and motility parameters of fresh semen in Holstein bulls, although its effect was not evident in frozen-thawed semen (Gholami et al. 2011). Despite the suggested structural role of polyunsaturated fatty acids (PUFA) in the sperm membrane, incorporation of docosahexaenoic acid or DHA into sperm could enhance a number of actions leading to the prevention of early apoptosis. This could promote a greater number of viable spermatozoa in the ejaculate (Gholami et al. 2011). In a study conducted in Upper Egypt, in an arid and dry area, essentially that of the desert, in which rainfall is almost negligible and the maximum ambient temperature and relative humidity ranged from 42 to 46 °C and 18 to 33%, respectively, during summer days, supplementation with probiotics in the ration of beef bulls improved growth performance and nutrients digestibility under heat stress conditions. Such improvement was due to a positive effect on blood metabolite parameters as a physiological response (Kassab et al. 2017).

Management alternatives and allowing animals to behave in ways that aid physiological homeostasis, such as the strategic use of shade, wind protection, sprinklers, and summer ventilation, allowing access to clean cool surfaces to lie, should always be considered to help livestock handle adverse conditions (Krishnan et al. 2017; Morrell 2020).

Cow cooling strategies described mainly for dairy cattle follow four general thermodynamic principles of heat transfer (radiation, conduction, convection, and evaporative cooling) (Ji et al. 2020).

Roof and shading constructions have the purpose of minimizing the solar load (radiation), consequently, evaporative, convective, or conductive cooling will be more able to provide adequate cooling (Sparke et al. 2001). Barn design is directly related to the mitigation of this solar load. Shoshani and Hetzroni (2013) reported some barn construction designs: the sliding roof, the louvered roof, the open ridge roof, and the pagoda (capped gable) roof. These authors mention that the orientation, width, height, and slope of the roof should be adjusted for different local conditions, with the wind and sunlight direction being the two key conditions. A review of several research studies (Kamal et al. 2018) shows that cows under heat stress spent less time standing, less time around the drinking trough, and more time ruminating when they had access to a shaded area. This fact shows that lying on a cool surface alleviates heat stress in cows.

In fact, cows lose heat by conduction when lying down, as long as the surface in contact with them is below body temperature. Cows may even sacrifice comfort and hygiene to seek conduction cooling when they are stressed by heat (Ji et al. 2020). The use of different bedding materials for this purpose has been studied, thus the use of refrigerated waterbeds (with 4.5 °C water) effectively mitigates heat stress in lactating dairy cows by decreasing rectal temperature (1.0 °C), respiration rate (18 breaths/min), and increasing milk yield (5%) and dry matter intake (14%) compared with control animals (Perano et al. 2015); other studies support the use of sand beds as a solution to heat stress situations, as cows transfer their surplus heat to this material quite effectively (Radoń et al. 2014). These heat exchangers are a viable complement to systems that employ fans, misters, and evaporative cooling methods to mitigate the effects of heat stress on dairy cows (Ortiz et al. 2015).

Regarding convection, ventilation is important because it helps to maintain a less stressful environment and to keep air quality inside the barn. Two common dairy barn designs are naturally ventilated barns (which have open sides to allow air to circulate freely) and tunnel-ventilated barns, which have closed sides with one end open and the other end with fans to move air through the barn (Ji et al. 2020). Stowell et al. (2001) demonstrated that the performance of the tunnel-ventilated barns is slightly better than that of the naturally ventilated barns, as the temperature differential between the inlet and outlet is lower (–0.4 to –0.3 °C). Regardless of the barn's design, adequate airspeed near the cows is important to control heat stress, with 2–2.5 m/s airspeed at cow level being the general recommendation (University of Wisconsin-Madison Dairyland Initiative 2019).

Evaporative cooling systems differ in the size of the water droplets. Misting and fogging systems produce smaller water droplets, which may have a negative effect by increasing the ambient humidity, which can lead to health problems such as mastitis and lameness (Nienaber and Hahn 2007). Sprinkler systems wet the cow with large water droplets and, therefore, are less dependent on good water-holding capacity in the air than misting and fogging systems. When a cow gets wet with a sprinkler, the water droplets become in direct contact with the cow's skin and evaporate through the body heat, cooling the animal (Ji et al. 2020).

The use of these strategies is enhanced when they are combined. Some authors showed that water soaking systems are most effective when combined with air movement along the feeding bunks and standing areas (Toledo et al. 2021). In another study, the performance of cows in tunnel-ventilated free stall barns was compared to free stall barns with sprinklers and fans, with results showing that tunnel-ventilated barns can cool cows better than barns with sprinklers and fans, with 2.3% less decrease in milk production (Dikmen et al. 2020). Grant and Miner (2015) summarized the benefits of cooling facilities in several studies and concluded likewise that cooling systems with fans only or shade only were less effective than their combination with sprinklers, fogging, or misting systems. All these strategies can be therefore implemented with the objective of preventing reduced reproductive performance through heat stress.

Other strategies described to improve fertility during the summer, are for example, the use of reproductive hormones such as GnRH and intravaginal progesterone devices, which induce cyclicity and the development of normal corpora lutea, leading to better fertility in summer, compared to the fertility of cows without these treatments (De Rensis and Scaramuzzi 2003). Another approach is embryo transfer of embryos produced during the cool seasons and transferred during the summer, which is an effective tool to keep fertility during heat stress conditions, as it avoids the damage caused by hyperthermia to the oocyte (Baruselli et al. 2020). However, the impact of the heat altering the uterine environment for the transferred embryos is not avoided with this strategy.

Under heat stress conditions, formulating diets containing slightly higher concentrations of neutral detergent fiber and acid detergent fiber can help to minimize the risk of ruminal acidosis associated with heat stress conditions. In addition, the inclusion of fat in higher fiber diets may help maintain energy intake under heat conditions (Staples 2007; Moallem et al. 2010).

Similarly to that described in bulls, cows, and heifers in temperate parts of the world, the introduction of females of breeds derived from *Bos indicus* subspecies and native breeds from tropical countries is an alternative to control the negative effect of high-temperature conditions. It has been reported that cattle with shorter hair, larger diameter hair, and lighter coat color are more adapted to hot environments than those with longer coats and darker colors (Bernabucci et al. 2010). This phenotype has been found in *B. taurus* in tropical environments, and this dominant gene is associated with a higher sweating rate, lower rectal temperature, and lower respiratory rate in homozygous cattle in hot conditions (Mariasegaram et al. 2007). Thus, heat shock protein genes associated with thermotolerance have been used as markers in selection breeding programs (Krishnan et al. 2017). The association of polymorphisms in thermotolerant genes has been described in several breeds, such as *HSP90AB1* in indigenous Thai cattle (Deb et al. 2014), or *HSF1* gene, *HSP70* A1 A gene, and *HSBP1* in Chinese Holstein cattle (Li et al. 2011a, b; Wang et al. 2013).

3.3   Cold Stress

It seems that cold stress has always been less important than heat stress, but it is a major issue in certain areas of the world, where it causes important economic losses in cattle farming (Kim et al. 2023).

Cold stress is one of the primary factors that reduces growth and enhances mortality in animals, resulting in significant economic losses to livestock worldwide, especially in the long, cold winter of the northern hemisphere (Hu et al. 2021). A response to cold stress in cattle will be activated when THI is below 38. Consequences of this response are (1) inhibition of the digestive system, making the increase in dry matter intake unable to keep up with the increase in heat production, resulting in a decreased growth; (2) enhancement of the body's energy metabolism and stress hormones imbalance (insulin, growth hormone, glucocorticoid, triiodothyronine, and tetraiodothyronine), which disrupt the normal growth and development of the animal's body (Wang et al. 2023).

Calves are much less able to adequately thermoregulate than adults do (Silva and Bittar 2019). Their thermoneutral zone ranges from 15 to 25 °C (Davis and Drackley 1998) and a cold environment has drastic, negative effects on calf health and survival (Davis and Drackley 1998), Carstens in 1994, showed a mortality range from 10 to 30% at birth when temperatures were below –10 °C. Therefore, efficiency and success in raising dairy calves in cold climates depend on high-level feeding management, proper hygienic practices, and disease prevention starting soon after birth. Feeding blood serum-derived proteins and fructooligosaccharides in addition to milk or milk replacers significantly increases survivorship (98 vs 84%) in preweaned dairy calves exposed to cold weather conditions (Pineda et al. 2016). One of the keys to maximizing benefits and keeping calves healthy in these regions is to create an environment that minimizes cold stress, as it contributes to poor performance and increased morbidity and mortality rates in the first weeks of life (Hulbert and Moisá 2016).

Strategies to reduce cold stress consequences are: (1) to assure the calf access to a dry, well-bedded shelter that provides protection from wind and extreme environmental conditions; (2) adequate colostrum administration, as colostrum is the most important energy source for newborns (Silva and Bittar 2019); and (3) increasing the daily volume of milk or milk replacer (8 L/day) or increasing the solids (17.5%) content (Raeth-Knight et al. 2009), but taking care of keeping an adequate osmolarity of the milk replacer administered to the calves.

Beef cattle exposed to cold stress modify their behavior (decreased dry matter intake and increased time standing) and blood parameters (increased blood cortisol, non-esterified fatty acids, and decreased glucose), as well as their physiological response to maintain homeostasis, independently of the age (Kim et al. 2023). Finally, as beef cattle are traditionally raised outdoors, with exposure to natural and variable environmental conditions, they are especially vulnerable not only to extreme weather conditions (heat and cold) but also to quick changes in these conditions (Mader 2003).

As stated before in bulls, the optimal environmental temperature for sperm production ranged from 15 to 18 °C for the entire spermatogenesis period (Parkinson 1987), and in a manner similar to heat stress, cold extreme conditions affect sires' reproductive performance, sperm production, and semen quality. As with heat stress, alterations in ejaculate characteristics caused by deleterious events may not be observed immediately. Possible damage to sperm cells may be observed 2–4 weeks after exposure to cold stress. Moreover, low temperatures and windy conditions can easily increase feed requirements by 25–30% above normal maintenance needs. Cold and wind can cause bull infertility in the following breeding season, but in addition, there is another valuable aspect related to cold stress, which is scrotal frostbite.

Scrotal frostbite can temporarily or permanently affect bull fertility, depending on the severity of the frostbite, causing testicular damage and thus altering semen production and quality (Faulkner et al. 1967). Lack of wind protection and bedding will increase the possibility of frostbite damage to the scrotum and testes. Scrotal frostbite appears a few days after exposure to cold, causing inflammation and swelling, and interfering with the cooling and warming mechanism of the testicles. The heat generated by the inflammation directly affects the spermatozoa stored in the epididymis. The damage caused may result in temporary or permanent sterility in the most severe cases. A scab may appear on the lower part of the scrotum because of scarring, but the absence of a scab does not indicate that frostbite injury has not occurred. Frostbite damage in the testis and epididymis can cause tissue adhesions, affecting scrotal mobility and circulation. A study conducted in western Canada found that most cases of frostbite occurred in the months of January and February and that the severity of frostbite correlated with semen quality (Barth and Waldner 2002). Thus, mild frostbite (one or two scrotal scabs at the base of the scrotum <2 cm in diameter) had a low effect on semen quality, however, severe frostbite (one or two scabs greater than 2 cm in diameter), significantly affected semen quality. Scrotal frostbite occurs quite frequently (5.9%) in beef bulls in western Canada. There was a correlation between age and the percentage of bulls with frostbite. Scrotal frostbite had a lower incidence in younger than in mature bulls because their scrotums are not as pendulous, so they have fewer thermoregulation problems. This retrospective analysis was carried out on 2110 beef bulls (*Bos taurus*) subjected to BBSE from January 1, 1986, through April 30, 1999, by the Western College of Veterinary Medicine (WCVM) in Saskatchewan (Canada). In this study, semen quality was categorized as satisfactory (good mass motility, individual sperm motility of 60%, and percentage of normal sperm of 70%), questionable (fair mass motility, individual sperm motility of 40–59%, and percentage of normal sperm of 50–69%) or unsatisfactory (poor mass motility, individual sperm motility lesser than 40%, and percentage of normal sperm lesser than 50%), in order to evaluate the impact of both physical anomalies and time of year on semen quality. Results showed that the proportion of physically normal bulls with satisfactory semen quality increased as spring and summer approached. Thus, the percentage of bulls with satisfactory semen quality increased from 55.1% in January to 74.6% in June. In bulls categorized as questionable according to semen quality, the percentage of sperm with midpiece defects was significantly higher from January to May than in June. In this study, only 65.8% of bulls were classified as satisfactory potential breeders, which can be considered low, and that the authors justify due to the stress effect of severe winters in western Canada.

Animal models have been used to evaluate the effects of cold stress on male fertility, focusing primarily on the effects on testes and sperm quality. Thus, the seminiferous epithelial area decreases in the testes of male rats exposed to immersion in cold water (15 min for 20 or 50 days), as well as body weight gain (Retana-Márquez et al. 2014). A reduction in litter size was also observed in females copulated with males subjected to cold stress for 50 days. This chronic stress increased corticosterone plasma levels that were higher in males stressed for 20 days than in males stressed for 50 days. In turn, testosterone decreased in cold-stressed males and it was lower in males subjected to thermal stress for 50 days. These results confirm HPA axis activation and HPG axis inactivation as possible mediators of this alteration. This study showed that chronic cold stress caused germ cell losses in the testes and a decrease in the prostatic and seminal epithelium, possibly because of decreased testosterone, which affected fertility. Corticotrophin-releasing hormone (CRH) secreted by the hypothalamus affects body weight by suppressing appetitive and feeding behavior in stressed males. In addition, chronic stress disrupted but did not suppress sexual behavior in sexually experienced males. A similar study in male rats subjected to chronic stress by daily immersion in cold water for 3, 8, 20, or 50 consecutive days, also showed that chronic stress causes germ cell loss and alterations in spermatogenesis. (Juárez-Rojas et al. 2017). Seminiferous tubules of cold-stressed rat males showed several degenerative signs, such as vacuoles in the basal epithelium, with pyknotic signs. Moderate to severe exfoliation of degenerative germ cells was also observed in the tubular lumen (mainly round spermatids and primary spermatocytes), thus disrupting spermatogenesis. All alterations appeared gradually and were related to a decrease in testosterone concentration, as well as a stress-induced increase in corticosterone.

4   Conclusion

Stress exerts a significant influence on reproductive procedures in livestock, affecting both male and female fertility, short-, medium- and long-term exposures to stress. Understanding, the mechanisms through which stress impacts the reproductive processes, veterinarians and producers can design and implement strategies to minimize it and to optimize the reproductive outcomes. Further research on stress management and its specific effect on reproductive efficiency is essential to improve the productivity of livestock systems.

References

Abbas Z, Hu L, Fang H, Sammad A, Kang L, Brito LF, Xu Q, Wang Y (2020) Association analysis of polymorphisms in the 5′ flanking region of the HSP70 gene with blood biochemical parameters of lactating Holstein cows under heat and cold stress. Animals 10(11):2016. https://doi.org/10.3390/ani10112016

Abdallah M, Joone C, Edwards S, Das S, Cavalieri J (2022) Comparison of the initial ovarian response, the synchrony of oestrus and ovulation and chronic stress response after administration of 100 or 250 μg of GnRH to randomly cycling Bos indicus cattle. Aust Vet J 100(10):479–491. https://doi.org/10.1111/avj.13196

Abeni F, Petrera F, Le Cozler Y (2019) Effects of feeding treatment on growth rates, metabolic profiles and age at puberty, and their relationships in dairy heifers. Animal 13(5):1020–1029. https://doi.org/10.1017/S1751731118002422

Aboagye IA, Frimpong JK, Ahunu BK (2015) Effect of drought on cattle production in Northern Ghana: Case study of the West Mamprusi District. J Sci Technol 35(3):94–106

Abril-Sánchez S, Freitas-de-Melo A, Damián JP, Giriboni J, Villagrá-García A, Ungerfeld R (2017) Ejaculation does not contribute to the stress response to electroejaculation in sheep. Reprod Domest Anim 52:403–408. https://doi.org/10.1111/rda.12922

Aitken RJ (1995) Free radicals, lipid peroxidation and sperm function. Reprod Fertil Dev 7(4):659–668. https://doi.org/10.1071/rd9950659

Aitken RJ, Krausz C (2001) Oxidative stress, DNA damage and the Y chromosome. Reproduction (Cambridge, England) 122(4):497–506. https://doi.org/10.1530/rep.0.1220497

Aitken RJ, Gordon E, Harkiss D, Twigg JP, Milne P, Jennings Z, Irvine DS (1998) Relative impact of oxidative stress on the functional competence and genomic integrity of human spermatozoa. Biol Reprod 59(5):1037–1046. https://doi.org/10.1095/biolreprod59.5.1037

Aitken RJ, Gibb Z, Baker MA, Drevet J, Gharagozloo P (2016) Causes and consequences of oxidative stress in spermatozoa. Reprod Fertil Dev 28(1–2):1–10. https://doi.org/10.1071/RD15325

Allen T, Pan Y, Hayman R (1963) The effect of evaporative heat loss and body temperature in Zebu and Jersey heifers. Aust J Agric Res 14(4):580. https://doi.org/10.1071/ar9630580

Alvarez L, Zerby HN, Starkey JD, Hunter RD (2000) Effect of handling stress and transport on intramuscular lipid and physical properties of the longissimus muscle of beef heifers. Meat Sci 55(3):313–318

Alves MBR, de Andrade AFC, de Arruda RP, Batissaco L, Florez-Rodriguez SA, Lançoni R, de Almeida TG (2015) An efficient technique to detect sperm reactive oxygen species: the CellRox deep red fluorescent probe. Biochem Physiol 4, 15

Amann RP (1983) Endocrine changes associated with onset of spermatogenesis in Holstein bulls. J Dairy Sci 66(12):2606–2622. https://doi.org/10.3168/jds.S0022-0302(83)82135-3

Armstrong DV (1994) Heat stress interaction with shade and cooling. J Dairy Sci 77(7):2044–2050. https://doi.org/10.3168/jds.S0022-0302(94)77149-6

Arnott G, Roberts D, Rooke JA, Turner SP, Lawrence AB, Rutherford KM (2012) Board invited review: The importance of the gestation period for welfare of calves: maternal stressors and difficult births. J Anim Sci 90(13):5021–5034. https://doi.org/10.2527/jas.2012-5463

Bach A, Calsamiglia S, Stern MD (2005) Nitrogen metabolism in the rumen. J Dairy Sci 88(Suppl 1):E9–E21. https://doi.org/10.3168/jds.S0022-0302(05)73133-7

Ball L, Furman JW (1972) Electroejaculation of the bull. Bovine Pract 1972(7):46–48. https://doi.org/10.21423/bovine-vol1972no7p46-48

Banks S, King SA, Irvine DS, Saunders PT (2005) Impact of a mild scrotal heat stress on DNA integrity in murine spermatozoa. Reproduction (Cambridge, England) 129(4):505–514. https://doi.org/10.1530/rep.1.00531

Barker DJ (1998) In utero programming of chronic disease. Clin Sci (London, England : 1979) 95(2):115–128

Barnett JL, Coleman GJ, Hemsworth PH, Newman EA, Fewings-Hall S, Ziini C (1999) Tail docking and beliefs about the practice in the Victorian dairy industry. Aust Vet J 77(11):742–747. https://doi.org/10.1111/j.1751-0813.1999.tb12919.x

Barroso FG, Alados CL, Boza J (2000) Social hierarchy in the domestic goat: effect on food habits and production. Appl Anim Behav Sci 69(1):35–53. https://doi.org/10.1016/s0168-1591(00)00113-1

Barrozo D, Buzanskas ME, Oliveira JA, Munari DP, Neves HH, Queiroz SA (2012) Genetic parameters and environmental effects on temperament score and reproductive traits of Nellore cattle. Animal 6(1):36–40. https://doi.org/10.1017/S1751731111001169

Baruselli PS, Ferreira RM, Vieira LM, Souza AH, Bó GA, Rodrigues CA (2020) Use of embryo transfer to alleviate infertility caused by heat stress. Theriogenology 155:1–11. https://doi.org/10.1016/j.theriogenology.2020.04.028

Barth AD (2012) Case-based studies of infertility in bulls. In: American Association of Bovine Practitioners conference, pp 50–59. https://doi.org/10.21423/aabppro20123871

Barth AD (2013) Bull breeding soundness, 3rd edn. Western Canadian Association of Bovine Practitioners, Saskatoon, SK, p 163

Barth AD, Bowman PA (1994) The sequential appearance of sperm abnormalities after scrotal insulation or dexamethasone treatment in bulls. Can Vet J 35(2):93–102

Barth AD, Waldner CL (2002) Factors affecting breeding soundness classification of beef bulls examined at the Western College of Veterinary Medicine. Can Vet J 43(4):274–284

Bartolome JA, van Leeuwen JJ, Thieme M, Sa'filho OG, Melendez P, Archbald LF, Thatcher WW (2009) Synchronization and resynchronization of inseminations in lactating dairy cows with the CIDR insert and the Ovsynch protocol. Theriogenology 72(6):869–878. https://doi.org/10.1016/j.theriogenology.2009.06.008

Bauché F, Fouchard MH, Jégou B (1994) Antioxidant system in rat testicular cells. FEBS Lett 349(3):392–396. https://doi.org/10.1016/0014-5793(94)00709-8

Bedford JM (2004) Enigmas of mammalian gamete form and function. Biol Rev Camb Philos Soc 79(2):429–460. https://doi.org/10.1017/s146479310300633x

Beggs DS (2013) Veterinary bull breeding soundness evaluation. Australian Cattle Veterinarians, Queensland, Australia, p 106

Beggs D, Bertram J, Chenoweth P, Entwistle K, Fordyce G, Johnston H et al (2013) Veterinary bull breeding soundness evaluation. Australian Veterinary Association, Brisbane, QLD

Bernabucci U, Lacetera N, Baumgard LH, Rhoads RP, Ronchi B, Nardone A (2010) Metabolic and hormonal acclimation to heat stress in domesticated ruminants. Animal 4(7):1167–1183. https://doi.org/10.1017/S175173111000090X

Bewley JM, Peacock AM, Lewis O, Boyce RE (2008) Application of the 2001 NRC model to estimate nutrient requirements of Holstein cows in commercial dairy herds. J Dairy Sci 91(8):3214–3222

Bhongade MB, Prasad S, Jiloha RC, Ray PC, Mohapatra S, Koner BC (2015) Effect of psychological stress on fertility hormones and seminal quality in male partners of infertile couples. Andrologia 47(3):336–342. https://doi.org/10.1111/and.12268

Bilby TR, Baumgard LH, Collier RJ, Zimbelman RB, Rhoads ML (2008) Heat stress effects on fertility: Consequences and possible solutions. In: South Western nutritional conference

Blazquez NB, Mallard GJ, Wedd SR (1988) Sweat glands of the scrotum of the bull. J Reprod Fertil 83(2):673–677. https://doi.org/10.1530/jrf.0.0830673

Brandão AP, Cooke RF (2021) Effects of temperament on the reproduction of beef cattle. Animals 11(11):3325. https://doi.org/10.3390/ani11113325

Brickell JS, McGowan MM, Pfeiffer DU, Wathes DC (2009) Mortality in Holstein-Friesian calves and replacement heifers, in relation to body weight and IGF-I concentration, on 19 farms in England. Animal 3(8):1175–1182. https://doi.org/10.1017/S175173110900456X

Brito LF, Silva AE, Barbosa RT, Unanian MM, Kastelic JP (2003) Effects of scrotal insulation on sperm production, semen quality, and testicular echotexture in Bos indicus and Bos indicus x Bos taurus bulls. Anim Reprod Sci 79(1-2):1–15. https://doi.org/10.1016/s0378-4320(03)00082-

Brito LF, Silva AE, Barbosa RT, Kastelic JP (2004) Testicular thermoregulation in Bos indi-
    cus, crossbred and Bos taurus bulls: relationship with scrotal, testicular vascular cone and
    testicular morphology, and effects on semen quality and sperm production. Theriogenology
    61(2–3):511–528. https://doi.org/10.1016/s0093-691x(03)00231-0
Brown EJ, Vosloo A (2017) The involvement of the hypothalamopituitary-adrenocortical
    axis in stress physiology and its significance in the assessment of animal welfare in cattle.
    Onderstepoort J Vet Res 84(1):e1–e9. https://doi.org/10.4102/ojvr.v84i1.1398
Burnett TA, Madureira AM, Silper BF, Nadalin A, Tahmasbi A, Veira DM, Cerri RL (2014) Short
    communication: Factors affecting hair cortisol concentrations in lactating dairy cows. J Dairy
    Sci 97(12):7685–7690. https://doi.org/10.3168/jds.2014-8444
Byrne CJ, Fair S, English AM, Urh C, Sauerwein H, Crowe MA, Lonergan P, Kenny DA (2017)
    Effect of breed, plane of nutrition and age on growth, scrotal development, metabolite con-
    centrations and on systemic gonadotropin and testosterone concentrations following a
    GnRH challenge in young dairy bulls. Theriogenology 96:58–68. https://doi.org/10.1016/j.
    theriogenology.2017.04.002
Byrne CJ, Keogh K, Kenny DA (2023) Review: Role of early life nutrition in regulating sexual devel-
    opment in bulls. Animal 17(Suppl 1):100802. https://doi.org/10.1016/j.animal.2023.100802
Cammack KM, Mesa H, Lamberson WR (2006) Genetic variation in fertility of heat-stressed male
    mice. Theriogenology 66(9):2195–2201. https://doi.org/10.1016/j.theriogenology.2006.06.011
Capela L, Leites I, Romão R, Lopes-da-Costa L, Pereira RMLN (2022) Impact of heat stress
    on bovine sperm quality and competence. Animals 12(8):975. https://doi.org/10.3390/
    ani12080975
Carroll JA, Forsberg NE (2007) Influence of stress and nutrition on cattle immunity. Vet Clin North
    Am Food Anim Pract 23(1):105–149. https://doi.org/10.1016/j.cvfa.2007.01.003
Carstens GE (1994) Cold thermoregulation in the newborn calf. Vet Clin North Am Food Anim
    Pract 10(1):69–106. https://doi.org/10.1016/s0749-0720(15)30590-9
Carter PD, Hamilton PA, Dufty JH (1990) Electroejaculation in goats. Aust Vet J 67(3):91–93.
    https://doi.org/10.1111/j.1751-0813.1990.tb07712.x
Cartmill JA, El-Zarkouny SZ, Hensley BA, Rozell TG, Smith JF, Stevenson JS (2001) An
    alternative AI breeding protocol for dairy cows exposed to elevated ambient temperatures
    before or after calving or both. J Dairy Sci 84(4):799–806. https://doi.org/10.3168/jds.
    S0022-0302(01)74536-5
Casady R, Myers R, Legates J (1953) The effect of exposure to high ambient temperature on
    spermatogenesis in the dairy bull. J Dairy Sci 36(1):14–23. https://doi.org/10.3168/jds.
    s0022-0302(53)91449-0
Chagas LM, Bass JJ, Blache D, Burke CR, Kay JK, Lindsay DR, Lucy MC, Martin GB, Meier S,
    Rhodes FM, Roche JR, Thatcher WW, Webb R (2007) Invited review: New perspectives on the
    roles of nutrition and metabolic priorities in the subfertility of high-producing dairy cows. J
    Dairy Sci 90(9):4022–4032. https://doi.org/10.3168/jds.2006-852
Chantziaras I, Dewulf J, Van Limbergen T, Klinkenberg M, Palzer A, Pineiro C, Aarestrup
    Moustsen V, Niemi J, Kyriazakis I, Maes D (2018) Factors associated with specific health,
    welfare and reproductive performance indicators in pig herds from five EU countries. Prev Vet
    Med 159:106–114. https://doi.org/10.1016/j.prevetmed.2018.09.006
Chernenko O, Prishedko V, Chernenko O, Mylostyvyi R, Shulzhenko N, Bordunova O (2023)
    Comparison of morphometric and histological properties of testicles and sperm production in
    breeding bulls with different reaction to stress. Veterinarska Stanica 54(2):193–209
Choshniak I, McEwan-Jenkinson D, Blatchford DR, Peaker M (1982) Blood flow and cate-
    cholamine concentration in bovine and caprine skin during thermal sweating. Comparative
    biochemistry and physiology C: comparative pharmacology 71C(1):37–42. https://doi.
    org/10.1016/0306-4492(82)90007-7
Cojkic A, Morrell JM (2023) Animal welfare assessment protocols for bulls in artificial insemina-
    tion centers: requirements, principles, and criteria. Animals 13(5):942. https://doi.org/10.3390/
    ani13050942

Collier RJ, Dahl GE, VanBaale MJ (2006) Major advances associated with environmental effects on dairy cattle. J Dairy Sci 89(4):1244–1253. https://doi.org/10.3168/jds.S0022-0302(06)72193-2

Collier RJ, Zimbelman RB, Rhoads RP, Rhoads ML, Baumgard LH (2011) A re-evaluation of the impact of temperature humidity index (THI) and black globe humidity index (BGHI) on milk production in high producing dairy cows. In Western dairy management conference, March 2011, Reno, NV, USA, pp 113–125

Collier RJ, Renquist BJ, Xiao Y (2017) A 100-year review: stress physiology including heat stress. J Dairy Sci 100(12):10367–10380. https://doi.org/10.3168/jds.2017-13676

Constable PD, Hinchcliff KW, Done SH, Grünberg W, Radostits OM (2017) Veterinary medicine: a textbook of the diseases of cattle, horses, sheep, pigs and goats. Can Vet J. Elsevier. https://doi.org/10.1016/B978-0-7020-5246-0.00008-5

Cook CW, Hunsaker PL, Coffey RE, Archive I (1997) Management and organizational behavior. Irwin, Chicago, IL. Internet archive. https://archive.org/details/managementorgani00cook

Cooke RF, Bohnert DW, Cappellozza BI, Mueller CJ, Delcurto T (2012) Effects of temperament and acclimation to handling on reproductive performance of Bos taurus beef females. J Anim Sci 90(10):3547–3555. https://doi.org/10.2527/jas.2011-4768

Cooke RF, Moriel P, Cappellozza BI, Miranda VFB, Batista LFD, Colombo EA, Ferreira VSM, Miranda MF, Marques RS, Vasconcelos JLM (2019) Effects of temperament on growth, plasma cortisol concentrations and puberty attainment in Nelore beef heifers. Animal 13(6):1208–1213. https://doi.org/10.1017/S1751731118002628

Costa JHC, von Keyserlingk MAG, Weary DM (2016) Invited review: Effects of group housing of dairy calves on behavior, cognition, performance, and health. J Dairy Sci 99(4):2453–2467. https://doi.org/10.3168/jds.2015-10144

Costa VGG, Vieira AD, Schneider A, Rovani MT, Gonçalves PBD, Gasperin BG (2018) Systemic inflammatory and stress markers in cattle and sheep submitted to different reproductive procedures. Ciência Rural 48(12)

Crookshank HR, Elissalde MH, White RG, Clanton DC, Smalley HE (1979) Effect of transportation and handling of calves upon blood serum composition. J Anim Sci 48(3):430–435. https://doi.org/10.2527/jas1979.483430x

Dado-Senn B, Skibiel AL, Fabris TF, Dahl GE, Laporta J (2019) Dry period heat stress induces microstructural changes in the lactating mammary gland. PLoS One 14(9):e0222120. https://doi.org/10.1371/journal.pone.0222120

Dado-Senn B, Laporta J, Dahl GE (2020) Carry over effects of late-gestational heat stress on dairy cattle progeny. Theriogenology 154:17–23. https://doi.org/10.1016/j.theriogenology.2020.05.012

Dahl GE, Skibiel AL, Laporta J (2019) In utero heat stress programs reduced performance and health in calves. Vet Clin North Am Food Anim Pract 35(2):343–353. https://doi.org/10.1016/j.cvfa.2019.02.005

Dahl GE, Tao S, Laporta J (2020) Heat stress impacts immune status in cows across the life cycle. Front Vet Sci 7:116. https://doi.org/10.3389/fvets.2020.00116

Damián JP, Ungerfeld R (2011) The stress response of frequently electroejaculated rams to electroejaculation: hormonal, physiological, biochemical, haematological and behavioural parameters. Reprod Domest Anim 46(4):646–650. https://doi.org/10.1111/j.1439-0531.2010.01722.x

Darwin (1871) The descent of man, and Selection in relation to sex, vol 1, 1st edn, John Murray, London. https://doi.org/10.1037/12293-000

Davis CL, Drackley JK (1998) The development, nutrition, and management of the young calf. Iowa State University Press, Ames, IA

Deb R, Sajjanar B, Singh U, Kumar S, Singh R, Sengar G, Sharma A (2014) Effect of heat stress on the expression profile of Hsp90 among Sahiwal (Bos indicus) and Frieswal (Bos indicus × Bos taurus) breed of cattle: a comparative study. Gene 536(2):435–440. https://doi.org/10.1016/j.gene.2013.11.086

de Castro LS, de Assis PM, Siqueira AF, Hamilton TR, Mendes CM, Losano JD, Nichi M, Visintin JA, Assumpção ME (2016) Sperm oxidative stress is detrimental to embryo development: a dose-dependent study model and a new and more sensitive oxidative status evaluation. Oxid Med Cell Longev 2016:8213071. https://doi.org/10.1155/2016/8213071

de Lamirande E, Jiang H, Zini A, Kodama H, Gagnon C (1997) Reactive oxygen species and sperm physiology. Rev Reprod 2(1):48–54. https://doi.org/10.1530/ror.0.0020048

De Rensis F, Scaramuzzi RJ (2003) Heat stress and seasonal effects on reproduction in the dairy cow—a review. Theriogenology 60(6):1139–1151. https://doi.org/10.1016/s0093-691x(03)00126-2

De Rensis F, Garcia-Ispierto I, López-Gatius F (2015) Seasonal heat stress: Clinical implications and hormone treatments for the fertility of dairy cows. Theriogenology 84(5):659–666. https://doi.org/10.1016/j.theriogenology.2015.04.021

de Vries M, Bokkers EA, van Schaik G, Engel B, Dijkstra T, de Boer IJ (2014) Exploring the value of routinely collected herd data for estimating dairy cattle welfare. J Dairy Sci 97(2):715–730. https://doi.org/10.3168/jds.2013-6585

Dikmen S, Hansen PJ (2009) Is the temperature-humidity index the best indicator of heat stress in lactating dairy cows in a subtropical environment? J Dairy Sci 92(1):109–116. https://doi.org/10.3168/jds.2008-1370

Dikmen S, Larson CC, De Vries A, Hansen PJ (2020) Effectiveness of tunnel ventilation as dairy cow housing in hot climates: rectal temperatures during heat stress and seasonal variation in milk yield. Trop Anim Health Prod 52(5):2687–2693. https://doi.org/10.1007/s11250-020-02309-3

Diskin MG, Kenny DA (2016) Managing the reproductive performance of beef cows. Theriogenology 86(1):379–387. https://doi.org/10.1016/j.theriogenology.2016.04.052

Diskin MG, Morris DG (2008) Embryonic and early foetal losses in cattle and other ruminants. Reprod Domest Anim 43 Suppl 2:260–267. https://doi.org/10.1111/j.1439-0531.2008.01171.x

Dobson H, Smith RF (2000) What is stress, and how does it affect reproduction? Anim Reprod Sci 60-61:743–752. https://doi.org/10.1016/s0378-4320(00)00080-4

Dobson H, Tebble JE, Smith RF, Ward WR (2001) Is stress really all that important? Theriogenology 55(1):65–73. https://doi.org/10.1016/s0093-691x(00)00446-5

Dobson H, Ghuman S, Prabhakar S, Smith R (2003) A conceptual model of the influence of stress on female reproduction. Reproduction (Cambridge, England) 125(2):151–163. https://doi.org/10.1530/rep.0.1250151

Dogan S, Vargovic P, Oliveira R, Belser LE, Kaya A, Moura A, Sutovsky P, Parrish J, Topper E, Memili E (2015) Sperm protamine-status correlates to the fertility of breeding bulls. Biol Reprod 92(4):92. https://doi.org/10.1095/biolreprod.114.124255

Drews C (1993) The concept and definition of dominance in animal behaviour. Behaviour 125(3–4):283–313. https://doi.org/10.1163/156853993x00290

DuBois PR, Williams DJ (1980) Increased incidence of retained placenta associated with heat stress in dairy cows. Theriogenology 13(2):115–121. https://doi.org/10.1016/0093-691x(80)90120-x

Durairajanayagam D, Agarwal A, Ong C (2015) Causes, effects and molecular mechanisms of testicular heat stress. Reprod Biomed Online 30(1):14–27. https://doi.org/10.1016/j.rbmo.2014.09.018

Ealy AD, Drost M, Hansen PJ (1993) Developmental changes in embryonic resistance to adverse effects of maternal heat stress in cows. J Dairy Sci 76(10):2899–2905. https://doi.org/10.3168/jds.S0022-0302(93)77629-8

Edwards JL, Ealy AD, Hansen PJ (2013) Environmental heat stress alters transcript abundances of genes that encode for key metabolic enzymes and potential markers of oocyte competence in Bos taurus. J Dairy Sci 96(2):1039–1050

Endo N (2022) Possible causes and treatment strategies for the estrus and ovulation disorders in dairy cows. J Reprod Dev 68(2):85–89. https://doi.org/10.1262/jrd.2021-125

Estevez I, Andersen I-L, Nævdal E (2007) Group size, density and social dynamics in farm animals. Appl Anim Behav Sci 103(3–4):185–204. https://doi.org/10.1016/j.applanim.2006.05.025

Færevik G, Jensen MB, Bøe KE (2006) Dairy calves social preferences and the significance of a companion animal during separation from the group. Appl Anim Behav Sci 99(3–4):205–221

Falk AJ, Waldner CL, Cotter BS, Gudmundson J, Barth AD (2001) Effects of epidural lidocaine anesthesia on bulls during electroejaculation. Can Vet J 42(2):116–120

Fatehi AN, Bevers MM, Schoevers E, Roelen BA, Colenbrander B, Gadella BM (2006) DNA damage in bovine sperm does not block fertilization and early embryonic development but induces apoptosis after the first cleavages. J Androl 27(2):176–188. https://doi.org/10.2164/jandrol.04152

Faulkner LC, Hopwood ML, Masken JF, Kingman HE Sr, Stoddard HL (1967) Scrotal frostbite in bulls. J Am Vet Med Assoc 151(5):602–605

Fernandes CE, Dode MA, Pereira D, Silva AE (2008) Effects of scrotal insulation in Nellore bulls (Bos taurus indicus) on seminal quality and its relationship with in vitro fertilizing ability. Theriogenology 70(9):1560–1568. https://doi.org/10.1016/j.theriogenology.2008.07.005

Fernandez-Novo A, Pérez-Garnelo SS, Villagrá A, Pérez-Villalobos N, Astiz S (2020) The effect of stress on reproduction and reproductive technologies in beef cattle—a review. Animals 10(11):2096. https://doi.org/10.3390/ani10112096

Fiol C, Aguerre M, Carriquiry M, Ungerfeld R (2019) Social dominance affects intake rate and behavioral time budget in pre-pubertal dairy heifers allocated in continuous competitive situations. Animal 13(6):1297–1303. https://doi.org/10.1017/S1751731118002835

Fisher AD, Crowe MA, Enright WJ (2001) The effects of road transport on subsequent fertility of heifers. Ir Vet J 54(6):289–293

Francis RC (1988) On the relationship between aggression and social dominance. Ethology 78:223–237. https://doi.org/10.1111/j.1439-0310.1988.tb00233.x

Franklin TB, Russig H, Weiss IC, Gräff J, Linder N, Michalon A, Vizi S, Mansuy IM (2010) Epigenetic transmission of the impact of early stress across generations. Biol Psychiatry 68(5):408–415. https://doi.org/10.1016/j.biopsych.2010.05.036

Funston RN, Martin JL, Larson DM, Roberts AJ (2012) Physiology and endocrinology symposium: nutritional aspects of developing replacement heifers. J Anim Sci 90(4):1166–1171. https://doi.org/10.2527/jas.2011-4569

Fuquay JW (1981) Heat stress as it affects animal production. J Anim Sci 52(1):164–174. https://doi.org/10.2527/jas1981.521164x

Galil KA, Setchell BP (1988) Effects of local heating of the testis on testicular blood flow and testosterone secretion in the rat. Int J Androl 11(1):73–85. https://doi.org/10.1111/j.1365-2605.1988.tb01218.x

Galindo F, Broom D (2000) The relationships between social behaviour of dairy cows and the occurrence of lameness in three herds. Res Vet Sci 69(1):75–79. https://doi.org/10.1053/rvsc.2000.0391

Gantner V, Mijić P, Kuterovac K, Solić D, Gantner R (2011) Temperature-humidity index values and their significance on the daily production of dairy cattle. Mljekarstvo: časopis za unaprjeđenje proizvodnje i prerade mlijeka 61(1):56–63

García-Ispierto I, López-Gatius F, Santolaria P, Yániz JL, Nogareda C, López-Béjar M, De Rensis F (2006) Relationship between heat stress during the peri-implantation period and early fetal loss in dairy cattle. Theriogenology 65(4):799–807. https://doi.org/10.1016/j.theriogenology.2005.06.011

Garcia-Oliveros LN, de Arruda RP, Batissaco L, Gonzaga VHG, Nogueira VJM, Florez-Rodriguez SA, Almeida FDS, Alves MBR, Pinto SCC, Nichi M, Losano JDA, Kawai GKV, Celeghini ECC (2020) Heat stress effects on bovine sperm cells: a chronological approach to early findings. Int J Biometeorol 64(8):1367–1378

Gaucher J, Reynoird N, Montellier E, Boussouar F, Rousseaux S, Khochbin S (2010) From meiosis to postmeiotic events: the secrets of histone disappearance. FEBS J 277(3):599–604. https://doi.org/10.1111/j.1742-4658.2009.07504.x

Gaughan JB, Bonner S, Loxton I, Mader TL, Lisle A, Lawrence R (2010a) Effect of shade on body temperature and performance of feedlot steers. J Anim Sci 88(12):4056–4067. https://doi.org/10.2527/jas.2010-2987

Gaughan JB, Mader TL, Holt SM, Sullivan ML, Hahn GL (2010b) Assessing the heat tolerance of 17 beef cattle genotypes. Int J Biometeorol 54(6):617–627. https://doi.org/10.1007/s00484-009-0233-4

Geary TW, Dahlen CR, Zezeski AL (2021) 254 Effects of nutrition on bull fertility. J Anim Sci 99(Suppl 3):136. https://doi.org/10.1093/jas/skab235.249

Gernand E, König S, Kipp C (2019) Influence of on-farm measurements for heat stress indicators on dairy cow productivity, female fertility, and health. J Dairy Sci 102(7):6660–6671. https://doi.org/10.3168/jds.2018-16011

Gholami H, Chamani M, Towhidi A, Fazeli MH (2011) Improvement of semen quality in holstein bulls during heat stress by dietary supplementation of omega-3 fatty acids. Int J Fertil Steril 4(4):160–167

Goyal HO (1983) Histoquantitative effects of orchiectomy with and without testosterone enanthate treatment on the bovine epididymis. Am J Vet Res 44(6):1085–1090

Grandin T (1997) Assessment of stress during handling and transport. J Anim Sci 75(1):249–257. https://doi.org/10.2527/1997.751249x

Grant R, Miner WH (2015) Economic benefits of improved cow comfort. Novus, St. Charles, MO

Grelet C, Vanden Dries V, Leblois J, Wavreille J, Mirabito L, Soyeurt H, Franceschini S, Gengler N, Brostaux Y, HappyMoo Consortium, Dehareng F (2022) Identification of chronic stress biomarkers in dairy cows. Animal 16(5):100502. https://doi.org/10.1016/j.animal.2022.100502

Gunn RMC (1936) Artificial production of seminal ejaculation and the characters of the spermatozoa contained therein. Council for Scientific and Industrial Research Bulletin 94

Guzeloglu A, Ambrose JD, Kassa T, Diaz T, Thatcher MJ, Thatcher WW (2001) Long-term follicular dynamics and biochemical characteristics of dominant follicles in dairy cows subjected to acute heat stress. Anim Reprod Sci 66(1–2):15–34. https://doi.org/10.1016/s0378-4320(01)00082-3

Gwazdauskas FC (1985) Effects of climate on reproduction in cattle. J Dairy Sci 68(6):1568–1578. https://doi.org/10.3168/jds.S0022-0302(85)80995-4

Gwazdauskas FC, Thatcher WW, Wilcox CJ (1973) Physiological, environmental, and hormonal factors at insemination which may affect conception. J Dairy Sci 56(7):873–877. https://doi.org/10.3168/jds.S0022-0302(73)85270-1

Handa RJ, Burgess LH, Kerr JE, O'Keefe JA (1994) Gonadal steroid hormone receptors and sex differences in the hypothalamo-pituitary-adrenal axis. Horm Behav 28(4):464–476. https://doi.org/10.1006/hbeh.1994.1044

Hansen PJ (2009) Effects of heat stress on mammalian reproduction. Philos Trans R Soc Lond Ser B Biol Sci 364(1534):3341–3350. https://doi.org/10.1098/rstb.2009.0131

Hansen PJ (2014) Current and future assisted reproductive technologies for mammalian farm animals. Adv Exp Med Biol 752:1–22. https://doi.org/10.1007/978-1-4614-8887-3_1

Hansen PJ (2020) The incompletely fulfilled promise of embryo transfer in cattle-why aren't pregnancy rates greater and what can we do about it? J Anim Sci 98(11):skaa288. https://doi.org/10.1093/jas/skaa288

Heinrichs AJ, Heinrichs BS, Harel O, Rogers GW, Place NT (2005) A prospective study of calf factors affecting age, body size, and body condition score at first calving of holstein dairy heifers. J Dairy Sci 88(8):2828–2835. https://doi.org/10.3168/jds.S0022-0302(05)72963-5

Hemsworth PH, Smith K, Karlen MG, Arnold NA, Moeller SJ, Barnett JL (2011) The choice behaviour of pigs in a Y maze: effects of deprivation of feed, social contact and bedding. Behav Process 87(2):210–217. https://doi.org/10.1016/j.beproc.2011.03.007

Hemsworth PH, Mellor DJ, Cronin GM, Tilbrook AJ (2015) Scientific assessment of animal welfare. N Z Vet J 63(1):24–30. https://doi.org/10.1080/00480169.2014.966167

Hochereau-de Reviers MT (1976) Variation in the stock of testicular stem cells and in the yield of spermatogonial divisions in ram and bull testes. Andrologia 8(2):137–146. https://doi.org/10.1111/j.1439-0272.1976.tb02122.x

Höckel M, Vaupel P (2001) Tumor hypoxia: definitions and current clinical, biologic, and molecular aspects. J Natl Cancer Inst 93(4):266–276. https://doi.org/10.1093/jnci/93.4.266

Hoflack G, Van Soom A, Maes D, de Kruif A, Opsomer G, Duchateau L (2006) Breeding soundness and libido examination of Belgian Blue and Holstein Friesian artificial insemination bulls in Belgium and The Netherlands. Theriogenology 66(2):207–216. https://doi.org/10.1016/j.theriogenology.2005.11.003

Hu L, Brito LF, Abbas Z, Sammad A, Kang L, Wang D, Wu H, Liu A, Qi G, Zhao M, Wang Y, Xu Q (2021) Investigating the short-term effects of cold stress on metabolite responses and metabolic pathways in inner-Mongolia Sanhe Cattle. Animals 11(9):2493. https://doi.org/10.3390/ani11092493

Hughes HD, Carroll JA, Burdick Sanchez NC, Richeson JT (2014) Natural variations in the stress and acute phase responses of cattle. Innate immunity 20(8):888–896. https://doi.org/10.1177/1753425913508993

Hulbert LE, Moisá SJ (2016) Stress, immunity, and the management of calves. J Dairy Sci 99(4):3199–3216. https://doi.org/10.3168/jds.2015-10198

Ideta A, Hayama K, Kawashima C, Urakawa M, Miyamoto A, Aoyagi Y (2009) Subjecting holstein heifers to stress during the follicular phase following superovulatory treatment may increase the female sex ratio of embryos. J Reprod Dev 55(5):529–533. https://doi.org/10.1262/jrd.20209

Iida T, Mine S, Fujimoto H, Suzuki K, Minami Y, Tanaka Y (2002) Hypoxia-inducible factor-1alpha induces cell cycle arrest of endothelial cells. Genes Cells 7(2):143–149. https://doi.org/10.1046/j.1356-9597.2001.00512.x

Ji B, Banhazi T, Perano K, Ghahramani A, Bowtell L, Wang C, Li B (2020) A review of measuring, assessing and mitigating heat stress in dairy cattle. Biosyst Eng 199:4–26. https://doi.org/10.1016/j.biosystemseng.2020.07.009

Johnson L (1995) Efficiency of spermatogenesis. Microsc Res Tech 32(5):385–422. https://doi.org/10.1002/jemt.1070320504

Johnson HD, Vanjonack WJ (1976) Effects of environmental and other stressors on blood hormone patterns in lactating animals. J Dairy Sci 59(9):1603–1617. https://doi.org/10.3168/jds.S0022-0302(76)84413-X

Johnson L, Varner DD, Roberts ME, Smith TL, Keillor GE, Scrutchfield WL (2000) Efficiency of spermatogenesis: a comparative approach. Anim Reprod Sci 60-61:471–480. https://doi.org/10.1016/s0378-4320(00)00108-1

Johnston JD, Buckland RB (1976) Response of male holstein calves from seven sires to four management stresses as measured by plasma corticoid levels. Can J Anim Sci 56(4):727–732. https://doi.org/10.4141/cjas76-087

Juárez-Rojas L, Vigueras-Villaseñor RM, Casillas F, Retana-Márquez S (2017) Gradual decrease in spermatogenesis caused by chronic stress. Acta Histochem 119(3):284–291. https://doi.org/10.1016/j.acthis.2017.02.004

Kadzere C, Murphy MR, Silanikove N, Maltz E (2002) Heat stress in lactating dairy cows: a review. Livest Prod Sci 77(1):59–91. https://doi.org/10.1016/s0301-6226(01)00330-x

Kamal R, Dutt T, Patel M, Dey A, Bharti PK, Chandran PC (2018) Heat stress and effect of shade materials on hormonal and behavior response of dairy cattle: a review. Trop Anim Health Prod 50(4):701–706. https://doi.org/10.1007/s11250-018-1542-6

Karimi MT, Ghorbani GR, Kargar S, Drackley JK (2015) Late-gestation heat stress abatement on performance and behavior of Holstein dairy cows. J Dairy Sci 98(10):6865–6875. https://doi.org/10.3168/jds.2014-9281

Kassab A, Hamdon H, Mohammed A (2017) Impact of probiotics supplementation on some productive performance, digestibility coefficient and physiological responses of beef bulls under heat stress conditions. Egypt J Nutr Feeds 20(1):29–39. https://doi.org/10.21608/ejnf.2017.75102

Kastelic JP, Coulter GH, Cook RB (1995) Scrotal surface, subcutaneous, intratesticular, and intraepididymal temperatures in bulls. Theriogenology 44(1):147–152

Kastelic JP, Cook RB, Coulter GH (1996) Contribution of the scrotum and testes to scrotal and testicular thermoregulation in bulls and rams. J Reprod Fertil 108(1):81–85. https://doi.org/10.1530/jrf.0.1080081

Kastelic JP, Cook RB, Coulter GH (1997) Scrotal/testicular thermoregulation and the effects of increased testicular temperature in the bull. Vet Clin North Am Food Anim Pract 13(2):271–282. https://doi.org/10.1016/s0749-0720(15)30340-6

Kastelic JP, Thundathil JC, Brito LF (2012) Bull BSE and semen analysis for predicting bull fertility. In: Proceedings of the Society for Theriogenology annual conference, Baltimore, MD, USA, pp 277–287

Kastelic JP, Rizzoto G, Thundathil J (2018) Review: testicular vascular cone development and its association with scrotal thermoregulation, semen quality and sperm production in bulls. Animal 12(s1):s133–s141. https://doi.org/10.1017/S175173111800116

Kastelic JP, Wilde RE, Bielli A, Genovese P, Rizzoto G, Thundathil J (2019) Hyperthermia is more important than hypoxia as a cause of disrupted spermatogenesis and abnormal sperm. Theriogenology 131:177–181. https://doi.org/10.1016/j.theriogenology.2019.03.040

Kaushik K, Kaushal N, Mittal PK, Kalla NR (2019) Heat induced differential pattern of DNA fragmentation in male germ cells of rats. J Therm Biol 84:351–356. https://doi.org/10.1016/j.jtherbio.2019.07.021

Kim WS, Ghassemi Nejad J, Lee HG (2023) Impact of cold stress on physiological, endocrinological, immunological, metabolic, and behavioral changes of beef cattle at different stages of growth. Animals 13(6):1073. https://doi.org/10.3390/ani13061073

Kipp C, Brügemann K, Zieger P, Mütze K, Möcklinghoff-Wicke S, König S, Halli K (2021) Across-generation effects of maternal heat stress during late gestation on production, female fertility and longevity traits in dairy cows. J Dairy Res 88(2):147–153. https://doi.org/10.1017/S0022029921000327

Knol BW (1991) Stress and the endocrine hypothalamus-pituitary-testis system: a review. Vet Q 13(2):104–114. https://doi.org/10.1080/01652176.1991.9694292

Koenneker K, Schulze M, Pieper L, Jung M, Schmicke M, Beyer F (2023) Comparative assessment of the stress response of cattle to common dairy management practices. Animals 13(13):2115. https://doi.org/10.3390/ani13132115

Koziol JH, Armstrong CL (2018) Manual for breeding soundness examination of bulls, 2nd edn. Society for Theriogenology, Providence, RI. 146

Krishnan G, Bagath M, Pragna P, Vidya MK, Aleena J, Archana PR, Sejian V, Bhatta R (2017) Mitigation of the heat stress impact in livestock reproduction. In: Theriogenology. InTech, eBooks, London. https://doi.org/10.5772/intechopen.69091

Lacasse P, Ollier S, Lollivier V, Boutinaud M (2016) New insights into the importance of prolactin in dairy ruminants. J Dairy Sci 99(1):864–874. https://doi.org/10.3168/jds.2015-10035

Lahn BT (2020) Social dominance hierarchy: toward a genetic and evolutionary understanding. Cell Res 30(7):560–561. https://doi.org/10.1038/s41422-020-0347-0

Lamb R (1976) Relationship between cow behavior patterns and management systems to reduce stress. J Dairy Sci 59(9):1630–1636. https://doi.org/10.3168/jds.s0022-0302(76)84416-5

Landaeta-Hernández AJ, Meléndez P, Bartolomé J, Rae DO, Archbald LF (2013) Effect of biostimulation and social organization on the interval from calving to resumption of ovarian cyclicity in postpartum Angus cows. Theriogenology 79(7):1041–1044. https://doi.org/10.1016/j.theriogenology.2013.01.020

Laporta J, Ferreira FC, Ouellet V, Dado-Senn B, Almeida AK, De Vries A, Dahl GE (2020) Late-gestation heat stress impairs daughter and granddaughter lifetime performance. J Dairy Sci 103(8):7555–7568. https://doi.org/10.3168/jds.2020-18154

Lay DC Jr, Randel RD, Friend TH, Jenkins OC, Neuendorff DA, Bushong DM, Lanier EK, Bjorge MK (1997) Effects of prenatal stress on suckling calves. J Anim Sci 75(12):3143–3151. https://doi.org/10.2527/1997.75123143x

Le Neindre P, Trillat G, Sapa J, Ménissier F, Bonnet JN, Chupin JM (1995) Individual differences in docility in Limousin cattle. J Anim Sci 73(8):2249–2253. https://doi.org/10.2527/1995.7382249x

Li QL, Ju ZH, Huang JM, Li JB, Li RL, Hou MH, Wang CF, Zhong JF (2011a) Two novel SNPs in HSF1 gene are associated with thermal tolerance traits in Chinese Holstein cattle. DNA Cell Biol 30(4):247–254. https://doi.org/10.1089/dna.2010.1133

Li QL, Han J, Du F, Ju Z, Huang J, Wang J, Li R, Wang C, Zhong J (2011b) Novel SNPs in HSP70A1A gene and the association of polymorphisms with thermo tolerance traits and tissue specific expression in Chinese Holstein cattle. Mol Biol Rep 38(4):2657–2663. https://doi. org/10.1007/s11033-010-0407-5

Liu YX (2010) Temperature control of spermatogenesis and prospect of male contraception. Front Biosci (Scholar edition) 2(2):730–755. https://doi.org/10.2741/s97

Llamas-Luceño N, Hostens M, Mullaart E, Broekhuijse M, Lonergan P, Van Soom A (2020) High temperature-humidity index compromises sperm quality and fertility of Holstein bulls in temperate climates. J Dairy Sci 103(10):9502–9514. https://doi.org/10.3168/jds.2019-18089

Loor JJ, Dann HM, Everts RE, Oliveira R, Green CA, Guretzky NA, Rodriguez-Zas SL, Lewin HA, Drackley JK (2005) Temporal gene expression profiling of liver from periparturient dairy cows reveals complex adaptive mechanisms in hepatic function. Physiological genomics 23(2):217–226. https://doi.org/10.1152/physiolgenomics.00132.2005

López-Gatius F (2012) Factors of a noninfectious nature affecting fertility after artificial insemination in lactating dairy cows. A review. Theriogenology 77(6):1029–1041. https://doi. org/10.1016/j.theriogenology.2011.10.014

Lucio AC, Alves BG, Alves KA, Martins MC, Braga LS, Miglio L, Alves BG, Silva TH, Jacomini JO, Beletti ME (2016) Selected sperm traits are simultaneously altered after scrotal heat stress and play specific roles in in vitro fertilization and embryonic development. Theriogenology 86(4):924–933. https://doi.org/10.1016/j.theriogenology.2016.03.015

Lucy MC (2019) Stress, strain, and pregnancy outcome in postpartum cows. Anim Reprod 16(3):455–464. https://doi.org/10.21451/1984-3143-AR2019-0063

Lynch EM (2010) Characterisation of physiological and immune-related biomarkers of weaning stress in beef cattle. PhD thesis, National University of Ireland Maynooth

Mader TL (2003) Environmental stress in confined beef cattle. J Anim Sci 81(14_suppl_2):E110–E119. https://doi.org/10.2527/2003.8114_suppl_2E110x

Mader TL, Davis MS, Brown-Brandl T (2006) Environmental factors influencing heat stress in feedlot cattle. J Anim Sci 84(3):712–719. https://doi.org/10.2527/2006.843712x

Mahmood SA, Ijaz A, Ahmad N, Rehman HU, Zaneb H, Farooq U (2013) Studies on libido and serum testosterone concentration of Cholistani AI bulls under stress free and stressful seasons. J Anim Plant Sci 23(6):1491–1495

Maia AS, daSilva RG, Battiston Loureiro CM (2005) Sensible and latent heat loss from the body surface of Holstein cows in a tropical environment. Int J Biometeorol 50(1):17–22. https://doi. org/10.1007/s00484-005-0267-1

Maines MD (1997) The heme oxygenase system: a regulator of second messenger gases. Annu Rev Pharmacol Toxicol 37:517–554. https://doi.org/10.1146/annurev.pharmtox.37.1.517

Marden WG (1954) New advances in the electroejaculation of the bull. J Dairy Sci 37(5):556–561. https://doi.org/10.3168/jds.s0022-0302(54)91298-9

Mariasegaram M, Chase CC Jr, Chaparro JX, Olson TA, Brenneman RA, Niedz RP (2007) The slick hair coat locus maps to chromosome 20 in Senepol-derived cattle. Anim Genet 38(1):54–59. https://doi.org/10.1111/j.1365-2052.2007.01560.x

Martinez MF, McLeod B, Tattersfield G, Smaill B, Quirke LD, Juengel JL (2015) Successful induction of oestrus, ovulation and pregnancy in adult ewes and ewe lambs out of the breeding season using a GnRH+progesterone oestrus synchronisation protocol. Anim Reprod Sci 155:28–35. https://doi.org/10.1016/j.anireprosci.2015.01.010

Mashoodh R, Habrylo IB, Gudsnuk KM, Pelle G, Champagne FA (2018) Maternal modulation of paternal effects on offspring development. Proc Biol Sci 285(1874):20180118. https://doi. org/10.1098/rspb.2018.0118

Mashoodh R, Habrylo IB, Gudsnuk K, Champagne FA (2023) Sex-specific effects of chronic paternal stress on offspring development are partially mediated via mothers. Horm Behav 152:105357. https://doi.org/10.1016/j.yhbeh.2023.105357

McDowell R, Hooven N, Camoens J (1976) Effect of climate on performance of holsteins in first lactation. J Dairy Sci 59(5):965–971. https://doi.org/10.3168/jds.s0022-0302(76)84305-6

Mee JF (2008) Prevalence and risk factors for dystocia in dairy cattle: a review. Vet J (London, England: 1997) 176(1):93–101. https://doi.org/10.1016/j.tvjl.2007.12.032

Mellor DJ, Murray L (1989) Effects of tail docking and castration on behaviour and plasma corti-sol concentrations in young lambs. Res Vet Sci 46:387–391

Minton JE, Parsons KM (2012) Heat stress in cattle: Basic and applied aspects of physiological responses. In: Livestock production and climate change, pp 179–204

Miranda-de la Lama GC, Pascual-Alonso M, Guerrero A, Alberti P, Alierta S, Sans P, Gajan JP, Villarroel M, Dalmau A, Velarde A, Campo MM, Galindo F, Santolaria MP, Sañudo C, María GA (2013) Influence of social dominance on production, welfare and the quality of meat from beef bulls. Meat Sci 94(4):432–437. https://doi.org/10.1016/j.meatsci.2013.03.026

Moallem U, Altmark G, Lehrer H, Arieli A (2010) Performance of high-yielding dairy cows sup-plemented with fat or concentrate under hot and humid climates. J Dairy Sci 93(7):3192–3202. https://doi.org/10.3168/jds.2009-2979

Moberg GP (2000) Biological response to stress: implications for animal welfare. In: Moberg GP, Mench JA (eds) The biology of animal stress: basic principles and implications for animal wel-fare. CABI, New York, pp 1–21. https://doi.org/10.1079/9780851993591.0001

Moberg GP, Mench JA (2000) The biology of animal stress: basic principles and implications for animal welfare. CABI eBooks, New York. https://doi.org/10.1079/9780851993591.0000

Moggy MA, Pajor EA, Thurston WE, Parker S, Greter AM, Schwartzkopf-Genswein KS, Campbell JR, Windeyer MC (2017) Management practices associated with stress in cattle on western Canadian cow-calf operations: A mixed methods study. J Anim Sci 95(4):1836–1844. https://doi.org/10.2527/jas.2016.1310

Molinari PCC, Davidson BD, Laporta J, Dahl GE, Sheldon IM, Bromfield JJ (2023) Prepartum heat stress in dairy cows increases postpartum inflammatory responses in blood of lactating dairy cows. J Dairy Sci 106(2):1464–1474. https://doi.org/10.3168/jds.2022-22405

Monteiro APA, Tao S, Thompson IMT, Dahl GE (2016) In utero heat stress decreases calf sur-vival and performance through the first lactation. J Dairy Sci 99(10):8443–8450. https://doi.org/10.3168/jds.2016-11072

Morignat E, Gay E, Vinard JL, Calavas D, Hénaux V (2015) Quantifying the influence of ambient temperature on dairy and beef cattle mortality in France from a time-series analysis. Environ Res 140:524–534. https://doi.org/10.1016/j.envres.2015.05.001

Mormède P, Andanson S, Aupérin B, Beerda B, Guémené D, Malmkvist J, Manteca X, Manteuffel G, Prunet P, van Reenen CG, Richard S, Veissier I (2007) Exploration of the hypothalamic-pituitary-adrenal function as a tool to evaluate animal welfare. Physiol Behav 92(3):317–339. https://doi.org/10.1016/j.physbeh.2006.12.003

Mormède P et al (2011a) Molecular genetics of hypothalamic-pituitary-adrenal axis activity and function. Ann N Y Acad Sci 1220:127–136

Mormède P, Foury A, Barat P, Corcuff JB, Terenina E, Marissal-Arvy N, Moisan MP (2011b) Molecular genetics of hypothalamic-pituitary-adrenal axis activity and function. Ann N Y Acad Sci 1220:127–136. https://doi.org/10.1111/j.1749-6632.2010.05902.x

Morrell JM (2020) Heat stress and bull fertility. Theriogenology 153:62–67. https://doi.org/10.1016/j.theriogenology.2020.05.014

Mosure WL, Meyer RA, Gudmundson J, Barth AD (1998) Evaluation of possible methods to reduce pain associated with electroejaculation in bulls. Can Vet J 39(8):504–506

Mujitaba MA, Egerszegi I, Kútvölgyi G, Nagy S, Vass N, Bodó S (2022) Alternative opportunities to collect semen and sperm cells for ex situ in vitro gene conservation in sheep. Agriculture 12:2001. https://doi.org/10.3390/agriculture12122001

Munkelwitz R, Gilbert BR (1998) Are boxer shorts really better? A critical analysis of the role of underwear type in male subfertility. J Urol 160(4):1329–1333

Murphy EM, Kelly AK, O'Meara C, Eivers B, Lonergan P, Fair S (2018) Influence of bull age, ejaculate number, and season of collection on semen production and sperm motility param-eters in Holstein Friesian bulls in a commercial artificial insemination centre. J Anim Sci 96(6):2408–2418. https://doi.org/10.1093/jas/sky130

Mychasiuk R, Harker A, Ilnytskyy S, Gibb R (2013) Paternal stress prior to conception alters DNA methylation and behaviour of developing rat offspring. Neuroscience 241:100–105. https://doi.org/10.1016/j.neuroscience.2013.03.025

Napolitano F, Arney D, Mota-Rojas D, De Rosa G (2020) Reproductive technologies and animal welfare. Reprod Technol Anim 275–286. https://doi.org/10.1016/b978-0-12-817107-3.00017-5

Naranjo-Gómez JS, Uribe-García HF, Herrera-Sánchez MP, Lozano-Villegas KJ, Rodríguez-Hernández R, Rondón-Barragán IS (2021) Heat stress on cattle embryo: gene regulation and adaptation. Heliyon 7(3):e06570. https://doi.org/10.1016/j.heliyon.2021.e06570

National Research Council (2000) Nutrient requirements of beef cattle, 7th edn. National Academies Press, Washington, DC

National Research Council (US) Subcommittee on Environmental Stress (1981) Effect of environment on nutrient requirements of domestic animals. National Academies Press (US), Washington, DC

Ndama PH, Entwistle KW, Lindsay JA (1983) Effect of protected protein supplements on some testicular traits in Brahman cross bulls. Theriogenology 20(6):639–650. https://doi.org/10.1016/0093-691X(83)90184-X

Netherton JK, Robinson BR, Ogle RA, Gunn A, Villaverde AISB, Colyvas K, Wise C, Russo T, Dowdell A, Baker MA (2022) Seasonal variation in bull semen quality demonstrates there are heat-sensitive and heat-tolerant bulls. Sci Rep 12(1):15322. https://doi.org/10.1038/s41598-022-17708-9

Nichi M, Rijsselaere T, Losano J, Angrimani D, Kawai G, Goovaerts I, Van Soom A, Barnabe VH, De Clercq J, Bols P (2017) Evaluation of epididymis storage temperature and cryopreservation conditions for improved mitochondrial membrane potential, membrane integrity, sperm motility and in vitro fertilization in bovine epididymal sperm. Reprod Domest Anim 52(2):257–263. https://doi.org/10.1111/rda.12888

Nienaber JA, Hahn GL (2007) Livestock production system management responses to thermal challenges. Int J Biometeorol 52(2):149–157. https://doi.org/10.1007/s00484-007-0103-x

O'Donnell L (2015) Mechanisms of spermiogenesis and spermiation and how they are disturbed. Spermatogenesis 4(2):e979623. https://doi.org/10.4161/21565562.2014.979623

O'Driscoll K, Gleeson D, O'Brien B, Boyle L, Brophy P (2017) A review of behavioural and physiological adaptations to social isolation in the domestic pig. J Appl Anim Welf Sci 20(3):302–320

Ong CN, Shen HM, Chia SE (2002) Biomarkers for male reproductive health hazards: are they available? Toxicol Lett 134(1–3):17–30. https://doi.org/10.1016/s0378-4274(02)00159-5

Orihuela A (2000) Some factors affecting the behavioural manifestation of oestrus in cattle: a review. Appl Anim Behav Sci 70(1):1–16. https://doi.org/10.1016/s0168-1591(00)00139-8

Orihuela A, Aguirre V, Hernandez C, Flores-Perez I, Vazquez R (2009a) Effect of anesthesia on welfare aspects of hair sheep (Ovis aries) during electro-ejaculation. J Anim Vet 8:305–308

Orihuela A, Virginio A, Carlos A, Carlos EH, Vazquez R, Flores-Perez I (2009b) Breaking down the effect of electro-ejaculation on the serum cortisol response, heart and respiratory rates in hair sheep (Ovis aries). J Anim Vet Adv 8(10):1968–1972

Ortiz XA, Smith JF, Rojano F, Choi CY, Bruer J, Steele T, Schuring N, Allen J, Collier RJ (2015) Evaluation of conductive cooling of lactating dairy cows under controlled environmental conditions. J Dairy Sci 98(3):1759–1771. https://doi.org/10.3168/jds.2014-8583

Ouellet V, Laporta J, Dahl GE (2020) Late gestation heat stress in dairy cows: Effects on dam and daughter. Theriogenology 150:471–479. https://doi.org/10.1016/j.theriogenology.2020.03.011

Ouellet V, Boucher A, Dahl GE, Laporta J (2021) Consequences of maternal heat stress at different stages of embryonic and fetal development on dairy cows' progeny. Anim Front 11(6):48–56. https://doi.org/10.1093/af/vfab059

Overton TR, Waldron MR (2004) Nutritional management of transition dairy cows: strategies to optimize metabolic health. J Dairy Sci 87:E105–E119. https://doi.org/10.3168/jds.s0022-0302(04)70066-1

Pacheco SE, Houseman EA, Christensen BC, Marsit CJ, Kelsey KT, Sigman M, Boekelheide K (2011) Integrative DNA methylation and gene expression analyses identify DNA packaging and epigenetic regulatory genes associated with low motility sperm. PLoS One 6(6):e20280. https://doi.org/10.1371/journal.pone.0020280

Palmer CW (2005) Welfare aspects of theriogenology: investigating alternatives to electroejaculation of bulls. Theriogenology 64(3):469–479. https://doi.org/10.1016/j.theriogenology.2005.05.032

Palmer CW (2016) Management and breeding soundness of mature bulls. Vet Clin North Am Food Anim Pract 32(2):479–495. https://doi.org/10.1016/j.cvfa.2016.01.014

Palmer CW, Brito LF, Arteaga AA, Söderquist L, Persson Y, Barth AD (2005) Comparison of electroejaculation and transrectal massage for semen collection in range and yearling feedlot beef bulls. Anim Reprod Sci 87(1–2):25–31. https://doi.org/10.1016/j.anireprosci.2004.09.004

Parker GA (1998) Sperm competition and the evolution of ejaculates: towards a theory base. In: Birkhead TR, Møller AP (eds) Sperm competition and sexual selection. Cambridge University Press, Cambridge, pp 3–54

Parkinson TJ (1987) Seasonal variations in semen quality of bulls: correlations with environmental temperature. Vet Rec 120(20):479–482. https://doi.org/10.1136/vr.120.20.479

Partida JA, Olleta JL, Campo MM, Sañudo C, María GA (2007) Effect of social dominance on the meat quality of young Friesian bulls. Meat Sci 76(2):266–273. https://doi.org/10.1016/j.meatsci.2006.11.008

Paul C, Murray AA, Spears N, Saunders PT (2008) A single, mild, transient scrotal heat stress causes DNA damage, subfertility and impairs formation of blastocysts in mice. Reproduction (Cambridge, England) 136(1):73–84. https://doi.org/10.1530/REP-08-0036

Paul C, Teng S, Saunders PT (2009) A single, mild, transient scrotal heat stress causes hypoxia and oxidative stress in mouse testes, which induces germ cell death. Biol Reprod 80(5):913–919. https://doi.org/10.1095/biolreprod.108.071779

Perano KM, Usack JG, Angenent LT, Gebremedhin KG (2015) Production and physiological responses of heat-stressed lactating dairy cattle to conductive cooling. J Dairy Sci 98(8):5252–5261. https://doi.org/10.3168/jds.2014-8784

Pérez-Crespo M, Pintado B, Gutiérrez-Adán A (2008) Scrotal heat stress effects on sperm viability, sperm DNA integrity, and the offspring sex ratio in mice. Mol Reprod Dev 75(1):40–47. https://doi.org/10.1002/mrd.20759

Perry GA (2012) Physiology and endocrinology symposium: harnessing basic knowledge of factors controlling puberty to improve synchronization of estrus and fertility in heifers. J Anim Sci 90(4):1172–1182. https://doi.org/10.2527/jas.2011-4572

Pineda A, Ballou MA, Campbell JM, Cardoso FC, Drackley JK (2016) Evaluation of serum protein-based arrival formula and serum protein supplement (Gammulin) on growth, morbidity, and mortality of stressed (transport and cold) male dairy calves. J Dairy Sci 99(11):9027–9039. https://doi.org/10.3168/jds.2016-11237

Preston BT, Stevenson IR, Pemberton JM, Wilson K (2001) Dominant rams lose out by sperm depletion. Nature 409(6821):681–682. https://doi.org/10.1038/35055617

Pugliese M, Monti S, Biondi V, Marino G, Passantino A (2022) Flashing lights, dark shadows, and future prospects of the current European Legislation for a better traceability and animal health requirements for movements of small animal germinal products. Front Vet Sci 9:852894. https://doi.org/10.3389/fvets.2022.852894

Radoń J, Bieda W, Lendelová J, Pogran Š (2014) Computational model of heat exchange between dairy cow and bedding. Comput Electron Agric 107:29–37. https://doi.org/10.1016/j.compag.2014.06.006

Raeth-Knight M, Chester-Jones H, Hayes S, Linn J, Larson R, Ziegler D, Ziegler B, Broadwater N (2009) Impact of conventional or intensive milk replacer programs on Holstein heifer performance through six months of age and during first lactation. J Dairy Sci 92(2):799–809. https://doi.org/10.3168/jds.2008-1470

Rahman MB, Vandaele L, Rijsselaere T, Maes D, Hoogewijs M, Frijters A, Noordman J, Granados A, Dernelle E, Shamsuddin M, Parrish JJ, Van Soom A (2011) Scrotal insulation and its relationship to abnormal morphology, chromatin protamination and nuclear shape of spermatozoa in Holstein-Friesian and Belgian Blue bulls. Theriogenology 76(7):1246–1257. https://doi.org/10.1016/j.theriogenology.2011.05.031

Rahman MB, Kamal MM, Rijsselaere T, Vandaele L, Shamsuddin M, Van Soom A (2014) Altered chromatin condensation of heat-stressed spermatozoa perturbs the dynamics of DNA methylation reprogramming in the paternal genome after in vitro fertilisation in cattle. Reprod Fertil Dev 26(8):1107–1116. https://doi.org/10.1071/RD13218

Rahman MB, Schellander K, Luceño NL, Van Soom A (2018) Heat stress responses in spermatozoa: Mechanisms and consequences for cattle fertility. Theriogenology 113:102–112. https://doi.org/10.1016/j.theriogenology.2018.02.012

Ren L, Li X, Weng Q, Trisomboon H, Yamamoto T, Pan L, Watanabe G, Taya K (2010) Effects of acute restraint stress on sperm motility and secretion of pituitary, adrenocortical and gonadal hormones in adult male rats. J Vet Med Sci 72(11):1501–1506. https://doi.org/10.1292/jvms.10-0113

Residiwati G, Tuska HSA, Budiono, Kawai GKV, Seifi-Jamadi A, Santoro D, Leemans B, Boccart C, Pascottini OB, Opsomer G, Van Soom A (2020) Practical methods to assess the effects of heat stress on the quality of frozen-thawed Belgian Blue semen in field conditions. Anim Reprod Sci 221:106572. https://doi.org/10.1016/j.anireprosci.2020.106572

Retana-Márquez S, Vigueras-Villaseñor RM, Juárez-Rojas L, Aragón-Martínez A, Torres GR (2014) Sexual behavior attenuates the effects of chronic stress in body weight, testes, sexual accessory glands, and plasma testosterone in male rats. Horm Behav 66(5):766–778. https://doi.org/10.1016/j.yhbeh.2014.09.002

Reynolds CK, Aikman PC, Lupoli B, Humphries DJ, Beever DE (2003) Splanchnic metabolism of dairy cows during the transition from late gestation through early lactation. J Dairy Sci 86(4):1201–1217. https://doi.org/10.3168/jds.S0022-0302(03)73704-7

Rhoads ML, Rhoads RP, VanBaale MJ, Collier RJ, Sanders SR, Weber WJ, Crooker BA, Baumgard LH (2009) Effects of heat stress and plane of nutrition on lactating Holstein cows: I. Production, metabolism, and aspects of circulating somatotropin. J Dairy Sci 92(5):1986–1997. https://doi.org/10.3168/jds.2008-1641

Rivera RM, Hansen PJ (2001) Development of cultured bovine embryos after exposure to high temperatures in the physiological range. Reproduction (Cambridge, England) 121(1):107–115

Rivier C, Rivest S (1991) Effect of stress on the activity of the hypothalamic-pituitary-gonadal axis: peripheral and central mechanisms. Biol Reprod 45(4):523–532. https://doi.org/10.1095/biolreprod45.4.523

Rizzoto G, Boe-Hansen G, Klein C, Thundathil JC, Kastelic JP (2020) Acute mild heat stress alters gene expression in testes and reduces sperm quality in mice. Theriogenology 158:375–381. https://doi.org/10.1016/j.theriogenology.2020.10.002

Rodríguez-Sánchez JA, Sanz A, Tamanini C, Casasús I (2015) Metabolic, endocrine, and reproductive responses of beef heifers submitted to different growth strategies during the lactation and rearing periods. J Anim Sci 93(8):3871–3885. https://doi.org/10.2527/jas.2015-8994

Roman-Ponce H, Thatcher WW, Caton D, Barron DH, Wilcox CJ (1978) Thermal stress effects on uterine blood flow in dairy cows. J Anim Sci 46(1):175–180. https://doi.org/10.2527/jas1978.461175x

Roth Z (2018) Symposium review: Reduction in oocyte developmental competence by stress is associated with alterations in mitochondrial function. J Dairy Sci 101(4):3642–3654. https://doi.org/10.3168/jds.2017-13389

Roth Z (2020) Reproductive physiology and endocrinology responses of cows exposed to environmental heat stress—experiences from the past and lessons for the present. Theriogenology 155:150–156. https://doi.org/10.1016/j.theriogenology.2020.05.040

Roth Z, Arav A, Bor A, Zeron Y, Braw-Tal R, Wolfenson D (2001a) Improvement of quality of oocytes collected in the autumn by enhanced removal of impaired follicles from previously heat-stressed cows. Reproduction (Cambridge, England) 122(5):737–744

Roth Z, Meidan R, Shaham-Albalancy A, Braw-Tal R, Wolfenson D (2001b) Delayed effect of heat stress on steroid production in medium-sized and preovulatory bovine follicles. Reproduction (Cambridge, England) 121(5):745–751

Russell JB, O'Connor JD, Fox DG, Van Soest PJ, Sniffen CJ (1992) A net carbohydrate and protein system for evaluating cattle diets: I. Ruminal fermentation. J Anim Sci 70(11):3551–3561. https://doi.org/10.2527/1992.70113551x

Ryan DP, Boland MP (1991) Frequency of twin births among Holstein-Friesian cows in a warm dry climate. Theriogenology 36(1):1–10. https://doi.org/10.1016/0093-691x(91)90428-g

Sabés-Alsina M, Johannisson A, Lundeheim N, Lopez-Bejar M, Morrell JM (2017) Effects of season on bull sperm quality in thawed samples in northern Spain. Vet Rec 180(10):251. https://doi.org/10.1136/vr.103897

Sakatani M, Alvarez NV, Takahashi M, Hansen PJ (2012) Consequences of physiological heat shock beginning at the zygote stage on embryonic development and expression of stress response genes in cattle. J Dairy Sci 95(6):3080–3091. https://doi.org/10.3168/jds.2011-4986

Sammad A, Umer S, Shi R, Zhu H, Zhao X, Wang Y (2020) Dairy cow reproduction under the influence of heat stress. J Anim Physiol Anim Nutr 104(4):978–986. https://doi.org/10.1111/jpn.13257

Santos JE, Bilby TR, Thatcher WW, Staples CR, Silvestre FT (2008) Long chain fatty acids of diet as factors influencing reproduction in cattle. Reprod Domest Anim 43 Suppl 2:23–30. https://doi.org/10.1111/j.1439-0531.2008.01139.x

Sarsaifi K, Rosnina Y, Ariff MO, Wahid H, Hani H, Yimer N, Vejayan J, Win Naing S, Abas MO (2013) Effect of semen collection methods on the quality of pre- and post-thawed Bali cattle (Bos javanicus) spermatozoa. Reprod Domest Anim 48(6):1006–1012. https://doi.org/10.1111/rda.12206

Savoini G, Farina G, Prandi A (2014) Social stress in dairy cows: A review. Ital J Anim Sci 13(3):363–372

Schütz KE, Stookey JM, Cox NR (2006) Effects of mixing beef cows from two different social groups on agonistic behavior, stress, and reproductive performance. J Anim Sci 84(11):2891–2901

Scientific Advisory Committee on Animal Health and Welfare (2022) Available online: http://www.fawac.ie/media/fawac/content/publications/scientificreports/LitReviewEE200717.pdf. Accessed 25 June 2023

Setchell BP (1998) The Parkes lecture. Heat and the testis. J Reprod Fertil 114(2):179–194. https://doi.org/10.1530/jrf.0.1140179

Setchell BP, Waites GMH, Till AR (1964) Variations in flow of blood within the epididymis and testis of the sheep and rat. Nature 203(4942):317–318. https://doi.org/10.1038/203317b0

Shabankareh HK, Kohram H, Shahneh AZ (2018) Heat stress effects on sperm cells: A review. Anim Reprod Sci 194:35–45

Shaha C, Tripathi R, Mishra DP (2010) Male germ cell apoptosis: regulation and biology. Philos Trans R Soc Lond Ser B Biol Sci 365(1546):1501–1515. https://doi.org/10.1098/rstb.2009.0124

Shiraishi K, Matsuyama H, Takihara H (2012) Pathophysiology of varicocele in male infertility in the era of assisted reproductive technology. Int J Urol 19(6):538–550. https://doi.org/10.1111/j.1442-2042.2012.02982.x

Shoshani E, Hetzroni A (2013) Optimal barn characteristics for high-yielding Holstein cows as derived by a new heat-stress model. Animal 7(1):176–182. https://doi.org/10.1017/S1751731112001085

Silva FLM, Bittar CMM (2019) Thermogenesis and some rearing strategies of dairy calves at low temperature—a review. J Appl Anim Res 47(1):115–122. https://doi.org/10.1080/09712119.2019.1580199

Sinclair KD, Watkins AJ (2013) Parental diet, pregnancy outcomes and offspring health: metabolic determinants in developing oocytes and embryos. Reprod Fertil Dev 26(1):99–114. https://doi.org/10.1071/RD13290

Skinner JD, Louw GN (1966) Heat stress and spermatogenesis in Bos indicus and Bos taurus cattle. J Appl Physiol 21(6):1784–1790. https://doi.org/10.1152/jappl.1966.21.6.1784

Solano J, Galindo F, Orihuela A, Galina CS (2004) The effect of social rank on the physiological response during repeated stressful handling in Zebu cattle (Bos indicus). Physiol Behav 82(4):679–683. https://doi.org/10.1016/j.physbeh.2004.06.005

Solymosi N, Torma C, Kern A, Maróti-Agóts A, Barcza Z, Könyves L, Berke O, Reiczigel J (2010) Changing climate in Hungary and trends in the annual number of heat stress days. Int J Biometeorol 54(4):423–431. https://doi.org/10.1007/s00484-009-0293-5

Sørensen H, Lambrechtsen J, Einer-Jensen N (1991) Efficiency of the countercurrent transfer of heat and 133Xenon between the pampiniform plexus and testicular artery of the bull under in-vitro conditions. Int J Androl 14(3):232–240. https://doi.org/10.1111/j.1365-2605.1991.tb01085.x

Sparke EJ, Young BA, Gaughan JB, Holt M, Goodwin PJ (2001) Heat load in feedlot cattle. Meat and Livestock Australia, North Sydney, NSW

Stafford KJ (1995) Electroejaculation: a welfare issue? Surveillance 22:15–17

Staples CR (2007) Nutrient and feeding strategies to enable cows to cope with heat stress conditions. In: 22nd annual southwest nutrition & management conference, Tempe, AZ. University of Arizona, Tempe, AZ, pp 93–108

Steger K, Cavalcanti MC, Schuppe HC (2011) Prognostic markers for competent human spermatozoa: fertilizing capacity and contribution to the embryo. Int J Androl 34(6 Pt 1):513–527. https://doi.org/10.1111/j.1365-2605.2010.01129.x

Steinberger E (1971) Hormonal control of mammalian spermatogenesis. Physiol Rev 51(1):1–22. https://doi.org/10.1152/physrev.1971.51.1.1

Stevenson JS, Britt JH (2017) A 100-Year Review: practical female reproductive management. J Dairy Sci 12:10292–10313. https://doi.org/10.3168/jds.2017-12959

Stewart WC, Wang SY, Soergel DA, Allgood G (2013) Reducing drought risk with irrigation in eastern Washington dryland wheat cropping systems. J Environ Qual 42(1):195–206

Stowell RR, Gooch CA, Inglis S (2001) Performance of tunnel ventilation for freestall dairy facilities as compared to natural ventilation with supplemental cooling fans. In: Livestock environment VI, Proceedings of the 6th international symposium 2001. American Society of Agricultural and Biological Engineers, p 29

St-Pierre N, Cobanov B, Schnitkey G (2003) Economic losses from heat stress by US livestock industries. J Dairy Sci 86:E52–E77. https://doi.org/10.3168/jds.s0022-0302(03)74040-5

Stull C, Reynolds J (2008) Calf welfare. Vet Clin North Am Food Anim Pract 24(1):191–203. https://doi.org/10.1016/j.cvfa.2007.12.001

Surani MA (1998) Imprinting and the initiation of gene silencing in the germ line. Cell 93(3):309–312. https://doi.org/10.1016/s0092-8674(00)81156-3

Suttle NF (2010) Mineral nutrition of livestock, 4th edn. CABI, Wallingford. https://doi.org/10.1079/9781845934729.0000

Tao S, Dahl GE (2013) Invited review: heat stress effects during late gestation on dry cows and their calves. J Dairy Sci 96(7):4079–4093. https://doi.org/10.3168/jds.2012-6278b

Tao S, Monteiro AP, Thompson IM, Hayen MJ, Dahl GE (2012) Effect of late-gestation maternal heat stress on growth and immune function of dairy calves. J Dairy Sci 95(12):7128–7136. https://doi.org/10.3168/jds.2012-5697

Thatcher WW, Bilby TR, Bartolome JA, Silvestre F, Staples CR, Santos JE (2006) Strategies for improving fertility in the modern dairy cow. Theriogenology 65(1):30–44. https://doi.org/10.1016/j.theriogenology.2005.10.004

Toledo IM, Dahl GE, De Vries A (2021) Dairy cattle management and housing for warm environments. Livest Sci 255:104802. https://doi.org/10.1016/j.livsci.2021.104802

Turner TT (1995) On the epididymis and its role in the development of the fertile ejaculate. J Androl 16(4):292–298

Turner AI, Hemsworth PH, Tilbrooka AJ (2002) Susceptibility of reproduction in female pigs to impairment by stress and the role of the hypothalamo-pituitary-adrenal axis. Reprod Fertil Dev 14(5–6):377–391. https://doi.org/10.1071/rd02012

University of Wisconsin-Madison Dairyland Initiative (2019) Ventilation and cooling in adult cattle facilities. https://thedairylandinitiative.vetmed.wisc.edu

Valeanu S, Johannisson A, Lundeheim N, Morrell J (2015) Seasonal variation in sperm quality parameters in Swedish red dairy bulls used for artificial insemination. Livest Sci 173:111–118. https://doi.org/10.1016/j.livsci.2014.12.005

Van Eetvelde M, Opsomer G (2017) Innovative look at dairy heifer rearing: Effect of prenatal and post-natal environment on later performance. Reprod Domest Anim 52 Suppl 3:30–36. https://doi.org/10.1111/rda.13019

Van Eetvelde M, Kamal MM, Vandaele L, Opsomer G (2017) Season of birth is associated with first-lactation milk yield in Holstein Friesian cattle. Animal 11(12):2252–2259. https://doi.org/10.1017/S1751731117001021

Van Soest PJ (1994) Nutritional ecology of the ruminant, 2nd edn. Cornell University Press, Ithaca, NY

Viljoen S (1968) Cattle psychology. Rhodesian Farmer 38(50):16

Vogler CJ, Saacke RG, Bame JH, Dejarnette JM, McGilliard ML (1991) Effects of scrotal insulation on viability characteristics of cryopreserved bovine semen. J Dairy Sci 74(11):3827–3835. https://doi.org/10.3168/jds.S0022-0302(91)78575-5

Vogler CJ, Bame JH, DeJarnette JM, McGilliard ML, Saacke RG (1993) Effects of elevated testicular temperature on morphology characteristics of ejaculated spermatozoa in the bovine. Theriogenology 40(6):1207–1219. https://doi.org/10.1016/0093-691x(93)90291-c

von Borell E, Dobson H, Prunier A (2007) Stress, behaviour and reproductive performance in female cattle and pigs. Horm Behav 52(1):130–138. https://doi.org/10.1016/j.yhbeh.2007.03.014

Walker J, Perry G, Olsen K (2008a) Bull nutrition. SDSU Extension Extra Archives 88. https://openprairie.sdstate.edu/extension_extra/88

Walker SL, Smith RF, Jones DN, Routly JE, Dobson H (2008b) Chronic stress, hormone profiles and estrus intensity in dairy cattle. Horm Behav 53(3):493–501. https://doi.org/10.1016/j.yhbeh.2007.12.003

Walters AH, Saacke RG, Pearson RE, Gwazdauskas FC (2006) Assessment of pronuclear formation following in vitro fertilization with bovine spermatozoa obtained after thermal insulation of the testis. Theriogenology 65(6):1016–1028. https://doi.org/10.1016/j.theriogenology.2005.07.005

Wang Y, Huang J, Xia P, He J, Wang C, Ju Z, Li J, Li R, Zhong J, Li Q (2013) Genetic variations of HSBP1 gene and its effect on thermal performance traits in Chinese Holstein cattle. Mol Biol Rep 40(6):3877–3882. https://doi.org/10.1007/s11033-012-1977-1

Wang Y, Chen ZP, Hu H, Lei J, Zhou Z, Yao B, Chen L, Liang G, Zhan S, Zhu X, Jin F, Ma R, Zhang J, Liang H, Xing M, Chen XR, Zhang CY, Zhu JN, Chen X (2021) Sperm microRNAs confer depression susceptibility to offspring. Sci Adv 7(7):eabd7605. https://doi.org/10.1126/sciadv.abd7605

Wang S, Li Q, Peng J, Niu H (2023) Effects of long-term cold stress on growth performance, behavior, physiological parameters, and energy metabolism in growing beef cattle. Animals 13(10):1619. https://doi.org/10.3390/ani13101619

Ward WS (2010) Function of sperm chromatin structural elements in fertilization and development. Mol Hum Reprod 16(1):30–36. https://doi.org/10.1093/molehr/gap080

Wathes DC, Fenwick M, Cheng Z, Bourne N, Llewellyn S, Morris DG, Kenny D, Murphy J, Fitzpatrick R (2007) Influence of negative energy balance on cyclicity and fertility in the high producing dairy cow. Theriogenology 68(Suppl 1):S232–S241. https://doi.org/10.1016/j.theriogenology.2007.04.006

Watters RD, Johnson RW, Johnson LP, Ellersieck MR (2013) Short communication: Adrenocorticotropic hormone challenge of Holstein replacement heifers administered an oral lipid infusion or intravenous glucose. J Dairy Sci 96(9):5845–5849

Wei Y, Yang CR, Wei YP, Zhao ZA, Hou Y, Schatten H, Sun QY (2014) Paternally induced transgenerational inheritance of susceptibility to diabetes in mammals. Proc Natl Acad Sci USA 111(5):1873–1878. https://doi.org/10.1073/pnas.1321195111

Weiss WP (2006) Nutritional management of dairy cows during the transition period. J Dairy Sci 89(Suppl):E31–E39

Welsh TH Jr, Johnson BH (1981) Stress-induced alterations in secretion of corticosteroids, progesterone, luteinizing hormone, and testosterone in bulls. Endocrinology 109(1):185–190. https://doi.org/10.1210/endo-109-1-185

Welsh TH, Randel RD, Johnson BH (1979) Temporal relationships among peripheral blood concentrations of corticosteroids, luteinizing hormone and testosterone in bulls. Theriogenology 12(3):169–179. https://doi.org/10.1016/0093-691x(79)90082-7

Whitlock BK, Coffman EA, Coetzee JF, Daniel JA (2012) Electroejaculation increased vocalization and plasma concentrations of cortisol and progesterone, but not substance P, in beef bulls. Theriogenology 78(4):737–746. https://doi.org/10.1016/j.theriogenology.2012.03.020

Wiltbank MC, Baez GM, Garcia-Guerra A, Toledo MZ, Monteiro PL, Melo LF, Ochoa JC, Santos JE, Sartori R (2016) Pivotal periods for pregnancy loss during the first trimester of gestation in lactating dairy cows. Theriogenology 86(1):239–253. https://doi.org/10.1016/j.theriogenology.2016.04.037

Winton TS, Nicodemus MC, Harvey KM (2024) Stressors inherent to beef cattle management in the United States of America and the resulting impacts on production sustainability: a review. Ruminants 4(2):227–240. https://doi.org/10.3390/ruminants4020016

Wolfenson D, Lew BJ, Thatcher WW, Graber Y, Meidan R (1997) Seasonal and acute heat stress effects on steroid production by dominant follicles in cows. Anim Reprod Sci 47(1–2):9–19. https://doi.org/10.1016/s0378-4320(96)01638-7

Wolfenson D, Roth Z, Meidan R (2000) Impaired reproduction in heat-stressed cattle: basic and applied aspects. Anim Reprod Sci 60-61:535–547. https://doi.org/10.1016/s0378-4320(00)00102-0

Xavier MJ, Nixon B, Roman SD, Scott RJ, Drevet JR, Aitken RJ (2019) Paternal impacts on development: identification of genomic regions vulnerable to oxidative DNA damage in human spermatozoa. Hum Reprod (Oxford, England) 34(10):1876–1890. https://doi.org/10.1093/humrep/dez153

Yadav P, Kumar J, Chauhan N, Chouksey S (2022) Effect of heat stress on reproduction and semen production: a review. Pharma Innov 11:630–635

Yanagimachi R (1994) Mammalian fertilization. In: Knobil E, Neill JD (eds) The physiology of reproduction. Raven, New York, pp 189–317

Yoshinaga K, Toshimori K (2003) Organization and modifications of sperm acrosomal molecules during spermatogenesis and epididymal maturation. Microsc Res Tech 61(1):39–45. https://doi.org/10.1002/jemt.10315

Zhang M, Jiang M, Bi Y, Zhu H, Zhou Z, Sha J (2012) Autophagy and apoptosis act as partners to induce germ cell death after heat stress in mice. PLoS One 7(7):e41412. https://doi.org/10.1371/journal.pone.0041412

Zhu BK, Setchell BP (2004) Effects of paternal heat stress on the in vivo development of preimplantation embryos in the mouse. Reprod Nutr Dev 44(6):617–629. https://doi.org/10.1051/rnd:2004064

Zhu B, Walker SK, Oakey H, Setchell BP, Maddocks S (2004) Effect of paternal heat stress on the development in vitro of preimplantation embryos in the mouse. Andrologia 36(6):384–394. https://doi.org/10.1111/j.1439-0272.2004.00635.x

The Effect of Stress on Equine Reproduction and Welfare

Gabriel Carreira Lencioni ⓘ, Ana Carolina Dierings Montec hese ⓘ, Yatta Boakari ⓘ, Maria Augusta Alonso ⓘ, Claudia Barbosa Fernandes ⓘ, and Amy Katherine McLean ⓘ

Abstract

Stress originating from various sources such as physical, metabolic, immuno-logical, and psychological factors can interact with and impair reproduction at all levels, from central systems in the brain to the functioning of reproductive organs. Though these effects have been demonstrated in both males and females in a range of species, stallions, and mares are still constantly exposed to stressful experiences and procedures, which may negatively affect their performance and predispose future offspring to negative health and behavioral outcomes. In don-keys, stress may relate to jacks not wanting to breed and dystocia in jennies. In this chapter, the effects of stress on stallions and mares' reproductive physiology, fetal programming, and development of foals' health and behavior will be dis-cussed, along with prevention or possible approaches to minimize these factors prior to and during breeding procedures, foaling, and weaning processes.

G. C. Lencioni · A. C. D. Montechese
Department of Preventive Veterinary Medicine and Animal Health, School of Veterinary Medicine and Animal Science (FMVZ), University of São Paulo, Sao Paulo, SP, Brazil

Y. Boakari
Department of Large Animal Clinical Sciences, College of Veterinary Medicine and Biomedical Sciences, Texas A & M University, College Station, TX, USA
e-mail: yboakari@tamu.edu

M. A. Alonso · C. B. Fernandes
Department of Animal Reproduction, School of Veterinary Medicine and Animal Science, University of Sao Paulo, Sao Paulo, SP, Brazil
e-mail: fernandescb@usp.br

A. K. McLean (✉)
Department of Animal Science, University of California Davis, Davis, CA, USA
e-mail: acmclean@ucdavis.edu

© The Author(s), under exclusive license to Springer Nature Switzerland AG 2024
J. C. Gardón, K. Satué Ambrojo (eds.), *Assisted Reproductive Technologies in Animals Volume 1*, https://doi.org/10.1007/978-3-031-73079-5_7

Keywords

Equine · Stress · Reproduction · Glucocorticoids · Corticosteroids · Pain · Distress
· Learning theory · Positive reinforcement

1   Introduction

Stress is a complex and significant factor influencing reproductive health in equines,
yet its precise impact and mechanisms remain relatively understudied. Understanding
how stressors such as exercise, transportation, pain, and social stress affect reproductive processes in horses is essential for optimizing breeding management and
enhancing equine welfare. Stress-induced release of glucocorticoids can disrupt the
delicate balance of reproductive hormones at various levels of the hypothalamic-
pituitary-gonadal axis, potentially leading to reduced fertility and reproductive inef-
ficiency. This paper aims to explore the interactions between stress and equine
reproduction, highlighting the physiological responses to stressors and their implications for reproductive success in horses. By elucidating these relationships, we
can identify strategies to mitigate stress-related reproductive disturbances and promote optimal breeding outcomes in equine.

2   Overall Effects of Stress on Reproduction

The effects of stress on reproductive physiology share a general pattern of inhibition
(Tilbrook et al. 2002; Joseph and Whirledge 2017; McCosh et al. 2022; Whirledge
and Cidlowski 2017; Whirledge and Cidlowski 2010; Rooney and Domar 2018;
Valsamakis et al. 2019), though the intensity of the impact appears to depend on the
nature of stress, for how long the stressor is maintained (Luo et al. 2016; Gao et al.
2016). Additionally, in females, the stage of the estrous cycle at which the exposure
to stress happens also makes a difference (Tilbrook et al. 2002). The interaction
between stress responses and reproduction is mediated by various factors such as
glucocorticoids (Tilbrook et al. 2000; Whirledge and Cidlowski 2010), catecholamines, endogenous opioids, and reproductive steroids (Valsamakis et al. 2019;
Tilbrook et al. 2002) which are modulated via multiple interacting pathways activated by the exposure to stressors.

   The stress response at the central nervous system broadly consists of rapid activation of the sympathetic adrenomedullary system (SAM) and relatively slower
engagement of the hypothalamic-pituitary-adrenal axis (HPA), which interact to
build both an immediate and a longer-lasting response to perceived stressors (Rasiah
et al. 2023; Godoy et al. 2018). The result of the mobilization of the SAM and HPA
axes is a cascade of systemic changes involving cardiovascular, immunologic, metabolic, reproductive, and digestive processes (Wadsworth et al. 2019). Additionally,
this mobilization requires a great amount of energy (Herman et al. 2016), which is
eventually resolved through mechanisms such as glucocorticoid-mediated feedback

loops once the stressor is no longer relevant in order to reestablish a neutral state (Rasiah et al. 2023; Godoy et al. 2018). Among the physiological circuits triggered by stressors, the HPA axis holds the most potential for influence over reproductive functions due to shared neuroanatomical features and hormonal sensibility (Wingfield and Sapolsky 2003; Whirledge and Cidlowski 2010).

Upon its activation following exposure to stressors, the HPA axis recruitment begins at the paraventricular nucleus of the hypothalamus (PVN) with the engagement of parvocellular neurosecretory cells (PNCs), which release corticotropin-releasing hormone (CRH) and other hypophysiotropic factors into the hypophyseal portal system. CRH stimulates the corticotroph cells in the anterior pituitary gland to produce and secrete proopiomelanocortin (POMC), which is converted into adrenocorticotropic hormone (ACTH) and released into circulation. Finally, in the *zona fasciculata* of the adrenal cortex, ACTH stimulates the secretion of corticosteroids, of which cortisol or corticosterone is the most widely regarded "stress hormone" (Rasiah et al. 2023; Godoy et al. 2018).

Corticosteroids are the main executors of HPA axis modulations of physiology, acting on processes such as glucose metabolism, cardiovascular functions, behavior, immunity, and reproduction. The pattern of HPA-mediated changes is a shift toward a catabolic state, increasing energy availability through favoring gluconeogenesis while also generally suppressing other processes that are considered nonessential to immediate survival, such as the functionality of the immune system and reproductive physiology. Corticosteroids also regulate the HPA axis through a classic endocrine feedback loop, inhibiting further release of CRH and ACTH (Rasiah et al. 2023; Godoy et al. 2018).

The HPA and HPG axes interact at various points, starting with their shared central circuitry (Valsamakis et al. 2019; Tilbrook et al. 2002; Whirledge and Cidlowski 2017). The CRH and POMC peptides are capable of inhibiting the secretion of gonadotropin-releasing hormone (GnRH) and its receptor (GnRHR) in the hypothalamus and anterior pituitary gland (Valsamakis et al. 2019), impairing further steps of the HPG axis such as the release of the gonadotropins luteinizing hormone (LH) and, to a smaller extent, follicle-stimulating hormone (FSH) (Wagenmaker and Moenter 2017; Barroso et al. 2001) from gonadotropes in the pituitary gland. ACTH administration also seems capable of disrupting LH secretion in response to GnRH (Dobson and Smith 2000; Biran et al. 2015).

Corticosteroids exert inhibitory effects on all levels of the HPG axis (Whirledge and Cidlowski 2010). Elevated circulating levels of corticosteroids can inhibit GnRH release from the hypothalamus and gonadotropins release from the pituitary through direct action (Whirledge and Cidlowski 2010, 2017) or mediation of other inhibitory neuropeptides, such as kisspeptin (KISS1) (McCosh et al. 2022) and gonadotropin-inhibitory hormone (GnIH) (Iwasa et al. 2017). These interactions will have downstream effects of impairment of steroid hormone secretion by the gonads, decreasing their sensitivity to gonadotropins and circulating levels of testosterone, estradiol, and progesterone (Whirledge and Cidlowski 2010, 2017). Moreover, CRH and corticosteroids can also act on testicular, ovarian, and uterine tissues, impairing processes such as spermatogenesis, ovulation, embryo

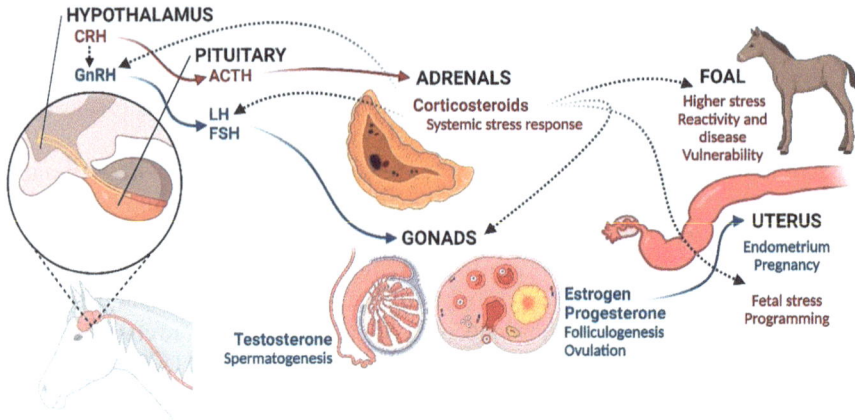

Fig. 1 Schematic representation of the HPA and HPG axes and their potential interactions, from shared neural circuitry to systemic effects and epigenetic programming phenomena. Text and arrows in red represent components and effects of the HPA axis; text and arrows in blue represent components and effects of the HPG axis and reproduction; dotted arrows in black represent points of influence from the HPA axis over the HPG axis and reproductive processes (created with BioRender.com)

implantation, placentation, and fetal programming, shaping the health and behavior of the offspring (Whirledge and Cidlowski 2010, 2017). A general representation of these interactions is presented in Fig. 1.

3 Effects on Stallions and Jacks

Improving our understanding of stress and how it affects stallions and jacks may help minimize the impact on reproduction. Further analyzing how stallions and jacks are managed daily as well as in the breeding shed may decrease stress, especially if modifications that respect their natural behavior are made. We know both are managed and bred extremely differently than expected in natural settings. Most breeding facilities keep both stallions and jacks in isolation, without any possibility of socialization, which is a fundamental need in the equine species (Fig. 2). Jacks may be found in a variety of management settings, including with other jacks, with a companion, or in solitude. Before being bred, many stallions come from performance backgrounds such as racing or showing. When the stallion is competing, it is not acceptable for him to behave like a stallion, and they are often reprimanded for such behavior. These management practices can lead to stress responses that may have a reproductive impact.

Fig. 2 Males turned out in a group to socialize

3.1   Impact on Libido

Many factors can play into the libido of stallions and jacks. Libido is closely linked to circulating testosterone, which can be decreased by the release of glucocorticoids when an animal is stressed. Management and training are key to helping decrease stress and decrease endocrine responses associated with stress, such as decreased testosterone and improved libido. When performance stallions are transitioned to the breeding phase of their career, counter-conditioning and retraining are often required. Classical conditioning, the association of a place or event, can easily be used to help decrease stress, increase circulating levels of testosterone, and improve the outcome. When being handled, it is not uncommon for stallions to be dealt with a chain placed across their nose or lips to help control or restrain them during the breeding or collection process. However, how such equipment is used can greatly determine the success and amount of stress in the breeding shed. Low libido is not uncommon in stallions and jacks, which can be related to high circulating levels of cortisol, which may block or reduce testosterone. This may be improved with exposure to more artificial light, exercise, a good nutrition program, and positive interactions in the breeding shed or process. If the stallion has been constantly reprimanded with positive punishment prior to breeding, then the stallion handler will have to retrain the stallion with classical conditioning that behaving like a stallion is acceptable and desired behavior. Otherwise, such stallions will have difficulty in achieving an erection and completing the process of ejaculation. Jacks may have low libido or decreased sexual desire based on how they are managed and teasing routines. Teasing can help increase libido by increasing testosterone and decreasing cortisol and estrogen if the mare or jenny is receptive to the jack or stallion (Fig. 3). More shy or timid breeders may improve their libido through visual stimulation and watching a more seasoned stallion breed. Other factors to consider are creating a routine when entering the breeding area, or shed and positive interactions. It's natural for a stallion to possibly paw, strike, vocalize, and express the flehmen response. Jacks, like stallions, are stimulated by visual, auditory, and olfactory senses. Tactile

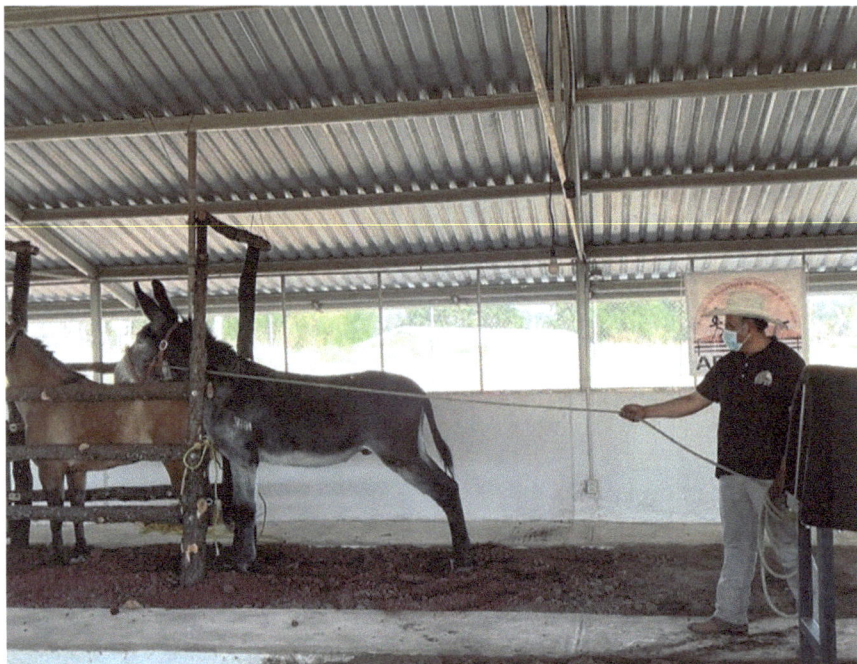

Fig. 3 Donkey being allowed to tease receptive female in a long line, with no chains being used

contact is additionally important for jacks when teasing. It's not uncommon for a jack to even kick out with a hind leg during the teasing and breeding process. Again, this behavior should be allowed and not prevented or punished to encourage a healthy libido and decrease stress associated with the process. Such behaviors should be rewarded and not punished. Over time, the stallion and jack will eventually learn that a breeding shed, or area, is a pleasurable and positive area, and libido will increase as cortisol decreases and time to erection will likely shorten. Other considerations when teasing include limiting the amount of teasing by breeding stallions and jacks. Both that are constantly used for teasing but never allowed to breed can also decrease libido due to the increased stress of constantly teasing mares but never allowed to breed.

Other factors associated with stress and decreased libido can be a change in where stallions and jacks are collected or bred. In some cases, a stallion or jack may be transported to the mares, or pasture breeding may be used, and after several unsuccessful attempts, a young stallion or jack may become disinterested. Different management strategies might need to be implemented to mitigate such a decrease in libido.

3.2 Spermatogenesis

Many stressors could affect sperm production, such as the stress of an infection (immune response) or increased body temperature for exercising stallions that increase body temperature, leading to lower or decreased sperm production. Nutrition can create metabolic stress that can reduce sperm production as well. Psychological stress may lead to an imbalance in hormones that could alter spermatogenesis. When cortisol levels are high, it is not uncommon to find high FSH or interstitial stimulating hormone levels. An imbalance in ISH can decrease testosterone levels and alter LH, which would, in turn, affect spermatozoa production and maturation. Stallions that are still training and exercising may have reduced spermatogenesis due to an increase in testicular temperature and an increase in stress associated with competing from travel to new environments, stress of competition, and constant environmental change and challenges. Stallions and jacks that lack stimulation and express signs of anxiety or stereotypic type behavior in their stalls and paddocks may have reduced sperm concentrations and reduced circulating GnRH. Frequency of collection and breeding may impact spermatogenesis and sperm concentration, but a regular collection or breeding schedule may reduce stress for some stallions; the same is true for access to turn out, exercise, and companion equine.

3.3 Acute and Chronic Stress Effects

Acute stress in stallions and jacks may be related to pain and nociceptive pain associated with the breeding process. Many performance horses enter the breeding phase due to a career-ending injury, but how the injury is maintained and treated may translate to decreased libido and reproductive efficiency related to acute stress. Once the pain is managed, the stallion or jack may continue without stress. Other examples of acute stress could be travel, immune responses, and anxiety. The change in the equine's environment is highly stressful, especially for a stallion who would naturally have a harem of mares and a jack who would live in a territorial social structure. The constant change can sometimes lead to frustrating behavior and even self-mutilation, where stallions bite at their flanks. Prolonged pain or anxiety can lead to many changes in the equids' body, from the limbic system to spermatozoa production, with the continued high levels of circulating stress hormones. One direct result is a change in heart rate and respiration rate. When stress is found to be chronic, many stallions will adapt by decreasing their libido, and a reduced quality of life may be associated with it (Hernandez-Avalos et al. 2021).

4   Effects on Mares and Jennies

Mares and jennies, depending on their use, may commonly be exposed to stressors in their daily lives, such as sports-related work or production (e.g., dairy donkeys) activities. Mares that are still competing and are being used for embryo transfer or ovum pick-up may experience the stress of competition, travel, new environments, and procedures related to being in estrus and breeding. Jennies that are working equids and/or dairy donkeys may experience the stress of carrying heavy loads, nursing a foal on her side, or having a foal separated for periods of time for milk purposes while maintaining a pregnancy and lactation. Each equid experiences different forms of stress that may impact their reproduction efficiency with high circulating levels of stress hormones creating an imbalance in gonadotrophic hormones and delaying folliculogenesis. Stress in the female equid, like in the male, can engage stress responses such as sympathetic pathways and the HPA axis (Malschitzky et al. 2015; Aurich and Aurich 2008; Berghold et al. 2007). Examples include physical demands of exercise and competition, physical restraint, transportation, separation from the social group, gynecological procedures, and human interferences around parturition (Melchert et al. 2019). Effects of stress over reproduction in females have been demonstrated in women, mice, rats (Whirledge and Cidlowski 2010, 2017), ewes, cows, pigs (Tilbrook et al. 2002), and mares (Malschitzky et al. 2015; Smith et al. 2012; Kelley et al. 2011; Mortensen et al. 2009), and though results suggest conserved mechanisms and consistent patterns of interaction between stress responses and female reproductive physiology across species (Tilbrook et al. 2002), some findings from studies in mares have been contradictory (Berghold et al. 2007; Baucus et al. 1990a, b).

4.1   Estrous Cycle and Fertility

Exposure to stressors is commonly capable of greatly altering the estrous cycle and fertility in women, rodents, and cattle (Whirledge and Cidlowski 2010; Tilbrook et al. 2002), though results have not been as consistent in mares when considering artificial and physiological increases in stress response mediators (Berghold et al. 2007; Hedberg et al. 2007). The administration of ACTH to intact and ovariectomized mares can stimulate the secretion of progesterone, testosterone, and cortisol, which may in turn impair the secretion of gonadotropins through feedback loops (Hedberg et al. 2007). Dexamethasone, which is a synthetic glucocorticoid, is also capable of reducing circulating LH concentrations, follicle size, and ovulation rate in intact mares (Asa and Ginther 1982) and suppressing sexual behavior in ovariectomized mares (Asa et al. 1980).

Several studies have investigated the potential effects of known stressors in mares on their fertility, especially transportation. In studies using different distances and durations of transportation as a stressor in mares, results vary between no effect over circulating concentrations of LH, estradiol, and progesterone and no impairment of estrous behavior and ovulation (Baucus et al. 1990a) to conflicting effects over concentrations of LH, FSH, and progesterone (Oikawa et al. 1996). When transportation was combined with other factors such as social stress, gynecological examinations, and artificial insemination protocols, researchers also failed to detect detrimental effects on mares' fertility parameters, such as duration of estrous and ovulation and pregnancy rates, even in mares with no previous history of exposure to breeding procedures who presented highest levels of cortisol (Berghold et al. 2007). It has been suggested that these results, which contrast with the findings from other species, may be attributed to the timing of the exposure to stressors during the cycle (Baucus et al. 1990a) or exposure to stressors which generate an insufficient mobilization of stress responses which does not have the ability to modulate the HPG axis, particularly regarding the increase of circulating cortisol (Berghold et al. 2007). It is also important to note that these contradictory findings from the mare studies may also have been due to the use of heterogeneous sample populations, by using mares ranging from 3 to 15 years of age without detailing their previous history (Baucus et al. 1990a), from different breeds (Berghold et al. 2007) or very low sample sizes (Oikawa et al. 1996). Further investigations into this phenomenon are still warranted, as previously highlighted by other authors (Campbell & Sandøe, 2015).

Exercise can also represent a stressor in horses and generate circulating cortisol increases which reflect its intensity and duration (Gordon et al. 2007; Hyyppä 2005; Williams et al. 2002), which have been reported to affect fertility parameters in mares (Kelley et al. 2009; Mortensen et al. 2009). Mares exposed to exercise may present increased interovulatory intervals and time from deviation to ovulation, decreased maximum circulating LH concentrations, a rise of estradiol levels earlier within the cycle (Kelley et al. 2011), ovulation from smaller follicles, delays in the effects of prostaglandin, and fewer and poorer quality embryos (Mortensen et al. 2009; Smith et al. 2012). Mares who were entered into more races during the year of mating or those racing after their first mating have been found to have worst foaling outcomes, though mares entered in up to ten races during the mating year or those who raced before mating had improvements in their fertility rates (Sairanen et al. 2011). The interaction between exercise and reproduction impairments is attributed to mediation by heat stress (Kelley et al. 2009; Mortensen et al. 2009), which is generated by exercise in horses (Mortensen et al. 2009) and upregulates the HPA axis and sympathetic nervous system (Gonzalez-Rivas et al. 2020; Wendt et al. 2007). Indeed, the effects caused by exercise on mares' estrous cycles and embryo recovery rates are compatible with reports from heat-stressed cows, suggesting a shared mechanism (Mortensen et al. 2009) (Fig. 4).

Investigations into the potential beneficial effects of positive management and handling techniques are still scarce, but one study has found that breeding mares kept under conditions considered less stressful, such as the absence of teasing practices, reduced or absent stall confinement, and maintenance of stable social groups of five to ten mares throughout the breeding season presented better fertility parameters, especially for barren mares. Mares managed in conditions that promoted better welfare, had lower rates of embryonic death, fewer embryo losses in foal heat pregnancies, and barren mares had higher pregnancy rates (Malschitzky et al. 2015).

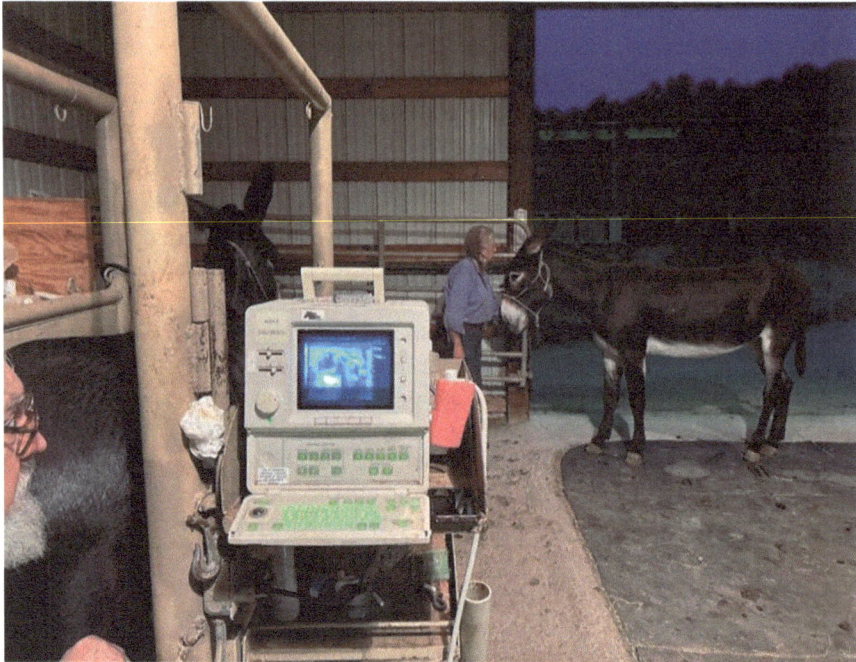

Fig. 4 Picture of a jenny contained in stocks for transrectal palpation and ultrasound exam with a companion within sight to reduce stress

Additionally, small management changes, such as having a companion female within sight during transrectal palpation in stocks, can reduce stress, especially in females who are not accustomed to being restrained in stocks.

4.2 Gestation and Foaling

It has been established that corticosteroids can modulate the uterine environment before and during gestation, influencing processes ranging from embryo implantation to fetal growth and epigenetic programming of offspring (Valsamakis et al. 2019; Whirledge and Cidlowski 2017; Kofman 2002; Maccari et al. 2003; Babenko et al. 2015; Rosenfeld 2021). These hormones may impair the growth and differentiation of uterine tissues induced by estrogen, resulting in reduced receptivity of the uterus to embryo implantation and the proper establishment and maintenance of pregnancy (Whirledge and Cidlowski 2017).

In pregnant mares, exposure to transportation stress at 3 or 5 weeks may induce changes in circulating levels of cortisol and progesterone but is not associated with early embryonic loss (EEL) (Baucus et al. 1990b), while heat stress seems capable of causing EEL due to the increased vulnerability of the embryo to environmental challenges in early pregnancy (Yu et al. 2022). Pain, psychogenic (emotional response) stress, and administration of exogenous steroids such as glucocorticoids have also been found to decrease plasma progestagen levels in mares and may be implicated in pregnancy loss during the first trimester of pregnancy (van Niekerk and Morgenthal 1982; Vaala and Sertich 1994). In later stages of pregnancy, exposure of mares to transportation stress may contribute to the advancement of parturition (Nagel et al. 2020), and disturbances during parturition, such as moving mares to an unfamiliar environment, may prolong the expulsive phase of foaling (Melchert et al. 2019) (Fig. 5).

Exposure to stress during pregnancy may also severely impact fetal development and long-term epigenetic programming (Kofman 2002; Maccari et al. 2003; Babenko et al. 2015; Rosenfeld 2021), resulting in higher stress reactivity (Dadds et al. 2015) and impaired functionality of multiple systems in the offspring (Maccari et al. 2003; Babenko et al. 2015; Collins et al. 2024). Though corticosteroids are critical to adequate development at various stages and tissues, including cardiovascular, respiratory, and nervous systems, the excessive supply to the fetus may be toxic and is usually prevented by protective mechanisms active in placental tissues, such as the conversion of excess cortisol to the inactive form cortisone by hydroxysteroid 11-beta dehydrogenase 2 (11HSDB2) (Davis et al. 2011; Appleton et al. 2015). Elevated levels of corticosteroids in the mother's system may exceed the protective capacity of the placenta and reach the developing fetus (Appleton et al. 2015), and this mechanism seems consistent even across species with different types of placentation, including sows which, like the mare, form epitheliochorial and diffuse placentas (Sarmiento et al. 2023). This phenomenon may potentially

Fig. 5 Mare post-foaling showing a relaxed demeanor by lying down next to the foal in a familiar environment

induce long-lasting epigenetic modulations which are often maladaptive and tend to predispose to higher vulnerability to stress as well as several pathological conditions (Oberlander et al. 2008; Dadds et al. 2015). There is also an interplay between maternal stress during pregnancy and the onset of maternal immune activation (MIA), which is also detrimental to development and may permanently affect several aspects of neural programming, immunity, and overall health (Goldstein et al. 2020; Minakova and Warner 2018).

4.3 Lactation

Ensuring proper milk supply to the foal (Fig. 6) is imperative for adequate development, especially during the first month of life, when foal growth is closely associated with milk intake (Schmidt et al. 2010). Exposure to stressors such as transportation to a novel environment (De Palo et al. 2018; Bruckmaier and Wellnitz 2008) and elevated environment temperatures (Tao et al. 2020) can impair processes involved in lactation, especially by interfering with the oxytocin supply (Bruckmaier and Wellnitz 2008) which is critical for contraction of myoepithelial cells in the mammary gland to promote milk ejection (Dewey 2001). Although these effects have not yet been investigated in mares, they have been proposed (Auclair-Ronzaud

Fig. 6 It is important that the foal be allowed to nurse and be close to the dam to reduce stress and guarantee normal development

et al. 2022) and previously observed in donkey jennies (De Palo et al. 2018). The disruption of lactation by stress can be corrected by habituation to stressors, which may be conducted by implementing a routine of gradually introducing the animals to any novel environments and procedures while also avoiding creating negative associations (e.g., minimizing stress and pain), resuming the adequate release of oxytocin (De Palo et al. 2018; Bruckmaier and Wellnitz 2008).

In dairy donkeys, cortisol is measured in their hair. Jennies in their first month of peri-parturition and lactation showed higher cortisol levels compared to jennies at 6 and 10 months. In dairy donkeys, the jennies must keep their foal on their side to continue to produce milk, yet the foal is removed for several hours each day for milking purposes. This study suggests that jennies are more stressed based on cortisol levels in their hair in the first month suggesting that jennies tested later in lactation may become adapted to the removal of their foal for several hours and the milking process. Understanding the level of stress may have a direct impact on their reproductive health, performance, and as well as sympathetic response (Salari et al. 2022).

4.4   Mare–Foal Relationship

In early life, the maternal figure occupies a vital role as the primary link between the environment and the developing offspring. Exposure of dams to stress has the capacity to interfere with their responsivity to offspring demands and patterns of maternal interaction, favoring the development of higher stress reactivity phenotypes in offspring and negatively affecting the development of cognition, emotional regulation, and health (Cameron et al. 2005; Champagne and Meaney 2006). Interestingly, the opposite is also true, and adequate patterns of maternal interaction can exert protective effects over offspring that may have been otherwise negatively programmed and predisposed to higher stress vulnerability (Cameron et al. 2005).

The bond between mares and foals has been extensively characterized by close proximity (Houpt 2002; Barber and Crowell-Davis 1994; Crowell-Davis 1986), especially in early life (Houpt 2002). Spatial relationships have also been explored in jennies and their foals. It's not uncommon in wild and domestic settings to find several generations of donkeys living together. As a donkey foal grows older, there's an increase in independence from its dam, yet after weaning, the spatial relationship returns, and the weanling and jenny remain close (French 1998). Similar behavior has been observed in Grevy zebras. Jennies have been noted to have close relationships with their foals with multiple tactile interactions. The critical period for the formation of the mare-foal bond is the first 30 min of the foal's life, which is characterized by intense maternal behavior such as investigation and licking of the foal. Therefore, unnecessary interventions should be avoided, especially at this moment, so as to not disrupt the establishment of the bond (Houpt 2002).

Under natural conditions, the mare-foal bond is maintained beyond weaning and birth of new foals (Henry et al. 2020), lasting up to 2.5 years (Boyd and Keiper 2005). Forced separation between mares and young foals, even for short durations, such as for reproductive interventions, is extremely stressful for the dams (Rogers et al. 2012) and represents a common occurrence as mares are often rebred shortly after foaling (Bosh et al. 2009) while the mare-foal bond is at its strongest (Crowell-Davis 1986). As previously outlined, exposure of the mare to such situations may feed into mechanisms of stress interference and negatively affect the supply of maternal care (Cameron et al. 2005) and milk to their current foal (De Palo et al. 2018), as well as potentially affect the success of the next breeding (Malschitzky et al. 2015) and the epigenetic makeup of the next foal (Kofman 2002; Maccari et al. 2003; Babenko et al. 2015).

5   Effects on Foals

Early life represents a critical developmental window in which long-term aspects of behavior and health are formed. Adequate care of foals, for example through providing them with opportunities for social learning (Crowell-Davis 1986), expression of play behaviors (McDonnell and Poulin 2002), employing less stressful management practices, and fostering positive human-animal interactions (Henry et al. 2005) may pay dividends in creating adults that are not fearful of humans, less reactive to other horses and novel situations and better equipped to deal with stressful challenges (Henry et al. 2009). Horses raised in such conditions may also express better performance in training, sports competitions, and reproduction and present higher resilience to disease and development of behavioral patterns deemed detrimental, such as stereotypies (Hausberger et al. 2007).

5.1   Neonatal Stage and Early Life

The formation of the mare-foal bond begins critically during the first 30 min of the foal's life, with intense expression of maternal behaviors, and disturbances of this process should be avoided as it may disrupt maternal recognition and pair bonding and interfere with the newborn's adaptation to the postnatal environment (Houpt 2002). Granted, in some cases, intervention may be necessary for dystocias, foals that have difficulty standing, and/or nursing. Foals must nurse early on to consume immunoglobins and begin to develop passive immunity. Mares who are first-time mothers may be more stressed with human handling and intervention than those mares who are multiparous. Granted, ensuring the foal, stands and nurses within the first few hours is imperative for the future of the foal's health.

Though there is a widespread practice in the horse industry to handle foals starting immediately after birth, with the intended goal of minimizing their reactivity to human contact and handling stress later in life, scientific evidence of the benefits of this practice is controversial (Williams et al. 2003; Simpson 2002; Spier et al. 2004; Lansade et al. 2005; Henry et al. 2009; Pereira-Figueiredo et al. 2024) and its welfare cost at short and long term is increasingly brought into discussion (Henry et al. 2009; Pereira-Figueiredo et al. 2024). Neonatal handling is demonstrably stressful for horse foals (Henry et al. 2009; Pereira-Figueiredo et al. 2024) and capable of exerting the exact opposite of desired effects by negatively modulating later responses to humans (Pereira-Figueiredo et al. 2024). Positive recorded results of this practice have been found to be temporary (Lansade et al. 2005), and the measured parameters of "reduced reactivity" especially meaning less expression of defensive reactions to subsequent human handling (Lansade et al. 2005; Spier et al. 2004) may actually be the result of learned helplessness, which represents a stress coping response (Pereira-Figueiredo et al. 2024; Henry et al. 2009; Simpson 2002). However, very little research has been focused on mule and donkey foals so the level of stress for early handling intervention may be specie dependent (McLean et al. 2019). Many suggest early foal handling for mules when compared to horse foals. Surveys with adult mules have shown decreased fear responses with routine husbandry and veterinary procedures with mules that did receive some form of early foal handling (McLean et al. 2019). One donkey study found that foals separated from their dams for 2 h showed little to no signs of stress, so each species may respond differently to handling and stress (De Souza Farias et al. 2021).

During the first 2 weeks of life, mares and foals spend more than 80% of their time within 5 meters of each other (Barber and Crowell-Davis 1994), and in natural conditions, a close spatial relationship is maintained even beyond weaning with foals still maintaining their dams as closest neighbors (Lansade et al. 2022). Interferences with this pattern of proximity through forced separations, even for short durations, are deeply stressful for foals (Moons et al. 2005; Henry et al. 2009; Heleski et al. 2002), and the exposure to such experiences in early life carries long-term consequences for their physical, emotional, and social development (Henry et al. 2009). Separation from the dam for even only 1 h after birth has been shown to intensely compromise at long-term the entire mare-foal bond and adequate development of behavior patterns by fostering higher dam dependency, impairing expression of play and exploration, and negatively modulating social interactions into patterns of higher withdrawal and aggression (Henry et al. 2009). In other species, exposure to stressors in early life, including neonatal handling and separation from the mother, is a known factor of developmental impairment and permanent vulnerability to higher stress reactivity, worse emotional regulation, disease, and even reproductive failure (Raineki et al. 2014).

Mares represent an important behavioral model for their foals and mediate processes of environmental knowledge acquisition and social facilitation (Nicol 1995; Henry et al. 2005). The mare is the key mediator to fostering positive, long-lasting human-animal relations with foals, as gentle handling of mares for short periods of time during the foals' first days of life promotes social interactions between foals and humans that last up to, and possibly longer than 1 year (Henry et al. 2005). Non-aversive handling of the dam, especially in the presence of the foal, while also avoiding known stressors such as separating the dyad (mare and foal behaviors) and forced foal handling, may represent an effective strategy for raising more physically and emotionally resilient foals.

5.2   Weaning

Artificial weaning is a common practice in the horse industry, which is conducted in precisely the opposite manner of how weaning occurs in the wild in horses (Henry et al. 2020). While naturally mares would gradually wean their foals between the ages of 8 and 12 months (Henry et al. 2020) and maintain the mare-foal bond and proximity in spatial relations for up to 2.5 years (Boyd and Keiper 2005), the ordinary procedure in commercial horse breeding is abrupt weaning as early as 4 months of age with complete and permanent rupture of the mare-foal bond and cessation of social contact between the dyad (Waran et al. 2008). For this reason, this procedure represents the most profoundly stressful experience in a foal's life (Henry et al. 2012, 2020; Waters et al. 2002; D'Eath 1987; Moons et al. 2005) and has been widely regarded as a critical contributor to several health and behavior challenges that may become permanent issues in the adult horse (Waran et al. 2008; Henry et al. 2020). In donkeys, most weaning is initiated by the jenny and depends on available resources (French 1998).

Weaning has also been considered the "single most important event in the development of stereotypies in horses" (D'Eath 1987), which are repetitive and unvarying patterns of behavior with no apparent function (Broom and Kennedy 1993) such as cribbing and weaving (Clegg et al. 2008). Up to $2/3$ of horses who perform stereotypies as adults start expressing these behavior patterns in the first month postweaning (Waters et al. 2002). Stereotypies in horses are broadly considered undesirable behaviors that may negatively impact their welfare, health (Henry et al. 2020; Waran et al. 2008), cognitive development, training performance (Hausberger et al. 2007), and even market value and sale potential (McBride and Long 2001). These behaviors are associated with long-term physiologic and anatomical alterations in the brain in horses (McBride and Hemmings 2005), suggesting that they may be permanent and potentially result from the interaction of genetic predisposition and environmental factors such as exposure to stress (Fureix et al. 2013). Moreover, horses that perform stereotypies are more likely to be exposed to further stress by owners and veterinarians through being socially isolated in an erroneous effort to prevent "social contagion" and subjected to invasive and painful procedures such as surgical interventions and electrical shock devices with the intention of physically preventing them from performing these behaviors (McBride and Long 2001; Nicol 1999).

Therefore, more appropriate weaning strategies have been suggested to minimize stress and later problems. Keeping weaning foals in social groups is an important measure, as the presence of an adult conspecific has been proven to alleviate signs of stress, aggression, and abnormal behaviors (Henry et al. 2012), and weaning foals with companions may foster positive interactions such as mutual grooming (Delank et al. 2023) (Fig. 7). Weaning foals in pastures as opposed to confinement inside stalls or barns has also been shown to have a positive effect, minimizing the expression of abnormal behaviors (Waters et al. 2002). Preventive measures such as ensuring foals are in sound health prior to weaning, providing a clean and safe

Fig. 7 Foals that were weaned as a group in a large paddock to minimize stress

environment, and habituating foals to human handling are also recommended to minimize the impact of weaning stress on foals (Apter and Householder 1996).

6   Low-Stress Handling and Management?

For horses, being restricted from socialization with other horses and not having enough forage and freedom of movement will lead to impaired levels of welfare (Krueger et al. 2021). Horses of any age, discipline, or status should be allowed access to those needs, which will be the foundation to build better welfare standards.

In addition to meeting the basic needs of horses, all handlers, trainers, veterinarians, and other people involved in the care of the animals should learn more about learning theory (Doherty et al. 2017). According to Mills (1998), most behavior problems can be prevented or corrected by accurate behavioral analysis and application of the techniques.

Applying learning theory and techniques such as operant conditioning, classical conditioning, habituation, counter conditioning, systematic desensitization, shaping, and always prioritizing methods that induce positive emotions will not only lead to better results with the animals but also reduce the risks and accidents with humans (Starling et al. 2016).

Some examples of situations where learning and applying those techniques to animals would be beneficial in breeding situations would be preparation for procedures and examinations, loading, general handling, traveling, early interactions, and many more (Carroll et al. 2022). Building positive associations with the animals in those situations could possibly benefit the individual and even bring positive effects on the offspring (Popescu et al. 2019).

References

Appleton AA, Lester BM, Armstrong DA, Lesseur C, Marsit CJ (2015) Examining the joint contribution of placental NR3C1 and HSD11B2 methylation for infant neurobehavior. Psychoneuroendocrinology 52:32–42. https://doi.org/10.1016/j.psyneuen.2014.11.004

Apter RC, Householder DD (1996) Weaning and weaning management of foals: A review and some recommendations. J Equine Vet Sci 16:428–435. https://doi.org/10.1016/S0737-0806(96)80208-5

Asa CS, Ginther OJ (1982) Glucocorticoid suppression of oestrus, follicles, LH and ovulation in the mare. J Reprod Fertil 32:247–251

Asa CS, Goldfoot DA, Garcia MC, Ginther OJ (1980) Dexamethasone suppression of sexual behavior in the ovariectomized mare. Horm Behav 14:55–64. https://doi.org/10.1016/0018-506x(80)90015-x

Auclair-Ronzaud J, Jaffrézic F, Wimel L, Dubois C, Laloë D, Chavatte-Palmer P (2022) Estimation of milk production in suckling mares and factors influencing their milk yield. Animal 16:100498. https://doi.org/10.1016/j.animal.2022.100498

Aurich C, Aurich J (2008) Effects of stress on reproductive functions in the horse. Pferdeheilkunde 24:99–102. https://doi.org/10.21836/pem20080121

Babenko O, Kovalchuk I, Metz GAS (2015) Stress-induced perinatal and transgenerational epigenetic programming of brain development and mental health. Neurosci Biobehav Rev 48:70–91. https://doi.org/10.1016/j.neubiorev.2014.11.013

Barber JA, Crowell-Davis SL (1994) Maternal behavior of Belgian (Equus caballus) mares. Appl Anim Behav Sci 41:161–189. https://doi.org/10.1016/0168-1591(94)90021-3

Barroso G, Oehninger S, Monzó A, Kolm P, Gibbons WE, Muasher SJ (2001) High FSH:LH ratio and low LH levels in basal cycle day 3: impact on follicular development and IVF outcome. J Assist Reprod Genet 18:499–505. https://doi.org/10.1023/a:1016601110424

Baucus KL, Ralston SL, Nockels CF, McKinnon AO, Squires EL (1990a) Effects of transportation on early embryonic death in mares. J Anim Sci 68:345–351. https://doi.org/10.2527/1990.682345x

Baucus KL, Squires EL, Ralston SL, McKinnon AO, Nett TM (1990b) Effect of transportation on the estrous cycle and concentrations of hormones in mares. J Anim Sci 68. https://doi.org/10.2527/1990.682419x

Berghold P, Möstl E, Aurich C (2007) Effects of reproductive status and management on cortisol secretion and fertility of oestrous horse mares. Anim Reprod Sci 102:276–285. https://doi.org/10.1016/j.anireprosci.2006.11.009

Biran D, Braw-Tal R, Gendelman M, Lavon Y, Roth Z (2015) ACTH administration during formation of preovulatory follicles impairs steroidogenesis and angiogenesis in association with ovulation failure in lactating cows. Domest Anim Endocrinol 53:52–59. https://doi.org/10.1016/j.domaniend.2015.05.002

Bosh KA, Powell D, Neibergs JS, Shelton B, Zent W (2009) Impact of reproductive efficiency over time and mare financial value on economic returns among Thoroughbred mares in central Kentucky. Equine Vet J 41:889–894. https://doi.org/10.2746/042516409x456059

Boyd LE, Keiper RR (2005) Behavioural ecology of feral horses. In: Mills DS, McDonnell SM (eds) The domestic horse. The evolution, development and management of its behaviour. Cambridge University Press, pp 55–82

Broom DM, Kennedy MJ (1993) Stereotypies in horses: their relevance to welfare and causation. Equine Vet Educ 5:151–154

Bruckmaier RM, Wellnitz O (2008) Induction of milk ejection and milk removal in different production systems. J Anim Sci 86:15–20. https://doi.org/10.2527/jas.2007-0335

Collins JM, Keane JM, Deady C, Khashan AS, Mccarthy FP, O'keeffe GW, Clarke G, Cryan JF, Caputi V, O'mahony SM (2024) Prenatal stress impacts foetal neurodevelopment: temporal windows of gestational vulnerability. Neurosci Biobehav Rev 164. https://doi.org/10.1016/j.neubiorev.2024.105793

Cameron NM, Champagne FA, Parent C, Fish EW, Ozaki-Kuroda K, Meaney MJ (2005) The programming of individual differences in defensive responses and reproductive strategies in the rat through variations in maternal care. Neurosci Biobehav Rev 29:843–865. https://doi.org/10.1016/j.neubiorev.2005.03.022

Campbell MLH, Sandøe P (2015) Welfare in horse breeding. Vet Rec 176:436. https://doi.org/10.1136/vr.102814

Carroll SL, Sykes BW, Mills PC (2022) Moving toward Fear-Free Husbandry and Veterinary Care for Horses. Animals 12(21):2907. https://doi.org/10.3390/ani12212907

Champagne FA, Meaney MJ (2006) Stress during gestation alters postpartum maternal care and the development of the offspring in a rodent model. Biol Psychiatry 59:1227–1235. https://doi.org/10.1016/j.biopsych.2005.10.016

Clegg HA, Buckley P, Friend MA, Mcgreevy PD (2008) The ethological and physiological characteristics of cribbing and weaving horses. Appl Anim Behav Sci 109:68–76. https://doi.org/10.1016/j.applanim.2007.02.001

Crowell-Davis SL (1986) Spatial relations between mares and foals of the Welsh pony (Equus caballus). Anim Behav 34:1007–1015. https://doi.org/10.1016/s0003-3472(86)80159-2

Dadds MR, Moul C, Hawes DJ, Diaz AM, Brennan J (2015) Individual differences in childhood behavior disorders associated with epigenetic modulation of the cortisol receptor gene. Child Dev 86:1311–1320. https://doi.org/10.1111/cdev.12391

Davis EP, Glynn LM, Waffarn F, Sandman CA (2011) Prenatal maternal stress programs infant stress regulation. J Child Psychol Psychiatry Allied Discip 52:119–129. https://doi.org/10.1111/j.1469-7610.2010.02314.x

D'Eath FM (1987) The behaviour of horses: In relation to their training and management by Dr Marthe Kiley-Worthington. Equine Vet J 19:325. https://doi.org/10.1111/j.2042-3306.1987.tb01421.x

Delank K, Reese S, Erhard M, Wöhr AC (2023) Behavioral and hormonal assessment of stress in foals (Equus caballus) throughout the weaning process. PLoS One 18. https://doi.org/10.1371/journal.pone.0280078

De Palo P, Maggiolino A, Albenzio M, Caroprese M, Centoducati P, Tateo A (2018) Evaluation of different habituation protocols for training dairy jennies to the milking parlor: Effect on milk yield, behavior, heart rate and salivary cortisol. Appl Anim Behav Sci 204:72–80. https://doi.org/10.1016/j.applanim.2018.05.003

De Souza Farias S, Dierings Montechese A, Bernardino T, Mazza Rodrigues P, de Araujo Oliveira C, Zanella A (2021) Two hours of separation prior to milking: is this strategy stressful for jennies and their foals. Animals (Basel) 11:178. https://doi.org/10.3390/ani11101078

Dewey KG (2001) Maternal and fetal stress are associated with impaired lactogenesis in humans. J Nutr 131:3012S–3015S. https://doi.org/10.1093/jn/131.11.3012s

Dobson H, Smith RF (2000) What is stress, and how does it affect reproduction? Anim Reprod Sci 60–61:743–752. https://doi.org/10.1016/s0378-4320(00)00080-4

Doherty O, McGreevy PD, Pearson G (2017) The importance of learning theory and equitation science to the veterinarian. Appl Anim Behav Sci 190:111–122

French J (1998) Mother-offspring relationships in donkeys. Appl Anim Behav Sci. https://doi.org/10.1016/S0168-1591(98)00173-7

Fureix C, Benhajali H, Henry S, Bruchet A, Prunier A, Ezzaouia M, Coste C, Hausberger M, Palme R, Jego P (2013) Plasma cortisol and faecal cortisol metabolites concentrations in stereotypic and non-stereotypic horses: do stereotypic horses cope better with poor environmental conditions? BMC Vet Res 9:3. https://doi.org/10.1186/1746-6148-9-3

Gao Y, Chen F, Kong Q-Q, Ning S-F, Yuan H-J, Lian H-Y, Luo M-J, Tan J-H (2016) Stresses on female mice Impair Oocyte Developmental Potential: Effects of stress severity and duration on oocytes at the growing follicle stage. Reprod Sci 23:1148–1157. https://doi.org/10.1177/1933719116630416

Godoy LD, Rossignoli MT, Delfino-Pereira P, Garcia-Cairasco N, De Lima Umeoka EH (2018) A comprehensive overview on stress neurobiology: basic concepts and clinical implications. Front Behav Neurosci 12. https://doi.org/10.3389/fnbeh.2018.00127

Goldstein JA, Gallagher K, Beck C, Kumar R, Gernand AD (2020) Maternal-fetal inflammation in the placenta and the developmental origins of health and disease. Front Immunol 11. https://doi.org/10.3389/fimmu.2020.531543

Gonzalez-Rivas PA, Chauhan SS, Ha M, Fegan N, Dunshea FR, Warner RD (2020) Effects of heat stress on animal physiology, metabolism, and meat quality: A review. Meat Sci 162:108025. https://doi.org/10.1016/j.meatsci.2019.108025

Gordon ME, McKeever KH, Betros CL, Filho HCM (2007) Exercise-induced alterations in plasma concentrations of ghrelin, adiponectin, leptin, glucose, insulin, and cortisol in horses. Vet J 173:532–540. https://doi.org/10.1016/j.tvjl.2006.01.003

Hausberger M, Gautier E, Müller C, Jego P (2007) Lower learning abilities in stereotypic horses. Appl Anim Behav Sci 107:299–306. https://doi.org/10.1016/j.applanim.2006.10.003

Hedberg Y, Dalin A-M, Forsberg M, Lundeheim N, Hoffmann B, Ludwig C, Kindahl H (2007) Effect of ACTH (tetracosactide) on steroid hormone levels in the mare. Anim Reprod Sci 100:73–91. https://doi.org/10.1016/j.anireprosci.2006.06.008

Heleski CR, Shelle AC, Nielsen BD, Zanella AJ (2002) Influence of housing on weanling horse behavior and subsequent welfare. Appl Anim Behav Sci 78:291–302. https://doi.org/10.1016/s0168-1591(02)00108-9

Henry S, Hemery D, Richard M-A, Hausberger M (2005) Human–mare relationships and behaviour of foals toward humans. Appl Anim Behav Sci 93:341–362. https://doi.org/10.1016/j.applanim.2005.01.008

Henry S, Richard-Yris M-A, Tordjman S, Hausberger M (2009) Neonatal handling affects durably bonding and social development. PLoS One 4:e5216. https://doi.org/10.1371/journal.pone.0005216

Henry S, Zanella AJ, Sankey C, Richard-Yris M-A, Marko A, Hausberger M (2012) Adults may be used to alleviate weaning stress in domestic foals (Equus caballus). Physiol Behav 106:428–438. https://doi.org/10.1016/j.physbeh.2012.02.025

Henry S, Sigurjónsdóttir H, Klapper A, Joubert J, Montier G, Hausberger M (2020) Domestic foal weaning: Need for re-thinking breeding practices? Animals 10:361. https://doi.org/10.3390/ani10020361

Herman JP, McKlveen JM, Ghosal S, Kopp B, Wulsin A, Makinson R, Scheimann J, Myers B (2016) Regulation of the hypothalamic-pituitary-adrenocortical stress response. Comp Physiol 603–621. https://doi.org/10.1002/cphy.c150015

Hernandez-Avalos I, Mota-Rojas D, Mendoza-Flores JE, Casas-Alvarado A, Flores-Padila K, Miranda-Cortes AE, Torres-Bernal F, Gomez-Prado J, Mora-Medina P (2021) Nociceptive pain and anxiety, in equines: physiological and behavioral alterations. Vet World 11:2984–2995. https://doi.org/10.14202/vetworld.2021.2984-2995

Houpt KA (2002) Formation and dissolution of the mare–foal bond. Appl Anim Behav Sci 78:319–328. https://doi.org/10.1016/s0168-1591(02)00111-9

Hyyppä S (2005) Endocrinal responses in exercising horses. Livest Prod Sci 92:113–121. https://doi.org/10.1016/j.livprodsci.2004.11.014

Iwasa T, Matsuzaki T, Yano K, Irahara M (2017) Gonadotropin-inhibitory hormone plays roles in stress-induced reproductive dysfunction. Front Endocrinol 8. https://doi.org/10.3389/fendo.2017.00062

Joseph DN, Whirledge S (2017) Stress and the HPA axis: Balancing homeostasis and fertility. Int J Mol Sci 18. https://doi.org/10.3390/ijms18102224

Kelley DE, Gibbons JR, Pratt SE, Smith RL, Mortensen CJ (2009) The effect of exercise on folliculogenesis in mares. Reprod Fertil Dev 21. https://doi.org/10.1071/rdv21n1ab173

Kelley DE, Gibbons JR, Smith R, Vernon KL, Pratt-Phillip. S. E, Mortensen CJ (2011) Exercise affects both ovarian follicular dynamics and hormone concentrations in mares. Theriogenology 76:615–622. https://doi.org/10.1016/j.theriogenology.2011.03.014

Kofman O (2002) The role of prenatal stress in the etiology of developmental behavioural disorders. Neurosci Biobehav Rev 26:457–470. https://doi.org/10.1016/s0149-7634(02)00015-5

Krueger K, Esch L, Farmer K, Marr I (2021) Basic Needs in Horses?—A Literature Review. Animals 11(6):1798. https://doi.org/10.3390/ani11061798

Lansade L, Lévy F, Parias C, Reigner F, Górecka-Bruzda A (2022) Weaned horses, especially females, still prefer their dam after five months of separation. Animal 16:(10). doi:https://doi.org/10.1016/j.animal.2022.100636

Lansade L, Bertrand M, Bouissou M-F (2005) Effects of neonatal handling on subsequent manageability, reactivity and learning ability of foals. Appl Anim Behav Sci 92:143–158. https://doi.org/10.1016/j.applanim.2004.10.014

Luo E, Stephens SBZ, Chaing S, Munaganuru N, Kauffman AS, Breen KM (2016) Corticosterone blocks ovarian cyclicity and the LH surge via decreased kisspeptin neuron activation in female mice. Endocrinology 157:1187–1199. https://doi.org/10.1210/en.2015-1711

Maccari S, Darnaudery M, Morley-Fletcher S, Zuena AR, Cinque C, Van Reeth O (2003) Prenatal stress and long-term consequences: implications of glucocorticoid hormones. Neurosci Biobehav Rev 27:119–127. https://doi.org/10.1016/s0149-7634(03)00014-9

Malschitzky E, Pimentel A, Garbade P, Jobim M, Gregory R, Mattos R (2015) Management strategies aiming to improve horse welfare Reduce embryonic death rates in mares. Reprod Domest Anim 50:632–636. https://doi.org/10.1111/rda.12540

McBride SD, Hemmings A (2005) Altered mesoaccumbens and nigro-striatal dopamine physiology is associated with stereotypy development in a non-rodent species. Behav Brain Res 159:113–118. https://doi.org/10.1016/j.bbr.2004.10.014

McBride SD, Long L (2001) Management of horses showing stereotypic behaviour, owner perception and the implications for welfare. Vet Rec 148:799–802. https://doi.org/10.1136/vr.148.26.799

McCosh RB, O'Bryne KT, Karsch FJ, Breen KM (2022) Regulation of the GnRH neuron during stress. J Neuroendocrinol 34. https://doi.org/10.1111/jne.13098

McDonnell SM, Poulin A (2002) Equid play ethogram. Appl Anim Behav Sci 78:263–290. https://doi.org/10.1016/s0168-1591(02)00112-0

McLean A, Varnum A, Ali A, Heleski C, Navas-Gonzalez F (2019) Comparing and contrasting knowledge on mules and hinnies as a tool to comprehend their behavior and improve their welfare. Animals 9:488. https://doi.org/10.3390/ani9080488

Melchert M, Aurich C, Aurich J, Gautier C, Nagel C (2019) External stress increases sympathoadrenal activity and prolongs the expulsive phase of foaling in pony mares. Theriogenology 128:110–115. https://doi.org/10.1016/j.theriogenology.2019.02.006

Mills DS (1998) Applying learning theory to the management of the horse: the difference between getting it right and getting it wrong. Equine Vet J 30:44–48. https://doi.org/10.1111/j.2042-3306.1998.tb05145.x

Minakova E, Warner BB (2018) Maternal immune activation, central nervous system development and behavioral phenotypes. Birth Defects Res 110:1539–1550. https://doi.org/10.1002/bdr2.1416

Moons CPH, Laughlin K, Zanella AJ (2005) Effects of short-term maternal separations on weaning stress in foals. Appl Anim Behav Sci 91:321–335. https://doi.org/10.1016/j.applanim.2004.10.007

Mortensen CJ, Choi YH, Hinrichs K, Ing NH, Kraemer DC, Vogelsang SG, Vogelsang MM (2009) Embryo recovery from exercised mares. Anim Reprod Sci 110:237–244. https://doi.org/10.1016/j.anireprosci.2008.01.015

Nagel C, Melchert M, Aurich J, Aurich C (2020) Road transport of Late-Pregnant mares advances the onset of foaling. J Equine Vet Sci 86:102894. https://doi.org/10.1016/j.jevs.2019.102894

Nicol CJ (1995) The social transmission of information and behaviour. Appl Anim Behav Sci 44:79–98. https://doi.org/10.1016/0168-1591(95)00607-t

Nicol C (1999) Understanding equine stereotypies. Equine Vet J 31:20–25. https://doi.org/10.1111/j.2042-3306.1999.tb05151.x

Oberlander TF, Weinberg J, Papsdorf M, Grunau R, Misri S, Devlin AM (2008) Prenatal exposure to maternal depression, neonatal methylation of human glucocorticoid receptor gene (NR3C1) and infant cortisol stress responses. Epigenetics 3:97–106. https://doi.org/10.4161/epi.3.2.6034

Oikawa M et al (1996) Effects of transport stress on concentrations of LH and FSH in plasma of mares: a preliminary study. J Equine Sci 7:1–5

Pereira-Figueiredo I, Rosa I, Sanchez CS (2024) Forced handling decreases emotionality but does not improve young horses' responses toward humans and their adaptability to stress. Animals 14:784. https://doi.org/10.3390/ani14050784

Popescu S, Lazar E, Borda C, Niculae M, Sandru C, Spinu M (2019) Welfare quality of breeding horses under different housing conditions. Animals 9:81. https://pubmed.ncbi.nlm.nih.gov/30841611/

Raineki C, Lucion AB, Weinberg J (2014) Neonatal handling: An overview of the positive and negative effects. Dev Psychobiol 56:1613–1625. https://doi.org/10.1002/dev.21241

Rasiah NP, Loewen SP, Bains JS (2023) Windows into stress: a glimpse at emerging roles for CRHPVN neurons. Physiol Rev 103:1667–1691. https://doi.org/10.1152/physrev.00056.2021

Rogers CW, Walsh V, Gee EK, Firth EC (2012) A preliminary investigation of the use of a foal image to reduce mare stress during mare–foal separation. J Vet Behav 7:49–54. https://doi.org/10.1016/j.jveb.2011.04.006

Rooney KL, Domar AD (2018) The relationship between stress and infertility. Dialogues Clin Neurosci 20:41–47. https://doi.org/10.31887/dcns.2018.20.1/klrooney

Rosenfeld CS (2021) The placenta-brain-axis. J Neurosci Res 99:271–283. https://doi.org/10.1002/jnr.24603

Sairanen J, Katila T, Virtala AM, Ojala M (2011) Effects of racing on equine fertility. Anim Reprod Sci 124:73–84. https://doi.org/10.1016/j.anireprosci.2011.02.010

Salari F, Mariti C, Altomonte I, Gazaano A, Martini M (2022) Impact of variability factors on hair cortisol, blood count and milk production in donkeys. Animals (Basel) 21. https://doi.org/10.3390/ani12213009

Sarmiento MP, Lanzoni L, Sabei L, Chincarini M, Palme R, Zanella AJ, Vignola G (2023) Lameness in pregnant sows alters placental stress response. Animals 13:1722. https://doi.org/10.3390/ani13111722

Schmidt B, Zeyner A, Kienzle E, Kirchhof S, Coenen M (2010) Milk intake and foal growth—a review of literature. In: Ellis AD, Longland AC, Coenen M, Miraglia N (eds) The impact of nutrition on the health and welfare of horses. Wageningen Academic, Wageningen, pp 43–46

Simpson BS (2002) Neonatal foal handling. Appl Anim Behav Sci 78:303–317. https://doi.org/10.1016/s0168-1591(02)00107-7

Smith RL, Vernon KL, Kelley DE, Gibbons JR, Mortensen CJ (2012) Impact of moderate exercise on ovarian blood flow and early embryonic outcomes in mares. J Anim Sci 90:3770–3777. https://doi.org/10.2527/jas.2011-4713

Spier SJ, Pusterla JB, Villarroel A, Pusterla N (2004) Outcome of tactile conditioning of neonates, or "imprint training" on selected handling measures in foals. Vet J 168:252–258. https://doi.org/10.1016/j.tvjl.2003.12.008

Starling M, McLean A, McGreevy P (2016) The contribution of equitation science to minimizing horse related risks to humans. Animals (Basel) 6. https://www.ncbi.nlm.nih.gov/pmc/articles/PMC4810043/

Tao S, Rivas RMO, Marins TN, Chen Y-C, Gao J, Bernard JK (2020) Impact of heat stress on lactational performance of dairy cows. Theriogenology 150:437–444. https://doi.org/10.1016/j.theriogenology.2020.02.048

Tilbrook AJ, Turner AI, Clarke IJ (2000) Effects of stress on reproduction in non-rodent mammals: the role of glucocorticoids and sex differences. Rev Reprod 5:105–113. https://doi.org/10.1530/ror.0.0050105

Tilbrook AJ, Turner AI, Clarke IJ (2002) Stress and reproduction: central mechanisms and differences in non-rodent species. Stress 5:83–100. https://doi.org/10.1080/10253890290027912

Vaala WE, Sertich PL (1994) Management strategies for mares at risk for periparturient complications. Vet Clin North Am Equine Pract 10:237–265. https://doi.org/10.1016/s0749-0739(17)30376-0

Valsamakis G, Chrousos G, Mastorakos G (2019) Stress, female reproduction and pregnancy. Psychoneuroendocrinology 100:48–57. https://doi.org/10.1016/j.psyneuen.2018.09.031

van Niekerk CH, Morgenthal JC (1982) Fetal loss and the effect of stress on plasma progestagen levels in pregnant Thoroughbred mares. J Reprod Fertil 32:453–457

Wadsworth ME, Broderick AV, Loughlin-Presnal JE, Bendezu JJ, Joos CM, Ahlkvist JA, Perzow SED, McDonald A (2019) Co-activation of SAM and HPA responses to acute stress: A review of the literature and test of differential associations with preadolescents' internalizing and externalizing. Dev Psychobiol 61:1079–1093. https://doi.org/10.1002/dev.21866

Wagenmaker ER, Moenter SM (2017) Exposure to acute psychosocial stress disrupts the luteinizing hormone surge independent of estrous cycle alterations in female mice. Endocrinology 158:2593–2602. https://doi.org/10.1210/en.2017-00341

Waran NK, Clarke N, Farnworth M (2008) The effects of weaning on the domestic horse (Equus caballus). Appl Anim Behav Sci 110:42–57. https://doi.org/10.1016/j.applanim.2007.03.024

Waters AJ, Nicol CJ, French NP (2002) Factors influencing the development of stereotypic and redirected behaviours in young horses: findings of a four year prospective epidemiological study. Equine Vet J 34:572–579. https://doi.org/10.2746/042516402776180241

Wendt D, Van Loon LJC, Van Marken Lichtenbelt WD (2007) Thermoregulation during exercise in the heat. Sports Med 37:669–682. https://doi.org/10.2165/00007256-200737080-00002

Whirledge S, Cidlowski JA (2010) Glucocorticoids, stress, and fertility. Minerva Endocrinol 35:109–125

Whirledge S, Cidlowski JA (2017) Glucocorticoids and reproduction: Traffic control on the road to reproduction. Trends Endocrinol Metab 28:399–415. https://doi.org/10.1016/j.tem.2017.02.005

Williams RJ, Marlin DJ, Smith N, Harris RC, Haresign W, Morel MCD (2002) Effects of cool and hot humid environmental conditions on neuroendocrine responses of horses to treadmill exercise. Vet J 164:54–63. https://doi.org/10.1053/tvjl.2002.0721

Williams JL, Friend TH, Collins MN, Toscano MJ, Sisto-burt A, Nevill CH (2003) Effects of imprint training procedure at birth on the reactions of foals at age six months. Equine Vet J 35:127–132. https://doi.org/10.2746/042516403776114126

Wingfield JC, Sapolsky RM (2003) Reproduction and Resistance to stress: When and how. J Neuroendocrinol 15:711–724. https://doi.org/10.1046/j.1365-2826.2003.01033.x

Yu K, Pfeiffer C, Burden C, Krekeler N, Marth C (2022) High ambient temperature and humidity associated with early embryonic loss after embryo transfer in mares. Theriogenology 188:37–42. https://doi.org/10.1016/j.theriogenology.2022.05.014

The Use of Assisted Reproductive Technologies to Improve Genetic Selection in Cattle

R. A. Chanaka Rabel ⊙, Elizabeth A. Bangert ⊙, Kenneth Wilson ⊙, and Matthew B. Wheeler ⊙

Abstract

Over the last 80 years, the US dairy industry saw the national milk production and individual cow milk yield increase by >80% and >400%, respectively, while the dairy cattle population and the carbon footprint of a glass of milk shrunk by >60% and 66%, respectively. Genetic improvements through artificial insemination were responsible for ~50% of these advancements. Although such dramatic statistics are not available for other assisted reproductive technologies (ARTs), they too have contributed to advancing dairy and beef industries in certain ways. Having been introduced in 2008, genomic selection has almost completely replaced phenotypic selection in US dairy cattle breeding while doubling the rate of genetic gain. Currently, it incorporates over 50 traits related to production, conformation, longevity, fertility, calving, health, and feed efficiency. With such comprehensiveness, genomic selection is expected to improve tomorrow's cattle even further. Gene editing is used for inserting/deleting specific genetic variations followed by precision breeding to obtain animals of favorable phenotypes. While being controversial in certain aspects, gene editing is expected to receive regulatory approval within the foreseeable future. ARTs can play a crucial role in

R. A. C. Rabel · E. A. Bangert · K. Wilson
Department of Animal Sciences, University of Illinois at Urbana-Champaign,
Urbana, IL, USA
e-mail: rrabel2@illinois.edu; eb21@illinois.edu; kwilson8@illinois.edu

M. B. Wheeler (✉)
Department of Animal Sciences, University of Illinois at Urbana-Champaign,
Urbana, IL, USA

Carl R. Woese Institute for Genomic Biology, University of Illinois at Urbana-Champaign,
Urbana, IL, USA
e-mail: mbwheele@illinois.edu

© The Author(s), under exclusive license to Springer Nature
Switzerland AG 2024
J. C. Gardón, K. Satué Ambrojo (eds.), *Assisted Reproductive Technologies in Animals Volume 1*, https://doi.org/10.1007/978-3-031-73079-5_8

increasing production efficiency and decreasing the carbon footprint of livestock industries, especially in the tropics whose ultimate impact will be felt globally.

Keywords

Assisted reproductive technologies · Selective breeding · Genetic improvement · Gene editing · Growth · Carcass merit · Fertility · Artificial insemination · Embryo transfer · Cloning · Thermotolerance · Sustainability

1   Introduction

1.1   Brief Background on Cattle Breeding and Genetics

The fact that a mere 80 female wild aurochs (*Bos primigenius*) domesticated ~10,500 years ago (Bollongino et al. 2012) in today's southeastern Turkey (Guliński 2021) has expanded the world over to reach a global cattle population of over 1.5 billion (Mueller and Van Eenennaam 2022; Kozicka et al. 2023) speaks volumes about the adaptability and reproductive prolificacy of modern cattle and their ancestors. Domestication and selective breeding practices have given rise to a tremendous amount of phenotypic and genetic diversity among modern dairy and beef cattle (Cesarani and Pulina 2021); for example, initial domestications in the Middle East/ Europe and the Indian subcontinent gave rise to the taurine and indicine lines of cattle, respectively (McTavish et al. 2013), and the intensification of animal husbandry practices after the industrial revolution gave rise to a variety of specialized breeds (Upadhyay et al. 2017).

The exact time point when humans started *selective breeding* of cattle is unknown. However, it is fair to assume that humans would have started this consciously or unconsciously at some point after their domestication. The first documented records of selective breeding come from the eighteenth century when Robert Bakewell adopted a systematic selective breeding approach to improve his beef cattle. Bakewell's selection criteria were essentially based on phenotypic characteristics such as *low-set, blocky, and quick-maturing* animals (Wykes 2004). Until the dawn of the twenty-first century, selection for better genetics was essentially carried out by selecting animals expressing desirable traits/phenotypes, even during relatively complex breeding strategies such as progeny testing. However, instead of relying directly on phenotypic data, the latter half of the twentieth century saw the development of estimated breeding values (EBVs) and expected progeny differences (EPDs), estimates of the true genetic merit of breeding animals.

Estimated breeding values, derived from performance data of an animal and its close relatives, enable producers to select sires that excel in traits of interest such as growth, carcass merit, fertility, and maternal characteristics (Dekkers et al. 2021; Johnsson 2023). By using Artificial Insemination (AI) sires with complementary strengths to their cowherd, producers could make accelerated progress toward their breeding objectives each generation. Expected progeny differences, and statistical estimates of genetic merit for a sire or dam for a specific trait, represent average performance differences in an individual and the reference group (Dekkers et al. 2021). Expected progeny differences have been a cornerstone in improving the genetics of beef cattle for nearly four decades. These predictions reflect the genetic transmitting ability of a parent to its offspring and can be used to guide selection decisions for desired traits within the herd. EPD values are calculated based on data from an animal's actual performance, progeny performance, and the performance of other relatives. A newer innovation in the realm of EPDs is the addition of genomic testing to improve their accuracy (Johnsson 2023).

The last quarter of the twentieth century saw the development of various genetics-related research and diagnostic applications that had the potential to select breeding animals based on actual genetic markers. These applications, together with advancements in genome sequencing, led to the transition from "phenotypic selection" to a "genomic selection" and was marked by the commercial release of the Illumina BovineSNP50 BeadChip in 2007, comprising 54,001 single nucleotide polymorphisms (SNPs) (reviewed by Weigel et al. 2017, Wiggans et al. 2017, Gutierrez-Reinoso et al. 2021). Since then, genomic selection has been widely adopted by major dairy-producing countries like the United States, Canada, Great Britain, Ireland, New Zealand, Australia, France, the Netherlands, Germany, and the Scandinavian countries in their dairy cattle breeding programs, leading to significant changes in the global dairy industry (Weller et al. 2017).

Genomic selection has enabled the upgrading of conventional EBVs to genomic breeding values (GBVs) which also capture hard-to-measure traits like feed efficiency (Pringle et al. 2019), disease resistance (Berry et al. 2011), and meat quality attributes (Dekkers et al. 2021). With DNA data being accessible with genomic selection, EPD accuracy can be increased through genomic-enhanced EPDs (GE-EPDs), which combine pedigree, performance, and genomic data. GE-EPDs provide the best estimate of an animal's genetic value as a parent, allowing more confident selection even for younger animals. Traits like feed efficiency, carcass quality, and reproductive performance benefit from this approach, amplifying genetic improvement in cattle.

Phenotypic and genomic selection, together with improved management practices, have transformed the livestock industries tremendously over the last century. For example, the average North American Holstein 305-day milk yield has transformed from producing ~2000 kg in the 1920s to producing over 10,000 kg today. However, this fivefold increase in milk production caused an overall decline in fertility and health traits. This was mainly because, until the twenty-first century, selective breeding was carried out almost exclusively based on milk production-related traits. To counteract this decline, selection goals have been made more comprehensive as new phenotypes become available and cost-effective to quantify (Miglior et al. 2017). Genomic selection is especially useful in this respect as it allows the evaluation of multiple genomic markers (correlated to various traits; e.g., health, reproduction, conformation, production efficiency, waste production, and gas emissions), in the calculation/prediction of overall breeding values (Weller et al. 2017).

1.2   Five Generations of Assisted Reproductive Technologies

Assisted Reproductive Technologies (ARTs) encompass a variety of techniques designed to enhance the reproductive process in animals. In the context of cattle breeding, ARTs play a crucial role in increasing reproductive efficiency, improving genetic traits, accelerating genetic gain, and preserving valuable genetic material. Overall, the use of ARTs in cattle breeding contributes to the development of more productive, resilient, and genetically diverse herds. These technologies play a vital role in meeting the increasing global demand for high-quality animal products.

Four generations of ARTs have been conventionally described for livestock: (1) AI and gamete and embryo freezing, (2) Multiple Ovulation and Embryo Transfer (MOET), (3) In Vitro Embryo Production (IVP), and (4) cloning by Somatic Cell Nuclear Transfer (SCNT) (Bertolini and Bertolini 2009).

The last decade has seen the successful integration of genetics/genomics-related techniques with conventional ARTs, giving rise to a fifth generation of ARTs. On the one hand, genomic selection is well established and is used for the selection of superior breeding animals with desirable traits. On the other hand, genome editing, a relatively new technology, is being used to create desirable traits in cattle, e.g., gene knockouts (gene inactivation), gene knock-ins, and allele introgression. Once precise edits are carried out, SCNT facilitates the generation of offspring carrying the genetic modifications/edits (reviewed by Fischer and Schnieke 2023). Such integrations have the potential to further improve the benefits of ARTs described previously, e.g., to accelerate genetic improvement through faster dissemination of desirable traits (Camargo et al. 2022; Mueller and Van Eenennaam 2022; Oliveira et al. 2023).

1.2.1   Artificial Insemination

Originated in Russia during the early 1900s, AI, the collection of semen from a male and its subsequent deposition into the reproductive tract of a female through artificial means, was the very first ART implemented in bovine species. Inspired by the Russian success, the first AI cooperative was established in Denmark in 1936, followed by the United States in 1938. Subsequent developments in methods for collection, evaluation, processing, cryopreservation, insemination, and sexing of semen, and prediction of bull breeding values have contributed tremendously to the enormous success of AI in modern-day cattle breeding (reviewed by Grosu et al. 2013; Moore and Hasler 2017; Weigel et al. 2017; Lonergan 2018). During the pre-AI era, a bull could breed ~60 cows in a 70-day breeding season each year. In contrast, with AI, a single ejaculate could be used to breed as many as 100 cows with the average bull fathering ~20,000 progeny. In an extreme example, the bull named

"Toystory" has fathered over 500,000 offspring thanks to AI (Zuidema et al. 2021). Today, the use of AI has grown to the extent that in countries like the Netherlands, Denmark, and the United Kingdom, more than 90% of dairy cows are bred using AI (Ombelet and Van Robays 2015).

1.2.2   Embryo Transfer

The first successful calf live birth from an embryo transfer (ET) took place in 1951 (Willett et al. 1951). However, it was not until the 1970s that ET transformed into a commercial industry, thanks to numerous developments in methods involving superovulation, embryo collection, embryo cryopreservation, and embryo transfer. Until the 1990s, all bovine ETs that took place commercially were essentially in vivo-derived (IVD) embryos.

1.2.3   In Vitro Production

The 1990s saw the development of in vitro embryo production (IVP) techniques in the bovine industry, i.e., fertilization of ova outside the body and subsequent culture of resulting embryos in a laboratory setting. Together with the development of ultrasound-guided ovum pick-up (OPU) techniques, ETs from IVP embryos gradually increased and in 2017, they exceeded ETs from IVD embryos (Viana 2018). Currently, ~1.5 million bovine embryos are being transferred annually (~75% IVP embryos) (reviewed by Seidel Jr. 1981b; Hasler 2014; Moore and Hasler 2017; Rabel et al. 2023; Viana 2023).

1.2.4   Cloning

Although not as commonly used, cloning involves producing genetically identical individuals. This can be done using techniques such as SCNT where the 2n nucleus of a somatic cell is transferred into an enucleated ovum. Alternatively, cloning can be done using embryonic cell nuclear transfer (ECNT) too. Cloning has enormous potential in replicating elite individuals with desirable traits. However, even after 25 years of producing the first mammalian clone, SCNT has a very poor efficiency to-date, attributed mainly to faulty or incomplete epigenetic reprogramming of the donor cell genome. The current applications of SCNT in the arena of reproduction are mainly limited to embryo production after genetic modification or gene editing.

1.2.5   Genomic Selection and Gene Editing

Genomic selection allows using genetic information to predict an individual's genetic merit or to estimate breeding value and to subsequently rank selection candidates. In contrast to phenotypic selection where you need to physically observe the adult animals' phenotype, with genomic selection, genetic signatures correlated with superior phenotypes could be identified at an early age, dramatically reducing the generation interval (average age of parents when offspring are born and impacts genetic improvement) and improving genetic gain. As a matter of fact, genomic selection has indeed doubled the rate of genetic gain in US dairy cattle since its introduction in 2008 (García-Ruiz et al. 2016).

Gene editing can be argued as the newest kid on the ART block. Gene editing allows specific genetic mutations/variations to be deleted or inserted and thereby confer desirable traits in a breed of choice. The different methodologies will be discussed in detail in subsequent sections.

2 Artificial Insemination in Cattle: A Gateway Tool for Genetic Advancement

AI was the gateway to the development and use of increasingly advanced ARTs. AI was first applied to cattle in the early 1900s (Ivanoff 1922). The first large-scale, bovine AI organization was established in 1936 in Denmark (Foote 2002). An in-depth timeline for major milestones in AI is shown in Fig. 1. This technique allows sperm cells collected from a bull to be deposited in the reproductive tract of an ovulating female and achieve a pregnancy without natural mating ever occurring. Prior to its adoption, there was much pushback both on moral grounds and from purebred breeders who believed it would take their business (Polge 2007). Since its emergence as an ART, AI has evolved significantly. Originally, procedures for semen collection and handling as well as the insemination procedure were quite time consuming and cumbersome. There was no protocol available for freezing bull semen in the early days and thus all animals had to be inseminated with freshly collected semen, or later on cooled semen (Edwards et al. 1938). Since 1949 and the development of bovine semen cryopreservation protocols (Polge et al. 1949) artificial insemination has become an indispensable tool for the enhancement of livestock genetics and the acceleration of selective breeding programs in cattle and other livestock species around the world (Polge and Rowson 1952).

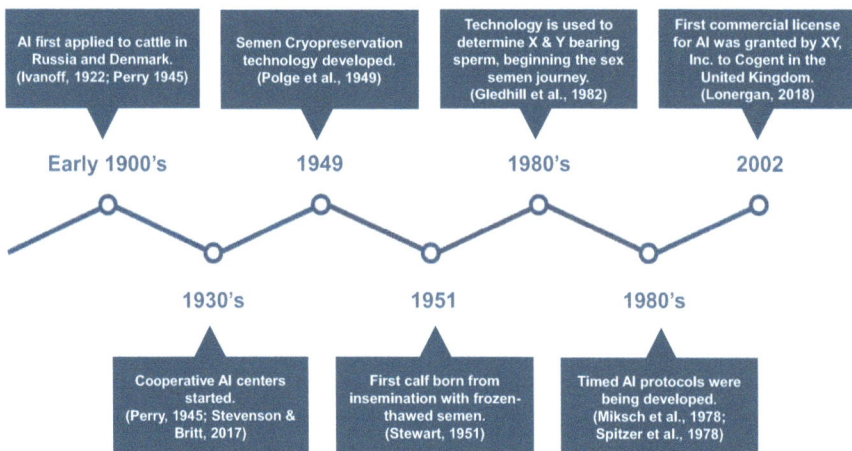

Fig. 1 Timeline for milestones in the development of artificial insemination technology (Ivanoff 1922; Perry 1945; Polge et al. 1949; Stewart 1951; Miksch et al. 1978; Spitzer et al. 1978; Gledhill et al. 1982; Stevenson and Britt 2017; Lonergan 2018)

2.1   Overview of the Process

The artificial insemination process is preceded by the process of semen collection and evaluation. Sires with superior genetics for characteristics such as milk production, meat quality, disease resistance, or other traits (usually with substantial economic value) are selected. Semen from these sires is collected, evaluated, and processed before being disseminated for breeding purposes. Evaluation of the semen sample is performed to ensure optimal fertilization potential of the dam. The semen is evaluated on the following criteria: sperm concentration, motility, and morphology. After passing the standards of evaluation the semen is processed and packaged into 0.25–0.5 cc straws. Today, semen is most frequently frozen and stored in liquid nitrogen until it is needed (Lieberman et al. 2016).

When using a frozen sample, the AI process begins with the thawing of the semen. A water bath or semen thaw unit set to 95 °F should be prepared for the thawing process (Looper 2000; Walker 2020). The straw of semen is then removed from the liquid nitrogen and placed in warm water for 30 s (Looper 2000; Walker 2020). Once the time has expired, the straw is removed quickly dried off, and loaded into the AI rod or gun for insemination. Prior to thawing the semen, the technician used paper towels or another method to clean the external surface of the vulva, removing any debris and fecal matter to the best of his/her ability. The technician then performs transrectal palpation to identify the uterus and cervix (Walker 2020). Once identified the technician will insert the AI rod (gun) into the vaginal canal. Then, placing the cervix in hand the technician guides the cervix gently over the AI rod carefully so as not to cause trauma to the uterine lining (Fig. 2a). Once the tip of the rod has reached the internal *os* of the cervix and the technician can palpate the end of the rod with his finger the semen the plunger of the AI gun is pressed slowly, and the semen is deposited at the mouth of the uterus (Fig. 2b). The AI gun is then removed from the animal, and the straw is examined to ensure proper deposition of the semen sample occurs.

The timing of artificial insemination is critical to obtaining pregnancy. To ensure successful conception rates, cows must be at the appropriate place within their

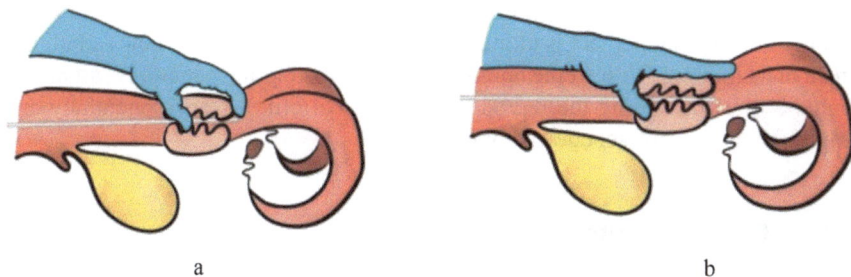

a   b

Fig. 2 Diagram of semen deposition in the bovine uterus during AI. Panel (**a**) depicts the passing of the AI rod through the cervix and into the uterus. Panel (**b**) shows where the semen is deposited in the uterus to complete the AI

estrous cycle. With data from approximately 44,000 head of cattle and unfrozen semen, R. H. Foote demonstrated that cows observed in estrus in the evening should be inseminated by noon the next day for the lowest return to estrus rate (Foote 1979). Today, the general rule of thumb for timing of AI is approximately 12 h post-observation of estrus. We refer to this as the AM:PM rule. If a cow is seen in estrus tonight, then inseminate her tomorrow morning. Likewise, if a cow is seen in estrus in the morning, then she should be inseminated that night. This rule holds true until you begin dealing with sexed sorted semen. Sexed semen typically results in a lower pregnancy per AI ratio than conventional semen (Sales et al. 2011). The typical timing of AI protocols is unfortunately not compatible with the use of sexed semen. Reasons for this incompatibility include a reduction in the lifespan of sperm within the reproductive tract (Maxwell et al. 2004) and lower sperm concentrations within an insemination dose (DeJarnette et al. 2008). For the fore mentioned reasons a decrease in the insemination to ovulation interval is needed to achieve the greatest conception rate possible.

2.2   Benefits of Selective Breeding with AI

Utilizing artificial insemination in a selective breeding program offers many benefits. These benefits include increasing genetic diversity within a population, accelerating genetic progress, precision breeding, and better control of reproductive diseases. AI enables worldwide access to superior genetics from across the globe. This makes the reality of implementing successful crossbreeding programs in areas where genetics for production or adaptability are poor much more attainable. The premise of AI is increasing selection intensity on the sire-to-progeny selection pathway, thus decreasing the time required for sire-related genetic progress (Murphy et al. 2018; Crowe et al. 2021). A clear demonstration of the power that artificial insemination holds to accelerate genetic progress can be seen when evaluating US dairy today. The dairy industry in the US produces over 80% more milk with 65% fewer cows than it did in 1944 (Mueller and Van Eenennaam 2022). Bertolini and Bertolini estimated in 2009 that approximately 50% of this gain in productivity could be attributed to the widespread adoption and use of artificial insemination in breeding programs (Bertolini and Bertolini 2009). When using natural mating the number of offspring per sire is ultimately limited by the number of cows a bull can breed in its lifetime. With AI the number of potential offspring per bull is essentially limitless, thus increasing the reproductive output of elite sires exponentially and amplifying the rate of genetic gain within a herd.

2.3   Other Benefits of AI

Increases in genetic potential are not where the benefits of AI stop. When using AI, breeders are given the opportunity to introduce new genetic material into their herds without the financial burden or biosecurity risk of bringing a new bull onsite. AI reduces the risk of transmitting sexual diseases between animals by minimizing direct animal-to-animal contact in the breeding process. Moreover, AI enables breeders to utilize semen from genetically superior sires that have been extensively disease tested, thus enhancing onsite biosecurity and safeguarding overall herd health. During the 1940s in the United Kingdom, AI was able to virtually eliminate the venereal pathogens *Tritrichomonas foetus* and *Campylobacter fetus* subsp. *Venerealis* (Parkinson and Morrell 2019). Financially, purchasing a straw of conventional beef semen priced at $15 as opposed to purchasing the sire himself makes more sense, especially for smaller breeders (Cabrera 2022). Reduction of the number of bulls on farms has also been a contributor to creating a safer working environment for farm employees (Foote 1981). Artificial insemination is the pioneer of first-generation assisted reproductive technologies. Its widespread adoption in the cattle industry represents a pivotal moment in the realm of ART. AI brings unparalleled opportunities for genetic advancement with limited equipment required. By harnessing the power of this technology breeders around the world have had the opportunity to experience expedited results of intensive breeding programs implemented to enhance the productivity and resilience of their herds contributing to the development of a more sustainable global livestock industry.

3   Embryo Production and Embryo Transfer

Embryo transfer (ET) is a reproductive technology that offers the greatest promise for advancement in genetic gain in the bovine industry. As one of the most common reproductive technologies in modern breeding programs, it has the practical advantage of enabling accelerated genetic gain, high reproductive efficiency, and conservation of valuable genetics (Galli et al. 2003; Hasler 2014). The use of multiple ovulation embryo transfer (MOET) allows the selection of superior genetics in the female and male with higher production of offspring than normal. Whereas in AI, the technique only allows intensive selection pressure on the male (Foote 2002; Galli et al. 2003).

3.1   Understanding Multiple Ovulation and Embryo Transfer

The process that is referred to as embryo transfer actually has two aspects that are sometimes confusing. The first aspect is the whole suite of techniques involved in embryo transfer (ET) technologies. These technologies include estrus synchronization, ovarian super-stimulation, embryo collection (via flushing of the uterus, or in vitro embryo production), embryo cryopreservation, and finally the transfer of the embryo itself. These techniques together are generally referred to as ET in cattle (Seidel Jr. 1981a). However, the more precise term is Multiple Ovulation and Embryo Transfer (MOET). If a beef or dairy producer is asked, "do you do ET in your herd?" and they answer yes, they are referring to the whole suite of methodologies. This, however, is a misnomer that has come to be accepted in the industry. The second aspect is the actual embryo transfer itself into a suitable prepared surrogate cow or heifer. This is the real embryo transfer. If you ask a practitioner, "do you do ET?" they are likely to answer yes, I transfer embryos.

3.2   Donor Selection and Ovarian Stimulation

3.2.1   Donor Selection

The process of Embryo Transfer (ET) usually begins by selecting females who are genetically superior for some specific trait or traits that have economic significance. Using these methods allows for more genetic selection pressure to be applied to the female (Seidel Jr. and Seidel 1991). Such females are referred to as "donors" or donor animals as they are the individuals that will donate the oocytes that contain the female contribution of DNA to the embryo. But using this method producers can rapidly expand desirable genetics in their herd (Seidel Jr. and Seidel 1991). Producers, geneticists, animal scientists, practitioners, and veterinarians involved in reproductive technologies will select donor cows with elite traits, such as high milk production, faster growth rate, and resistance to common livestock diseases (Hasler 2014; Bo and Mapletoft 2018).

3.2.2   Ovarian Stimulation

The next step is to increase the number of oocytes that the donor cow produces so that more than one embryo is produced per reproductive cycle. This process is most precisely known as ovarian super-stimulation but is more commonly known as "superovulation." This method uses reproductive hormones to increase the number of oocyte-containing ovarian follicles that can ovulate and can then be fertilized (Seidel Jr. 1981b; Hasler 2014). In the 1970s and 1980s, superovulation was achieved using Pregnant Mare Serum Gonadotropin (PMSG), but the outcome was inconsistent (Slenning and Wheeler 1989; Seidel Jr. and Seidel 1991). Today, super-ovulation is commonly initiated using a Follicle-Stimulating Hormone (FSH) treatment to stimulate the production of multiple ovulatory follicles (Moore and Hasler 2017). This method can be used for both in vivo embryo collection or for use with OPU used for in vitro embryo production, which will be discussed in a later section. With the ovulation of more follicles, the potential yield of embryos recovered is increased (Seidel Jr. and Seidel 1991; Hasler 2014; Bo and Mapletoft 2018).

3.3   Embryo Collection and In Vitro Embryo Production

3.3.1   Embryo Collection

In the 1960s, 1970s, and early 1980s, bovine embryo collection was a surgical process (Betteridge 1977; Mapletoft 2013). The surgery was relatively time consuming (15–60 min), invasive, and relatively expensive (Betteridge 1977) It required special facilities and could not easily be performed on a farm. In the mid-1970s, a non-surgical method was developed that was basically a uterine lavage where a modified

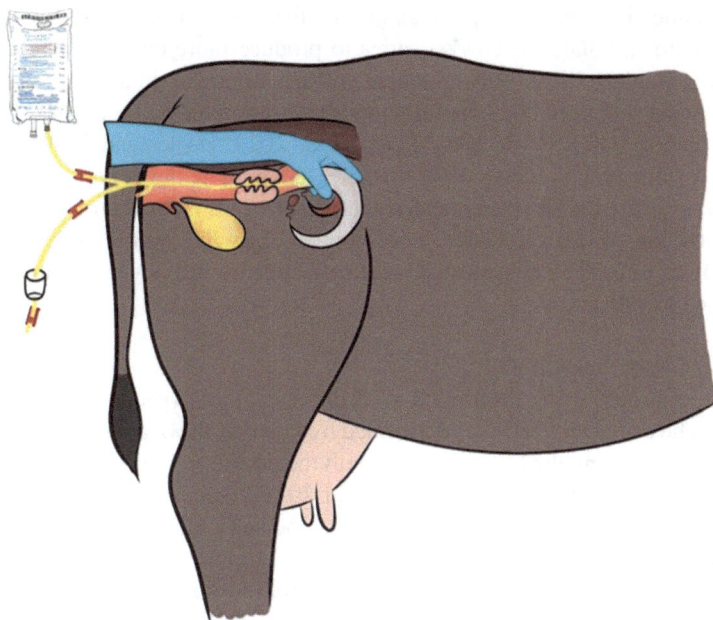

Fig. 3 Diagram of in vivo embryo recovery from the bovine uterus via lavage also known as "uterine flushing"

saline solution was introduced into the uterus, and when recovered it contained the embryos (Elsden et al. 1976; Rowe et al. 1976). This non-surgical technique was far easier on both the practitioners and donor animals (Seidel Jr. and Seidel 1991). Furthermore, non-surgical recovery methods are less technology demanding and lower in cost (Moore and Hasler 2017). Superovulated donors are bred artificially to elite sires to allow for the preferred genetics to be passed on to the offspring. Seven days after artificial insemination, embryos are non-surgically collected from the donor's uterus by manual lavage (Fig. 3). During embryo recovery, the embryos are evaluated based on the stage of development, morphology, and cell viability (Seidel Jr. 1981a; Seidel Jr. and Seidel 1991; Dahlen et al. 2014; Hasler 2014; Rabel et al. 2023).

3.3.2   In Vitro Embryo Production

While AI has revolutionized breeding practices for decades, In Vitro Fertilization (IVF) offers distinct advantages for specific breeding goals. In cattle breeding, IVF is a powerful tool alongside traditional methods like AI and conventional embryo collection. Bovine IVF involves a controlled laboratory environment to fertilize an oocyte with sperm, mimicking natural conception outside the cow's body (Galli et al. 2003; Ferre et al. 2020; Hasler and Barfield 2021). The process of IVF is not simple, as there are many steps/tasks involved that must be performed in the correct order with the correct timing. The first step is similar to that done with conventional embryo collection, that is, superovulation of the donor. The donor is given FSH treatments to stimulate the cow's ovaries to produce more ovarian follicles rather than just a single follicle. This treatment will usually result in a higher number of oocytes being produced and hopefully more embryos than would be produced by a natural mating (Galli et al. 2003, Ferre et al. 2020, Hasler and Barfield 2021). However, some practitioners do not use FSH to stimulate the donors, especially in *Bos indicus* cattle for OPU (Fernandes et al. 2014). This is not surprising since *Bos indicus* breeds typically produce more oocytes than *Bos taurus* breeds of cattle (Baruselli et al. 2021). The next step is the collection of the oocytes, commonly now performed by OPU.

3.3.2.1   Ovum Pick-Up (OPU)

OPU is performed using ultrasound-guided needles on a special aspiration probe (Fig. 4). The oocyte collection is performed by piercing the ovarian follicle with the needle and aspirating the follicular fluid with the oocyte from the ovary (Galli et al. 2003, Ferre et al. 2020, Hasler and Barfield 2021). Once the oocyte is recovered from the ovary the next step is to mature the oocyte in vitro so that it is ready for fertilization (Hasler and Barfield 2021). Oocyte maturation is the progression of the nucleus of the oocyte from its meiotic arrest phase into metaphase II of the cell cycle (Wassarman et al. 1979). Metaphase II is the stage where the oocyte is ready to be fertilized (Wassarman et al. 1979; Hasler and Barfield 2021), Collected oocytes are placed into maturation media to further mature in a controlled laboratory environment until the fertilization step (Galli et al. 2003, Ferre et al. 2020, Hasler and Barfield 2021).

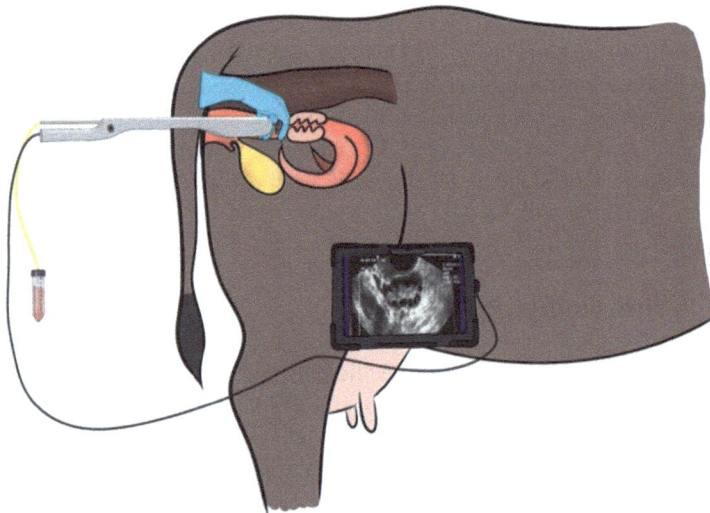

Fig. 4 Diagram of ovum pick-up (OPU) from the bovine ovary via ultrasound-guided follicular aspiration

3.3.2.2   Semen Preparation

Cryopreserved semen must be thawed and processed to separate the motile sperm from the surrounding seminal fluid and ensure high-quality, fertile sperm candidates for fertilization prior to the collection (Parrish et al. 1995; Galli et al. 2003; Rubessa et al. 2011; Ferre et al. 2020; Hasler and Barfield 2021). Frozen bull semen samples, usually pre-tested for IVF, are thawed at 37 °C for ~30–40 s. The sperm cells are then separated by centrifugation for 15–25 min at 300 × g using a discontinuous Percoll gradient ranging from 40 to 90% concentration. The resulting pellet of sperm cells will be resuspended in 1–2 mL of IVF medium and centrifuged again for 5 min. Finally, the sperm cell pellet will be diluted with IVF medium to a concentration of one million sperm cells per milliliter, and this diluted sperm sample will be added to the fertilization wells. There are a number of variations to the sperm preparation protocol but they have the same basic steps (Parrish et al. 1995; Rubessa et al. 2011, 2016, 2020; Ferre et al. 2020; Kandel et al. 2020; Hasler and Barfield 2021).

3.3.2.3   In Vitro Fertilization

Once In Vitro Maturation (IVM) is complete, mature oocytes and prepared sperm are co-incubated in fertilization media for up to 8 days to produce viable embryos (Rubessa et al. 2016; Hasler and Barfield 2021). Briefly, gametes are co-incubated for 18–22 h at 39 °C, in 5% CO_2 in air, after which presumptive zygotes are vortexed for 2–3 min to remove cumulus cells in HEPES-TCM-199 medium with 5% bovine serum (BS) and washed twice in the same medium.

3.3.2.4   In Vitro Culture

Presumptive zygotes are placed in a modified Synthetic Oviduct Fluid (mSOF) medium (Tervit et al. 1972) with 30 μL/mL essential amino acids, 10 μL/mL non-essential amino acids, and 5% BS where they are incubated in a humidified mixture of 5% CO_2, 6% O_2, and 89% N_2 at 39 °C (Rubessa et al. 2016). The fertilized oocytes (zygotes) undergo cell division and develop into blastocyst stage embryos (~100–200 cells) over the next 6–8 days in a controlled temperature, humidity, and gas atmosphere incubator (Rubessa et al. 2016; Ferre et al. 2020; Hasler and Barfield 2021). These IVF embryos may be transferred as fresh embryos or cryopreserved at −196 °C for transfer at a later date (Hasler and Barfield 2021).

3.3.2.5   Recipient Selection and Preparation

Recipients (surrogates) are selected based on criteria including age, reproductive history, estrous cyclicity, and health status to maximize the likelihood of successful embryo implantation and subsequent pregnancy establishment. Proper synchronization of donors and recipients is crucial to ensure optimal timing for embryo transfer as they both must be in the same stage of the estrous cycle. Synchronization is typically achieved through hormonal treatments and estrous synchronization protocols (Seidel Jr. 1984; Dahlen et al. 2014; Hasler and Barfield 2021). Recipient animals can either be naturally synchronized with the estrous cycles of donor animals, or their estrous cycles can be artificially synchronized using hormones like progestins or prostaglandins without negatively impacting pregnancy rates (Seidel Jr. 1984; Seidel Jr. and Seidel 1991).

After estrus synchrony has been confirmed in the recipient, she is restrained, and the feces are evacuated from the rectum. Shortly after, the presence and side of a functional corpus luteum (CL) is confirmed by transrectal ultrasonography or rectal palpation. The vulva is cleaned, and an epidural anesthetic is administered. The embryo is loaded into a 0.25 or 0.5 cc straw between at least two air bubbles and two columns of medium. The straw is then placed into an embryo transfer pipette (gun). A sterile sheath is placed over the ET gun that will protect the straw as it is passed through the vulva and the vagina up to insertion into the cervix. Some practitioners cover the ET sheath with a plastic sleeve, known as a chemise, to provide additional protection for the embryo containing straw up to the cervix. This is the process for the currently widespread non-surgical embryo transfer in cattle (Bo and Mapletoft 2018). When estrus in the donors and recipients occurred within 24 h of each other, transferring quality embryos in cattle should result in high pregnancy rates (Seidel Jr. 1984; Bo and Mapletoft 2018). Nutrition and postpartum intervals are the two management factors that determine the success or failure of estrus synchronization programs (Dahlen et al. 2014; Bo and Mapletoft 2018).

3.3.2.6   Embryo Transfer

During the 1970s, bovine embryos were primarily transferred surgically via midline incision under general anesthesia (Betteridge 1977; Seidel Jr. 1981a, b, 1984; Seidel Jr. and Seidel 1991). The animal was anesthetized, placed in dorsal recumbency, the incision site disinfected and then an incision was made on the midline (*Linea alba*) of the abdomen. Next uterus was exteriorized, and a small hole was made, with the blunt end of a 1/2 circular #2-4 suture needle, in the anterior tip of the uterine horn near the oviduct. The embryo was then transferred through the hole in the uterus into the uterine lumen by a fine glass pipette. Today, specialized embryo transfer pipettes are used for non-surgical transfers (Moore and Hasler 2017; Bo and Mapletoft 2018). With this method, the embryo is loaded into a 0.25 or 0.5 cc artificial insemination straw, which is then loaded into the ET pipette (a.k.a ET gun) and then that is covered with a sheath to protect the embryo and maintain sterility while passing it through the vulva, vagina, and the cervix into the uterus (Moore and Hasler 2017; Bo and Mapletoft 2018). The ET gun will pass through the cervix up into the uterine horn ipsilateral (on the same side) to the ovary bearing the CL. The embryo will be deposited as close to the tip of the uterine horn as possible (Fig. 5). The plunger on the ET gun will be pressed slowly and firmly to transfer the embryos into the uterus.

3.3.2.7   Pregnancy Diagnosis and Management

There are three major pregnancy detection methods used: palpation through the rectum, ultrasound scanning, or blood tests for pregnancy-associated glycoproteins (PAGs) (Dahlen et al. 2014). Rectal palpation is the most common and inexpensive method for pregnancy diagnosis in cattle (Romano et al. 2007). The use of ultrasonography has become popular with veterinarians and other practitioners over the

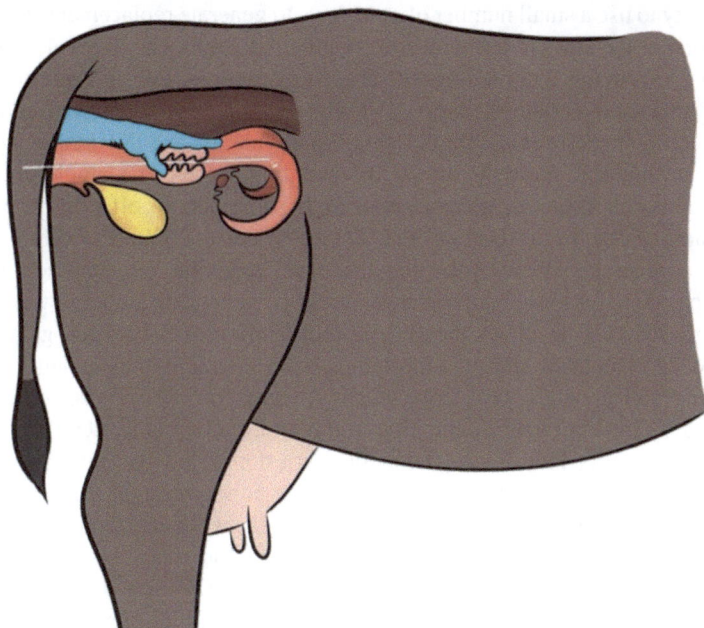

Fig. 5 Diagram of non-surgical embryo transfer into the bovine

past few years. As the cost of the equipment has decreased the use of this technology has become more widespread. The use of ultrasound allows for more precise measurement, more precise dating of the fetus, and even sex determination (Lamb et al. 2003), The final method is a blood test for PAGs. These tests are highly accurate, between 95 and 99% (Lamb et al. 2003). However, they do not provide an immediate pregnancy diagnosis and the producers must resort the cattle later. If veterinarians are not available or accessible then this method is a viable alternative (Lamb et al. 2003). Once pregnancy is confirmed the recipient cow must receive proper care, nutrition, and management throughout gestation.

3.4   Advantages and Applications

Implementation of embryo transfer into cattle production systems is the quickest way to change the genetic base of a herd using existing females. Within herds, females with suboptimal genetic potential may serve as a surrogate (recipient) to carry a calf with superior genetics. The combination of ET technology and the use of semen (conventional or gender selected) from superior bulls to produce the embryos can exponentially increase the genetic improvement within a herd. The most common use of sexed semen in the dairy industry is to increase the number of female animals in purebred operations. Additionally, sexed semen provides the opportunity to use a small number of elite cows to generate replacements while mating the rest of the cows to terminal sires (Galli et al. 2003; Dahlen et al. 2014).

While AI remains a cornerstone of breeding practices, IVF in cattle offers specific advantages in certain scenarios. IVF allows a single cow to have multiple offspring each year (Ferre et al. 2020). Using superovulation and IVF makes it possible to generate multiple embryos during the breeding season of a single elite donor female. This can allow the generation of higher numbers of offspring from select female animals each year (Dahlen et al. 2014; Ferre et al. 2020). IVF offers the ability to maximize the spread of the genetics of a single elite female much the same way AI has done for the spread of the genetics of a genetically superior bull (Bo and Mapletoft 2018). IVF allows the genetic preservation of the female gametes like what frozen ejaculated and post-mortem recovered epididymal semen allows for preserving bull genetics. Ova from recently deceased cows can be removed and preserved by freezing for insemination in the future (Galli et al. 2003). The use of IVF allows the breeding of a high genetic merit donor cow that may have fertility issues (Nowicki 2021). IVF can be an option for cows with non-genetic reproductive tract damage from disease or trauma that prevents natural conception or successful AI. Furthermore, there is the potential for more biological control, "biosecurity," with IVF. According to Givens et al. (2007), there are critical controls for handling raw or processed materials of animal origin to prevent the spread of disease. With IVF we have the possibility to test all raw or processed materials of animal origin for the presence of contaminating agents. Further, the washing of oocytes and washing and treating developed embryos with trypsin will decrease disease spread (Givens et al. 2007). Next, the appropriate use of antimicrobial substances, potentially including antiviral agents and the testing of samples of all media/cells from IVM, IVF, and in vitro culture (IVC) for the presence of contaminating microorganisms will also reduce the spread of disease from in vitro embryo production. Lastly, the establishment of minimal sanitary standards not only for IVF laboratories but also for the abattoirs from which raw materials are collected will improve the biosecurity of IVF technologies (Givens et al. 2007).

4   Technologies for Genetic Selection in Cattle Breeding

4.1   Genetic Screening and Genomic Selection: Identifying the "Best" Animals to Propagate

One of the very first genetic screening-related technologies used in the reproduction of cattle was for embryo sexing, first using in situ hybridization (Leonard et al. 1987) followed by PCR (Macháty et al. 1993) also termed pre-implantation genetic diagnosis (PGD). Since then, PCR (especially quantitative real-time PCR) has been used extensively for genomic screening (embryo genotyping) of in vitro-derived and in vivo-produced embryos, from a research perspective (reviewed by Rabel et al. 2023). However, obtaining biopsies from pre-implantation embryos, while maintaining embryo viability, can be challenging and impractical, especially at a mass scale. Therefore, embryo genomic screening is unlikely to have commercial applications soon.

However, genomic selection, a type of genetic screening, has tremendous applications and in fact, has already revolutionized how cattle are selected for breeding. In the United States, genomic selection carried out using SNP chips, was introduced in 2008 for dairy cattle. The first official genomic evaluations for Holsteins and Jerseys were released in January 2009. As of 2022, the number of genotypes has reached 6.6 million. Initially, the focus was on bulls; however, the last few years have seen a rapid increase in genotyping of females. Currently, genomic evaluations take into consideration over 50 traits related to (1) production; e.g., milk yield, fat yield, and percentage, protein yield and percentage, milking speed; (2) conformation; (3) longevity; e.g., productive life, cow livability, heifer livability, and birth to first calving; (4) fertility; e.g., daughter pregnancy rate, cow conception rate, calving to first insemination, gestation length, and early first calving; (5) calving; e.g., calving ability index; (6) health; e.g., somatic cell score and evaluations for displaced abomasum, ketosis, mastitis, metritis, milk fever, and retained placenta; and (7) feed efficiency; e.g., feed saved (García-Ruiz et al. 2016; Weigel et al. 2017; Wiggans and Carrillo 2022).

Even though over a million American Angus cattle have been genotyped (Retallick et al. 2022), in contrast to dairy cattle, implementing genomic selection in beef cattle poses a few notable challenges. For example, the accuracies of estimated breeding values are likely to be low to moderate given the limited size of the reference populations, a limited number of progeny-tested bulls, and the variety of beef breeds dominating different regions of the world. As a result, genomic selection has been adopted to a much lesser extent than in dairy cattle (reviewed by Esrafili Taze Kand Mohammaddiyeh et al. 2023).

4.2   Gene Editing and Genetic Modification

Conventional breeding is essentially a means of "randomly" extracting "desirable" genes from the two parental breeds. However, we have no control over the genes that will be inherited by the offspring. On the one hand, it is essentially dependent on a random assortment of genes and crossover of chromosomes, sub-cellular activities we have no control over. On the other hand, in addition to the "desirable and intended" genes, many hundreds, if not thousands of "undesirable and unintended" genes will be inherited by the offspring. Genetic modification and genome editing technologies have the potential to revolutionize the above slow and inefficient conventional selective breeding process. In contrast to crossbreeding, where there is a wholesale transfer of genes including those that may be deleterious for production, fertility, and/or health, genome editing makes it possible to transfer targeted genes as per specific requirements (Hansen 2020). This chapter will focus on gene editing modifications as genetic modification has been recently reviewed in cattle (Monzani et al. 2016; Monzani et al. 2022).

Genome editing enables the manipulation of gene function through deletion, addition, or base change at a desired and specific location. Different techniques have been used to-date for genome editing, i.e., Zinc Finger Nucleases (ZFN; Smith et al. 2000, DeFrancesco 2011), Translation Activator-Like Effector Nucleases (TALENs; Cermak et al. 2011), Meganucleases (MegNs; Arnould et al. 2007), and Clustered Regularly Interspaced Short Palindromic Repeats (CRISPR)/CRISPR-associated nuclease (Cas) 9 (CRISPR-Cas9; Barrangou et al. 2007). Briefly, these techniques first induce a double-stranded break (DSB) at a targeted sequence through a site-directed nuclease, and secondly facilitate targeted genetic modifications by repairing the DSB via non-homologous end-joining (NHEJ) or homology-directed repair (HDR). Such genetic modifications are considered superior to conventional transgenesis because exogenous DNA is not introduced (Ishii 2017). Out of these techniques, CRISPR-Cas9 is considered the "system of choice" for genetic engineering in livestock because it overcomes many limitations in previous technologies (Fischer and Schnieke 2023).

Genome editing can have tremendous applications in agriculture, including dairy and beef industries. For example, using gene editing, alleles associated with production, disease resistance, or heat tolerance could be transferred from one breed to another without the need for crossbreeding. As mentioned previously, gene editing is superior to conventional crossbreeding in transferring such genes between breeds; for example, crossbreeding (1) will consume decades given the relatively long generation interval in cattle and (2) will transfer not only the favorable traits/genes but also unfavorable traits/genes. In combination with conventional ARTs and genomic selection, gene editing can contribute to faster dissemination of desirable traits by either reducing the generation interval, increasing selection intensity and accuracy, and/or by increasing genetic variation thereby accelerating the genetic gain for the desired trait(s) (reviewed by Camargo et al. 2022). Once gene editing is carried out in a donor cell, gene-edited embryos are produced using SCNT, making precision livestock breeding a reality.

To-date, gene editing has been used in cattle to modify genes related to thermotolerance, disease resistance, growth, allergenicity, and welfare (Table 1). In addition to the overall benefits of gene editing mentioned earlier, modifications related to the above categories will lead to a plethora of benefits. For example, genome-edited disease-resistant livestock will result in improved animal welfare, improved sustainability of production, and decreased use of antimicrobials in agriculture; genome-edited thermotolerant livestock will result in more efficient feed conversion leading to a decreased carbon footprint, and improved production and reproduction.

Thus far, these remain as proof-of-concept, feasibility studies. Adoption of these precision breeding technologies at the commercial level is very likely soon after regulatory review. In fact, genome-edited SLICK cattle have been identified as "low-risk" by the FDA (2022).

Table 1 How gene editing has been used in cattle to-date

Gene editing technology used	Gene modification	ART used to produce gene-edited offspring	Remarks on gene-edited offspring	Reference(s)
Lentiviral transduction	Introgression of bovine β-casein promoter and human insulin sequences	SCNT	Presence of human insulin in milk implies potential for using transgenic animals as bioreactors to produce proteins of pharmaceutical interest	Monzani et al. (2024)
CRISPR/ Cas9-mediated homology-directed repair (HDR)	BVDV binding domain of CD46 was modified with $A_{83}LPTFS_{88}$ substitution	SCNT	Reduced susceptibility to BVDV infection upon exposure to a persistently infected (BVDV-PI) calf	Workman et al. (2023)
Single guide RNA (sgRNA)/ Cas9	*MSTN* knockout	IVP	*MSTN* knockout Korean beef calf showed double muscling	Gim et al. (2022)
Guide RNA (gRNA)/Cas9	Naturally occurring mutation p.Leu18del in the *PMEL* gene (from Galloway and Highland cattle) was introgressed into Holstein Friesian cattle	SCNT	Black coat markings replaced with silvery gray markings implying increased thermotolerance	Laible et al. (2021)
gRNA/Cas9	Naturally occurring SLICK mutation of *PRLR* (from Senepol cattle) was introgressed into Red Angus cattle	IVP	SLICK mutation detected implying increased thermotolerance	Rodriguez-Villamil et al. (2021)

(continued)

Table 1 (continued)

Gene editing technology used	Gene modification	ART used to produce gene-edited offspring	Remarks on gene-edited offspring	Reference(s)
CRISPR-mediated homology-mediated end-joining (HMEJ)	Natural resistance-associated macrophage protein-1 (*NRAMP1*) gene was inserted to bovine *ROSA26* locus	SCNT	Rate of *M. bovis* multiplication in monocyte-derived macrophages (MDMs) was lower than in controls implying reduced susceptibility to tuberculosis	Yuan et al. (2021)
TALEN	The putative P_C Celtic *POLLED* allele was introgressed	SCNT	Three homozygous (P_CP_C) and one heterozygous (P_Cp) polled calves were produced. Breeding a homozygous bull to horned cows (pp) produced six polled calves (P_Cp)	Carlson et al. (2016), Young et al. (2020)
TALEN	Disruption in *LGB* by creating a premature stop codon in the original reading frame	IVP	Milk was devoid of mature beta-lactoglobulin (BLG) implying hypoallergenicity	Wei et al. (2018)
CRISPR/single Cas9 nickase (Cas9n)-mediated HDR	*NRAMP1* gene was inserted to bovine *FSCN1-ACTB* (F-A) locus	*SCNT*	Rate of *M. bovis* growth in MDMs was lower than in controls. In vivo experiments also showed increased resistance to *M. bovis*	Gao et al. (2017)
ZFNs	*CD18* gene was edited to cause a single amino acid substitution of the leukocyte surface signal peptide CD18	SCNT	Leukocyte population was resistant to *Mannheimia* (*Pasteurella*) *haemolytica* leukotoxin-induced cytolysis implying reduced susceptibility to bovine respiratory disease	Shanthalingam et al. (2016)
TALEN	Introgression of SP110 nuclear body protein gene (SP110)	SCNT	Reduced growth and multiplication of *M. bovis* in in vitro and in vivo assays implying reduced susceptibility to tuberculosis	Wu et al. (2015)

(continued)

Table 1 (continued)

Gene editing technology used	Gene modification	ART used to produce gene-edited offspring	Remarks on gene-edited offspring	Reference(s)
TALEN	Direct injection of TALEN mRNA into zygotes to edit *MSTN*	IVP	Increased muscle mass implying increased carcass yield	Proudfoot et al. (2015)
ZFN	Human lysozyme (hLYZ) gene was inserted into bovine β-casein locus	SCNT	Milk had the ability to kill *Staphylococcus aureus* implying reduced susceptibility to mastitis	Liu et al. (2014)
ZFN	Transfection with *MSTN*-specific ZFN mRNA	SCNT	Double muscling implies increased carcass yield	Luo et al. (2014)
ZFNickases	Lysostaphin gene was inserted into the endogenous β-casein (CSN2) locus	SCNT	Secreted lysostaphin in milk upon being induced to lactate. In vitro assays demonstrated the milk's ability to kill *Staphylococcus aureus* implying reduced susceptibility to mastitis	Liu et al. (2013)

5   Applications of ART for Improving Traits Related to Production, Reproduction, Health, and Thermotolerance

ARTs, together with improved management practices have resulted in massive improvements in dairy cow milk production, the carbon footprint of the dairy industry, fertility, and heat tolerance of temperate breeds.

5.1   Dairy Industry

For example, in the United States, the dairy cattle population has plummeted from 25.6 to 9 million heads from 1944 to the present (a 65% reduction). Yet the national production has increased by over 80% with a more than fourfold rise in individual cow milk yield, increasing from 2000 kg/cow in 1944 to over 10,000 kg/cow today. It has been estimated that the genetic improvements brought about by AI are responsible for ~50% of these developments. As a direct result of the increased productivity, the carbon footprint from a glass of milk has come down to approximately one-third of its 1944 equivalent (reviewed by Mueller and Van Eenennaam 2022), an invaluable side-effect of ARTs.

Conventional selective breeding programs were focused on profitability and easily quantifiable traits such as milk yield, and milk fat composition of dairy cows. They largely overlooked health and reproduction-related traits. As a result, over the last quarter of the twentieth century, a decline in the health and reproductive status of dairy cows was observed in contrast to the massive improvements in milk yield-related traits (Berry et al. 2014). However, current selection indices such as Lifetime Net Merit Dollars (NM$) take into account as many as 50 traits related to (1) milk production, (2) conformation, (3) longevity, (4) fertility, (5) calving ability (Schmitt et al. 2019), (6) health, and (7) feed efficiency (Wiggans et al. 2017; Schmitt et al. 2019; Wiggans and Carrillo 2022). Other commonly examined traits analyzed include somatic cell count, fat, and protein yield, nonreturn rate, days open, calving interval, udder health, milking speed, and conformation of feet and legs among others (Cole and VanRaden 2018). Some of these traits are described in Table 2. Identification of these economically important traits has enabled the selection of breeders that are superior not only in traits related to production but also to reproduction and health. Reproduction-related traits generally have a low heritability. A 2016 study (García-Ruiz et al. 2016) that evaluated the impact of genomic selection on different traits found that since being introduced in 2008, genomic selection has

Table 2 Important traits for genetic selection and dairy breeding programs

Trait[a]	Abbreviation	Consequence
Milk production		
Milk	Milk	Increased milk production
Fat	Fat	Indicator of fat percentage in milk
Protein	Protein	Indicator of protein percentage in milk
Somatic cell score	SCS	
Fertility and maternal traits		
Daughter pregnancy rate	DPR	Decrease in days open and improved herd productivity
Heifer conception rate	HCR	Cycling and capability to conceive early
Cow conception rate	CCR	Increase fertile longevity of mature cows
Calving ease sire/ daughter	SCE/DCE	Low dystocia for an easier calving process
Confirmation traits		
Body weight composite	BWC	Indicates the overall size and efficiency of a cow
Udder composite	UC	Improved udder confirmation and functionality
Feet and legs composite	FLC	Evaluator of overall foot and leg confirmation
Other functional traits		
Productive life	PL	Indicator of longevity and productive lifespan within the herd
Cow livability	LIV	Indicator of growth at later ages and potential feed maintenance requirements

[a]Berry et al. (2011, 2014); García-Ruiz et al. (2016); Wiggans et al. (2017); Cole and VanRaden (2018); Schmitt et al. (2019); Wiggans and Carrillo (2022)

produced dramatic increases in dairy cow fertility and health traits. The greatest response to genomic selection was observed for daughter pregnancy rate, productive life, and somatic cell score.

5.2   Beef Industry

Compared to the dairy industry, the beef industry has been slow to adopt ARTs. For example, only ~12% of beef producers use AI, and ~7% use estrus synchronization in the United States. In Northern Australia, AI is used in <1% of the breeding herds. However, with the advent of Timed AI (TAI), a significant improvement in the use of AI was seen in the beef industries of countries such as Brazil, Argentina, and Uruguay. For example, it has been estimated that TAI returns over US\$ 0.5 billion/year to the Brazilian beef industry attributed to genetic improvements in growth and carcass merit, as compared to natural service (Baruselli et al. 2019; Mueller and Van Eenennaam 2022). Crossbreeding systems facilitated by AI can also boost productivity through effects like heterosis (Leroy et al. 2018; Lamberson and Thomas 2021). However, crossbreeding programs require the use of a large herd and more breeding pastures if several breeds of bulls are to be used. Maximum heterosis can also be achieved by using AI in crossbreeding programs as precise mating with multiple breed bulls can be achieved relatively easily (Lamberson and Thomas 2021). The last few years have seen a significant rise in the popularity of "beef on dairy operations" where sexed semen from a beef breed is inseminated into dairy cows (Fuerniss et al. 2023). This industry is predicted to develop by many a fold within the next few years and AI will no doubt play a big role in this development.

5.2.1   Leveraging Assisted Reproductive Techniques in Beef Cattle Breeding for Enhanced Meat Quality

For commercial beef producers, genetic improvement of their herds is the key to driving productivity, efficiency, and profitability. While conventional breeding programs can make steady genetic progress over many generations, ARTs offer powerful tools to accelerate the rate of genetic gain. Over the years, research has shown evidence of the effectiveness of ARTs in rapidly multiplying high-ranking genetics within a herd. By giving producers more control over which animals contribute genetics to the next generation, ARTs enable more intensive selection pressure and more rapid dissemination of valuable genetics. While some of these, like AI, have been widely adopted, others are currently limited to more advanced breeding programs. But all have potential applications for amplifying genetic improvement in beef herds.

5.2.2   Building an Integrated System for Genetic Improvement

Realizing the full value of ARTs requires integrating them into a cohesive genetic improvement program with well-defined breeding objectives and intense genetic selection at multiple levels. The objectives of a breeding program are often multi-faceted requiring the use of multiple ARTs to accomplish. Determining how to use the different ARTs to achieve your objectives is an integral part of program development. Each technique has its own unique advantages that can be applied to the herd. Granleese et al. demonstrated this need for synergism between ARTs by determining there to be an extra 25–60% gain when multiple ovulation embryo transfer was added to artificial insemination or natural breeding programs (Granleese et al. 2015). It was also noted that this gain could be increased even more by the implementation of genomic selection.

When selecting for genetic improvement in a beef cattle breeding program, there are several important traits to consider. Table 3 lists many of these traits as well as a brief description of their consequence to the beef industry.

The specific traits emphasized depend on the production system, breeding objectives, and market targets. Tools like EPDs, genomics, and indexes allow multi-trait selection. Crossbreeding can also complement strengths across breeds.

Continued research and technology refinements are working to enhance biological efficiencies, streamline protocols, and reduce expenses associated with advanced ARTs. This makes implementation more accessible and economically feasible for a wider swath of producers and enterprises pursuing accelerated genetic improvement (Lamb et al. 2016; Moore and Hasler 2017). Assisted reproductive technologies like embryo transfer (ET) and IVF have been effectively used to select for improved carcass traits in beef cattle breeding programs. There have been numerous studies examining the use of ARTs in conjunction with the selection of economically important traits to accelerate genetic improvement in beef cattle populations. Some examples of these traits are listed in Table 4.

These types of studies demonstrate how ARTs, coupled with intensive selection of economically relevant traits, can greatly accelerate the rate of genetic improvement and dissemination of superior genetics compared to conventional breeding schemes alone.

5.2.3 Challenges and Considerations

While ART techniques hold immense promise for improving meat quality in beef cattle, several challenges and considerations must be addressed to maximize their effectiveness. These include the selection of appropriate sires and donor females based on their genetic merit for meat quality traits, the optimization of reproductive management practices to ensure high conception rates (Dahlen et al. 2014; Timlin 2020), and the integration of genetic selection with other management strategies aimed at enhancing meat quality throughout the production chain.

Furthermore, factors such as cost-effectiveness and regulatory compliance may influence the adoption of ART techniques in beef cattle breeding. ARTs provide powerful capabilities but also require significant investment in facilities, equipment, technical expertise, and operating costs. Their economic viability depends on the projected returns from more rapid genetic gain outweighing the implementation costs (Lamb et al. 2016; Crowe et al. 2021). For commercial beef producers, factors like herd size, production system, breeding objectives, and point on the supply chain factor into determining whether adopting ART makes financial sense (Seidel Jr. 2014). By employing rigorous genetic selection and breeding practices informed by

Table 3 Important traits for genetic selection and beef breeding programs

Trait	Abbreviation for EPD or EBV	Consequence	Related sources
Growth traits			
Birth weight	BW	Moderate birth weights for calving ease but acceptable growth potential	Lykins et al. (2000)
Weaning weight	WW	Increased pre-weaning growth and milk production	Lykins et al. (2000)
Yearling weight	YW	Improved post-weaning growth and feed efficiency	Hough et al. (1985)
Carcass merit and feed efficiency			
Carcass weight	CW	Increased saleable meat yield	Crews (2002)
Rib eye area	REA/RE	Indicator of muscle expression and cutability	Cundiff et al. (1998)
Back fat thickness	Fat	Optimum level for quality grade and yield	Do et al. (2016), Naserkheil et al. (2021)
Marbling score	Mrb	Intramuscular fat for improved eating quality	Pringle et al. (2019)
Tenderness	SHR	Meat quality attributes like shear force and tenderness	Akanno et al. (2014)
Lean yield	YG	Dressing percentage and ratio of lean-to-fat yield	Fukuhara et al. (1989)
Residual feed intake	RADG/RFI	Improved feed conversion and efficiency	Pringle et al. (2019)
Fertility traits			
Scrotal circumference	SC/SCR	Indicator of bull fertility	Eler et al. (2004), Evans et al. (1999)
Heifer pregnancy rate	HP/HPG	Cycling and capability to conceive early	Evans et al. (1999), Eler et al. (2004), Moorey and Biase (2020)
Stayability	STAY	Longevity and reproductive lifespan in the herd	Sánchez-Castro et al. (2017)
Maternal traits			
Milk production	Milk	Adequate milking ability for calf pre-weaning growth	Mallinckrodt et al. (1993)
Calving ease direct/maternal	CED/CEM	Low dystocia for an easier calving process	Nugent et al. (1991)
Maternal weaning weight	MWW	Maternal influences on calf pre-weaning gain	Frazier et al. (1999)
Other functional traits			
Docility	Doc	Calmer temperament for handling and management	Beckman et al. (2007), Beckman (2008)
Disease resistance	N/A	Animal health and disease resistance	Berry et al. (2011)
Structural soundness	Claw/angle	Locomotion, feet, leg, udder conformation	Goldsmith (2022)

Table 4 Impact of ARTs on genetic improvement of beef cattle

Growth and carcass traits	A study by Cundiff et al. (1998) utilized embryo transfer to produce calves from elite donor dams and sires selected for increased muscularity and carcass cutability traits. The resulting progeny showed significant improvements in ribeye area and red meat yield compared to unselected controls (Cundiff et al. 1998)
	A recent study explored the use of IVF and genomic selection to produce embryos with high genomic breeding values for carcass traits in cattle (Fujii et al. 2021). This group previously examined the use of high genomic breeding values for carcass traits like carcass weight, ribeye area, and marbling score and fat. Subsequent progeny testing confirmed higher genetic merit for these traits (Fujii et al. 2019)
Reproductive traits	Researchers in Brazil have recently shown that using genomic selection in combination with in vitro embryo production (IVP) using oocytes from heifer calves is a powerful technology that reduces the generation interval and significantly increases the rate of genetic gain in cattle (Baruselli et al. 2021). Selection for fertility traits remains a complex and challenging problem (Kertz et al. 2023)
Feed efficiency	Heifers with better feed efficiency were found to be leaner and reached puberty later than those with lower feed efficiency (Randel and Welsh 2013). Conversely, another study showed that heifers with good feed efficiency reached puberty earlier than those with poorer feed efficiency (Canal et al. 2020)
Multi-trait selection	Researchers in Japan used genomic selection on biopsied embryonic cells from parents with high breeding values for carcass traits like carcass weight, ribeye area, and marbling score. Subsequent progeny testing of the ET calves confirmed their higher genetic merit for these economically relevant carcass characteristics (Fujii et al. 2019). This work was subsequently followed by a similar study using IVF embryos (Fujii et al. 2021)

scientific research, breeders can overcome many of these challenges and drive continuous improvement in both meat quality and quantity, ultimately meeting the demands of consumers for high-quality beef products.

In summary, assisted reproductive techniques represent powerful tools for enhancing meat quality traits in beef cattle breeding, offering breeders unprecedented opportunities to selectively propagate desirable genetic traits and drive genetic progress in regard to meat quality within their herds. By strategically leveraging AI, ET, and IVF in conjunction with the results of scientific research, breeders can accelerate the genetic improvement of meat quality traits, strengthen the competitiveness of the beef industry, and meet the evolving demands of consumers for premium beef products.

5.3   Thermotolerance of Temperate Breeds: From Cross Breeding to Gene Editing

Tropical and sub-tropical regions of the world are home to more than 80% of the world's cattle population, the majority of whom are *Bos indicus*-influenced breeds.

Bos indicus are well adapted to the warm and parasitic conditions prevailing in the tropics; however, compared to *Bos taurus*, *Bos indicus* are inferior in many facets related to feed efficiency, production, and reproduction. As a result, they are relatively inefficient, less economical, and have a higher carbon footprint compared to *Bos taurus* (Madalena 2002; Cooke et al. 2020; Marchioretto et al. 2023). Even though taurine cattle are superior in terms of production and reproduction-related traits, they are less tolerant to the heat and prevailing parasitic conditions of the tropics. Therefore, it was recognized that crossbreeding of *Bos taurus* and *Bos indicus* cattle could be used to produce thermotolerant synthetic/composite breeds that would exhibit superior milk production and fertility traits of *B. taurus* ancestors.

One of the very first attempts to produce such a thermotolerant composite breed came from Jamaica, as early as 1910. After decades of trial and error, the synthetic breed Jamaica Hope was finally declared a breed in 1952 (Wellington and Mahadevan 1977; Schneeberger et al. 1982). Other examples of *Bos taurus* × *Bos indicus* thermotolerant, tropically adapted composite breeds that were successfully developed are Australian Milking Zebu (AMZ), Australian Friesian Sahiwal (AFS), Girolando, and Guzerá. Development of these composite breeds involved extensive use of ARTs such as AI and MOET (Tierney 1992; Alexander and Tierney 1996; Campolina Diniz Peixoto et al. 2022). Systematic evaluations of the contribution of ARTs toward the development of composite breeds are rare. However, one study concluded that MOET had contributed significantly to the phenotypic and genetic progress of milk production-related traits in the Guzerá breed (Campolina Diniz Peixoto et al. 2022).

The abovementioned crossbreeding programs took multiple decades to develop a satisfactory synthetic breed. Yet, they were far from ideal and as a result only a very few of the above breeds, e.g., Girolando, are being commercially used today. As mentioned earlier, on the one hand, crossbreeding does not allow the luxury of handpicking thermotolerance-related genes, on the other hand, it allows hundreds, if not thousands of unintended and possibly deleterious genes to "leak" through to the offspring. In contrast to crossbreeding, genome editing makes it possible to handpick the genetic variations/mutations as per specific requirements, e.g., thermotolerance-related genes (Hansen 2020) thermotolerance-related genes (Hansen 2020; de Almeida Camargo and Pereira 2022).

To-date, genome editing has been used to transfer/introduce thermotolerance-related mutations, SLICK (p.Leu462* mutation in the prolactin receptor gene; PRLR; Rodriguez-Villamil et al. 2021) and p.Leu18del (mutation in the pre-melanosomal protein 17 gene; PMEL; Laible et al. 2021) in cattle. The SLICK/p. Leu462* mutation of *PRLR* results in a short, smooth coat increasing heat dissipation and thereby conferring thermotolerance. In contrast, p.Leu18del mutation of *PMEL* results in a lighter-colored coat reducing heat absorption and thereby conferring thermotolerance.

The SLICK mutation is a naturally occurring gene in Senepol cattle found in the Caribbean Island of St. Croix. The mutation confers thermotolerance to cattle by virtue of being responsible for a smooth, shiny, and short coat, which allows superior heat dissipation. Senepol cattle have been crossbred to Holstein to produce

Holsteins carrying the SLICK mutation; these Holsteins also have short and sleek coats and demonstrate superior thermoregulatory properties compared to non-SLICK counterparts with a rough and thick hair coat. As a result, it was shown that Holsteins carrying the SLICK mutation experience a diminished decline in milk yield under hot environmental conditions (Dikmen et al. 2014). Rectal temperature, respiration rate, skin temperature, and sweating rate of these cattle support the notion that the SLICK mutation contributes to thermotolerance. Several countries including the United States and Puerto Rico have incorporated the SLICK mutation into their Holstein breeding programs. Even commercial entities involved in AI are actively marketing Holstein bulls carrying the SLICK variant (Reviewed by Worku et al. 2023). Recently, the SLICK variant was successfully introgressed into Red Angus beef cattle through gene editing, specifically, using gRNA/Cas9 and embryo microinjection (Rodriguez-Villamil et al. 2021).

As mentioned earlier, p.Leu18del mutation of *PMEL*, a naturally occurring mutation found in Galloway and Highland cattle, confers thermotolerance by reducing heat absorption. Animals with black hair are known to absorb twice as much solar radiation compared to those with white hair. Therefore, animals with black coat colors are subjected to higher levels of heat stress compared to those with lighter-colored coats. In hot summers and in tropical countries, primarily black dairy cows, e.g., Holstein Friesian, exhibit a reduced ability to regulate body temperature. This has negative effects on both milk production levels and reproductive performance.

Mutations of the pre-melanosomal protein 17 gene (*PMEL*) are known to be associated with color dilution effects. A recent study took advantage of gene editing technology, specifically gRNA/Cas9, to introduce the naturally occurring p.Leu18del mutation of *PMEL* into Holstein Friesian cattle (Laible et al. 2021). Black coat markings of calves homozygous for the *PMEL* mutation were replaced with silvery gray markings whereas white areas remained unaffected. The thermotolerance levels of these calves had not been analyzed at the time, however, Holsteins with such a coat color are expected to absorb less heat and be highly heat tolerant.

Heat-tolerant Holsteins with the SLICK/p.Leu462* mutation of *PRLR* and p.Leu18del mutation of *PMEL* would be a great asset for the tropics as well as hot summers in temperate regions. Especially in the tropics, even modest improvements in feed/production efficiency would translate into substantial environmental benefits on a global scale. Recent estimates show that ~80% of the world's cattle reside in tropical or sub-tropical regions (Cooke et al. 2020, Marchioretto et al. 2023).

6   Ethical Considerations

Despite their massive potential, certain ARTs face many practical and ethical bottlenecks in terms of obtaining social acceptance for livestock breeding. This is especially true for genome editing-related applications. Some of them are related to animal welfare while others are ethical concerns not directly related to animal welfare (reviewed by Ishii 2017; Jans et al. 2018; Ritter et al. 2019; Campbell 2021).

6.1   Animal Welfare Issues Associated with ART Procedures

Most animal husbandry practices result in various levels of stress and pain due to animal handling, sorting, restraint, and social isolation. However, certain ARTs result in additional welfare-related concerns as summarized in Table 5.

Advocates of ARTs argue that ARTs have certain positive effects on animal welfare as summarized in Table 6.

Many of the ARTs are perceived as unnatural and therefore have poor social acceptance (Ishii 2017; Ritter et al. 2019). As such, several issues not directly related to animal welfare have been raised (reviewed by Ormandy et al. 2011); particularly related to gene editing, SCNT, and IVP, as they result in genetic and epigenetic modifications. It has been argued that these ARTs (1) produce "unnatural" animals with altered physiology and behavior, (2) violate species' integrity, (3) disregard the inherent value of animals, (4) disturb the natural "power balance" between humans and animals, (5) upset the natural balance of the ecosystem, and (6) generate livestock/products unfit for human consumption. In fact, the European Union has banned the sale of cloned livestock, their offspring, and products derived from them (Vogel 2015).

Table 5 Major concerns of ARTs on animal welfare

ART	Welfare-related matter
AI	• Increased risk of mass spreading of infectious and genetic diseases
Semen collection	• Pain associated with electro-ejaculation
Timed AI, ovulation induction, superovulation for OPU and MOET, surrogate synchronization in ET	• Pain and stress • Injection-site injuries/lesions • Unintended effects of exogenous hormone administration
Embryo flushing/collection/transfer	• Pain and stress
OPU	• Repeated OPU at short intervals • Using older females past their breeding age
SCNT	• Placental abnormalities • Fetal abnormalities • Dystocia • Higher perinatal mortality rates
IVP	• Dystocia due to large calves • Longer gestation periods • Higher pregnancy losses • Fetal abnormalities
AI, ET, IVP	• Denial of expression of natural mating behavior
Genetic selection	• Pain and stress during tissue sampling
Gene editing	• Sacrifice/waste of animals with unsuccessful edits

Table 6 Positive impacts of ARTs on animal welfare

ART	Welfare-related concerns
AI	• Reduces risk of spreading infectious and genetic conditions • Protects both males and females from injuries that may result from natural mating • Use of semen from polled animals eliminates the need for disbudding and dehorning

Sex-sorted semen	• Reduces male calf births subsequently reducing incidence of:
	  – Ill-treatment of male calves
	  – Dystocia due to feto-maternal disproportion

Genetic selection	• Selection against unfavorable traits, e.g., teat and hoof confirmation, reduces incidence of mastitis and lameness, respectively, in future generations
Gene editing	• Cattle with higher resistance to mastitis, tuberculosis, BVD, etc. • Polled cattle that do not need to undergo invasive dehorning procedures • Thermotolerant cattle with diminished heat stress • Compared to phenotypic selection, far fewer animals are required to generate a desired genotype

A counterargument to the above beliefs is that gene editing is simply a logical continuation of phenotype-based selective breeding, a practice carried out over many centuries; therefore, the final products of gene editing are as "natural" as those of conventional selective breeding. As a matter of fact, there are over 86.5 million known genetic variants among different breeds of cattle, including 84 million single nucleotide variants. Further, genetic variations do not pose a hazard in terms of food safety (Van Eenennaam 2017; Hayes and Daetwyler 2019). Supporting this notion, several countries including Japan, Brazil, and Australia have recently announced that genome-edited livestock with simple indel mutations will not be considered Genetically Modified Organisms (GMOs) because the technology does not involve the integration of exogenous DNA (Gim et al. 2023).

7 Future Directions

The different generations of ARTs, especially, AI and genomic selection, have contributed to the tremendous growth of the dairy and beef industries. However, there is room for improvement especially in the tropics in terms of production, feed efficiency, disease resistance, fertility, thermotolerance, and carbon footprint. While the benefits of improved production, feed efficiency, disease resistance, fertility, and thermotolerance will be mainly felt locally, the impact of the reduced carbon footprint will be felt globally. Therefore, every effort must be made to encourage the adoption of ARTs by tropical countries in addition to fine-tuning them for even further improvements for use in temperate countries.

7.1   AI, IVP, ET, and Genomic Selection: Improve Adoption in the Tropics and Further Development

Even though most temperate countries with well-developed dairy industries use AI to breed >80% of their dairy cows, the same cannot be said about many tropical countries. The use of sex-sorted semen is even lower in these developing countries. Therefore, widespread adoption of AI, preferably using sex-sorted semen, can make a great difference in the production efficiency, and subsequently, the carbon footprint. As mentioned above, the entire world will benefit from the reduced greenhouse gas emissions from the tropics. As such temperate countries can be expected to play an active role in developing breeds suitable for the tropics in the future (Marchioretto et al. 2023). Even in developed countries where AI coverage is >80%, there is potential for improvement of AI-associated technologies such as semen processing, semen cryopreservation, semen sorting, and semen analysis. Improvements in these areas will result in improved sire fertility and semen shelf-life, which in turn can have a positive impact on the dairy industry (Zuidema et al. 2021). The beef industry in developing countries would also benefit from these improvements in AI technology in their specific situation.

Despite the massive advancements in every aspect of IVP, on average, only ~27% of the transferred embryos will produce a live calf. Reasons for the ~73% pregnancy failures include deficiencies related to surrogate cow synchronization, IVP methodology (reviewed by Ealy et al. 2019), and pre-transfer embryo evaluation (Rabel et al. 2023). Every effort must be made to pinpoint and rectify mechanisms behind these deficiencies so that IVP-ET success rates can be improved soon. Artificial intelligence-related technologies are currently being tested for the evaluation of semen, oocytes, and embryos (reviewed by Rabel et al. 2023). Artificial intelligence is expected to result in major improvements in both AI and IVP-ET in the coming decade. The use of IVP-ET in the tropics is even lower than that of AI, given the intricacies of the technology. It is well known that ET results in higher pregnancy rates compared to AI under warm conditions such as those found in the tropics. Therefore, adoption of ET (IVP or IVD) should be encouraged in the tropics.

Genomic selection has already resulted in major improvements in genetic gain in the dairy industry. With the number of available genotypes and analyzed traits increasing, genetic gain will further increase resulting in improvements in traits related to production, health, and reproduction. Whether tropical countries are ready for genomic selection or not is debatable. Many tropical countries lack the underlying framework needed for genomic selection, e.g., milk recording. As such implementing genomic selection under such circumstances can be challenging.

The next great horizon of ART methods is the use of "in vitro breeding." In vitro breeding (VB) integrates two techniques: genomic selection and deriving germ cells from pluripotent stem cells. VB involves differentiating germ cells (which give rise to eggs and sperm) from embryonic stem cells grown in the lab. This method combined with genomic selection will dramatically increase the progress of genetically enhancing livestock in breeding programs (Goszczynski et al. 2018). These methods currently being applied to livestock arise from the groundbreaking work in mice where oogonia were produced from induced pluripotent stem cells (Yamashiro et al. 2018). VB holds great promise, but many technical issues need to be resolved before it becomes a practical method for improving genetics in livestock production. Only time will tell if this will be a viable suite of tools for assisted reproduction.

7.2   Gene Editing: Social Acceptance and Regulatory Approval

Winning social acceptance and regulatory approval for gene editing will be the major challenges that lie ahead in relation to gene editing. Even if gene editing gets the green light from a few countries, the lack of global harmony in terms of regulatory approval will be a challenge for the global trade of gene-edited animals and their products. Regulatory approval and social acceptance of gene editing will allow mass production of cattle with increased productivity, increased feed efficiency, increased disease resistance, increased fertility, increased thermotolerance, and decreased carbon footprint driving dairy and beef industries forward.

From a health perspective, gene editing can be used to produce not only disease-resistant cattle (as summarized in Table 1) but also virulence-attenuated parasites that can be used for producing vaccines (Hakimi et al. 2019; Pal and Dam 2022). Together, these will decrease the incidence of infectious diseases, which in turn reduces the use of antimicrobials and subsequently promotes improved animal health and public health.

It has been estimated that improving livestock production efficiencies in the ten countries with the largest emission reduction potential (i.e., Madagascar, Morocco, Niger, South Africa, Tanzania, China, India, Iran, Turkey, and Brazil) could decrease global livestock emissions by as much as 60–65% (Chang et al. 2021). This goal can be achieved by producing thermotolerant and feed-efficient cattle using ARTs. For example, cattle gene-edited for thermotolerance can be mass-bred using AI and IVP. Global warming gives even more importance to thermotolerant breeds. On the one hand, cattle will be subjected to even more heat stress in the tropics where it is already too warm for optimal performance. On the other hand, it has been estimated that global warming will lead to milk production losses of up to $1.7 billion/year by 2050 in North America. It has been estimated that milk production diminishes by 1.15 kg/day for every 1 °C rise in temperature above the thermal comfort zone. In addition to reduced milk production, heat stress will also affect meat and milk quality as well as reproductive efficiency. As such, the engineering of thermotolerant breeds is of utmost importance in the face of global warming (de Almeida Camargo and Pereira 2022). In this context, gene-edited cattle with the SLICK mutation in *PRLR* and p.Leu18del mutation in *PMEL* can be expected to play a significant role. Gene editing can also be used to insert certain known mutations associated with higher milk production. An example is the F279Y mutation of the growth hormone receptor (*GHR*) gene, associated with increased production of milk, casein, and lactose. It is known to be present in Holstein, Jersey, and Ayrshire breeds, but not in *Bos indicus* cattle. Introducing such mutations to *Bos indicus* cows using gene editing will help improve the productivity and feed efficiency of the dairy industry of the tropics (reviewed by de Almeida Camargo and Pereira 2022).

Producing polled cattle through gene editing is a way of tackling a major animal welfare issue at the genetic level. It has the potential to prevent welfare concerns associated with dehorning of cattle. As such, from a welfare point of view, introgression of the Celtic *POLLED* allele (P_C) in naturally horned breeds should receive high priority.

7.3   ARTs as Solutions to Tackle Current and Future Global Problems

Two of the biggest problems humans will face during the twenty-first century are (1) feeding the ever-expanding population and (2) global warming. On one hand, ARTs have the potential to improve quantities of dairy and beef products by improving the efficiency of production, so that the expanding population's animal protein requirements can be satisfied. On the other hand, by virtue of improving production efficiency, ARTs have the potential to reduce the carbon footprint of dairy and beef industries minimizing their contribution to global warming.

More than 60% of the livestock greenhouse gas emissions come from the dairy and beef industries. If this impact is to be reduced, higher-efficiency cattle (in terms of feed conversion, production, reproduction, disease resistance, and heat tolerance) need to be used in the dairy and beef industries. With regard to the dairy industry, 75% of the global greenhouse gas emissions come from developing/emerging countries, most of which are in the tropical and sub-tropical regions. Due to the relatively poor genetics and harsh environmental conditions, cattle in these regions are generally inefficient in feed conversion, growth, milk production, and reproduction and subsequently have a greater carbon footprint. Therefore, if greenhouse gas emissions from cattle are to be controlled, the tropics cannot be neglected. ARTs have the potential to provide sustainable solutions to this problem in the tropics. For example, on one hand, the widespread adoption of AI and IVP-ET will help rapidly disperse existing superior genetics within the region. On the other hand, gene editing will allow the creation of new genotypes that are more adapted to the tropics in terms of thermotolerance and disease resistance.

Feeding the rising global population by 2050 will require global food production to be ramped up by ~50%. Half of the world's habitable land is used for agriculture of which more than three-quarters is used for livestock (Cole 2019). Therefore, increasing land use for dairy and beef industries is not a practical solution. What can be done instead is to improve the production efficiency of dairy and beef cattle so that production per unit of land area or production per animal could be improved. Genomic selection and gene editing can be used to identify and incorporate genetic polymorphisms/mutations that are responsible for superior feed conversion, disease resistance, fertility, and heat resistance and then IVP and SCNT can be used for mass production of animals carrying such genetics culminating in overall improved production per unit land area.

8 Conclusions

ARTs, together with improved management practices have resulted in massive improvements in livestock industries. The US dairy industry is the perfect example where the national milk production increased by over 80% while the dairy cattle population plunged from 25.6 to 9 million heads (a 65% reduction) from 1944 to date. This was made possible because individual cow milk yield increased from 2000 kg/cow to over 10,000 kg/cow today. As a direct result of this, today, the carbon footprint from a glass of milk is estimated to be approximately one-third of its 1944 equivalent. It has been estimated that genetic improvements brought about by AI are responsible for ~50% of the above advancements.

Most of the above improvements took place during the pre-genomic selection era. Genomic selection is an extremely powerful technology that is still very new. Regardless, it has already doubled the rate of genetic gain in US dairy cattle since its introduction in 2008. Genomic selection is improving and evolving by the day and now incorporates over 50 traits related to (1) production, (2) conformation, (3) longevity, (4) fertility, (5) calving, (6) health, and (7) feed efficiency. With such a degree of comprehensiveness, genomic selection is expected to improve the above aspects even further.

In the field of livestock breeding, gene editing can be argued as the next best thing after sliced bread. Specifically, CRISPR-Cas9 technology has revolutionized gene editing and has been used to produce cattle with increased disease resistance (to BVDV, mastitis, tuberculosis, and BRD) increased carcass yield, increased thermotolerance, polledness, and hypoallergenic milk. Some of these gene-edited cattle have already been declared by the FDA as "low risk" and, therefore, are expected to receive complete regulatory approval within the foreseeable future.

There is a lot of potential for developing livestock industries, especially in the tropics where ARTs do not have widespread adoption yet. Increased productivity leading to reduced greenhouse gas emissions from the tropical and sub-tropical regions will help counter challenges related to global nutrition and global warming.

References

Akanno EC, Plastow G, Woodward BW, Bauck S, Okut H, Wu X-L, Sun C, Aalhus JL, Moore SS, Miller SP, Wang Z, Basarab JA (2014) Reliability of molecular breeding values for Warner-Bratzler shear force and carcass traits of beef cattle: An independent validation study. J Anim Sci 92(7):2896–2904. https://doi.org/10.2527/jas.2013-7374

Alexander GI, Tierney MJ (1996) Improved tropical dairy production. In: ACIAR proceedings

Arnould S, Perez C, Cabaniols J-P, Smith J, Gouble A, Grizot S, Epinat J-C, Duclert A, Duchateau P, Pâques F (2007) Engineered I-CreI derivatives cleaving sequences from the human XPC gene can induce highly efficient gene correction in mammalian cells. J Mol Biol 371(1):49–65. https://doi.org/10.1016/j.jmb.2007.04.079

Barrangou R, Fremaux C, Deveau H, Richards M, Boyaval P, Moineau S, Romero DA, Horvath P (2007) CRISPR provides acquired resistance against viruses in prokaryotes. Science 315(5819):1709–1712. https://doi.org/10.1126/science.1138140

Baruselli PS, Catussi BLC, Ângelo L, de Abreu F, Elliff M, Garcia L, da Silva, and Emiliana de Oliveira Santana Batista. (2019) Challenges to increase the AI and ET markets in Brazil. Anim Reprod 16:364–375. https://doi.org/10.21451/1984-3143-AR2019-0050

Baruselli PS, Rodrigues CA, Ferreira RM, Sales JNS, Elliff FM, Silva LG, Viziack MP, Factor L, D'Occhio MJ (2021) Impact of oocyte donor age and breed on in vitro embryo production in cattle, and relationship of dairy and beef embryo recipients on pregnancy and the subsequent performance of offspring: A review. Reprod Fertil Dev 34(2):36–51. https://doi.org/10.1071/rd21285

Beckman DW (2008) Docility EPD: A tool for temperament. In: Beef Improvement Federation—40th annual meeting, Calgary, Alberta, Canada, June 30–July 3, 2008

Beckman DW, Enns RM, Speidel SE, Brigham BW, Garrick DJ (2007) Maternal effects on docility in Limousin cattle. J Anim Sci 85(3):650–657. https://doi.org/10.2527/jas.2006-450

Berry DP, Bermingham ML, Good M, More SJ (2011) Genetics of animal health and disease in cattle. Ir Vet J 64(1):5. https://doi.org/10.1186/2046-0481-64-5

Berry DP, Wall E, Pryce JE (2014) Genetics and genomics of reproductive performance in dairy and beef cattle. Animal 8(Suppl 1):105–121. https://doi.org/10.1017/s1751731114000743

Bertolini M, Bertolini LR (2009) Advances in reproductive technologies in cattle: from artificial insemination to cloning. Revista de la Facultad de Medicina Veterinaria y de Zootecnia 56(III):184–194

Betteridge KJ (1977) Embryo transfer in farm animals: a review of techniques and applications. Health of Animals Branch Agriculture Canada Animal Pathology Division, Animal Diseases Research Institute (Eastern)

Bo GA, Mapletoft RJ (2018) Embryo transfer technology in cattle. In: Niemann H, Wrenzycki C (eds) Animal biotechnology 1: Reproductive biotechnologies. Springer, Cham, pp 107–133

Bollongino R, Burger J, Powell A, Mashkour M, Vigne J-D, Thomas MG (2012) Modern taurine cattle descended from small number of near-Eastern founders. Mol Biol Evol 29(9):2101–2104. https://doi.org/10.1093/molbev/mss092

Cabrera VE (2022) Economics of using beef semen on dairy herds. JDS Commun 3(2):147–151. https://doi.org/10.3168/jdsc.2021-0155

Camargo LSA, Saraiva NZ, Oliveira CS, Carmickle A, Lemos DR, Siqueira LGB, Denicol AC (2022) Perspectives of gene editing for cattle farming in tropical and subtropical regions. Anim Reprod 19(4):e20220108. https://doi.org/10.1590/1984-3143-ar2022-0108

Campbell MLH (2021) Ethics: use and misuse of assisted reproductive techniques across species. Reprod Fertil 2(3):C23–c28. https://doi.org/10.1530/raf-21-0004

Campolina Diniz Peixoto MG, Carrara ER, Lopes PS, Tomita Bruneli FÂ, Penna VM (2022) The contribution of a MOET nucleus scheme for the improvement of Guzerá (Bos indicus) cattle for milk traits in Brazil. Front Genet 13:982858. https://doi.org/10.3389/fgene.2022.982858

Canal LB, Fontes PLP, Sanford CD, Mercadante VRG, DiLorenzo N, Lamb GC, Oosthuizen N (2020) Relationships between feed efficiency and puberty in Bos taurus and Bos indicus-influenced replacement beef heifers. J Anim Sci 98(10). https://doi.org/10.1093/jas/skaa319

Carlson DF, Lancto CA, Zang B, Kim E-S, Walton M, Oldeschulte D, Seabury C, Sonstegard TS, Fahrenkrug SC (2016) Production of hornless dairy cattle from genome-edited cell lines. Nat Biotechnol 34(5):479–481. https://doi.org/10.1038/nbt.3560

Cermak T, Doyle EL, Christian M, Wang L, Zhang Y, Schmidt C, Baller JA, Somia NV, Bogdanove AJ, Voytas DF (2011) Efficient design and assembly of custom TALEN and other TAL effector-based constructs for DNA targeting. Nucleic Acids Res 39(12):e82. https://doi.org/10.1093/nar/gkr218

Cesarani A, Pulina G (2021) Farm animals are long away from natural behavior: open questions and operative consequences on animal welfare. Animals (Basel) 11(3). https://doi.org/10.3390/ani11030724

Chang J, Peng S, Yin Y, Ciais P, Havlik P, Herrero M (2021) The key role of production efficiency changes in livestock methane emission mitigation. Agu Adv 2(2):e2021AV000391. https://doi.org/10.1029/2021AV000391

Cole J (2019) Agriculture: Land use, food systems and biodiversity. In: Cole J (ed) Planetary health: human health in an era of global environmental change. CABI, p 69. https://doi.org/10.1079/9781789241655.0069

Cole JB, VanRaden PM (2018) Symposium review: Possibilities in an age of genomics: The future of selection indices1. J Dairy Sci 101(4):3686–3701. https://doi.org/10.3168/jds.2017-13335

Cooke RF, Daigle CL, Moriel P, Smith SB, Tedeschi LO, Vendramini JMB (2020) Cattle adapted to tropical and subtropical environments: social, nutritional, and carcass quality considerations. J Anim Sci 98(2). https://doi.org/10.1093/jas/skaa014

Crews DH (2002) The relationship between beef sire carcass EPD and progeny phenotype. Can J Anim Sci 82(4):503–506. https://doi.org/10.4141/a02-037

Crowe AD, Lonergan P, Butler ST (2021) Use of assisted reproduction techniques to accelerate genetic gain and increase value of beef production in dairy herds. J Dairy Sci 104(12):12189–12206. https://doi.org/10.3168/jds.2021-20281

Cundiff LV, Gregory KE, Koch RM (1998) Germplasm evaluation in beef cattle-cycle IV: birth and weaning traits2. J Anim Sci 76(10):2528–2535. https://doi.org/10.2527/1998.76102528x

Dahlen C, Larson J, Cliff Lamb G (2014) Impacts of reproductive technologies on beef production in the United States. In: Cliff Lamb G, DiLorenzo N (eds) Current and future reproductive technologies and world food production. Springer, New York, pp 97–114

de Almeida Camargo LS, Pereira JF (2022) Genome-editing opportunities to enhance cattle productivity in the tropics. CABI Agric Biosci 3(1):8. https://doi.org/10.1186/s43170-022-00075-w

DeFrancesco L (2011) Move over ZFNs: a new technology for genome editing may put the zinc finger nuclease franchise out of business, some believe. Not so fast, say the finger people. Nat Biotechnol 29(8):681–685

DeJarnette JM, Nebel RL, Marshall CE, Moreno JF, McCleary CR, Lenz RW (2008) Effect of sex-sorted sperm dosage on conception rates in holstein heifers and lactating cows. J Dairy Sci 91(5):1778–1785. https://doi.org/10.3168/jds.2007-0964

Dekkers JCM, Hailin S, Cheng J (2021) Predicting the accuracy of genomic predictions. Genet Sel Evol 53(1):55. https://doi.org/10.1186/s12711-021-00647-w

Dikmen S, Khan FA, Huson HJ, Sonstegard TS, Moss JI, Dahl GE, Hansen PJ (2014) The SLICK hair locus derived from Senepol cattle confers thermotolerance to intensively managed lactating Holstein cows. J Dairy Sci 97(9):5508–5520. https://doi.org/10.3168/jds.2014-8087

Do C, Park B, Kim S, Choi T, Yang B, Park S, Song H (2016) Genetic parameter estimates of carcass traits under national scale breeding scheme for beef cattle. Asian Australas J Anim Sci 29(8):1083–1094. https://doi.org/10.5713/ajas.15.0696

Ealy AD, Wooldridge LK, McCoski SR (2019) Board invited review: post-transfer consequences of in vitro-produced embryos in cattle. J Anim Sci 97(6):2555–2568. https://doi.org/10.1093/jas/skz116

Edwards J, Walton A, Siebenga J (1938) On the exchange of bull semen between England and Holland. J Agric Sci 28(3):503–508. https://doi.org/10.1017/S0021859600050929

Eler JP, Silva JAIIV, Evans JL, Ferraz JBS, Dias F, Golden BL (2004) Additive genetic relationships between heifer pregnancy and scrotal circumference in Nellore cattle. J Anim Sci 82(9):2519–2527. https://doi.org/10.2527/2004.8292519x

Elsden RP, Hasler JF, Seidel GE (1976) Non-surgical recovery of bovine eggs. Theriogenology 6(5):523–532. https://doi.org/10.1016/0093-691X(76)90120-5

Esrafili Taze Kand Mohammaddiyeh M, Rafat SA, Shodja J, Javanmard A, Esfandyari H (2023) Selective genotyping to implement genomic selection in beef cattle breeding. Front Genet 14:1083106. https://doi.org/10.3389/fgene.2023.1083106

Evans JL, Golden BL, Bourdon RM, Long KL (1999) Additive genetic relationships between heifer pregnancy and scrotal circumference in Hereford cattle. J Anim Sci 77(10):2621–2628. https://doi.org/10.2527/1999.77102621x

FDA (2022) Risk assessment summary—V-006378 PRLR-SLICK cattle. Food and Drug Administration, Washington, DC

Fernandes CADC, Miyauchi TM, de Figueiredo ACS, Palhão MP, Varago FC, Nogueira ESC, Neves JP, Miyauchi TA (2014) Hormonal protocols for in vitro production of Zebu and taurine embryos. Pesq Agrop Brasileira 49(10):813–817. https://doi.org/10.1590/S0100-204X2014001000008

Ferre LB, Kjelland ME, Stroebech LB, Hyttel P, Mermillod P, Ross PJ (2020) Review: Recent advances in bovine in vitro embryo production: reproductive biotechnology history and methods. Animal 14(5):991–1004. https://doi.org/10.1017/s1751731119002775

Fischer K, Schnieke A (2023) How genome editing changed the world of large animal research. Front Genome Ed 5. https://doi.org/10.3389/fgeed.2023.1272687

Foote RH (1979) Time of artificial insemination and fertility in dairy cattle. J Dairy Sci 62(2):355–358. https://doi.org/10.3168/jds.S0022-0302(79)83248-8

Foote RH (1981) New technologies in animal breeding, 1st edn. Academic, New York

Foote RH (2002) The history of artificial insemination: Selected notes and notables. J Anim Sci 80:1–10

Frazier EL, Sprott LR, Sanders JO, Dahm PF, Crouch JR, Turner JW (1999) Sire marbling score expected progeny difference and weaning weight maternal expected progeny difference associations with age at first calving and calving interval in Angus beef cattle. J Anim Sci 77(6):1322–1328. https://doi.org/10.2527/1999.7761322x

Fuerniss LK, Young JD, Hall JR, Wesley KR, Benitez OJ, Corah LR, Rathmann RJ, Johnson BJ (2023) Beef embryos in dairy cows: calfhood growth of Angus-sired calves from Holstein, Jersey, and crossbred beef dams. Transl Anim Sci 7(1):txad096. https://doi.org/10.1093/tas/txad096

Fujii T, Naito A, Hirayama H, Kashima M, Yoshino H, Hanamure T, Domon Y, Hayakawa H, Watanabe T, Moriyasu S, Kageyama S (2019) Potential of preimplantation genomic selection for carcass traits in Japanese Black cattle. J Reprod Dev 65(3):251–258. https://doi.org/10.1262/jrd.2019-009

Fujii T, Naito A, Moriyasu S, Kageyama S (2021) Potential of preimplantation genomic selection using the blastomere separation technique in bovine in vitro fertilized embryos. J Reprod Dev 67(2):155–159. https://doi.org/10.1262/jrd.2020-153

Fukuhara R, Moriya K, Harada H (1989) Estimation of genetic parameters and sire evaluation for carcass characteristics with special reference to yield grade of the new beef carcass grading standards. Jpn J Zootech Science 60(12):1128–1134

Galli C, Duchi R, Crotti G, Turini P, Ponderato N, Colleoni S, Lagutina I, Lazzari G (2003) Bovine embryo technologies. Theriogenology 59(2):599–616. https://doi.org/10.1016/S0093-691X(02)01243-8

Gao Y, Haibo W, Wang Y, Liu X, Chen L, Li Q, Cui C, Liu X, Zhang J, Zhang Y (2017) Single Cas9 nickase induced generation of NRAMP1 knockin cattle with reduced off-target effects. Genome Biol 18(1):13. https://doi.org/10.1186/s13059-016-1144-4

García-Ruiz A, Cole JB, VanRaden PM, Wiggans GR, Ruiz-López FJ, Van Tassell CP (2016) Changes in genetic selection differentials and generation intervals in US Holstein dairy cattle as a result of genomic selection. Proc Natl Acad Sci USA 113(28):E3995-4004. https://doi.org/10.1073/pnas.1519061113

Gim G-M, Kwon D-H, Eom K-H, Moon JH, Park J-H, Lee W-W, Jung D-J, Kim D-H, Yi J-K, Ha J-J, Lim K-Y, Kim J-S, Jang G (2022) Production of MSTN-mutated cattle without exogenous gene integration using CRISPR-Cas9. Biotechnol J 17(7):2100198. https://doi.org/10.1002/biot.202100198

Gim GM, Eom KH, Kwon DH, Jung DJ, Kim DH, Yi JK, Ha JJ, Lee JH, Lee SB, Son WJ, Yum SY, Lee WW, Jang G (2023) Generation of double knockout cattle via CRISPR-Cas9 ribonucleoprotein (RNP) electroporation. J Anim Sci Biotechnol 14(1):103. https://doi.org/10.1186/s40104-023-00902-8

Givens DM, Gard JA, Stringfellow DA (2007) Relative risks and approaches to biosecurity in the use of embryo technologies in livestock. Theriogenology 68(3):298–307. https://doi.org/10.1016/j.theriogenology.2007.04.004

Gledhill BL, Pinkel D, Garner DL (1982) Identifying X- and Y-chromosome-bearing sperm by DNA content: retrospective perspectives and prospective opinions. In: Conference on prospects of sexing mammalian sperm, Denver, CO, USA, March 18, 1982

Goldsmith TJ (2022) A new angle on the bovine foot. In: Proceedings of the fifty-fifth annual conference, American Association of Bovine Practitioners, Long Beach, CA, USA, July 17, 2023

Goszczynski DE, Cheng H, Demyda-Peyrás S, Medrano JF, Wu J, Ross PJ (2018) In vitro breeding: application of embryonic stem cells to animal production. Biol Reprod 100(4):885–895. https://doi.org/10.1093/biolre/ioy256

Granleese T, Clark SA, Swan AA, van der Werf JHJ (2015) Increased genetic gains in sheep, beef and dairy breeding programs from using female reproductive technologies combined with optimal contribution selection and genomic breeding values. Genet Sel Evol 47(1):70. https://doi.org/10.1186/s12711-015-0151-3

Grosu HB, Lungu SA, Oltenacu PA (2013) History of genetic evaluation methods in dairy cattle I. Daughter-dam comparisons. The Publishing House of the Romanian Academy, Bucharest

Guliński P (2021) Cattle breeds—contemporary views on their origin and criteria for classification: a review. Acta Scientiarum Polonorum Zootechnica 20(2):3–18. https://doi.org/10.21005/asp.2021.20.2.01

Gutierrez-Reinoso MA, Aponte PM, Garcia-Herreros M (2021) Genomic analysis, progress and future perspectives in dairy cattle selection: a review. Animals (Basel) 11:(3). https://doi.org/10.3390/ani11030599

Hakimi H, Ishizaki T, Kegawa Y, Kaneko O, Kawazu S-i, Asada M (2019) Genome editing of Babesia bovis using the CRISPR/Cas9 system. mSphere 4(3). https://doi.org/10.1128/msphere.00109-19

Hansen PJ (2020) Prospects for gene introgression or gene editing as a strategy for reduction of the impact of heat stress on production and reproduction in cattle. Theriogenology 154:190–202. https://doi.org/10.1016/j.theriogenology.2020.05.010

Hasler JF (2014) Forty years of embryo transfer in cattle: A review focusing on the journal Theriogenology, the growth of the industry in North America, and personal reminisces. Theriogenology 81(1):152–169. https://doi.org/10.1016/j.theriogenology.2013.09.010

Hasler JF, Barfield JP (2021) In vitro fertilization. In: Hopper RM (ed) Bovine reproduction. Wiley Blackwell, Hoboken, NJ, pp 1124–1141. Original edition, 2014

Hayes BJ, Daetwyler HD (2019) 1000 Bull Genomes Project to map simple and complex genetic traits in cattle: applications and outcomes. Annu Rev Anim Biosci 7:89–102. https://doi.org/10.1146/annurev-animal-020518-115024

Hough JD, Benyshek LL, Mabry JW (1985) Direct and correlated response to yearling weight selection in hereford cattle using nationally evaluated sires. J Anim Sci 61(6):1335–1344. https://doi.org/10.2527/jas1985.6161335x

Ishii T (2017) Genome-edited livestock: Ethics and social acceptance. Anim Front 7(2):24–32. https://doi.org/10.2527/af.2017.0115

Ivanoff EI (1922) On the use of artificial insemination for zootechnical purposes in Russia. J Agric Sci 12(3):244–256. https://doi.org/10.1017/S002185960000530X

Jans V, Dondorp W, Goossens E, Mertes H, Pennings G, de Wert G (2018) Balancing animal welfare and assisted reproduction: ethics of preclinical animal research for testing new reproductive technologies. Med Health Care Philos 21(4):537–545. https://doi.org/10.1007/s11019-018-9827-0

Johnsson M (2023) Genomics in animal breeding from the perspectives of matrices and molecules. Hereditas 160. https://doi.org/10.1186/s41065-023-00285-w

Kandel ME, Rubessa M, He YR, Schreiber S, Meyers S, Matter Naves L, Sermersheim MK, Sell GS, Szewczyk MJ, Sobh N, Wheeler MB, Popescu G (2020) Reproductive outcomes predicted by phase imaging with computational specificity of spermatozoon ultrastructure. Proc Natl Acad Sci USA 117(31):18302–18309. https://doi.org/10.1073/pnas.2001754117

Kertz NC, Banerjee P, Dyce PW, Diniz WJS (2023) Harnessing genomics and transcriptomics approaches to improve female fertility in beef cattle—a review. Animals (Basel) 13(20). https://doi.org/10.3390/ani13203284

Kozicka K, Žukovskis J, Wójcik-Gront E (2023) Explaining global trends in cattle population changes between 1961 and 2020 directly affecting methane emissions. Sustainability 15(13):10533. https://doi.org/10.3390/su151310533

Laible G, Cole SA, Brophy B, Wei J, Leath S, Jivanji S, Littlejohn MD, Wells DN (2021) Holstein Friesian dairy cattle edited for diluted coat color as a potential adaptation to climate change. BMC Genomics 22(1):856. https://doi.org/10.1186/s12864-021-08175-z

Lamb GC, Dahlen CR, Brown DR (2003) Reproductive ultrasonography for monitoring ovarian structure development, fetal development, embryo survival, and twins in beef cows. Presented at the managing reproduction in beef cattle symposium as a part of the 2002 Midwest ASAS and ADSA Regional Meeting in Des Moines, IA in March 2002. Prof Anim Sci 19(2):135–143. https://doi.org/10.15232/S1080-7446(15)31392-9

Lamb GC, Mercadante VRG, Henry DD, Fontes PLP, Dahlen CR, Larson JE, DiLorenzo N (2016) Advantages of current and future reproductive technologies for beef cattle production. Prof Anim Sci 32(2):162–171. https://doi.org/10.15232/pas.2015-01455

Lamberson W, Thomas J (2021) Crossbreeding systems for small herds of beef cattle. University of Missouri Extension, Columbia, MO

Leonard M, Kirszenbaum M, Cotinot M, Chesne P, Heyman Y, Stinnakre MG, Bishop C, Delouis C, Vaiman M, Fellous M (1987) Sexing bovine embryos using Y chromosome specific DNA probe. Theriogenology 27(1):248

Leroy G, Boettcher P, Scherf B, Hoffmann I, Notter DR (2018) Breeding of animals. In: Reference module in life sciences. Elsevier

Lieberman D, McClure E, Harston S, Madan D (2016) Maintaining semen quality by improving cold chain equipment used in cattle artificial insemination. Sci Rep 6(1):28108. https://doi.org/10.1038/srep28108

Liu X, Wang Y, Guo W, Chang B, Liu J, Guo Z, Quan F, Zhang Y (2013) Zinc-finger nickase-mediated insertion of the lysostaphin gene into the beta-casein locus in cloned cows. Nat Commun 4(1):2565. https://doi.org/10.1038/ncomms3565

Liu X, Wang Y, Tian Y, Yu Y, Gao M, Hu G, Su F, Pan S, Luo Y, Guo Z, Quan F, Zhang Y (2014) Generation of mastitis resistance in cows by targeting human lysozyme gene to β-casein locus using zinc-finger nucleases. Proc R Soc B Biol Sci 281(1780):20133368. https://doi.org/10.1098/rspb.2013.3368

Lonergan P (2018) Review: Historical and futuristic developments in bovine semen technology. Animal 12:s4–s18. https://doi.org/10.1017/S175173111800071X

Looper M (2000) Proper semen handling improves conception rates of dairy cows. New Mexico State University College of Agriculture, Home Economics Cooperative Extension Service and U.S. Department of Agriculture Cooperating, Las Cruces, NM

Luo J, Song Z, Shengli Y, Cui D, Wang B, Ding F, Li S, Dai Y, Li N (2014) Efficient generation of myostatin (MSTN) biallelic mutations in cattle using zinc finger nucleases. PLoS One 9(4):e95225. https://doi.org/10.1371/journal.pone.0095225

Lykins LE Jr, Bertrand JK, Baker JF, Kiser TE (2000) Maternal birth weight breeding value as an additional factor to predict calf birth weight in beef cattle. J Anim Sci 78(1):21–26. https://doi.org/10.2527/2000.78121x

Macháty Z, Páldi A, Csáki T, Varga Z, Kiss I, Bárándi Z, Vajta G (1993) Biopsy and sex determination by PCR of IVF bovine embryos. J Reprod Fertil 98(2):467–470. https://doi.org/10.1530/jrf.0.0980467

Madalena FE (2002) Dairy animals I Bos indicus breeds and Bos indicus × Bos taurus crosses. In: Roginski H (ed) Encyclopedia of dairy sciences. Elsevier, Oxford, pp 576–585

Mallinckrodt CH, Bourdon RM, Golden BL, Schalles RR, Odde KG (1993) Relationship of maternal milk expected progeny differences to actual milk yield and calf weaning weight. J Anim Sci 71(2):355–362. https://doi.org/10.2527/1993.712355x

Mapletoft RJ (2013) History and perspectives on bovine embryo transfer. Anim Reprod 10(3):168–173

Marchioretto PV, Chanaka Rabel RA, Allen CA, Ole-Neselle MMB, Wheeler MB (2023) Development of genetically improved tropical-adapted dairy cattle. Anim Front 13(5):7–15. https://doi.org/10.1093/af/vfad050

Maxwell WMC, Evans G, Hollinshead FK, Bathgate R, De Graaf SP, Eriksson BM, Gillan L, Morton KM, O'Brien JK (2004) Integration of sperm sexing technology into the ART toolbox. Anim Reprod Sci 82-83:79–95. https://doi.org/10.1016/j.anireprosci.2004.04.013

McTavish EJ, Decker JE, Schnabel RD, Taylor JF, Hillis DM (2013) New world cattle show ancestry from multiple independent domestication events. Proc Natl Acad Sci USA 110(15):E1398–E1406. https://doi.org/10.1073/pnas.1303367110

Miglior F, Fleming A, Malchiodi F, Brito LF, Martin P, Baes CF (2017) A 100-year review: identification and genetic selection of economically important traits in dairy cattle. J Dairy Sci 100(12):10251–10271. https://doi.org/10.3168/jds.2017-12968

Miksch ED, LeFever DG, Mukembo G, Spitzer JC, Wiltbank JN (1978) Synchronization of estrus in beef cattle. II. Effect of an injection of norgestomet and an estrogen in conjunction with a norgestomet implant in heifers and cows. Theriogenology 10(2-3):201–221. https://doi.org/10.1016/0093-691X(78)90020-1

Monzani PS, Adona PR, Ohashi OM, Meirelles FV, Wheeler MB (2016) Transgenic bovine as bioreactors: Challenges and perspectives. Bioengineered 7(3):123–131. https://doi.org/10.1080/21655979.2016.1171429

Monzani PS, Adona PR, Long SA, Wheeler MB (2022) Cows as bioreactors for the production of nutritionally and biomedically significant proteins. In: Guoyao W (ed) Recent advances in animal nutrition and metabolism. Springer, Cham, pp 299–314

Monzani PS, Sangalli JR, Sampaio RV, Guemra S, Zanin R, Adona PR, Berlingieri MA, Cunha-Filho LFC, Mora-Ocampo IY, Pirovani CP, Meirelles FV, Wheeler MB, Ohashi OM (2024) Human proinsulin production in the milk of transgenic cattle. Biotechnol J 19(3):2300307. https://doi.org/10.1002/biot.202300307

Moore SG, Hasler JF (2017) A 100-year review: Reproductive technologies in dairy science. J Dairy Sci 100(12):10314–10331. https://doi.org/10.3168/jds.2017-13138

Moorey SE, Biase FH (2020) Beef heifer fertility: importance of management practices and technological advancements. J Anim Sci Biotechnol 11(1):97. https://doi.org/10.1186/s40104-020-00503-9

Mueller ML, Van Eenennaam AL (2022) Synergistic power of genomic selection, assisted reproductive technologies, and gene editing to drive genetic improvement of cattle. CABI Agric Biosci 3(1):13. https://doi.org/10.1186/s43170-022-00080-z

Murphy EM, Kelly AK, O'Meara C, Eivers B, Lonergan P, Fair S (2018) Influence of bull age, ejaculate number, and season of collection on semen production and sperm motility parameters in Holstein Friesian bulls in a commercial artificial insemination centre. J Anim Sci 96(6):2408–2418. https://doi.org/10.1093/jas/sky130

Naserkheil M, Lee D, Chung K, Park MN, Mehrban H (2021) Estimation of genetic correlations of primal cut yields with carcass traits in Hanwoo beef cattle. Animals (Basel) 11(11). https://doi.org/10.3390/ani11113102

Nowicki A (2021) Embryo transfer as an option to improve fertility in repeat breeder dairy cows. J Vet Res 65(2):231–237. https://doi.org/10.2478/jvetres-2021-0018

Nugent RA III, Notter DR, Beal WE (1991) Body measurements of newborn calves and relationship of calf shape to sire breeding values for birth weight and calving ease. J Anim Sci 69(6):2413–2421. https://doi.org/10.2527/1991.6962413x

Oliveira CS, Camargo LSA, da Silva M, Saraiva NZ, Quintão CC, Machado MA (2023) Embryo biopsies for genomic selection in tropical dairy cattle. Anim Reprod 20(2):e20230064. https://doi.org/10.1590/1984-3143-ar2023-0064

Ombelet W, Van Robays J (2015) Artificial insemination history: hurdles and milestones. Facts Views Vis Obgyn 7(2):137–143

Ormandy EH, Dale J, Griffin G (2011) Genetic engineering of animals: ethical issues, including welfare concerns. Can Vet J 52(5):544–550

Pal S, Dam S (2022) CRISPR-Cas9: Taming protozoan parasites with bacterial scissor. J Parasit Dis 46(4):1204–1212. https://doi.org/10.1007/s12639-022-01534-x

Parkinson TJ, Morrell JM (2019) 43—Artificial insemination. In: Noakes DE, Parkinson TJ, England GCW (eds) Veterinary reproduction and obstetrics, 10th edn. W.B. Saunders, St. Louis, MO, pp 746–777

Parrish JJ, Krogenaes A, Susko-Parrish JL (1995) Effect of bovine sperm separation by either swim-up or Percoll method on success of in vitro fertilization and early embryonic development. Theriogenology 44(6):859–869. https://doi.org/10.1016/0093-691x(95)00271-9

Perry EJ (1945) Historical background. In: Perry EJ (ed) The artificial insemination of farm animals. Rutgers University Press, New Brunswick, NJ, pp 3–12. Original edition, 1945

Polge C (2007) The work of the Animal Research Station, Cambridge. Stud Hist Philos Biol Biomed Sci 38(2):511–520. https://doi.org/10.1016/j.shpsc.2007.03.011

Polge C, Rowson LEA (1952) Fertilizing capacity of bull spermatozoa after freezing at −79° C. Nature 169(4302):626–627. https://doi.org/10.1038/169626b0

Polge C, Smith AU, Parkes AS (1949) Revival of spermatozoa after vitrification and dehydration at low temperatures. Nature 164(4172):666–666. https://doi.org/10.1038/164666a0

Pringle TD, Segers J, Wells J, Detweiler R, Rekaya R, Gilleland H, Thinguldstad H (2019) The impact of selection using residual average daily gain and marbling EPDs on growth performance and carcass traits in angus cattle. Meat Muscle Biol 1(3)

Proudfoot C, Carlson DF, Huddart R, Long CR, Pryor JH, King TJ, Lillico SG, Mileham AJ, McLaren DG, Whitelaw CB, Fahrenkrug SC (2015) Genome edited sheep and cattle. Transgenic Res 24(1):147–153. https://doi.org/10.1007/s11248-014-9832-x

Rabel RAC, Marchioretto PV, Bangert EA, Wilson K, Milner DJ, Wheeler MB (2023) Pre-implantation bovine embryo evaluation—from optics to omics and beyond. Animals 13(13):2102. https://doi.org/10.3390/ani13132102

Randel RD, Welsh TH Jr (2013) Joint Alpharma-Beef Species Symposium: interactions of feed efficiency with beef heifer reproductive development. J Anim Sci 91(3):1323–1328. https://doi.org/10.2527/jas.2012-5679

Retallick KJ, Lu D, Garcia A, Miller SP (2022) Genomic selection in the US: where it has been and where it is going? In: Proceedings of 12th World Congress on Genetics Applied to Livestock Production (WCGALP) technical and species orientated innovations in animal breeding, and contribution of genetics to solving societal challenges: 1795–1798

Ritter C, Beaver A, von Keyserlingk MAG (2019) The complex relationship between welfare and reproduction in cattle. Reprod Domest Anim 54(S3):29–37. https://doi.org/10.1111/rda.13464

Rodriguez-Villamil P, Ongaratto FL, Bostrom JR, Larson S, Sonstegard T (2021) 13 Generation of SLICK beef cattle by embryo microinjection: A case report. Reprod Fertil Dev 33(2):114–114. https://doi.org/10.1071/RDv33n2Ab13

Romano JE, Thompson JA, Kraemer DC, Westhusin ME, Forrest DW, Tomaszweski MA (2007) Early pregnancy diagnosis by palpation per rectum: Influence on embryo/fetal viability in dairy cattle. Theriogenology 67(3):486–493. https://doi.org/10.1016/j.theriogenology.2006.08.011

Rowe RF, Del Campo MR, Eilts CL, French LR, Winch RP, Ginther OJ (1976) A single cannula technique for nonsurgical collection of ova from cattle. Theriogenology 6(5):471–483. https://doi.org/10.1016/0093-691X(76)90114-X

Rubessa M, Boccia L, Campanile G, Longobardi V, Albarella S, Tateo A, Zicarelli L, Gasparrini B (2011) Effect of energy source during culture on in vitro embryo development, resistance to cryopreservation and sex ratio. Theriogenology 76(7):1347–1355. https://doi.org/10.1016/j.theriogenology.2011.06.004

Rubessa M, Ambrosi A, Gonzalez-Pena D, Polkoff KM, Denmark SE, Wheeler MB (2016) Non-invasive analysis of bovine embryo metabolites during in vitro embryo culture using nuclear magnetic resonance. AIMS Bioeng 3(4):538–551. https://doi.org/10.3934/bioeng.2016.4.538

Rubessa M, Kandel ME, Schreiber S, Meyers S, Beckd DH, Popescu G, Wheeler MB (2020) Morphometric analysis of sperm used for IVP by three different separation methods with spatial light interference microscopy. Syst Biol Reprod Med. https://doi.org/10.1080/1939636 8.2019.1701139

Sales JNS, Neves KAL, Souza AH, Crepaldi GA, Sala RV, Fosado M, Campos Filho EP, de Faria M, Sá Filho MF, Baruselli PS (2011) Timing of insemination and fertility in dairy and beef cattle receiving timed artificial insemination using sex-sorted sperm. Theriogenology 76(3):427–435. https://doi.org/10.1016/j.theriogenology.2011.02.019

Sánchez-Castro MA, Boldt RJ, Thomas MG, Enns RM, Speidel SE (2017) Expected progeny differences for stayability in Angus cattle using a random regression model. J Anim Sci 95(suppl_4):88–89. https://doi.org/10.2527/asasann.2017.179

Schmitt MR, VanRaden PM, De Vries A (2019) Ranking sires using genetic selection indices based on financial investment methods versus lifetime net merit. J Dairy Sci 102(10):9060–9075. https://doi.org/10.3168/jds.2018-16081

Schneeberger CP, Wellington KE, McDowell RE (1982) Performance of Jamaica Hope cattle in commercial dairy herds in Jamaica. J Dairy Sci 65(7):1364–1371. https://doi.org/10.3168/jds. S0022-0302(82)82354-0

Seidel GE Jr (1981a) Critical review of embryo transfer procedures with cattle. In: Mastroianni L, Biggers JD (eds) Fertilization and embryonic development in vitro. Springer, Boston, MA, pp 323–353

Seidel GE Jr (1981b) Superovulation and embryo transfer in cattle. Science 211(4480):351–358

Seidel GE Jr (1984) Applications of embryo transfer and related technologies to cattle. J Dairy Sci 67(11):2786–2796. https://doi.org/10.3168/jds.S0022-0302(84)81635-5

Seidel GE Jr (2014) Beef cattle in the year 2050. In: Lamb GC, DiLorenzo N (eds) Current and future reproductive technologies and world food production. Springer, New York, NY, pp 239–244

Seidel GE Jr, Seidel SM (1991) Training manual for embryo transfer in cattle, vol 77. FAO, Rome, Italy

Shanthalingam S, Tibary A, Beever JE, Kasinathan P, Brown WC, Srikumaran S (2016) Precise gene editing paves the way for derivation of Mannheimia haemolytica leukotoxin-resistant cattle. Proc Natl Acad Sci USA 113(46):13186–13190. https://doi.org/10.1073/pnas.1613428113

Slenning BD, Wheeler MB (1989) Risk evaluation for bovine embryo transfer services using computer simulation and economic decision theory. Theriogenology 31(3):653–673. https://doi.org/10.1016/0093-691x(89)90249-5

Smith J, Bibikova M, Whitby FG, Reddy AR, Chandrasegaran S, Carroll D (2000) Requirements for double-strand cleavage by chimeric restriction enzymes with zinc finger DNA-recognition domains. Nucleic Acids Res 28(17):3361–3369. https://doi.org/10.1093/nar/28.17.3361

Spitzer JC, Burrell WC, LeFever DG, Whitman RW, Wiltbank JN (1978) Synchronization of estrus in beef cattle. I. Utilization of a norgestomet implant and injection of estradiol valerate. Theriogenology 10(2-3):181–200. https://doi.org/10.1016/0093-691X(78)90019-5

Stevenson JS, Britt JH (2017) A 100-year review: Practical female reproductive management. J Dairy Sci 100(12):10292–10313. https://doi.org/10.3168/jds.2017-12959

Stewart DL (1951) Storage of bull spermatozoa at low temperatures. Vet Rec 65:65–66

Tervit HR, Whittingham DG, Rowson LE (1972) Successful culture in vitro of sheep and cattle ova. J Reprod Fertil 30(3):493–497. https://doi.org/10.1530/jrf.0.0300493

Tierney ML (1992) The AFS—a tropical dairy cattle export resource. In: Proceedings of the Australian Association of Animal Breeding and Genetics

Timlin CL (2020) Strategies for improving reproductive efficiency of beef cattle with assisted reproductive technologies. Ph.D., Animal and Poultry Science, Virginia Polytechnic Institute and State University

Upadhyay MR, Chen W, Lenstra JA, Goderie CRJ, MacHugh DE, Park SDE, Magee DA, Matassino D, Ciani F, Megens HJ, van Arendonk JAM, Groenen MAM, Marsan PA, Balteanu V, Dunner S, Garcia JF, Ginja C, Kantanen J, Consortium European Cattle Genetic Diversity, and Rpma Crooijmans (2017) Genetic origin, admixture and population history of aurochs (Bos primigenius) and primitive European cattle. Heredity 118(2):169–176. https://doi.org/10.1038/hdy.2016.79

Van Eenennaam AL (2017) Genetic modification of food animals. Curr Opin Biotechnol 44:27–34. https://doi.org/10.1016/j.copbio.2016.10.007

Viana JHM (2018) 2017 Statistics of embryo production and transfer in domestic farm animals: Is it a turning point? In 2017 more in vitro-produced than in vivo-derived embryos were transferred worldwide. Embryo Technol Newsl 36:46

Viana JHM (2023) 2022 Statistics of embryo production and transfer in domestic farm animals: The main trends for the world embryo industry still stand. Embryo Technol Newsl 41:48

Vogel G (2015) EU parliament votes to ban cloning of farm animals. Science News. Retrieved from http://www.sciencemag.org/news/2015/09/eu-parliament-votes-ban-cloning-farm-animals

Walker C (2020) Investigating cattle artificial insemination technique on farm. Livestock 25(1):13–18. https://doi.org/10.12968/live.2020.25.1.13

Wassarman PM, Schultz RM, Letourneau GE, LaMarca MJ, Josefowicz WJ, Bleil JD (1979) Meiotic maturation of mouse oocytes in vitro. In: Channing CP, Marsh JM, Sadler WA (eds) Ovarian follicular and corpus luteum function. Springer, Boston, MA, pp 251–268

Wei J, Wagner S, Maclean P, Brophy B, Cole S, Smolenski G, Carlson DF, Fahrenkrug SC, Wells DN, Laible G (2018) Cattle with a precise, zygote-mediated deletion safely eliminate the major milk allergen beta-lactoglobulin. Sci Rep 8(1):7661. https://doi.org/10.1038/s41598-018-25654-8

Weigel KA, VanRaden PM, Norman HD, Grosu H (2017) A 100-year review: methods and impact of genetic selection in dairy cattle—from daughter–dam comparisons to deep learning algorithms. J Dairy Sci 100(12):10234–10250. https://doi.org/10.3168/jds.2017-12954

Weller JI, Ezra E, Ron M (2017) Invited review: A perspective on the future of genomic selection in dairy cattle. J Dairy Sci 100(11):8633–8644. https://doi.org/10.3168/jds.2017-12879

Wellington KE, Mahadevan P (1977) Development of the Jamaica hope breed of dairy cattle. FAO Anim Prod Health Pap 1:67–72

Wiggans GR, Carrillo JA (2022) Genomic selection in United States dairy cattle. Front Genet 13. https://doi.org/10.3389/fgene.2022.994466

Wiggans GR, Cole JB, Hubbard SM, Sonstegard TS (2017) Genomic selection in dairy cattle: The USDA experience. Annu Rev Anim Biosci 5(1):309–327. https://doi.org/10.1146/annurev-animal-021815-111422

Willett EL, Black WG, Casida LE, Stone WH, Buckner PJ (1951) Successful transplantation of a fertilized bovine ovum. Science 113(2931):247–247

Workman AM, Heaton MP, Vander Ley BL, Webster DA, Sherry L, Bostrom JR, Larson S, Kalbfleisch TS, Harhay GP, Jobman EE, Carlson DF, Sonstegard TS (2023) First gene-edited calf with reduced susceptibility to a major viral pathogen. PNAS Nexus 2(5). https://doi.org/10.1093/pnasnexus/pgad125

Worku D, Hussen J, De Matteis G, Schusser B, Alhussien MN (2023) Candidate genes associated with heat stress and breeding strategies to relieve its effects in dairy cattle: a deeper insight into the genetic architecture and immune response to heat stress. Front Vet Sci 10. https://doi.org/10.3389/fvets.2023.1151241

Wu H, Wang Y, Zhang Y, Yang M, Lv J, Liu J, Zhang Y (2015) TALE nickase-mediated SP110 knockin endows cattle with increased resistance to tuberculosis. Proc Natl Acad Sci USA 112(13):E1530-9. https://doi.org/10.1073/pnas.1421587112

Wykes D (2004) Robert Bakewell (1725-1795) of Dishley: Farmer and livestock improver. Agric Hist Rev 52:38–55

Yamashiro C, Sasaki K, Yabuta Y, Kojima Y, Nakamura T, Okamoto I, Yokobayashi S, Murase Y, Ishikura Y, Shirane K, Sasaki H, Yamamoto T, Saitou M (2018) Generation of human oogonia from induced pluripotent stem cells in vitro. Science 362(6412):356–360. https://doi.org/10.1126/science.aat1674

Young AE, Mansour TA, McNabb BR, Owen JR, Trott JF, Titus Brown C, Van Eenennaam AL (2020) Genomic and phenotypic analyses of six offspring of a genome-edited hornless bull. Nat Biotechnol 38(2):225–232. https://doi.org/10.1038/s41587-019-0266-0

Yuan M, Zhang J, Gao Y, Yuan Z, Zhu Z, Wei Y, Wu T, Han J, Zhang Y (2021) HMEJ-based safe-harbor genome editing enables efficient generation of cattle with increased resistance to tuberculosis. J Biol Chem 296. https://doi.org/10.1016/j.jbc.2021.100497

Zuidema D, Kerns K, Sutovsky P (2021) An exploration of current and perspective semen analysis and sperm selection for livestock artificial insemination. Animals (Basel) 11(12). https://doi.org/10.3390/ani11123563

Part IV

The Importance of Male Factors in ART

Assessment of Boar Sperm Quality: New Diagnostic Techniques

Jon Romero-Aguirregomezcorta ⓘ, Laura Abril Parreño ⓘ,
Armando Quintero Montero ⓘ,
and Joaquín Gadea Mateos ⓘ

Abstract

Evaluation of boar semen quality is essential for diagnosing reproductive problems and optimizing pig production. Traditional parameters such as sperm concentration, motility, and morphology are often used, but they do not always accurately predict fertility. Computer Assisted Semen Analysis (CASA) technology improves semen evaluation by providing automated and accurate measurements of sperm motility, concentration, and morphology. Over the years, flow cytometry has gained popularity as a powerful tool. It facilitates high-throughput analysis, allowing the rapid and accurate examination of large numbers of sperm cells. Flow cytometry rapidly evaluates sperm viability by assessing the integrity of the plasma membrane, and it also analyzes DNA integrity, which is critical for fertilization and healthy embryo development. The advent of omics technologies has accelerated research into accurate predictors of male fertility. Single nucleotide polymorphisms (SNPs) and genome-wide association studies (GWAS) have identified genetic markers associated with sperm quality. In addition, transcriptomic, proteomic, and metabolomic analyses have identified molecular biomarkers associated with fertility. Validation of these biomarkers can improve swine production by increasing reproductive efficiency through improved selection for boar semen quality. This chapter examines both conventional and cutting-edge

J. Romero-Aguirregomezcorta · L. A. Parreño · A. Q. Montero
Department of Physiology, University of Murcia, Murcia, Spain
e-mail: jon.romero@um.es; laura.abril2@um.es; armando.quintero@um.es

J. G. Mateos (✉)
Department of Physiology, International Excellence Campus for Higher Education and
Research "Campus Mare Nostrum", University of Murcia, Murcia, Spain

Institute for Biomedical Research of Murcia (IMIB-Arrixaca), Murcia, Spain
e-mail: jgadea@um.es

© The Author(s), under exclusive license to Springer Nature
Switzerland AG 2024
J. C. Gardón, K. Satué Ambrojo (eds.), *Assisted Reproductive Technologies in
Animals Volume 1*, https://doi.org/10.1007/978-3-031-73079-5_9

methods for evaluating boar semen, emphasizing the importance of combining these approaches to improve the overall effectiveness of breeding programs.

Keywords

Spermiogram · Flow cytometry · Computer assessment · Omics technologies

1   Introduction

The evaluation of boar semen quality is a fundamental process in the diagnosis and management of reproductive problems in pig production (Maside et al. 2023). Semen quality is currently assessed using several conventional parameters such as sperm concentration, motility, and morphology. However, these parameters are not always sufficient to fully assess the quality of boar semen, and in some cases, they may not accurately predict the fertility potential of the semen (Gadea 2005).

The evaluation of sperm parameters, traditionally referred to as the spermiogram, is critical to the assessment of reproductive health in both humans and animals. In humans, the World Health Organization (WHO) recognizes the importance of sperm parameters in semen analysis for the diagnosis of infertility and provides support with a defined protocol for the evaluation of semen and defined cut-off values for the characterization of human sperm samples. Similarly, in animal breeding programs, evaluation of semen parameters is essential for the selection of superior breeding candidates. However, there is no defined worldwide standard for the methodology of application for semen evaluation in different species, including swine.

For this reason, in the pig industry stakeholders in different countries have implemented standardized quality control and quality assurance systems in semen processing programs. These standards in semen quality are proposals for autoregulation of the sector in different countries. Some experiences are the German–Austrian–Swiss model named ZDS standard (Riesenbeck et al. 2015), the United Kingdom, and the Spanish model (ANPSTAND) (Gadea 2019). These guidelines include the technical and health requirements in the use of boars, the methodology to be used for sperm evaluation, and the minimum values in some specific semen parameters that the semen samples must comply with before being offered to customers (motility, number of spermatozoa per semen dose, morphological abnormalities, bacterial contamination, etc.).

Although traditional semen analysis remains the most widely used technique, significant progress has been made in recent years in the use of new diagnostic techniques that provide more detailed and accurate information on semen quality. These techniques include Computer Assisted Sperm Analysis (CASA), flow cytometry, genomics, transcriptomics, proteomics, and metabolomics, which allow a more detailed and accurate assessment of boar semen quality (Quintero-Moreno et al. 2004; Broekhuijse et al. 2012b; Babamoradi et al. 2015; Soler et al. 2017; Boe-Hansen and Satake 2019).

Omics technologies have revolutionized biological research by enabling scientists to monitor the dynamics of living systems at multiple molecular levels simultaneously. This allows a more comprehensive understanding of biological systems and the underlying mechanisms that regulate them. Genomics involves the study of the entire genome, including the DNA sequences of all genes and non-coding regions. Transcriptomics focuses on the study of gene expression, including the measurement of mRNA levels and alternative splicing patterns. Proteomics is the study of the complete set of proteins expressed by cells. Metabolomics focuses on the study of the small molecules present in a biological system, including metabolites, lipids, and small peptides.

On the other hand, these new techniques provide a wealth of information that must be processed and analyzed to extract valuable insights and useful knowledge using big data techniques. These techniques make it possible to handle and manage large amounts of data that are too complex for traditional data analysis tools (Pineiro et al. 2019). In this sense, computational biology has become an essential tool for identifying potential markers to improve pig reproduction.

The integration of data from different technologies allows the construction of networks of interactions between genes, metabolites, and proteins that can be analyzed using computational biology methods. These methods allow the identification of potential markers for semen quality and male fertility (Llavanera et al. 2022).

In this chapter, we will review the basic principles of each of these techniques and their applications in the field of andrology for the evaluation of boar semen quality.

2   Basic Principles of Boar Sperm Evaluation

The classical method of semen evaluation is based on the use of a battery of tests that are easy, quick to perform, and relatively inexpensive, based on the examination of cell structure and sperm functionality (Kondracki et al. 2006b).

Typically, the spermiogram includes the measurement of (1) ejaculate volume, as the total amount of semen produced in a single ejaculation. (2) Sperm concentration: This determines the number of sperm cells present per milliliter of semen. (3) Total number of sperm in the ejaculate: This is calculated by multiplying the ejaculate volume by the sperm concentration to give the total number of sperm cells in the entire sample. (4) Progressive motility: This assesses the percentage of sperm that are actively moving in a forward direction, which is critical for fertilization. (5) Percentage of viable cells: This assesses the proportion of live sperm cells in the sample, as opposed to dead or non-viable cells. (6) Morphology: This involves examining the shape and structure of the sperm cells to identify any abnormalities that may affect fertility.

Although these traditional techniques are valuable for detecting very poor-quality samples that can then be discarded, they generally do not provide reliable predictions about the future fertility of the sperm (Gadea and Matas 2000; Gadea 2005). The ability to identify suboptimal samples ensures that only semen of acceptable quality is used, thereby maintaining the overall effectiveness of breeding programs (Broekhuijse et al. 2012a).

2.1   Volume, Sperm Concentration, and Total Number of Sperm in the Ejaculate

Assessing ejaculate volume, sperm concentration, and the total number of sperm in boar semen is crucial for evaluating reproductive health, optimizing semen use, and ensuring high-fertility rates in artificial insemination programs.

Ejaculate volume provides insight into boar reproductive health as a marker of reproductive gland function and assists in proper semen dilution. Total boar semen volume can be efficiently measured using graduated cylinders or by weight. Weight is more commonly used in large boar semen centers because it facilitates the semen dosing process.

Sperm concentration determines sperm density, essential for ensuring each insemination dose contains enough sperm cells for successful fertilization. Direct microscopic observation and counting using hematocytometric chambers (Neubauer, Thoma, Bürker, Makler, etc.) is an inexpensive and simple option but this method is rather time-consuming and limited by its subjectivity (Maes et al. 2010). Also, there are differences in the accuracy of the measurements according to the kind of chamber used (Christensen et al. 2005).

Several alternative methods can also provide accurate and reliable results, such as spectrophotometry, which measures the absorbance or optical density of a semen sample at specific wavelengths. The absorbance reading correlates with sperm concentration (Camus et al. 2011). The use of flow cytometry and CASA systems has improved the accuracy of sperm concentration determination and favored automation processes in centers producing porcine semen doses (Jung et al. 2015). On the other hand, the Nucleoconter system is used as a gold standard for the validation of evaluation systems due to its high repeatability (Hansen et al. 2006).

The accuracy and repeatability of the different methodologies for measuring sperm concentration have been evaluated and compared (Christensen et al. 2005; Hansen et al. 2006; Paulenz et al. 2007; Maes et al. 2010; Murgas et al. 2010; Valverde-Abarca et al. 2019; Grossfeld et al. 2022). The precision of a measurement device can be described by the variation in the results of repeated measurements. The coefficient of variation is usually the parameter of choice to describe the precision of an instrument. For sperm concentration assessment, CVs as low as 3% have been achieved for flow cytometers and Nucleoconter (Hansen et al. 2002, 2006).

2.2   Sperm Motility

Motility assessment is the most widely used test because it is simple, fast, and inexpensive. It is a good indicator of membrane integrity and functionality. Although semen motility is considered an important parameter to validate the quality of the ejaculate processed, its relationship to the fertility results is significant but limited (Gadea et al. 2004; Broekhuijse et al. 2012d). Despite severe limitations, motility seems to be an efficient seminal parameter because it is significantly related to farrowing rate and total number of piglets born, so it is included as a significant component in all multivariate models (Gadea et al. 2004).

Evaluation of sperm motility in porcine semen is a critical aspect of assessing semen quality and suitability for artificial insemination (AI). Several factors can affect the accuracy and reliability of assessing sperm motility in porcine semen. The manner in which semen is collected, processed, and stored can significantly affect sperm motility. Proper handling techniques, including temperature control and timely processing, are essential to preserve sperm motility (Lopez Rodriguez et al. 2012; Lopez Rodriguez et al. 2017; Schulze et al. 2018; Balogun and Stewart 2021). On the one hand, the composition of the extender used to dilute semen can affect sperm motility. Extenders containing appropriate nutrients, buffers, and antioxidants help maintain sperm motility during storage (Gadea 2003; Vyt et al. 2004a). On the other hand, the ratio of semen to the extender and the dilution technique used can affect sperm motility (Centurion et al. 2003; Schulze et al. 2017).

When observing sperm motility, the quality and calibration of the microscope and associated equipment such as thermal plates, slides or chambers, and micropipettes used to evaluate sperm motility can affect the accuracy of the results. A high-quality contrast-phase microscope with appropriate magnification (x100 or x200) and a well-calibrated thermal plate will ensure an accurate assessment of sperm motility.

Although, sperm motility is a subjective measure that is dependent on the individual observer. Standardized evaluation protocols and training programs help minimize variability and ensure consistent results. CASA system has been developed to help minimize the subjectivity of the process, increasing the repeatability (Vyt et al. 2004b). However, the CASA system is not free from other technical and human errors (Amann and Katz 2004). To solve this problem another alternative has been reported as the use of confocal-based methodologies to evaluate motility (Kim et al. 2017).

2.3   Sperm Morphology

Assessing sperm morphology involves examining the details of sperm cells, including their size, shape, and structural integrity. Assessment of sperm morphology plays a critical role in providing valuable information about male fertility potential and reproductive success. In animal breeding programs, evaluation of sperm morphology is essential, guiding the selection of elite breeding candidates. The presence of morphological anomalies in porcine sperm can significantly affect the fertilization potential and reproductive success of boars. Normal sperm morphology is essential for successful fertilization and embryonic development. Abnormalities in sperm shape and structure can impair sperm motility, capacitation, and the ability to penetrate the oocyte, ultimately resulting in reduced fertility rates and compromised reproductive outcomes. Consequently, evaluation of sperm morphology is included in the quality control of commercial semen doses for porcine semen (Gadea 2005).

Several studies in pigs have shown an inverse relationship between sperm morphological abnormalities and subsequent fertility (Waberski et al. 1994; Alm et al. 2006; Tsakmakidis et al. 2010). The importance of morphology has also been emphasized in the productive context of discarded boar semen and boar culling in AI centers (Feitsma 2009; Schulze et al. 2014; Knecht et al. 2017; Wang et al. 2019).

Sperm morphology and morphometry are two aspects of sperm analysis that provide insight into the shape and size of sperm cells. While there are similarities between the two terms, there are also some differences. Sperm morphology is the evaluation of the size, shape, and structural characteristics of sperm cells. It involves examining the overall appearance of the sperm under a microscope to determine whether the cells have a normal shape or abnormalities. Sperm morphometry, on the other hand, focuses on the quantitative measurement of various sperm parameters, such as length, width, and ratios of different parts of the sperm cell. Morphometric aspects will be reviewed in another section of this chapter (CASA measurements).

When evaluating the morphology of porcine spermatozoa, various common morphological abnormalities can be observed. These abnormalities have been characterized and classified according to the localization of the change, including head abnormalities, midpiece abnormalities, tail abnormalities, cytoplasmic droplets, and acrosome abnormalities (Bonet 1990; Bonet and Briz 1991; Kondracki et al. 2006a; Briz and Fàbrega 2013). Additionally, morphological anomalies could be classified according to their origin, either testicular (primary anomalies) or epididymal (secondary anomalies) origin (Lasley and Bogart 1944; Briz et al. 1996). For example, a coiled-tail sperm defect in a boar was found to originate from the epididymis (Holt 1982), and the presence and localization of cytoplasmic droplets serve as indicators of sperm maturation and epididymal function (Kaplan et al. 1984). Another classification, based on the bovine methodology (Blom 1983) categorizes spermatozoa into three groups: those with major sperm defects (abnormalities of sperm head and acrosome, coiled sperm tail), presence of proximal cytoplasmic droplets, and those with minor sperm defects (simple bent tail, loose sperm head) (Sutkeviciene et al. 2005; Juonala et al. 2007).

A variety of techniques can be utilized to evaluate porcine sperm morphology, ranging from traditional manual methods to advanced automated systems. Here is an overview of both traditional and advanced techniques commonly used for assessing porcine sperm morphology.

Traditional techniques include microscopic evaluation, which involves examining sperm morphology under a microscope. The samples can be fixed using substances such as formaldehyde or glutaraldehyde (Hancock 1957; Pursel and Johnson 1974), and wet-fixed samples can be visualized using phase contrast or interdifferential contrast. This technique utilizes phase-contrast optics to enhance the contrast and visibility of sperm structures. It allows for a better assessment of head morphology, acrosome integrity, tail abnormalities, and the presence of cytoplasmic droplets, that are assessed manually. An alternative method involves utilizing air-fix samples, which are subsequently stained and observed under a high-magnification light microscope. Various staining techniques can be employed to enhance the visibility of specific structures and abnormalities in sperm morphology. Some commonly used stains include eosin-nigrosin, Giemsa, hematoxylin-eosin, Papanicolaou, Bengal Rose, Panoptic, Sperm Blue, Hemacolor, or Diff-Quik (Hancock 1957; van der Horst and Maree 2009; Oberlender et al. 2012; Czubaszek et al. 2019).

Only a limited number of studies have compared the results of morphological sperm evaluations between different techniques. One such study conducted by (Oberlender et al. 2012) compared the proportion of morphological abnormalities in wet-fixed samples and Bengal Rosa-stained smears. The researchers found no significant difference in morphological alterations, except for tail abnormalities, which were higher in the stained smears. These comparative studies provide valuable insights into the consistency and effectiveness of different techniques in evaluating sperm morphology. However, further research is needed to conduct more comprehensive comparisons and assess the agreement between various evaluation methods.

Transmission electron microscopy (TEM) and scanning electron microscopy (SEM) are powerful techniques that offer high-resolution imaging of sperm morphology. These advanced microscopy methods provide detailed information about the ultrastructural features of sperm cells, including the acrosome, mitochondria, and tail abnormalities. They enable researchers to visualize and analyze the fine details of sperm morphology at a microscopic level. In the case of pigs, several studies have employed TEM and SEM to investigate and characterize sperm morphology (Jones 1971; Holt 1982; Kaplan et al. 1984; Bonet 1990; Bonet et al. 1993; Gu et al. 2010). These studies have contributed to a deeper understanding of porcine sperm morphology by revealing specific structural features and abnormalities that may impact fertility and reproductive processes (Gu et al. 2019).

Various new techniques for assessing boar sperm morphology have emerged, including the Quantitative Phase Microscope (Iglesias and Vargas-Martin 2013), Pyramid Phase Microscope (Iglesias 2016), Digital Holographic Microscopy (Di Caprio et al. 2010), or Atomic Force Microscopy (Ellis et al. 2002; Jones et al. 2008; Rubessa et al. 2020). Each has its own advantages and limitations, and the choice depends on factors like cost, equipment availability, and specific research or diagnostic needs. Typically, a combination of traditional and advanced methods is used to gain a comprehensive understanding of porcine sperm morphology.

In recent years, advancements in machine and deep learning have introduced a novel approach to assessing morphological parameters, such as acrosome integrity, using cost-effective equipment, such as bright-field microscopes and computers (Keller and Kerns 2022; Park et al. 2023).

3   New Diagnostic Techniques for Boar Sperm Evaluation

3.1   Computer Assisted Sperm Analysis (CASA): Principles, Advantages, and Limitations

In most pig-producing countries, reproduction is mainly performed by traditional or post-cervical artificial insemination (AI), using semen cooled between 15 and

18 °C. Computer Assisted Semen Analysis (CASA) technology allows automated and accurate measurement of individual sperm motility patterns (CASA-Mot) in the ejaculate (Quintero-Moreno et al. 2004; Amann and Waberski 2014), in addition to the measurement of morphometry (CASA-Morph) in live cells in suspension (Soler et al. 2015) or fixed and stained with dyes (Thurston et al. 2001; Kondracki et al. 2005; Peña et al. 2005; Quintero-Moreno et al. 2009; Yaniz et al. 2016). Another application of CASA systems is the assessment of sperm concentration. Several studies have demonstrated the accuracy of this parameter compared to other methods such as cell counting chambers, flow cytometry systems, or colorimetry (Hansen et al. 2002, 2006; Gaczarzewicz 2015; Sevilla et al. 2023).

The main aspects that influence the accuracy of the values provided in the evaluation of porcine semen by the CASA technology will be described below. First, a brief analysis of the technical conditions and limitations of the CASA technology will be made, and then the review will focus on the descriptors of motility by CASA-Mot and sperm morphology by CASA-Morph.

The use of CASA is becoming increasingly common on commercial farms to evaluate sperm motility in boars (Holt et al. 1997; Didion 2008; Broekhuijse et al. 2012b). It is known that the handling of freshly collected boar semen responds very well when diluted with commercial semen extenders, as evidenced by the observation of optimal and uniform movement of most spermatozoa. This is explained by the fact that 90% of the sperm in an ejaculate have the same linear motion pattern and optimal average velocity (Quintero-Moreno et al. 2004). However, as with most technologies, CASA measurements are subject to external biases associated with the processing of semen for use in AI or for proper evaluation. The main sources of variation that affect sperm performance after sample evaluation include sample incubation time, incubation temperature, and the time interval between slide loading and analysis (Gaczarzewicz 2015). It is also important to consider some other factors, such as the dilution rates (Centurion et al. 2003) and the dilution protocol (Schulze et al. 2017), the type and depth of the counting chamber to be used (Bompart et al. 2018; Valverde and Madrigal-Valverde 2019), and the technical conditions of the video acquisition, including capture time and frame rate frequency (Valverde et al. 2019a, b, 2021). Another aspect to consider is the technician performing the evaluation (Yeste et al. 2018). The user's experience with the CASA system could determine changes in the results, so the training and learning process is a crucial step for new operators, showing that after proper learning, the coefficient of variation of CASA measurements was reduced (Broekhuijse et al. 2011; Ehlers et al. 2011).

CASA technology was developed in the late 1970s (Dott and Foster 1979; Amann and Katz 2004) and commercialized in the late 1980s but it was not until the 1990s that universities and some leading swine companies acquired this technology, which objectively provides sperm characteristics of an ejaculate (Holt 1995; Holt et al. 1997). In the early years, however, only the sperm motility module and, to a lesser extent, the concentration and morphometry modules were routinely used. The main components of a CASA system are a microscope equipped with a heating plate, a negative phase contrast optical device, and an attached video camera with good resolution quality. In addition, and of great importance, there is a computer that must have been equipped with specific software that allows various analyses to be carried out.

3.1.1   Sperm Motility Measured by CASA (CASA-Mot)

The CASA-Mot module generates a video file with many spermatozoa, which generates a robust data structure and offers many possibilities for generating information with great statistical validity, in addition to the possibility of discriminating the data, creating and eliminating trajectories and even atypical morphometric measurements (Martinez-Pastor et al. 2011).

Total and progressive motility, as well as the percentage of fast, medium, slow, and static spermatozoa, are the basic parameters provided by CASA systems (Holt et al. 1996). Sperm kinematics involves measuring the distance between each point on a sperm head along its trajectory at a given time. The main descriptors of motility relate to sperm oscillation, where curvilinear velocity (VCL, μm/s) represents the actual measured sperm motion by adding the distance between the head positions frame by frame, divided by the elapsed time. Straight line velocity (VSL, μm/s) is a descriptor that is estimated by measuring the distance between the first and the last point traveled by the sperm, divided by the elapsed time. Average trajectory velocity (VAP, μm/s) represents the distance traveled by the sperm along its average trajectory, which is determined by an algorithm. Straightness index (STR, %) is the ratio of VSL/VAP to measure the compactness of the path. Linearity index (LIN, %) is the linear progression calculated by estimating the VSL/VCL ratio. Wobble index (WOB,%) represents the oscillation of the actual trajectory with respect to the trajectory of the VAP/LCV ratio; amplitude of lateral head displacement (ALH, μm) is the amplitude of the sinusoidal oscillation made by the sperm head (from one side to the other) in its curvilinear path adjusted to the average path, and beating frequency (BCF, Hz) is the frequency with which the sperm head crosses half the length of the path (Quintero-Moreno 2003).

CASA-Mot measurements are highly susceptible to sample variability and vary depending on hardware and/or software (Holt et al. 1994; Yeste et al. 2018). Therefore, it is necessary to standardize the CASA measurement for a number of parameters, such as sample temperature, dilution factor, sample mixing, chamber slide depth, frame rate, and image acquisition time, among others, in order to reduce this variability (Fair and Romero-Aguirregomezcorta 2019).

Heating the semen sample to the same temperature (37, 37.5, or 38 °C), both in the preheating container and in the microscope where the sample is placed, in addition to being consistent with the time elapsed until its evaluation (Del Gallego et al. 2017), the number of fields analyzed must be constant (Broekhuijse et al. 2011). In the evaluation of motility, it is suggested to capture at least three fields (preferably 5) and increase to more fields if the image shows few sperm. It is also important to specify the type of camera (Bompart et al. 2018), the video capture time (Valverde et al. 2019a) and to adjust the frame rate (Castellini et al. 2011; Valverde et al. 2019b). In the early days of CASA technology, frame rates as low as 16 or 25 frames per second (fps) were used (Holt et al. 1997), but more recently an acquisition frequency of 50–60 Hz has been used (Valverde et al. 2020). When cells exhibit high speed and low linearity, such as hyperactivated spermatozoa, it is recommended to increase the frame rate to properly assess their movements (Valverde et al. 2019b). According to Valverde et al. (2021), the optimal frame rate for analyzing boar sperm kinetic variables with a CASA-Mot system is 200 fps. A lower frame rate may result in a loss of accuracy.

Although there is no worldwide guide for boar semen evaluation, we appreciate that the majority of laboratories in universities and research centers with boar semen and semen production centers for AI have uniform criteria for processing and evaluating sperm motility over time, a fact that allows a more objective comparison of results, both within the same laboratory and between them, allowing quantitative differences in semen parameters to be detected with greater precision. These criteria are used to find differences between treatments designed for research or field trials, or between boars participating in the same experiment (Verstegen et al. 2002).

Several studies have reported a positive correlation between sperm kinematic parameters assessed by computer assisted sperm analysis (CASA), and field fertility after artificial insemination (AI) (Holt et al. 1997; Broekhuijse et al. 2012b; Schulze et al. 2013). Although sperm motility accounts for only 6% of the total variation in a sow's fertility (Broekhuijse et al. 2011), it is not a percentage to underestimate, especially since this variable is still very important in deciding whether to use a sow's ejaculate (Broekhuijse et al. 2012b). The review by Fair and Romero-Aguirregomezcorta (2019) refers to these positive associations between sperm motility and fertility in the sow. In some studies, VAP and VSL have been related to fertility and litter size (Holt et al. 1997; Broekhuijse et al. 2012b), while in some other cases relationship is low or non-existent (Hirai et al. 2001; Quintero-Moreno et al. 2004).

3.1.2   Sperm Morphology and Morphometry Measurements with CASA (CASA-Morph)

In the pig industry, CASA systems are mainly used for evaluating motility and motion parameters in the production of commercial seminal doses. Another interesting technical demand is the direct morphological analysis using CASA without the need for staining the preparation. Despite the introduction of various sperm morphology analysis systems by different companies, there is currently limited scientific information available to fully understand the advantages these systems provide. Only a few studies have been published, and in those studies, no significant differences have been observed when comparing the evaluation of stained smears under microscopic assessment (Antonczyk et al. 2012; Suarez-Trujillo et al. 2022). For example, one study compared sperm morphology in semen samples from teratospermic boars using CASA and Giemsa-stained smears and found no differences between the two methods (Antonczyk et al. 2012). Recently, a smartphone-based device has been reported to accurately measure morphology, with results similar to contrast-phase microscopy evaluation (Suarez-Trujillo et al. 2022). These studies highlight the potential of alternative approaches for morphological assessment, including classical CASA-based systems and new smartphone-based devices, which offer convenient and reliable methods for evaluating sperm morphology.

Most of the published studies on sperm morphometry focus on the sperm head (Thurston et al. 2001; Kondracki et al. 2005; Peña et al. 2005; Quintero-Moreno et al. 2009; Morales et al. 2012; Yaniz et al. 2016) and to a lesser extent, other researchers have also measured the acrosome, the intermediate piece, the entire flagellum and even the nucleus (Thurston et al. 1999; Yaniz et al. 2015; Kondracki et al. 2017). CASA-morph provides several biometric descriptors to characterize the sperm, however, the most common and accepted are primary parameters that provide information about the dimensions of the sperm head and generally include length (L, μm), width (W, μm), area (A, μm^2) and perimeter (P, μm), and parameters that describe the silhouette or shape of the head through a series of mathematical formulas, including ellipticity = L/W, roughness = $4\pi A/P2$, elongation or exit of roundness of the sperm = $(L - W)/(L + W)$, regularity = $\pi LW/4$, and grayscale. It should be noted that ellipticity and elongation provide redundant information and describe the same phenomenon, since they associate the relationship between elongation and widening of the sperm head (Varea Sanchez et al. 2013), therefore, because the biological importance of these variables is unknown, only research with the first four variables mentioned is reported and only the importance of sperm elongation has been emphasized.

Several publications analyzed sperm head dimension parameters over the last 20 years; using different sperm stains and commercial software (Table 1 summarizes the most important studies). From this information, the mean ± SEM and confidence intervals (IC95%) were calculated, which could be used as reference values for sperm morphometry in porcine spermatozoa. It is important to note that the coefficient of variation (CV) of these parameters is below 10%, which implies that there is 90% homogeneity in the information provided by the referenced experiments.

Morphometric results can be influenced by both internal and external factors in boars. Intrinsic factors include genetic or environmental factors, including sampling frequency, as well as the age and sexual maturity of the boar (Kondracki et al. 2005; Quintero-Moreno et al. 2009). The dimensions of the boar's sperm head vary from the beginning of its reproductive life until it reaches 2 and a half years of age; pigs older than 18 months produce more elongated, less wide sperm and have a smaller area and perimeter in 79% of their ejaculate (Quintero-Moreno et al. 2009). There is a new technique called Trumorph which describes the observation of unstained spermatozoa in suspension immobilized by a brief 60 °C thermal shock under a contrast-phase microscope (Soler et al. 2015). On the other hand, sperm staining techniques in boars and other species are previously fixed with solutions that dehydrate the cell and cause changes in its dimensions (Quintero-Moreno et al. 2015); permeates the background with the dye and the silhouette of the sperm it is captured with great difficulty in most protocols (Quintero-Moreno et al. 2015; Czubaszek et al. 2019).

It is worth noting that the dimensions of boar sperm heads are significantly influenced by sample management and staining methods (Czubaszek et al. 2019; Szablicka et al. 2022). Standardizing the morphometric evaluation process is

Table 1 Morphometric dimensions of boar spermatozoa (mean ± SEM) with the CASA-Morp system

CASA-Morph software	Area, µm²	Perimeter, µm	Length, µm	Width, µm	Reference
Minitube®	35.70 ± 0.20	–	9.27 ± 0.05	4.66 ± 0.02	Hirai et al. (2001)
SM.v. 4.1	41.17 ± 2.68	–	9.37 ± 0.44	4.88 ± 0.28	Kondracki et al. (2005)
ISAS®	36.20 ± 1.71	26.60 ± 1.23	9.10 ± 0.34	4.60 ± 0.19	Saravia et al. (2007)
Microptic®	32.39 ± 0.06	26.33 ± 0.04	8.95 ± 0.01	4.32 ± 0.009	Quintero et al. (2009)
Microptic®	31.62 ± 0.05	22.49 ± 0.02	8.50 ± 0.10	4.24 ± 0.02	Morales et al. (2012)
SM v.4.1	33.96 ± 3.74	27.69 ± 1.85	8.92 ± 0.50	4.48 ± 0.33	Kondracki et al. (2017)
ISAS®	37.98 ± 4.67	23.23 ± 3.54	9.88 ± 1.20	5.20 ± 0.38	Yaniz et al. (2016)
SM.v. 4.1	38.97 ± 4.27	23.44 ± 1.09	9.12 ± 0.43	4.86 ± 0.32	Górski et al. (2017)
ISAS®	34.20 ± 2.47	23.84 ± 1.13	8.62 ± 0.60	4.46 ± 0.23	Barquero et al. (2021)
MSI	27.79 ± 1.17	23.39 ± 0.98	8.29 ± 0.33	4.26 ± 0.19	Banaszewska and Andraszek (2021)
NIS-D5®	37.72 ± 4.28	24.84 ± 1.85	9.11 ± 0.10	4.80 ± 0.72	Szablicka et al. (2022)
Mean ± SEM	35.24 ± 1.14	24.65 ± 0.59	9.01 ± 0.13	4.61 ± 0.09	–
IC95%	37.78–32.70	23.25–26.04	9.30–8.71	4.41–4.81	–
CV %	10.74	7.35	4.87	6.51	–

SM, screen measurement v. 4.1; NIS-D%, NIS-elements D5 image analysis software; MSI, MultiScan image analysis system

essential to address this issue (Garcia-Herreros et al. 2006; van der Horst and Maree 2009; Vicente-Fiel et al. 2013).

3.1.3 Determination of Sperm Concentration by CASA

CASA systems offer numerous advantages over traditional manual methods for assessing semen concentration in pigs. These benefits include speed and efficiency. CASA can quickly process and analyze a large number of samples, making it ideal for large-scale farms and research studies. The data generated from the analysis is integrated into management software that facilitates storage, retrieval, and analysis of large data sets. While the initial investment in CASA systems can be high, the long-term benefits include reduced labor costs and improved efficiency.

An important issue is the accuracy of the measurement. Reports on the accuracy of CASA for measuring sperm concentration are inconsistent and limited in number for boar semen (Vianna et al. 2004; Kuster 2005; Hansen et al. 2006; Gaczarzewicz 2015; Valverde-Abarca et al. 2019; Grossfeld et al. 2022; Sevilla et al. 2023). The accuracy of CASA systems tends to be higher than hematocytometric assessment and lower than nucleoconter or flow cytometric assessment of sperm concentration. In this sense, and according to Hansen et al. (2006), the CVs for some CASA systems ranged 5.4–8.1%, while nucleoconter and flow cytometry were close to 3%. Recently, a new study compared CV ranging 3.6–7.6% for two CASA systems compared to a CV of 1.6% for Nucleoconter (Grossfeld et al. 2022). A recent investigation concluded that there are large differences in sperm concentration when comparing four different software, finding variations between them (Sevilla et al. 2023).

3.2 Flow Cytometry: Principles, Advantages, Limitations, and Applications

Over the years, several diagnostic techniques have emerged to assess boar semen quality, with flow cytometry increasingly gaining popularity as a powerful tool in the swine industry. Its application enables more effective selection of breeding boars, enhancing reproductive efficiency and offspring quality (Broekhuijse et al. 2012c). In addition, flow cytometry facilitates high-throughput analysis, allowing the examination of a large number of sperm cells in a short period of time with excellent accuracy and sensitivity, even for samples with low sperm concentrations (Spano and Evenson 1993).

One crucial aspect of boar sperm analysis is evaluating sperm viability, which flow cytometry accomplishes rapidly and accurately by assessing plasma membrane integrity and specific intracellular enzymes. DNA integrity is another parameter analyzed using flow cytometry, which is essential as DNA damage can affect sperm's ability to fertilize and produce healthy embryos. By identifying and quantifying DNA damage, flow cytometry allows the selection of sires with lower levels of DNA damage, thereby improving the quality of semen used in artificial insemination.

Flow cytometry allows the rapid analysis of multiple sperm parameters in thousands of cells within seconds, enhancing accuracy, reliability, and repeatability (Gonzalez-Castro et al. 2022). While traditional assessments of sperm motility, morphology, DNA fragmentation, acrosome, and membrane integrity are crucial for evaluating boar fertility, they fall short in predicting potential fertility due to the complexity of fertilization. By enabling the simultaneous multiparametric assessment of various sperm attributes, flow cytometry provides a more precise estimation of sperm fertilizing potential (Babamoradi et al. 2015; Ortega-Ferrusola et al. 2017; Jakel et al. 2021; Quirino et al. 2022). The following sections will delve into the principles, advantages, and limitations of using flow cytometry as a diagnostic tool for evaluating boar sperm.

3.2.1 Principles of Flow Cytometry

Flow cytometry is a technique that enables the rapid and simultaneous analysis of multiple parameters of individual cells or particles. It employs principles of hydrodynamic focusing, laser-induced fluorescence, and optical and electronic detection systems. Hydrodynamic focusing is a crucial principle of flow cytometry that ensures the cells or particles are aligned and pass through the laser beam single file, leading to accurate detection and measurement. In this process, the sample is aspirated into a sheath fluid, creating a laminar flow. The sheath fluid surrounds the sample fluid and creates a narrow, central core where the cells or particles are confined. This hydrodynamic focusing enhances the precision of positioning the cells or particles in the laser beam, allowing for accurate analysis.

Laser-induced fluorescence is another principle of flow cytometry that enables the detection and analysis of specific cellular features or components. A laser beam is directed to the focused stream of cells or particles, causing the emission of fluorescent light. Depending on the specific characteristics of the cells or particles being analyzed, fluorescent dyes or probes can be used to label certain components or structures within the cells. These labeled components emit fluorescence when excited by the laser, allowing for their detection and measurement.

The optical and electronic detection systems in flow cytometry are responsible for capturing and analyzing the emitted fluorescent signals. The light scattered and emitted from each cell or particle is collected by a series of lenses and filters, which separate the different wavelengths of light based on their characteristics. The emitted fluorescence is typically filtered through bandpass filters specific to the fluorochromes used, allowing only the relevant wavelengths to reach the detectors. Multiple detectors are positioned to collect the emitted fluorescence at different wavelengths, enabling the simultaneous analysis of multiple parameters.

3.2.2   Applications of Flow Cytometry in Boar Sperm Analysis

3.2.2.1   Nuclear Counterstaining
Traditionally, sperm selection in flow cytometry involves evaluating cellular granulation and size. This is achieved using the forward scatter (FSC) and side scatter (SSC) channels. However, in order to achieve a precise differentiation between spermatozoa and other cellular elements and debris, various cell-permeable dyes are employed combined with other staining agents to pinpoint the sperm cell and its distinct characteristics (Robles and Martínez-Pastor 2013). Nuclear counterstaining stands out as the predominant method for this purpose. In boar semen analysis, SYBR-14 and Hoechst 33342, both membrane-permeant DNA fluorochromes, are widely used as standards (Boe-Hansen and Satake 2019).

3.2.2.2   Assessment of Sperm Viability
The integrity of the sperm plasma membrane holds critical importance for optimal sperm function and fertility prediction (Yeste et al. 2010). Moreover, changes induced by liquid storage and cryopreservation can alter membrane lipid and protein organization, affecting membrane permeability (Johnson et al. 2000; Suo et al. 2024). Hence, an accurate assessment of plasma membrane integrity is crucial.

Initially, fluorochromes such as 6-carboxymethylfluorescein diacetate (CFDA) and propidium iodide (PI) were used to evaluate the plasma membrane integrity of pig sperm in the 1980s (Garner et al. 1986). Nowadays, the widely used stain for pig sperm viability is SYBR-14, a membrane-permeable DNA fluorochrome, combined with PI to identify membrane-damaged sperm (Garner and Johnson 1995). This combination, excited by a 488-nm laser, using two optical filters, distinguishes viable cells emitting green fluorescence from damaged cells stained in red fluorescence (Gonzalez-Castro et al. 2022).

Newly developed DNA stains, such as SYTO dyes, offer alternatives to SYBR14 and/or PI, providing varied fluorochrome characteristics for comprehensive assessment. They come in a variety of colors, each with different excitation and emission spectra, allowing researchers to choose the dye that best suits their experimental needs. Additionally, some SYTO dyes are cell permeant, allowing them to penetrate cell membranes and stain nucleic acids in live cells, while others are cell impermeant and can be used to selectively stain nucleic acids in fixed cells (Boe-Hansen and Satake 2019).

Annexin-V is a protein that has a high affinity for phosphatidylserine (PS), a phospholipid normally located on the inner leaflet of the cell membrane. During apoptosis, a process of programmed cell death, PS is translocated from the inner to the outer leaflet of the cell membrane. This "flipping" of PS to the outer surface is an early marker of apoptosis. To detect PS exposure on the cell surface, Annexin-V is often conjugated with a fluorophore, such as fluorescein isothiocyanate (FITC), which emits green fluorescence when excited by a specific wavelength of light. When Annexin-V-FITC binds to exposed PS on the outer surface of cells undergoing apoptosis, it generates a fluorescent signal, indicating the presence of apoptotic cells (Peña et al. 2003).

Sperm apoptosis can be influenced by various factors such as oxidative stress, temperature changes, or exposure to toxins, and it can have implications for fertility and reproductive health. By using Annexin-V-FITC staining, researchers can quantify the percentage of spermatozoa undergoing early apoptosis, providing valuable insights into sperm quality and potential fertility issues (Maside et al. 2023).

Fixable stains offer the advantage of delaying the determination of sperm viability and mitochondrial membrane potential (Boe-Hansen and Satake 2019), to do so, amine-reactive dyes, or live/dead fixable viability stains, allow for the identification of dead cells in fixed samples. These dyes differ from other viability dyes because they react with free amines in the cytoplasm. Live cells exclude these dyes because their cell membranes are intact and free dye is washed away after staining. Moreover, the reaction is irreversible so when cells are fixed the bound dye remains associated with the dead cells. As an example, the Zombie Aqua dye is employed in horse sperm to detect non-viable/damaged sperm in fixed samples (Peña et al. 2018). But to the authors' knowledge, no studies reporting these dyes have been employed yet in porcine.

3.2.2.3 Evaluation of Acrosome Integrity

Assessing acrosome integrity often involves utilizing lectins labeled with fluorescein, such as *Pisum sativum* agglutinin (PSA) or *Arachis hypogaea* agglutinin (PNA), combined with the viability probe PI (Flesch et al. 1998). These lectins specifically interact with glycosidic residues on damaged or exocytosed acrosomal membranes. The choice between PSA and PNA depends on the specificity of the labeling desired. PNA-FITC binding is restricted to β-galactose moieties on the outer acrosomal membrane, offering a precise assessment of acrosome integrity. In contrast, PSA-FITC binds to α-mannose and α-galactose moieties, staining not only the acrosomal matrix but also the flagella and head, resulting in less specific labeling. However, it is important to note that while lectin staining indicates membrane damage, it does not elucidate the molecules crucial for fertilization. Studies in boars frequently employ the PNA-FITC/PI and PSA-FITC/PI combinations to distinguish sperm subpopulations with intact plasma and acrosomal membranes (Matas et al. 2007; Martin-Hidalgo et al. 2011). Recent advancements include the availability of PNA conjugated with Alexa Fluor 488 (Alvarez-Rodriguez et al. 2018) or Alexa Fluor 647 (Murphy et al. 2017; Jakel et al. 2021), offering several options for fluorescent labeling in acrosome assessment.

3.2.2.4 Sperm Capacitation
Sperm capacitation is indeed a critical process in fertilization, encompassing various changes that prepare sperm for interaction with the oocyte. Among these changes are alterations in membrane fluidity, calcium levels, mitochondrial membrane potential, intracellular pH, and protein phosphorylation patterns (Harrison et al. 1996; Tardif et al. 2001; Petrunkina et al. 2005; Visconti 2009; Visconti et al. 2011). In the following sections, this general topic will be split into each specific event involved in preparing sperm for fertilization.

Evaluation of Membrane Fluidity
Membrane fluidity refers to the degree to which the lipid molecules in a cell membrane can move within the membrane structure. It is an important aspect of membrane function because it affects various cellular processes like signal transduction, protein trafficking, and membrane fusion. One common method to evaluate membrane fluidity involves using fluorescent dyes such as Merocyanine 540 (M-540) in combination with Yo-Pro-1(Harrison et al. 1996; Gadea et al. 2005).

Measurement of Mitochondrial Function
Mitochondrial activity holds pivotal importance in sperm pathophysiology and structure. It provides the energy required for motility, generates reactive oxygen species (ROS), and participates in calcium signaling, hyperactivation, and fertilization processes (Vertika et al. 2020).

Different fluorescent probes have been used to detect changes in sperm mitochondrial membrane potential (MMP) using flow cytometry, discriminating cells with active and non-active mitochondria (Graham et al. 1990). The commonly used fluorescent dyes to detect sperm mitochondrial activity in boar spermatozoa include Rhodamine 123 (Fraser et al. 2002), 5,5′,6,6′-tetrachloro-1,1′,3,3′ tetraethylbenzimidazolylcarbocyanine iodide (JC-1) (Guthrie and Welch 2006), and increasingly MitoTracker Deep Red 633 (Caamano et al. 2021).

The probes may be suitable for discriminating spermatozoa with deteriorated mitochondria, and MMP serves as a crucial indicator of cellular health and functional status. As an example, the cyanine dye JC-1 enables the differentiation between energized and de-energized mitochondria. This is achieved because the typically green fluorescent dye shifts to red fluorescence when it forms aggregates in energized mitochondria due to their higher membrane potential (Perelman et al. 2012).

JC-1 has found application in porcine spermatozoa studies for characterizing MMP. In combination with SiR700-DNA, which identifies DNA-containing events, and Calbryte 630, utilized for assessing intracellular calcium levels, JC-1 provides a comprehensive approach to probing mitochondrial function and its interplay with other cellular processes (Jakel et al. 2021).

Mitochondria play a crucial role in generating the energy required for sperm movement, and changes in MMP are closely associated with sperm capacitation—the process through which sperm become capable of fertilizing an egg. The positive correlation between sperm motility and mitochondrial activity, as highlighted by (Althouse and Hopkins 1995), suggested that the efficiency of mitochondrial function directly influences the ability of sperm to move effectively. Moreover, the findings from Guo et al. (2017) provided additional evidence by demonstrating that motile boar spermatozoa exhibit higher mitochondrial activity compared to immotile sperm.

Calcium Levels Measurement

Measurement of calcium levels in pig sperm, both before and after in vitro capacitation, has been conducted using the flow cytometry technique (Harrison et al. 1993; Yeste et al. 2015). For this purpose, the fluorescent probe Fluo-3 AM is commonly used. Fluo-3, a green fluorochrome excited by a standard 488 nm laser, emits fluorescence intensity directly proportional to the intracellular calcium levels.

The use of Fluo-3 AM allows for the assessment of calcium uptake in specific regions of sperm cells, such as the midpiece region in boar spermatozoa. This method provides valuable insights into the responsiveness of spermatozoa to capacitation stimuli and can also be used to study the function of ion channels, such as the CatSper channel, in relation to sperm capacitation (Vicente-Carrillo et al. 2017).

Measurement of Intracellular pH

The intracellular pH (pHi) serves as an important indicator of sperm viability and function, with associations noted between pHi levels, sperm motility, and capacitation (Nishigaki et al. 2014; Soriano-Ubeda et al. 2019). The fluorescent probe 2′,7′-bis-(2-carboxyethyl)-5-(and-6)-carboxyfluorescein, acetoxymethyl ester (BCECF AM) is commonly utilized to measure pHi in spermatozoa (Harrison et al. 1993). BCECF AM penetrates the sperm plasma membrane and is hydrolyzed by esterases released into the cytoplasm. Upon excitation, the fluorescence intensity emitted by BCECF is directly dependent on the pH of the intracellular environment. This method allows the assessment of pHi dynamics in sperm cells, providing insights into their metabolic activity, viability, and functional status (Harrison et al. 1993; Soriano-Ubeda et al. 2019; Martin-Hidalgo et al. 2024).

Detection of Protein Tyrosine Phosphorylation

In boar semen, the detection of protein tyrosine phosphorylation, which in this species has been associated with capacitation (Flesch et al. 1999), has been used to identify subpopulations by flow cytometry using an anti-phosphotyrosine antibody conjugated to FITC (Torres et al. 2016).

Evaluation of Oxidative Stress

Oxidative stress occurs when there is an imbalance between ROS and the antioxidant defense mechanisms. While ROS at normal physiological levels play crucial roles in sperm processes like maturation, capacitation, acrosome reaction, and fertilization, elevated ROS concentrations are linked to decreased sperm motility, sperm membrane lipid peroxidation, and DNA damage (Brouwers et al. 2005).

The use of fluorescent probes like $2',7'$-dichlorodihydrofluorescein diacetate (H2DCFDA) and dihydroethidium (DHE) provides valuable insights into intracellular ROS levels in sperm (Guthrie and Welch 2006). For instance, H2DCFDA is extensively employed to detect total ROS in pig sperm (Gadea et al. 2005). When ROS are present H2DCFDA is oxidized into dichlorofluorescein (DCF), emitting green fluorescence. This probe allows the assessment of overall ROS levels within the cell. DHE, on the other hand, serves as a specific probe for superoxides. In the presence of superoxides, DHE is oxidized into ethidium, which binds to DNA and emits red fluorescence at 610 nm. Therefore, the intensity of ethidium fluorescence indicates the levels of superoxides in the cell. These probes can be combined with viability dyes such as PI or YO-PRO-1 to distinguish between viable and non-viable sperm cells while assessing ROS levels (Maside et al. 2023). Additionally, MitoSOX Red, a derivative of DHE, is specifically designed to detect mitochondrial ROS levels. This probe provides a focused assessment of ROS specifically within the mitochondria, shedding light on the oxidative stress status of this organelle (Guo et al. 2017).

Lipid peroxidation of sperm plasma membrane due to ROS can be detected using a dual fluorescence probe like BODIPY581/59 (Guthrie and Welch 2006). The probe incorporates into cells and undergoes a spectral emission shift, which is uniquely very small, when attacked by reactive oxygen metabolites, indicating exposure of phospholipids to ROS and quantifying lipid peroxidation. This probe emits different wavelengths according to the oxidation status (non-oxidized membrane in red and peroxidized membrane in green). This probe can be combined with viability stains to assess lipid peroxidation in viable and non-viable sperm (Boe-Hansen and Satake 2019).

Reactive nitrogen species (RNS) are considered a subgroup of ROS and include nitric oxide (NO), peroxynitrite, and nitroxyl ion, among others (Sikka 2001). RNS can also affect sperm quality, excessive NO or peroxynitrite levels have detrimental effects on sperm function and fertilization ability (Romero-Aguirregomezcorta et al. 2021). NO and peroxynitrite levels can be assessed using fluorescent probes like 4,5-Diaminofluorescein diacetate (DAF-2 DA), which becomes highly fluorescent when it reacts with NO, and dihydrorhodamine 123 (DHR 123), oxidized by peroxynitrite to rhodamine 123 (Serrano et al. 2020).

A list of different fluorochromes and applications is given in Table 2.

Table 2 Main fluorochromes employed in boar sperm flow cytometry, emission color, and functional parameters assessed

Target/parameter assessed	Stain	Emission color	References
Acrosome integrity	PNA-FITC	Green	Martín-Hidalgo et al. (2011)
Acrosome integrity	PNA-Alexa Fluor 488	Green	Alvarez-Rodriguez et al. (2018); Jakel et al. (2021)
Acrosome integrity	PNA-Alexa Fluor 647	Red	Murphy et al. (2017)
Plasma membrane integrity	Propidium iodide	Red	Garner et al. (1986, 1996); Gonzalez-Castro et al. (2022); Jakel et al. (2021)
Plasma membrane integrity	6-Carboxymethylfluorescein diacetate (CFDA)	Green	Garner et al. (1986)
DNA presence	Hoechst 33342	Blue	Torres et al. (2016)
DNA presence	SYBR-14	Green	Gonzalez-Castro et al. (2022)
DNA presence	SiR700-DNA	Red	Jakel et al. (2021)
Nucleic acids presence	YO-PRO-1	Green	Alvarez-Rodriguez et al. (2018); Martin-Hidalgo et al. (2011)
Phosphatidylserine exposure (Apoptosis)	Annexin-V-FITC	Green	Peña et al. (2003)
Membrane fluidity	Merocyanine 540	Red	Martin-Hidalgo et al. (2011)
Mitochondrial status	Mitotracker Deep Red	Red	Alvarez-Rodriguez et al. (2018); Jakel et al. (2021); Li et al. (2023)
Mitochondrial status	Rhodamine 123	Green	Fraser et al. (2002)
Mitochondrial superoxide indicator	MitoSOX Red	Red	Alvarez-Rodriguez et al. (2018); Guo et al. (2017)
Mitochondrial transmembrane potential	5,5′,6,6′-Tetrachloro-1,1′,3,3′ tetraethylbenzimidazolylcarbocyanine iodide (JC-1)	Green / Red	Jakel et al. (2021); Martin-Hidalgo et al. (2011)
Intracellular calcium levels	Fluo-3 AM	Green	Caballero et al., (2009); Vicente-Carrillo et al. (2017)

(continued)

Table 2 (continued)

Target/parameter assessed	Stain	Emission color	References
Intracellular calcium levels	Calbryte 630	Red	Jakel et al. (2021)
Intracellular pH (pHi)	2′,7′-bis-(2-carboxyethyl)-5-(and-6)-carboxyfluorescein, acetoxymethyl ester (BCECF AM)	Green	Li et al. (2023)
Protein tyrosine phosphorylation	Anti-phosphotyrosine-FITC	Green	Torres et al. (2016)
Total ROS detection	2′,7′-Dichlorodihydrofluorescein diacetate (H2DCFDA)	Green	Guthrie and Welch (2006)
Superoxide detection	Dihydroethidium (DHE)	Red	Guthrie and Welch (2006)
Lipid peroxidation	BODIPY581/59	Red	Guthrie and Welch (2006)
NO levels detection	4,5-Diaminofluorescein diacetate (DAF-2 DA)	Green	Serrano et al. (2020)
Peroxynitrite levels detection	Dihydrorhodamine 123 (DHR 123)	Green	Serrano et al. (2020)

3.2.2.5 Multiparametric Analysis

The advancement of flow cytometers with multiple lasers and optical channels enables simultaneous multiwavelength analyses, enhancing sperm evaluation by measuring several parameters at once. This approach improves the prediction of fertilizing ability and allows for a deeper understanding of the relationship between different sperm functional parameters.

By determining multiple sperm attributes on a single-cell basis, it becomes possible to visualize sperm subpopulations, reflecting the physiological heterogeneity of semen samples. Additionally, links between structural and functional sperm attributes can be established. Multiparametric flow cytometry eliminates the need to split semen samples into multiple aliquots for analysis, potentially aiding in identifying ideal spermatozoa for successful fertilization and embryo development.

Recent advancements allow the joint evaluation of sperm viability, acrosome integrity, and mitochondrial activity using a four-color protocol in pigs (Gonzalez-Castro et al. 2022). In brief, spermatozoa were incubated at 37 °C for 20 min using a four-color protocol compound of 8 µg/ml Hoechst 33342, 5.6 µg/ml PNA-Alexa Fluor™ 488 conjugate, 10 µM PI, and 64 nM MitoTracker™ Deep Red FM as final concentrations. Furthermore, a multifluorochrome panel consisting of SiR700-DNA to identify DNA-containing events, JC-1 to characterize the MMP, and Calbryte 630 to assess the intracellular calcium level, allowed the simultaneous assessment of functional sublethal alterations related to calcium homeostasis and mitochondrial function (Jakel et al. 2021)

Complex multiparametric studies using numerous colors need an increase in the number of 2D plots for each marker combination. To manage the intricate data analysis involved, automated computational methods have been devised to identify populations in multidimensional flow cytometry (Mair et al. 2016). Overall, the implementation of multiparametric and computational flow cytometry holds great promise for advancing knowledge and discoveries in semen assessment (Boe-Hansen and Satake 2019).

4   The Use of Omics Technologies for Boar Sperm Evaluation

In animal reproduction, the correlation between methods to assess semen quality and fertility is often limited, with sub-fertile males being a particularly challenging case. The development of the new era of "omics" technology has stimulated research into the identification of accurate predictors of male fertility. In this context, the fertility status of males is frequently unknown until sexual maturity is reached. Therefore, the validation of biochemical and molecular markers (i.e., genes, transcripts, proteins, and metabolites) associated with male subfertility phenotypes has the potential to improve pig production by increasing reproductive efficiency.

4.1   Genomic Approaches to Predict Sperm Quality

4.1.1   Single Nucleotide Polymorphism Genotyping

Single nucleotide polymorphisms (SNPs) represent the most prevalent type of genomic variation, being these single-base substitutions in the DNA code frequently associated with known genetic traits. Advances in molecular genetics during the past two decades have allowed the identification of specific SNPs associated with boar sperm quality, which has increased the interest in understanding the molecular mechanisms underlying sperm phenotype traits. For example, it has been identified that several putative candidate genes for hormone receptors (GnRHR), as well as the inhibitory gonadal hormone inhibin (*INHBA; INHBB*), possess SNPs that have been demonstrated to impact boar semen quality and sperm function (Lin et al. 2006).

Examples of SNPs associated with reduced boar fertility include genes encoding sperm proteins such as phospholipase C zeta (*PLCZ*), which partially controls oocyte activation during fertilization (Kaewmala et al. 2012), *DAZL* (deleted in azoospermia-like) involved in the regulation of spermatogenesis (Ma et al. 2013) and the *CD9* antigen, which is involved in epididymal sperm maturation (Kaewmala et al. 2011). Regarding spermatogenesis and epididymal sperm maturation, Sironen et al. (2010) also found a SNP in the gene encoding for ubiquitin ligase *HECW2* associated with heritable knobbed acrosome defect in spermatozoa from Finnish Yorkshire boars.

Recently, several candidate SNPs related to post-thaw semen quality have been identified (Mańkowska et al. 2020; Brym et al. 2021), which might be used to improve the selection procedure of boars for cryopreservation. In conclusion, the detection of polymorphisms in genes expressed in spermatozoa offers potential candidate factors that are associated with boar semen quality although the evaluation of various sperm traits rather than a single trait analysis provides a better fertility prediction of boar semen quality.

4.1.2   Genome-Wide Association Studies

Commercial SNP arrays are now available for all major livestock species, as well as a collection of phenotypic traits such as male reproductive and semen production traits. Genome-wide association studies (GWAS) identify quantitative trait loci (QTL) and specific regions of the genome as genetic markers, which are associated with phenotypic data.

A GWA study in boar identified genomic regions and candidate genes that were associated with abnormal sperm morphology (Zhao et al. 2020). In addition, an enrichment analysis of those candidate genes revealed pathways associated with sperm concentration, sperm motility, sperm progression, and abnormal morphology. A recent study has presented candidate genes with functions involved in spermatogenesis and sperm motility in young Pietrain boars (Reyer et al. 2024). Also, they identified genes with functions related to the sperm coating and morphology (i.e., *SUCLA2* and *CMAH*) that might contribute to differences in the sperm motility and environmental resistance of sperm. Interestingly, some of these genes have been previously associated with fertility traits in livestock species or related to infertility in human patients (*CEP78* and *GPER1*). Therefore, these studies provide useful information about the genetic architecture of complex phenotypes such as boar semen production and quality traits.

4.2   The Boar Sperm Transcriptome

Although sperm have been considered solely as vehicles of paternal DNA, it is now accepted that mature spermatozoa retained RNAs and these are delivered to the oocyte. However, the biological role of these transcripts largely remains unknown (Johnson et al. 2011). Some of the proposed functions included the regulation of early embryonic gene expression and stabilization of the nuclear matrix (Godia et al. 2018). In addition, growing evidence indicates that, beyond messenger RNA (mRNA), small non-coding RNAs, including microRNA (miRNA) and piwi-interacting RNA (piRNA), may also play a role in the initial stages of embryonic development and could potentially function as important biomarkers for male fertility (Jodar et al. 2015).

Transcriptome analysis is achieved by screening isolated RNA from biological samples such as tissues, fluids, and cells. The analysis of RNA sequencing data includes the acquisition of raw reads, adapter trimming, read alignment, gene quantification, and quality checkpoints after each step of the analysis (Fig. 1). While microarrays can only interrogate the expression of the target RNA molecules, RNA sequencing has the potential to analyze the expression of all the RNA molecules present in a sample. RNA sequencing also allows the detection of novel genes and splicing isoforms and the identification of transcribed genetic variants (Wang et al.

2009). The principal advantages of the RNA sequencing are the higher dynamic range, specificity, and sensitivity. However, RNA sequencing requires a complex computational processing of large data sets and the presence of a significant cDNA library, possible bias through RNA fragmentation methods, and repeated identical reads (Wang et al. 2009).

In the evaluation of boar sperm, transcriptomics has been used to identify molecular biomarkers associated with semen quality traits (Godia et al. 2020; Mańkowska et al. 2020; Ablondi et al. 2021), some of these are shown in Table 3. The transcriptome profile of high and low fertile boar revealed a high number of differentially expressed genes (Alvarez-Rodriguez et al. 2020), among which *CATSPERG* (CatSper channel auxiliary subunit gamma) was upregulated, and *CATSPERB* (CatSper channel auxiliary subunit beta) was downregulated in pigs with high-fertility pigs. In addition, the differentially expressed genes annotating to the family of zinc finger nucleases (ZFNs) such as *ZNRF4, PLAGL2, ZFN25, LOC100739821*, and *ZDHHC7* were primarily upregulated in high-fertility boars. Furthermore, Godia et al. (2019) analyzed the sperm transcriptome of boar ejaculates obtained during both summer and winter seasons, periods associated with lower and higher semen quality and fertility, respectively. Notably, 34 transcripts exhibited altered levels between the summer and winter samples, likely in response to heat stress. These transcripts predominantly pertained to oxidative stress, sperm membrane and DNA damage, and autophagy pathways.

The distinctive roles of microRNAs (miRNA) in RNA silencing and the post-transcriptional control of gene expression, as highlighted by Ambros (2004) make them highly interesting potential biomarkers for male fertility. Consequently, miRNAs have been extensively investigated in various mammalian species, including pigs. MicroRNAs are a class of small and non-coding ribonucleic acids that play a crucial role in spermatogenesis. As an example, miRNAs have been reported to modulate expression during the different stages of sperm maturation (Moazed 2009). The differential expression of miRNAs in porcine spermatozoa has been also associated with sperm morphology and motility. Curry et al. (2011) found that the expression of four miRNAs (let-7a, -7d, -7e, and miR-22) was increased in sperm samples with abnormal morphology compared to normal sperm, whereas miR-15b was decreased. In addition to being upregulated in the abnormal morphology group,

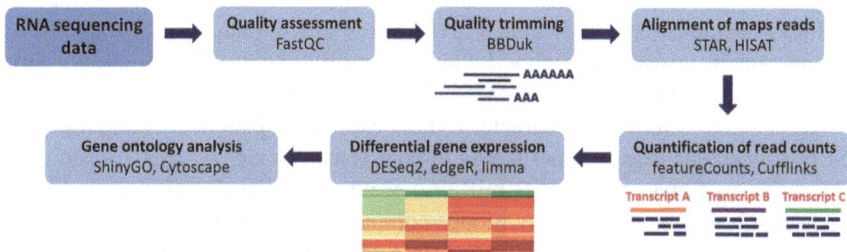

Fig. 1 Overview of RNA sequencing data analysis pipeline and examples of software that can be used in each step from quality assessment to gene ontology analysis

Table 3 List of important transcripts reported in pig spermatozoa by RNA sequencing analysis

Type of sample	Phenotype	RNAs	Finding	Reference
Sperm	Summer vs. winter fresh ejaculates	*MCM8, STARD9, OSGIN1, miR-34c, miR-1249, miR-106b*	Oxidative stress and autophagy related to summer periods	Godia et al. (2019)
Sperm	Fresh vs. frozen sperm	*PLCZ1, CRISP2, VEGFA, CDKN1B, PRKD2, SC5D*	Sperm response to environment stimuli, apoptosis, and metabolic activities	Dai et al. (2019)
Sperm	High fertile vs. low fertile	*CATSPERG, ZNRF4, CATSPERB, PLAGL2, ZFN25, miR-621, miR-221*	Capacitation and immune regulation response	Alvarez-Rodriguez et al. (2020)
Sperm	Cauda epididymal vs. ejaculated	*let-7a, miR-92a*	Calcium and cAMP signaling	Chang et al. (2016)
Sperm	Fresh sperm ejaculates	*CSNK1G2, SIL1, PSMF1*	Spermatogenesis	Ablondi et al. (2021)
Sperm	Capacitated and non-capacitated sperm	*miR-1343, miR-151-3p, miR-1285, miR-127*	PI3K-Akt, MAPK, cAMP-PKA, and Ca^{2+} signaling pathways	Li et al. (2018)
Seminal plasma	RNAs from extracellular vesicles	*let-7a, ssc-miR-21-5p, Ssc-miR-148a-3p*	Spermatogenesis, immune functions	Xu et al. (2020)

let-7d and -7e were also significantly increased in the low motility group, suggesting a fundamental role of the let-7s in mature sperm function.

Interestingly, the expression of two miRNAs (miR-34c and miR-1249) has been also differentially expressed in boar sperm between the summer and the winter months (Godia et al. 2019). For example, the most abundant miRNA in the study was miR-34c, which was downregulated in the winter samples. In contrast, miR-1249 was upregulated in the winter group, which was also found to be altered in the semen of bulls with moderate fertility (Fagerlind et al. 2015). This suggests a potential correlation between the seasonal variation in semen quality and the expression levels of these two miRNAs.

Capacitated and non-capacitated sperm also appeared to have different miRNA profiles as demonstrated by Li et al. (2018). These miRNA targets play a significant role in protein tyrosine phosphorylation, cell proliferation, differentiation, sperm capacitation, sperm motility, hyperactivation, and acrosome reaction.

Recent studies have also characterized the expression pattern of seminal plasma (SP) extracellular vesicle small RNAs in boar semen (Xu et al. 2020). Zhao et al. (2024) showed that 14 and 10 miRNAs were upregulated and downregulated (respectively) in the low motility groups compared to sperm with high motility. In conclusion, the identification of seminal RNAs as non-invasive biomarkers for male infertility might be useful to incorporate in selection programs and thus have a high economic impact on the livestock industry.

4.3 Proteomic Identification of Potential Biomarkers

Proteomics encompasses the comprehensive study of proteins, encompassing aspects such as quantitative expression, post-translational modifications, and protein interactions on a large scale. Compared to genomic and transcriptomic analyses, proteomics is suggested to offer a potentially more accurate means of identifying biomarkers. This is because gene expression varies significantly across cells, and mRNA does not always translate into protein, as noted by Kovac et al. (2013). Since the first proteomics approach in 1980, advancements in protein extraction techniques and liquid chromatography have notably enhanced the resolution and sensitivity of mass spectrometers. This progress has significantly augmented the outcomes of proteomic analyses conducted on intricate biological entities, such as SP and spermatozoa (Jimenez et al. 2007).

Generally during sample preparation for analysis, the entire semen is typically subjected to low-speed centrifugation (700–2000 × g) to separate the SP from the spermatozoa content and debris (Fig. 2). For the analysis of SP proteins, the supernatant is then further centrifuged at high speed (10,000–12,000 × g) for one hour at 4 °C. Alternatively, if the goal is to analyze proteins from spermatozoa, the pellet obtained after the initial centrifugation is washed in phosphate buffer to eliminate SP constituents. Subsequently, the cell pellet is resuspended in a lysis buffer, and mechanical lysis methods, such as sonication and centrifugation, are employed to lyse the cells. The resulting supernatant comprises the whole spermatozoa protein extract (Bustamante-Filho et al. 2022). After isolating proteins, they are typically digested with trypsin or other proteases, leading to the generation of a highly complex mixture of peptides. These peptides are then separated using liquid chromatography. Subsequently, the peptides are ionized and introduced into an LC-coupled mass spectrometer via electron spray injection.

Spectra generated by the mass spectrometry (MS) (m/z values, intensities of parent peptide and its associated fragment ions) are used for protein identification using various software and databases such as UniProtKB, Ensebml/UniProt/NCBI (Kumar et al. 2017). Protein data can be also used for enrichment analysis (Chen et al. 2020), which identifies proteins that share common functions and pathways and helps understand the physiological relevance of the identified changes in the proteome. Bioinformatics is the last essential step of proteomic analysis. Some of these online databases include DAVID, STRING, PANTHER, KEGG pathway, and Reactome, among others.

In boar, SP represents the major portion of the ejaculate volume (approximately 95%) and has been related to sperm quality, membrane integrity, and ejaculate resistance to cooling and freezing. Proteins are one of the most important components of the boar SP, with concentrations between 30 and 60 g/L (Rodriguez-Martinez et al. 2009). Druart et al. (2013) conducted the first MS-based proteomics investigation of boar SP, where they highlight expected spermadhesins, specifically PSP-I, PSP-II, AQN-1, AQN-3, AWN-1, but also fibronectin (FN1) among others. Several studies have reported that pig SP is rich in spermadhesins (75–90% of total SP protein content) (Rodriguez-Martinez et al. 2011; Gonzalez-Cadavid et al. 2014), a highly

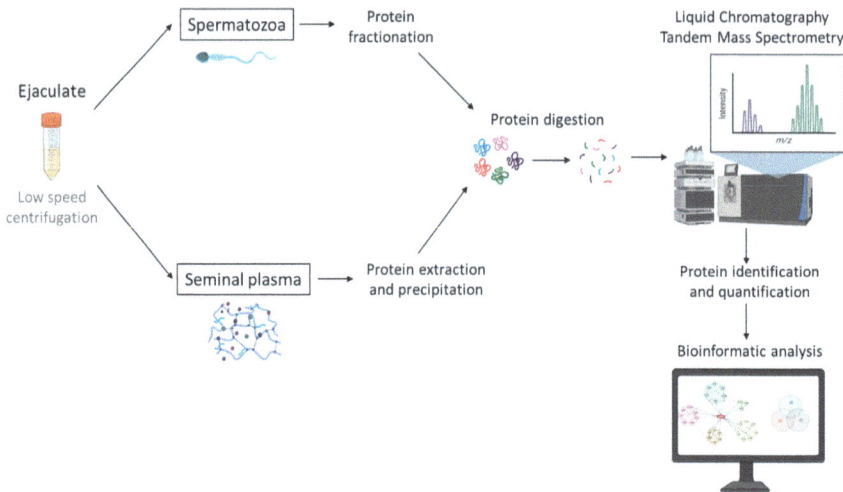

Fig. 2 Schematic overview of a typical workflow for high-throughput proteomics for spermatozoa and seminal plasma samples. The semen is initially centrifuged at low speed to distinguish the seminal plasma from the spermatozoa content. For spermatozoa samples, a lysis buffer is applied, and mechanical lysis techniques like sonication and centrifugation are employed to acquire the protein fraction. In contrast, to isolate seminal plasma proteins, the supernatant is subjected to further centrifugation at high speed, yielding a supernatant enriched with seminal proteins. The purified proteins from each sample are analyzed by liquid chromatography and mass spectrometry analyses. The resulting data allows protein identification, quantification as well as enrichment pathway analysis

versatile family of glycoproteins categorized based on their capacity to bind (AQN-1, AQN-3, and AWN) or not (PSP-I and PSP-II) to heparin. Furthermore, various spermadhesins have been linked to the fertilization capability of sperm in vivo (Caballero et al. 2012)

Differences in protein concentration and composition have been also observed between the sperm-rich fraction (SRF) and the post-SRF (with the largest SP volume). Perez-Patino et al. (2016) identified 34 proteins that were differentially expressed between the SRF and post-SRF, some of them were previously related to sperm capacitation, the acrosome reaction, zona pellucida binding, membrane stability and permeability (HEXB, GP2, ARSA, GLB1L3). In addition, deoxyribonuclease-2-alpha, an acid endonuclease with known bactericidal activity was found in greater quantity in the post-SRF. This suggests that spermadhesins may protect spermatozoa during their journey through the female reproductive tract.

Comparative proteomics has been also used to study ejaculate resistance to cooling. De Lazari et al. (2020) identified that spermadhesin PSP-I, cathepsin B, epididymal secretory protein E1 precursor, and IgG Fc binding protein were more abundant in ejaculates with higher resistance to cooling at 17 °C. A recent study identified exclusive proteins in the SP of boars showing lower resistance to seminal cooling at 5 °C, which were involved in the proteolytic activation of metalloproteinases (i.e., cathepsin B, cystatin, and clusterin) and proteins related to immune and inflammatory modulation (prostaglandin-H2 D-isomerase, complement regulator factor H; Ig-like domain-containing protein and MHCI-like Ag-recog domain-containing protein; ribonuclease A family member 9 and peptidyl-prolyl cis-trans isomerase) (Menezes et al. 2020).

Similarly to the SP, studies have recently been performed on the complete proteome of boar spermatozoa under physiological and capacitation conditions, providing valuable information to identify potentially useable biomarkers of sperm fertilizing capacity. The first in-depth study of the boar ejaculated spermatozoa-proteome was done by Feugang et al. (2018), where they identified 2728 proteins in mature spermatozoa using a shotgun approach. In addition, Perez-Patino et al. (2019) demonstrated that the proteome of pig spermatozoa, even those classified as mature within the cauda epididymis, undergoes remodeling upon ejaculation. This remodeling primarily occurs due to sperm interactions with SP, resulting in quantitative changes in proteins implicated in sperm functionality.

Several studies have shown that fertility-related proteins identified in capacitated spermatozoa are able to predict litter size more accurately than when these proteins are studied in non-capacitated spermatozoa (Kwon et al. 2015a, b). Recently, Zigo et al. (2023) compared the sperm surface proteome of non-capacitated and in vitro capacitated boar spermatozoa, with a particular focus on the substrates of sperm proteasomes. They found 141 proteins that were more abundant, while 91 proteins were less abundant on the surface of the in vitro capacitated group compared to the control group. Also, they demonstrated that the ubiquitin–proteasome system regulates the capacitation-associated sperm surface remodeling by targeting at least 14 individual sperm surface proteins.

Taken together, proteomics allows a "shotgun" analysis by the identification of thousands of proteins at the same time. Moreover, studying multiple biomarkers for one trait could increase the overall accuracy and sensitivity of male infertility predictions, and thus, could be considered for field application. Before these findings can be effectively applied in the field, a validation step is mandatory to confirm that the identified proteins can be reliably used as biomarkers of male fertility. The employment of specific antibodies for techniques such as western blotting, enzyme-linked immunosorbent assay (ELISA), and immunolocalization will enhance the validity and utility of specific proteins as biomarkers. However, as semen is a dynamic fluid whose proteins can be influenced by many factors; other techniques are needed to validate whether the expression level of protein is stable in time to better elucidate the physiological implications. In addition, the application of machine learning and artificial intelligence to manage large data (protein data and fertility indexes) would accelerate the validation steps and identification of semen biomarkers.

4.4   Metabolomic Analysis

Metabolomics is an emerging technology that studies small (<1500 Da), low-molecular weight compounds, which are the products of sperm metabolism and might control signaling pathways such as sperm motility, hyperactivation, and energy acquisition-related signaling directly or indirectly. Metabolomics is a useful technique since reflects downstream of gene and protein expression (Kovac et al. 2013). Although metabolomics research has developed rapidly in recent years, nuclear magnetic resonance (NMR) and MS are the two main measurement techniques. Nuclear Magnetic Resonance technology offers remarkable reproducibility and precise quantitative analysis. It also provides unequivocal data essential for comprehending compound structures through non-invasive measurements (Markley et al. 2017). On the other hand, MS delivers comprehensive and accurate mass determination, aiding in structural identification and high-sensitivity assessment of contents. Additionally, MS instrumentation can also be accessible on the level of analytical resolution at a low price. Due to these characteristics, MS has emerged as the predominant technique in the field of metabolomics.

Metabolomics has just begun to be used in the search for biomarkers in boar sperm and SP to predict male fertility (Mateo-Otero et al. 2021; Zhang et al. 2021). Focusing on the pig SP metabolomic profile, Mateo-Otero et al. (2020) revealed for the first time intra-ejaculate variability in the metabolite composition of the SP, observing different relative abundances in choline, glycerophosphocholine, and glycine metabolites. They also showed variability in the metabolite composition of SP between the different portions of the ejaculate. A further study by the same group found specific relationships between certain metabolites and sperm variables (Mateo-Otero et al. 2021). For example, some fresh sperm quality and functionality parameters were related to glutamate, methanol, trimethylamine N-oxide, carnitine, and isoleucine. In addition, sperm's ability to survive liquid storage was seen to be related to the presence of leucine, hypotaurine, carnitine, and isoleucine metabolites.

Other studies have also shown that the metabolome of boar sperm significantly differs between poor and good freezability sperm samples. Torres et al. (2022) revealed a number of metabolites associated with the energy production pathway (including inosine, hypoxanthine, creatine, ADP, niacinamide, spermine, and 2-methylbutyrylcarnitine) with the improvement of ejaculate freezability. Sui et al. (2023) showed lower expression of the metabolite L-citrulline in sperm samples with bad freezability. This is in agreement with previous studies where they demonstrated that L-citrulline supplementation mitigates oxidative stress which resulted in enhanced ram sperm freezability (Zhao et al. 2023).

In conclusion, the use of metabolomics allows the identification and quantification of metabolites that could be used as biomarkers of sperm quality. As metabolites are the final products of metabolism, the changes in the composition of which can reflect the state of sperm and individual metabolic timely. However, there are still some features that would need to be improved such as the development of more real-time and universal detection methods based on the existing independent analysis platforms; establishing a new metabolome analysis platform to perform the parallel analysis and data integration of multiple analysis platforms and further integrating metabolomics data with other omics data to improve our understanding of biological processes.

5 Conclusions

Throughout this chapter, the critical importance of evaluating boar semen quality for diagnosing reproductive problems and optimizing pig production has been highlighted. Traditional semen analysis parameters such as sperm concentration, motility, and morphology are commonly used, but often fall short in accurately predicting fertility. The introduction of CASA technology has improved semen evaluation by providing automated, accurate measurements of some of these parameters. However, flow cytometry has emerged as a complementary tool, offering high-throughput analysis that quickly and accurately assesses sperm viability, plasma membrane integrity, and DNA integrity, which are critical for fertilization and healthy embryo development. The advent of omics technologies, such as genomics, transcriptomics, proteomics, and metabolomics, has furthered the understanding of male fertility by identifying molecular biomarkers associated with sperm quality.

There is a clear need to combine traditional and advanced methods to improve the overall effectiveness of boar semen evaluation and selection in breeding programs. The integration of omics technologies has revolutionized the field by providing comprehensive insights into the molecular basis of sperm quality and fertility. By validating these biomarkers, it will be possible to improve reproductive efficiency in swine production in the near future. It must be emphasized that while traditional techniques remain valuable, the application of CASA and flow cytometry, along with the insights gained from omics research, represents a significant advancement in the field, allowing more accurate prediction of fertility and better selection of superior breeding candidates.

References

Ablondi M, Godia M, Rodriguez-Gil JE, Sanchez A, Clop A (2021) Characterisation of sperm piRNAs and their correlation with semen quality traits in swine. Anim Genet 52(1):114–120. https://doi.org/10.1111/age.13022

Alm K, Peltoniemi OA, Koskinen E, Andersson M (2006) Porcine field fertility with two different insemination doses and the effect of sperm morphology. Reprod Domest Anim 41(3):210–213. https://doi.org/10.1111/j.1439-0531.2005.00670.x

Althouse GC, Hopkins SM (1995) Assessment of boar sperm viability using a combination of two fluorophores. Theriogenology 43(3):595–603. https://doi.org/10.1016/0093-691x(94)00065-3

Alvarez-Rodriguez M, Vicente-Carrillo A, Rodriguez-Martinez H (2018) Hyaluronan improves neither the long-term storage nor the cryosurvival of liquid-stored CD44-bearing AI boar spermatozoa. J Reprod Dev 64(4):351–360. https://doi.org/10.1262/jrd.2017-141

Alvarez-Rodriguez M, Martinez C, Wright D, Barranco I, Roca J, Rodriguez-Martinez H (2020) The transcriptome of pig spermatozoa, and its role in fertility. Int J Mol Sci 25(5):1572. https://doi.org/10.3390/ijms21051572

Amann RP, Katz DF (2004) Reflections on CASA after 25 years. J Androl 25(3):317–325. https://doi.org/10.1002/j.1939-4640.2004.tb02793.x

Amann RP, Waberski D (2014) Computer-assisted sperm analysis (CASA): capabilities and potential developments. Theriogenology 81(1):5–17e11–13. https://doi.org/10.1016/j.theriogenology.2013.09.004

Ambros V (2004) The functions of animal microRNAs. Nature 431(7006):350–355. https://doi.org/10.1038/nature02871

Antonczyk A, Nizanski W, Partyka A, Ochota M, Mila H (2012) The usefulness of real time morphology software in semen assessment of teratozoospermic boars. Syst Biol Reprod Med 58(6):362–368. https://doi.org/10.3109/19396368.2012.715229

Babamoradi H, Amigo JM, van den Berg F, Petersen MR, Satake N, Boe-Hansen G (2015) Quality assessment of boar semen by multivariate analysis of flow cytometric data. Chemom Intell Lab Syst 142:219–230. https://doi.org/10.1016/j.chemolab.2015.02.008

Balogun KB, Stewart KR (2021) Effects of air exposure and agitation on quality of stored boar semen samples. Reprod Domest Anim 56(9):1200–1208. https://doi.org/10.1111/rda.13975

Banaszewska D, Andraszek K (2021) Assessment of the morphometry of heads of normal sperm and sperm with the dag defect in the semen of duroc boars. J Vet Res 65(2):239–244

Barquero V, Roldan ERS, Soler C, Yaniz JL, Camacho M, Valverde A (2021) Predictive capacity of boar sperm morphometry and morphometric sub-populations on reproductive success after artificial insemination. Animals (Basel) 11(4)

Blom E (1983) [Pathological conditions in genital organs and sperm as a cause for the rejection of breeding bulls for import into and export from Denmark (an andrologic retrospective, 1958-1982)]. Nord Vet Med 35(3):105–130

Boe-Hansen GB, Satake N (2019) An update on boar semen assessments by flow cytometry and CASA. Theriogenology 137:93–103. https://doi.org/10.1016/j.theriogenology.2019.05.043

Bompart D, Garcia-Molina A, Valverde A, Caldeira C, Yaniz J, Nunez de Murga M, Soler C (2018) CASA-Mot technology: how results are affected by the frame rate and counting chamber. Reprod Fertil Dev 30(6):810–819. https://doi.org/10.1071/RD17551

Bonet S (1990) Immature and aberrant spermatozoa in the ejaculate of Sus domesticus. Anim Reprod Sci 22(1):67–80. https://doi.org/10.1016/0378-4320(90)90039-i

Bonet S, Briz M (1991) New data on aberrant spermatozoa in the ejaculate of Sus domesticus. Theriogenology 35(4):725–730. https://doi.org/10.1016/0093-691x(91)90413-8

Bonet S, Briz M, Fradera A (1993) Ultrastructural abnormalities of boar spermatozoa. Theriogenology 40(2):383–396. https://doi.org/10.1016/0093-691x(93)90276-b

Briz M, Fàbrega A (2013) The Boar spermatozoon. In: Bonet C, Holt Y (eds) Boar reproduction: fundamentals and new biotechnological trends. Springer, Berlin, pp 3–47. https://doi.org/10.1007/978-3-642-35049-8_1

Briz MD, Bonet S, Pinart B, Camps R (1996) Sperm malformations throughout the boar epididymal duct. Anim Reprod Sci 43(4):221–239

Broekhuijse ML, Sostaric E, Feitsma H, Gadella BM (2011) Additional value of computer assisted semen analysis (CASA) compared to conventional motility assessments in pig artificial insemination. Theriogenology 76(8):1473–1486. e1471. https://doi.org/10.1016/j.theriogenology.2011.05.040

Broekhuijse ML, Feitsma H, Gadella BM (2012a) Artificial insemination in pigs: predicting male fertility. Vet Q 32(3–4):151–157. https://doi.org/10.1080/01652176.2012.735126

Broekhuijse ML, Sostaric E, Feitsma H, Gadella BM (2012b) Application of computer-assisted semen analysis to explain variations in pig fertility. J Anim Sci 90(3):779–789. https://doi.org/10.2527/jas.2011-4311

Broekhuijse ML, Sostaric E, Feitsma H, Gadella BM (2012c) Relationship of flow cytometric sperm integrity assessments with boar fertility performance under optimized field conditions. J Anim Sci 90(12):4327–4336. https://doi.org/10.2527/jas.2012-5040

Broekhuijse ML, Sostaric E, Feitsma H, Gadella BM (2012d) The value of microscopic semen motility assessment at collection for a commercial artificial insemination center, a retrospective study on factors explaining variation in pig fertility. Theriogenology 77(7):1466–1479. e1463. https://doi.org/10.1016/j.theriogenology.2011.11.016

Brouwers JF, Silva PF, Gadella BM (2005) New assays for detection and localization of endogenous lipid peroxidation products in living boar sperm after BTS dilution or after freeze-thawing. Theriogenology 63(2):458–469. https://doi.org/10.1016/j.theriogenology.2004.09.046

Brym P, Wasilewska-Sakowska K, Mogielnicka-Brzozowska M, Mankowska A, Paukszto L, Pareek CS, Kordan W, Kondracki S, Fraser L (2021) Gene promoter polymorphisms in boar spermatozoa differing in freezability. Theriogenology 166:112–123. https://doi.org/10.1016/j.theriogenology.2021.02.018

Bustamante-Filho IC, Pasini M, Moura AA (2022) Spermatozoa and seminal plasma proteomics: Too many molecules, too few markers. The case of bovine and porcine semen. Anim Reprod Sci 247:107075. https://doi.org/10.1016/j.anireprosci.2022.107075

Caamano JN, Tamargo C, Parrilla I, Martinez-Pastor F, Padilla L, Salman A, Fueyo C, Fernandez A, Merino MJ, Iglesias T, Hidalgo CO (2021) Post-thaw sperm quality and functionality in the autochthonous pig breed Gochu Asturcelta. Animals (Basel) 11(7). https://doi.org/10.3390/ani11071885

Caballero I, Vazquez JM, Mayor GM, Alminana C, Calvete JJ, Sanz L, Roca J, Martinez EA (2009) PSP-I/PSP-II spermadhesin exert a decapacitation effect on highly extended boar spermatozoa. Int J Androl 32(5):505–513

Caballero I, Parrilla I, Alminana C, del Olmo D, Roca J, Martinez EA, Vazquez JM (2012) Seminal plasma proteins as modulators of the sperm function and their application in sperm biotechnologies. Reprod Domest Anim 47 Suppl 3(s3):12–21. https://doi.org/10.1111/j.1439-0531.2012.02028.x

Camus A, Camugli S, Leveque C, Schmitt E, Staub C (2011) Is photometry an accurate and reliable method to assess boar semen concentration? Theriogenology 75(3):577–583. https://doi.org/10.1016/j.theriogenology.2010.09.025

Castellini C, Dal Bosco A, Ruggeri S, Collodel G (2011) What is the best frame rate for evaluation of sperm motility in different species by computer-assisted sperm analysis? Fertil Steril 96(1):24–27. https://doi.org/10.1016/j.fertnstert.2011.04.096

Centurion F, Vazquez JM, Calvete JJ, Roca J, Sanz L, Parrilla I, Garcia EM, Martinez EA (2003) Influence of porcine spermadhesins on the susceptibility of boar spermatozoa to high dilution. Biol Reprod 69(2):640–646. https://doi.org/10.1095/biolreprod.103.016527

Chang Y, Dai DH, Li Y, Zhang Y, Zhang M, Zhou GB, Zeng CJ (2016) Differences in the expression of microRNAs and their predicted gene targets between cauda epididymal and ejaculated boar sperm. Theriogenology 86(9):2162–2171

Chen C, Hou J, Tanner JJ, Cheng J (2020) Bioinformatics methods for mass spectrometry-based proteomics data analysis. Int J Mol Sci 21(8):2873. https://doi.org/10.3390/ijms21082873

Christensen P, Stryhn H, Hansen C (2005) Discrepancies in the determination of sperm concentration using Burker-Turk, Thoma and Makler counting chambers. Theriogenology 63(4):992–1003. https://doi.org/10.1016/j.theriogenology.2004.05.026

Curry E, Safranski TJ, Pratt SL (2011) Differential expression of porcine sperm microRNAs and their association with sperm morphology and motility. Theriogenology 76(8):1532–1539. https://doi.org/10.1016/j.theriogenology.2011.06.025

Czubaszek M, Andraszek K, Banaszewska D, Walczak-Jedrzejowska R (2019) The effect of the staining technique on morphological and morphometric parameters of boar sperm. PLoS One 14(3):e0214243. https://doi.org/10.1371/journal.pone.0214243

Dai DH, Qazi IH, Ran MX, Liang K, Zhang Y, Zhang M, Zhou GB, Angel C, Zeng CJ (2019) Exploration of miRNA and mRNA Profiles in Fresh and Frozen-Thawed Boar Sperm by Transcriptome and Small RNA Sequencing. Int J Mol Sci 20(4):802

De Lazari FL, Sontag ER, Schneider A, Araripe Moura AA, Vasconcelos FR, Nagano CS, Dalberto PF, Bizarro CV, Mattos RC, Mascarenhas Jobim MI, Bustamante-Filho IC (2020) Proteomic identification of boar seminal plasma proteins related to sperm resistance to cooling at 17 degrees C. Theriogenology 147:135–145. https://doi.org/10.1016/j.theriogenology.2019.11.023

Del Gallego R, Sadeghi S, Blasco E, Soler C, Yaniz JL, Silvestre MA (2017) Effect of chamber characteristics, loading and analysis time on motility and kinetic variables analysed with the CASA-mot system in goat sperm. Anim Reprod Sci 177:97–104. https://doi.org/10.1016/j.anireprosci.2016.12.010

Di Caprio G, Gioffrè MA, Saffioti NA, Grilli SA, Ferraro PA, Puglisi RA, Balduzzi DA, Galli AA, Coppola GA (2010) Quantitative label-free animal sperm imaging by means of digital holographic microscopy. IEEE J Selected Top Quantum Electron 16(4):833–840. https://doi.org/10.1109/jstqe.2009.2036741

Didion BA (2008) Computer-assisted semen analysis and its utility for profiling boar semen samples. Theriogenology 70(8):1374–1376. https://doi.org/10.1016/j.theriogenology.2008.07.014

Dott HM, Foster GC (1979) The estimation of sperm motility in semen, on a membrane slide, by measuring the area change frequency with an image analysing computer. J Reprod Fertil 55(1):161–166. https://doi.org/10.1530/jrf.0.0550161

Druart X, Rickard JP, Mactier S, Kohnke PL, Kershaw-Young CM, Bathgate R, Gibb Z, Crossett B, Tsikis G, Labas V, Harichaux G, Grupen CG, de Graaf SP (2013) Proteomic characterization and cross species comparison of mammalian seminal plasma. J Proteome 91:13–22. https://doi.org/10.1016/j.jprot.2013.05.029

Ehlers J, Behr M, Bollwein H, Beyerbach M, Waberski D (2011) Standardization of computer-assisted semen analysis using an e-learning application. Theriogenology 76(3):448–454. https://doi.org/10.1016/j.theriogenology.2011.02.021

Ellis DJ, Shadan S, James PS, Henderson RM, Edwardson JM, Hutchings A, Jones R (2002) Post-testicular development of a novel membrane substructure within the equatorial segment of ram, bull, boar, and goat spermatozoa as viewed by atomic force microscopy. J Struct Biol 138(3):187–198. https://doi.org/10.1016/s1047-8477(02)00025-4

Fagerlind M, Stalhammar H, Olsson B, Klinga-Levan K (2015) Expression of miRNAs in bull spermatozoa correlates with fertility rates. Reprod Domest Anim 50(4):587–594. https://doi.org/10.1111/rda.12531

Fair S, Romero-Aguirregomezcorta J (2019) Implications of boar sperm kinematics and rheotaxis for fertility after preservation. Theriogenology 137:15–22. https://doi.org/10.1016/j.theriogenology.2019.05.032

Feitsma H (2009) Artificial insemination in pigs, research and developments in The Netherlands, a review. Acta Sci Vet 37(Supl 1):s61–s71

Feugang JM, Liao SF, Willard ST, Ryan PL (2018) In-depth proteomic analysis of boar spermatozoa through shotgun and gel-based methods. BMC Genomics 19(1):62. https://doi.org/10.1186/s12864-018-4442-2

Flesch FM, Voorhout WF, Colenbrander B, van Golde LM, Gadella BM (1998) Use of lectins to characterize plasma membrane preparations from boar spermatozoa: a novel technique for monitoring membrane purity and quantity. Biol Reprod 59(6):1530–1539. https://doi.org/10.1095/biolreprod59.6.1530

Flesch FM, Colenbrander B, van Golde LM, Gadella BM (1999) Capacitation induces tyrosine phosphorylation of proteins in the boar sperm plasma membrane. Biochem Biophys Res Commun 262(3):787–792. https://doi.org/10.1006/bbrc.1999.1300

Fraser L, Lecewicz M, Strzezek J (2002) Fluorometric assessments of viability and mitochondrial status of boar spermatozoa following liquid storage. Pol J Vet Sci 5(2):85–92

Gaczarzewicz D (2015) Influence of chamber type integrated with computer-assisted semen analysis (CASA) system on the results of boar semen evaluation. Pol J Vet Sci 18(4):817–824. https://doi.org/10.1515/pjvs-2015-0106

Gadea J (2003) Review: Semen extenders used in the artificial insemination of swine. Span J Agric Res 1(2):17–27

Gadea J (2005) Sperm factors related to in vitro and in vivo porcine fertility. Theriogenology 63(2):431–444. https://doi.org/10.1016/j.theriogenology.2004.09.023

Gadea J (2019) Desarrollo de un estándar de calidad para centros de inseminación porcina en España. La propuesta ANPSTAND Suis 154:16–20

Gadea J, Matas C (2000) Sperm factors related to in vitro penetration of porcine oocytes. Theriogenology 54(9):1343

Gadea J, Selles E, Marco MA (2004) The predictive value of porcine seminal parameters on fertility outcome under commercial conditions. Reprod Domest Anim 39(5):303–308. https://doi.org/10.1111/j.1439-0531.2004.00513.x

Gadea J, Gumbao D, Matas C, Romar R (2005) Supplementation of the thawing media with reduced glutathione improves function and the in vitro fertilizing ability of boar spermatozoa after cryopreservation. J Androl 26(6):749–756. https://doi.org/10.2164/jandrol.05057

Garcia-Herreros M, Aparicio IM, Baron FJ, Garcia-Marin LJ, Gil MC (2006) Standardization of sample preparation, staining and sampling methods for automated sperm head morphometry analysis of boar spermatozoa. Int J Androl 29(5):553–563. https://doi.org/10.1111/j.1365-2605.2006.00696.x

Garner DL, Johnson LA (1995) Viability assessment of mammalian sperm using SYBR-14 and propidium iodide. Biol Reprod 53(2):276–284. https://doi.org/10.1095/biolreprod53.2.276

Garner DL, Pinkel D, Johnson LA, Pace MM (1986) Assessment of spermatozoal function using dual fluorescent staining and flow cytometric analyses. Biol Reprod 34(1):127–138. https://doi.org/10.1095/biolreprod34.1.127

Garner DL, Dobrinsky JR, Welch GR, Johnson LA (1996) Porcine sperm viability, oocyte fertilization and embryo development after staining spermatozoa with SYBR-14. Theriogenology 45(6):1103–1113

Godia M, Swanson G, Krawetz SA (2018) A history of why fathers' RNA matters. Biol Reprod 99(1):147–159. https://doi.org/10.1093/biolre/ioy007

Godia M, Estill M, Castello A, Balasch S, Rodriguez-Gil JE, Krawetz SA, Sanchez A, Clop A (2019) A RNA-Seq analysis to describe the boar sperm transcriptome and its seasonal changes. Front Genet 10:299. https://doi.org/10.3389/fgene.2019.00299

Godia M, Castello A, Rocco M, Cabrera B, Rodriguez-Gil JE, Balasch S, Lewis C, Sanchez A, Clop A (2020) Identification of circular RNAs in porcine sperm and evaluation of their relation to sperm motility. Sci Rep 10(1):7985. https://doi.org/10.1038/s41598-020-64711-z

Gonzalez-Cadavid V, Martins JA, Moreno FB, Andrade TS, Santos AC, Monteiro-Moreira AC, Moreira RA, Moura AA (2014) Seminal plasma proteins of adult boars and correlations with sperm parameters. Theriogenology 82(5):697–707. https://doi.org/10.1016/j.theriogenology.2014.05.024

Górski K, Kondracki S, Wysokińska A (2017) Effects of season on semen parameters and relationships between selected semen characteristics in Hypor boars. Turk J Vet Anim Sci 41:563–569

Gonzalez-Castro RA, Peña FJ, Herickhoff LA (2022) Validation of a new multiparametric protocol to assess viability, acrosome integrity and mitochondrial activity in cooled and frozen thawed boar spermatozoa. Cytom B Clin Cytom 102(5):400–408. https://doi.org/10.1002/cyto.b.22058

Graham JK, Kunze E, Hammerstedt RH (1990) Analysis of sperm cell viability, acrosomal integrity, and mitochondrial function using flow cytometry. Biol Reprod 43(1):55–64. https://doi.org/10.1095/biolreprod43.1.55

Grossfeld R, Pable J, Jakop U, Simmet C, Schulze M (2022) Comparison of NUCLEOCOUNTER, ANDROVISION with Leja chambers and the newly developed ANDROVISION eFlow for sperm concentration analysis in boars. Sci Rep 12(1):11943. https://doi.org/10.1038/s41598-022-16280-6

Gu A, Ji G, Shi X, Long Y, Xia Y, Song L, Wang S, Wang X (2010) Genetic variants in Piwi-interacting RNA pathway genes confer susceptibility to spermatogenic failure in a Chinese population. Hum Reprod 25(12):2955–2961. https://doi.org/10.1093/humrep/deq274

Gu NH, Zhao WL, Wang GS, Sun F (2019) Comparative analysis of mammalian sperm ultrastructure reveals relationships between sperm morphology, mitochondrial functions and motility. Reprod Biol Endocrinol 17(1):66. https://doi.org/10.1186/s12958-019-0510-y

Guo H, Gong Y, He B, Zhao R (2017) Relationships between mitochondrial DNA content, mitochondrial activity, and boar sperm motility. Theriogenology 87:276–283. https://doi.org/10.1016/j.theriogenology.2016.09.005

Guthrie HD, Welch GR (2006) Determination of intracellular reactive oxygen species and high mitochondrial membrane potential in Percoll-treated viable boar sperm using fluorescence-activated flow cytometry. J Anim Sci 84(8):2089–2100. https://doi.org/10.2527/jas.2005-766

Hancock JL (1957) The morphology of boar spermatozoa. J R Microsc Soc 76(3):84–97. https://doi.org/10.1111/j.1365-2818.1956.tb00443.x

Hansen C, Christensen P, Stryhn H, Hedeboe AM, Rode M, Boe-Hansen G (2002) Validation of the FACSCount AF system for determination of sperm concentration in boar semen. Reprod Domest Anim 37(6):330–334. https://doi.org/10.1046/j.1439-0531.2002.00367.x

Hansen C, Vermeiden T, Vermeiden JP, Simmet C, Day BC, Feitsma H (2006) Comparison of FACSCount AF system, Improved Neubauer hemocytometer, Corning 254 photometer, SpermVision, UltiMate and NucleoCounter SP-100 for determination of sperm concentration of boar semen. Theriogenology 66(9):2188–2194. https://doi.org/10.1016/j.theriogenology.2006.05.020

Harrison RA, Mairet B, Miller NG (1993) Flow cytometric studies of bicarbonate-mediated Ca2+ influx in boar sperm populations. Mol Reprod Dev 35(2):197–208. https://doi.org/10.1002/mrd.1080350214

Harrison RA, Ashworth PJ, Miller NG (1996) Bicarbonate/CO2, an effector of capacitation, induces a rapid and reversible change in the lipid architecture of boar sperm plasma membranes. Mol Reprod Dev 45(3):378–391. https://doi.org/10.1002/(SICI)1098-2795(199611)45:3<378::AID-MRD16>3.0.CO;2-V

Hirai M, Boersma A, Hoeflich A, Wolf E, Foll J, Aumuller TR, Braun J (2001) Objectively measured sperm motility and sperm head morphometry in boars (Sus scrofa): relation to fertility and seminal plasma growth factors. J Androl 22(1):104–110

Holt WV (1982) Epididymal origin of a coiled-tail sperm defect in a boar. J Reprod Fertil 64(2):485–489. https://doi.org/10.1530/jrf.0.0640485

Holt C (1995) An investigation of boar sperm motility using a novel computerized analysis system. PhD, University of London, University College London, London, UK

Holt W, Watson P, Curry M, Holt C (1994) Reproducibility of computer-aided semen analysis: comparison of five different systems used in a practical workshop. Fertil Steril 62(6):1277–1282. https://doi.org/10.1016/s0015-0282(16)57201-x

Holt C, Holt WV, Moore HD (1996) Choice of operating conditions to minimize sperm subpopulation sampling bias in the assessment of boar semen by computer-assisted semen analysis. J Androl 17(5):587–596

Holt C, Holt WV, Moore HD, Reed HC, Curnock RM (1997) Objectively measured boar sperm motility parameters correlate with the outcomes of on-farm inseminations: results of two fertility trials. J Androl 18(3):312–323

Iglesias I (2016) Tomographic imaging of transparent biological samples using the pyramid phase microscope. Biomed Opt Express 7(8):3049–3055. https://doi.org/10.1364/BOE.7.003049

Iglesias I, Vargas-Martin F (2013) Quantitative phase microscopy of transparent samples using a liquid crystal display. J Biomed Opt 18(2):26015. https://doi.org/10.1117/1.JBO.18.2.026015

Jakel H, Henning H, Luther AM, Rohn K, Waberski D (2021) Assessment of chilling injury in hypothermic stored boar spermatozoa by multicolor flow cytometry. Cytom A 99(10):1033–1041. https://doi.org/10.1002/cyto.a.24301

Jimenez CR, Piersma S, Pham TV (2007) High-throughput and targeted in-depth mass spectrometry-based approaches for biofluid profiling and biomarker discovery. Biomark Med 1(4):541–565. https://doi.org/10.2217/17520363.1.4.541

Jodar M, Sendler E, Moskovtsev SI, Librach CL, Goodrich R, Swanson S, Hauser R, Diamond MP, Krawetz SA (2015) Absence of sperm RNA elements correlates with idiopathic male infertility. Sci Transl Med 7(295):295re296. https://doi.org/10.1126/scitranslmed.aab1287

Johnson LA, Weitze KF, Fiser P, Maxwell WMC (2000) Storage of boar semen. Anim Reprod Sci 62(1):143–172

Johnson GD, Lalancette C, Linnemann AK, Leduc F, Boissonneault G, Krawetz SA (2011) The sperm nucleus: chromatin, RNA, and the nuclear matrix. Reproduction 141(1):21–36. https://doi.org/10.1530/REP-10-0322

Jones RC (1971) Studies of the structure of the head of boar spermatozoa from the epididymis. J Reprod Fertil Suppl 13(Suppl 13):51–64

Jones R, James PS, Oxley D, Coadwell J, Suzuki-Toyota F, Howes EA (2008) The equatorial subsegment in mammalian spermatozoa is enriched in tyrosine phosphorylated proteins. Biol Reprod 79(3):421–431. https://doi.org/10.1095/biolreprod.107.067314

Jung M, Rudiger K, Schulze M (2015) In vitro measures for assessing boar semen fertility. Reprod Domest Anim 50(Suppl 2):20–24. https://doi.org/10.1111/rda.12533

Juonala T, Lintukangas S, Nurttila T, Andersson M (2007) Relationship between semen quality and fertility in 106 AI-boars. Reprod Domest Anim 33(3–4):155–158. https://doi.org/10.1111/j.1439-0531.1998.tb01334.x

Kaewmala K, Uddin MJ, Cinar MU, Grosse-Brinkhaus C, Jonas E, Tesfaye D, Phatsara C, Tholen E, Looft C, Schellander K (2011) Association study and expression analysis of CD9 as candidate gene for boar sperm quality and fertility traits. Anim Reprod Sci 125(1–4):170–179. https://doi.org/10.1016/j.anireprosci.2011.02.017

Kaewmala K, Uddin MJ, Cinar MU, Grosse-Brinkhaus C, Jonas E, Tesfaye D, Phatsara C, Tholen E, Looft C, Schellander K (2012) Investigation into association and expression of PLCz and COX-2 as candidate genes for boar sperm quality and fertility. Reprod Domest Anim 47(2):213–223. https://doi.org/10.1111/j.1439-0531.2011.01831.x

Kaplan M, Russell LD, Peterson RN, Martan J (1984) Boar sperm cytoplasmic droplets: their ultrastructure, their numbers in the epididymis and at ejaculation and their removal during isolation of sperm plasma membranes. Tissue Cell 16(3):455–468. https://doi.org/10.1016/0040-8166(84)90063-6

Keller A, Kerns K (2022) Deep learning, artificial intelligence methods to predict boar sperm acrosome health. Anim Reprod Sci 247:107110. https://doi.org/10.1016/j.anireprosci.2022.107110

Kim SW, Ki MS, Kim CL, Hwang IS, Jeon IS (2017) A simple confocal microscopy-based method for assessing sperm movement. Dev Reprod 21(3):229–235. https://doi.org/10.12717/DR.2017.21.3.229

Knecht D, Jankowska-Makosa A, Duzinski K (2017) Analysis of the lifetime and culling reasons for AI boars. J Anim Sci Biotechnol 8(1):49. https://doi.org/10.1186/s40104-017-0179-z

Kondracki S, Bonaszewska D, Mielnicka C (2005) The effect of age on the morphometric sperm traits of domestic pigs (Sus scrofa domestica). Cell Mol Biol Lett 10(1):3–13

Kondracki S, Banaszewska D, Wysokinska A, Chomicz J (2006a) Sperm morphology of cattle and domestic pigs. Reprod Biol 6(Suppl 2):99–104

Kondracki S, Wysokinska A, Banaszewska D, Wozniak E (2006b) Evaluation of males spermiogram in domestic pigs. Reprod Biol 6(Suppl 2):93–98

Kondracki S, Wysokinska A, Kania M, Gorski K (2017) Application of two staining methods for sperm morphometric evaluation in domestic pigs. J Vet Res 61(3):345–349. https://doi.org/10.1515/jvetres-2017-0045

Kovac JR, Pastuszak AW, Lamb DJ (2013) The use of genomics, proteomics, and metabolomics in identifying biomarkers of male infertility. Fertil Steril 99(4):998–1007. https://doi.org/10.1016/j.fertnstert.2013.01.111

Kumar D, Yadav AK, Dash D (2017) Choosing an optimal database for protein identification from tandem mass spectrometry data. In: Keerthikumar S, Mathivanan S (eds) Proteome bioinformatics. Springer, New York, pp 17–29. https://doi.org/10.1007/978-1-4939-6740-7_3

Kuster C (2005) Sperm concentration determination between hemacytometric and CASA systems: why they can be different. Theriogenology 64(3):614–617. https://doi.org/10.1016/j.theriogenology.2005.05.047

Kwon WS, Oh SA, Kim YJ, Rahman MS, Park YJ, Pang MG (2015a) Proteomic approaches for profiling negative fertility markers in inferior boar spermatozoa. Sci Rep 5:13821. https://doi.org/10.1038/srep13821

Kwon WS, Rahman MS, Lee JS, Yoon SJ, Park YJ, Pang MG (2015b) Discovery of predictive biomarkers for litter size in boar spermatozoa. Mol Cell Proteomics MCP 14(5):1230–1240. https://doi.org/10.1074/mcp.M114.045369

Lasley JF, Bogart R (1944) A comparative study of epididymal and ejaculated spermatozoa of the boar. J Anim Sci 3(4):360–370. https://doi.org/10.2527/jas1944.34360x

Li Y, Li RH, Ran MX, Zhang Y, Liang K, Ren YN, He WC, Zhang M, Zhou GB, Qazi IH, Zeng CJ (2018) High throughput small RNA and transcriptome sequencing reveal capacitation-related microRNAs and mRNA in boar sperm. BMC Genomics 19(1):736. https://doi.org/10.1186/s12864-018-5132-9

Li J, Zhao W, Zhu J, Ju H, Liang M, Wang S, Chen S, Ferreira-Dias G, Liu Z (2023) Antioxidants and oxidants in boar spermatozoa and their surrounding environment are associated with AMPK activation during liquid storage. Vet Sci 10(3)

Lin CL, Ponsuksili S, Tholen E, Jennen DG, Schellander K, Wimmers K (2006) Candidate gene markers for sperm quality and fertility of boar. Anim Reprod Sci 92(3–4):349–363. https://doi.org/10.1016/j.anireprosci.2005.05.023

Llavanera M, Delgado-Bermudez A, Ribas-Maynou J, Salas-Huetos A, Yeste M (2022) A systematic review identifying fertility biomarkers in semen: a clinical approach through Omics to diagnose male infertility. Fertil Steril 118(2):291–313. https://doi.org/10.1016/j.fertnstert.2022.04.028

Lopez Rodriguez A, Rijsselaere T, Vyt P, Van Soom A, Maes D (2012) Effect of dilution temperature on boar semen quality. Reprod Domest Anim 47(5):e63–e66. https://doi.org/10.1111/j.1439-0531.2011.01938.x

Lopez Rodriguez A, Van Soom A, Arsenakis I, Maes D (2017) Boar management and semen handling factors affect the quality of boar extended semen. Porcine Health Manag 3:15. https://doi.org/10.1186/s40813-017-0062-5

Ma C, Li J, Tao H, Lei B, Li Y, Tong K, Zhang X, Guan K, Shi Y, Li F (2013) Discovery of two potential DAZL gene markers for sperm quality in boars by population association studies. Anim Reprod Sci 143(1–4):97–101. https://doi.org/10.1016/j.anireprosci.2013.10.002

Maes D, Rijsselaere T, Vyt P, Sokolowska A, Deley W, Van Soom A (2010) Comparison of five different methods to assess the concentration of boar semen. Vlaams Diergeneeskundig Tijdschrift 79(1):42–47

Mair F, Hartmann FJ, Mrdjen D, Tosevski V, Krieg C, Becher B (2016) The end of gating? An introduction to automated analysis of high dimensional cytometry data. Eur J Immunol 46(1):34–43. https://doi.org/10.1002/eji.201545774

Mańkowska A, Brym P, Paukszto Ł, Jastrzębski JP, Fraser L (2020) Gene polymorphisms in boar spermatozoa and their associations with post-thaw semen quality. Int J Mol Sci 21(5):1902

Markley JL, Bruschweiler R, Edison AS, Eghbalnia HR, Powers R, Raftery D, Wishart DS (2017) The future of NMR-based metabolomics. Curr Opin Biotechnol 43:34–40. https://doi.org/10.1016/j.copbio.2016.08.001

Martinez-Pastor F, Tizado EJ, Garde JJ, Anel L, de Paz P (2011) Statistical series: opportunities and challenges of sperm motility subpopulation analysis. Theriogenology 75(5):783–795. https://doi.org/10.1016/j.theriogenology.2010.11.034

Martin-Hidalgo D, Baron FJ, Bragado MJ, Carmona P, Robina A, Garcia-Marin LJ, Gil MC (2011) The effect of melatonin on the quality of extended boar semen after long-term storage at 17 degrees C. Theriogenology 75(8):1550–1560. https://doi.org/10.1016/j.theriogenology.2010.12.021

Martin-Hidalgo D, Solar-Malaga S, Gonzalez-Fernandez L, Zamorano J, Garcia-Marin LJ, Bragado MJ (2024) The compound YK 3-237 promotes pig sperm capacitation-related events. Vet Res Commun 48(2):773–786. https://doi.org/10.1007/s11259-023-10243-6

Maside C, Recuero S, Salas-Huetos A, Ribas-Maynou J, Yeste M (2023) Animal board invited review: An update on the methods for semen quality evaluation in swine—from farm to the lab. Animal 17(3):100720. https://doi.org/10.1016/j.animal.2023.100720

Matas C, Decuadro G, Martinez-Miro S, Gadea J (2007) Evaluation of a cushioned method for centrifugation and processing for freezing boar semen. Theriogenology 67(5):1087–1091. https://doi.org/10.1016/j.theriogenology.2006.11.010

Mateo-Otero Y, Fernandez-Lopez P, Gil-Caballero S, Fernandez-Fuertes B, Bonet S, Barranco I, Yeste M (2020) (1)H nuclear magnetic resonance of pig seminal plasma reveals intra-ejaculate variation in metabolites. Biomolecules 10(6):906. https://doi.org/10.3390/biom10060906

Mateo-Otero Y, Fernandez-Lopez P, Delgado-Bermudez A, Nolis P, Roca J, Miro J, Barranco I, Yeste M (2021) Metabolomic fingerprinting of pig seminal plasma identifies in vivo fertility biomarkers. J Anim Sci Biotechnol 12(1):113. https://doi.org/10.1186/s40104-021-00636-5

Menezes TA, Bustamante-Filho IC, Paschoal AFL, Dalberto PF, Bizarro CV, Bernardi ML, Ulguim RDR, Bortolozzo FP, Mellagi APG (2020) Differential seminal plasma proteome signatures of boars with high and low resistance to hypothermic semen preservation at 5 degrees C. Andrology 8(6):1907–1922. https://doi.org/10.1111/andr.12869

Moazed D (2009) Small RNAs in transcriptional gene silencing and genome defence. Nature 457(7228):413–420. https://doi.org/10.1038/nature07756

Morales B, Quintero-Moreno A, Osorio-Meléndez C, Rubio-Guillén J (2012) Valoración de la biometría de la cabeza del espermatozoide mediante análisis computarizado en semen de cerdo recién colectado y refrigerado. Rev Fac Agron 29:413–431

Murgas LDS, Lima D, Alvarenga ALN, Zangeronimo MG, Oberlender G, Oliveira SL (2010) Estudo comparativo de diferentes técnicas de avaliação da concentração espermática em suínos. Archivos de Zootecnia 59:463–466

Murphy EM, Stanton C, Brien CO, Murphy C, Holden S, Murphy RP, Varley P, Boland MP, Fair S (2017) The effect of dietary supplementation of algae rich in docosahexaenoic acid on boar fertility. Theriogenology 90:78–87. https://doi.org/10.1016/j.theriogenology.2016.11.008

Nishigaki T, Jose O, Gonzalez-Cota AL, Romero F, Trevino CL, Darszon A (2014) Intracellular pH in sperm physiology. Biochem Biophys Res Commun 450(3):1149–1158. https://doi.org/10.1016/j.bbrc.2014.05.100

Oberlender G, Murgas L, Zangeronimo M, Silva A, Pereira L, Muzzi R (2012) Comparison of two different methods for evaluating boar semen morphology. Arch Med Vet 44:201–205

Ortega-Ferrusola C, Gil MC, Rodriguez-Martinez H, Anel L, Peña FJ, Martin-Munoz P (2017) Flow cytometry in spermatology: a bright future ahead. Reprod Domest Anim 52(6):921–931. https://doi.org/10.1111/rda.13043

Park M, Yoon H, Kang BH, Lee H, An J, Lee T, Cheong HT, Lee SH (2023) Deep learning-based precision analysis for acrosome reaction by modification of plasma membrane in boar sperm. Animals (Basel) 13(16):2622. https://doi.org/10.3390/ani13162622

Paulenz H, Grevle IS, Tverdal A, Hofmo PO, Berg KA (2007) Precision of the Coulter® counter for routine assessment of boar-sperm concentration in comparison with the haemocytometer and spectrophotometer. Reprod Domest Anim 30(3):107–111. https://doi.org/10.1111/j.1439-0531.1995.tb00614.x

Peña FJ, Johannisson A, Wallgren M, Rodríguez-Martínez H (2003) Assessment of fresh and frozen-thawed boar semen using an Annexin-V assay: a new method of evaluating sperm membrane integrity. Theriogenology 60(4):677–689

Peña FJ, Saravia F, Garcia-Herreros M, Nunez-martinez I, Tapia JA, Johannisson A, Wallgren M, Rodriguez-Martinez H (2005) Identification of sperm morphometric subpopulations in two different portions of the boar ejaculate and its relation to postthaw quality. J Androl 26(6):716–723. https://doi.org/10.2164/jandrol.05030

Peña FJ, Ball BA, Squires EL (2018) A new method for evaluating stallion sperm viability and mitochondrial membrane potential in fixed semen samples. Cytom B Clin Cytom 94(2):302–311. https://doi.org/10.1002/cyto.b.21506

Perelman A, Wachtel C, Cohen M, Haupt S, Shapiro H, Tzur A (2012) JC-1: alternative excitation wavelengths facilitate mitochondrial membrane potential cytometry. Cell Death Dis 3(11):e430. https://doi.org/10.1038/cddis.2012.171

Perez-Patino C, Barranco I, Parrilla I, Valero ML, Martinez EA, Rodriguez-Martinez H, Roca J (2016) Characterization of the porcine seminal plasma proteome comparing ejaculate portions. J Proteome 142:15–23. https://doi.org/10.1016/j.jprot.2016.04.026

Perez-Patino C, Parrilla I, Li J, Barranco I, Martinez EA, Rodriguez-Martinez H, Roca J (2019) The proteome of pig spermatozoa is remodeled during ejaculation. Mol Cell Proteomics MCP 18(1):41–50. https://doi.org/10.1074/mcp.RA118.000840

Petrunkina AM, Volker G, Brandt H, Topfer-Petersen E, Waberski D (2005) Functional significance of responsiveness to capacitating conditions in boar spermatozoa. Theriogenology 64(8):1766–1782. https://doi.org/10.1016/j.theriogenology.2005.04.007

Pineiro C, Morales J, Rodriguez M, Aparicio M, Manzanilla EG, Koketsu Y (2019) Big (pig) data and the internet of the swine things: a new paradigm in the industry. Anim Front 9(2):6–15. https://doi.org/10.1093/af/vfz002

Pursel VG, Johnson LA (1974) Glutaraldehyde fixation of boar spermatozoa for acrosome evaluation. Theriogenology 1(2):63–68. https://doi.org/10.1016/0093-691x(74)90008-9

Quintero-Moreno A (2003) Estudio sobre la dinámica de poblaciones espermáticas en semen de caballo, cerdo y conejo. Doctoral thesis, Universitat Autònoma de Barcelona, Barcelona, Spain

Quintero-Moreno A, Rigau T, Rodríguez-Gil J (2004) Regression analyses and motile sperm sub-population structure study as improving tools in boar semen quality analysis. Theriogenology 61(4):673–690

Quintero-Moreno A, González-Villalobos D, López-Brea JJ, Esteso MC, Fernández-Santos MR, Carvalho-Crociata JL, Mejía-Silva W, León-Atencio G (2009) Valoración morfométrica de la cabeza del espermatozoide del cerdo doméstico según su edad. Revista Científica 19:153–158

Quintero-Moreno A, Ramirez M, Nava-Trujillo H, Hidalgo M (2015) Comparison of two histologic stains in the evaluation of sperm head morphometric measurements in frozen-thawed bull semen. Acta Microsc 24:103

Quirino M, Jakop U, Mellagi APG, Bortolozzo FP, Jung M, Schulze M (2022) A 5-color flow cytometry panel to assess plasma membrane integrity, acrosomal status, membrane lipid organization and mitochondrial activity of boar and stallion spermatozoa following liquid semen storage. Anim Reprod Sci 247:107076. https://doi.org/10.1016/j.anireprosci.2022.107076

Reyer H, Abou-Soliman I, Schulze M, Henne H, Reinsch N, Schoen J, Wimmers K (2024) Genome-wide association analysis of semen characteristics in Piétrain boars. Genes 15(3):382

Riesenbeck A, Schulze M, Rudiger K, Henning H, Waberski D (2015) Quality control of boar sperm processing: implications from European AI centres and two spermatology reference laboratories. Reprod Domest Anim 50(Suppl 2):1–4. https://doi.org/10.1111/rda.12573

Robles V, Martínez-Pastor F (2013) Flow cytometric methods for sperm assessment. In: Carrell A (ed) Spermatogenesis: methods and protocols. Humana, Totowa, NJ, pp 175–186. https://doi.org/10.1007/978-1-62703-038-0_16

Rodriguez-Martinez H, Kvist U, Saravia F, Wallgren M, Johannisson A, Sanz L, Peña FJ, Martinez EA, Roca J, Vazquez JM, Calvete JJ (2009) The physiological roles of the boar ejaculate. Soc Reprod Fertil Suppl 66:1–21

Rodriguez-Martinez H, Kvist U, Ernerudh J, Sanz L, Calvete JJ (2011) Seminal plasma proteins: what role do they play? Am J Reprod Immunol 66(Suppl 1):11–22. https://doi.org/10.1111/j.1600-0897.2011.01033.x

Romero-Aguirregomezcorta J, Soriano-Ubeda C, Matas C (2021) Involvement of nitric oxide during in vitro oocyte maturation, sperm capacitation and in vitro fertilization in pig. Res Vet Sci 134:150–158. https://doi.org/10.1016/j.rvsc.2020.12.011

Rubessa M, Feugang JM, Kandel ME, Schreiber S, Hessee J, Salerno F, Meyers S, Chu I, Popescu G, Wheeler MB (2020) High-throughput sperm assay using label-free microscopy: morphometric comparison between different sperm structures of boar and stallion spermatozoa. Anim Reprod Sci 219:106509. https://doi.org/10.1016/j.anireprosci.2020.106509

Saravia F, Nunez-Martinez I, Moran JM, Soler C, Muriel A, Rodriguez-Martinez H, Pena FJ (2007) Differences in boar sperm head shape and dimensions recorded by computer-assisted sperm morphometry are not related to chromatin integrity. Theriogenology 68(2):196–203. https://doi.org/10.1016/j.theriogenology.2007.04.052

Schulze M, Ruediger K, Mueller K, Jung M, Well C, Reissmann M (2013) Development of an in vitro index to characterize fertilizing capacity of boar ejaculates. Anim Reprod Sci 140(1–2):70–76. https://doi.org/10.1016/j.anireprosci.2013.05.012

Schulze M, Buder S, Rudiger K, Beyerbach M, Waberski D (2014) Influences on semen traits used for selection of young AI boars. Anim Reprod Sci 148(3–4):164–170. https://doi.org/10.1016/j.anireprosci.2014.06.008

Schulze M, Ammon C, Schaefer J, Luther AM, Jung M, Waberski D (2017) Impact of different dilution techniques on boar sperm quality and sperm distribution of the extended ejaculate. Anim Reprod Sci 182:138–145. https://doi.org/10.1016/j.anireprosci.2017.05.013

Schulze M, Bortfeldt R, Schafer J, Jung M, Fuchs-Kittowski F (2018) Effect of vibration emissions during shipping of artificial insemination doses on boar semen quality. Anim Reprod Sci 192:328–334. https://doi.org/10.1016/j.anireprosci.2018.03.035

Serrano R, Garrido N, Cespedes JA, Gonzalez-Fernandez L, Garcia-Marin LJ, Bragado MJ (2020) Molecular mechanisms involved in the impairment of boar sperm motility by peroxynitrite-induced nitrosative stress. Int J Mol Sci 21(4):1208. https://doi.org/10.3390/ijms21041208

Sevilla F, Soler C, Araya-Zúñiga I, Barquero V, Roldan ERS, Valverde A (2023) Are there differences between methods used for the objective estimation of boar sperm concentration and motility? Animals 13(10):1622

Sikka SC (2001) Relative impact of oxidative stress on male reproductive function. Curr Med Chem 8(7):851–862. https://doi.org/10.2174/0929867013373039

Sironen A, Uimari P, Nagy S, Paku S, Andersson M, Vilkki J (2010) Knobbed acrosome defect is associated with a region containing the genes STK17b and HECW2 on porcine chromosome 15. BMC Genomics 11(1):699. https://doi.org/10.1186/1471-2164-11-699

Soler C, Garcia-Molina A, Contell J, Silvestre MA, Sancho M (2015) The Trumorph(R) system: The new universal technique for the observation and analysis of the morphology of living sperm [corrected]. Anim Reprod Sci 158:1–10. https://doi.org/10.1016/j.anireprosci.2015.04.001

Soler C, Valverde A, Bompart D, Fereidounfar S, Sancho M, Yániz JL, García-Molina A, Korneenko-Zhilyaev YA (2017) New methods of semen analysis by CASA. Sel'skokhozyaistvennaya Biologiya 52(2):232–241. https://doi.org/10.15389/agrobiology.2017.2.232eng

Soriano-Ubeda C, Romero-Aguirregomezcorta J, Matas C, Visconti PE, Garcia-Vazquez FA (2019) Manipulation of bicarbonate concentration in sperm capacitation media improves in vitro fertilisation output in porcine species. J Anim Sci Biotechnol 10(1):19. https://doi.org/10.1186/s40104-019-0324-y

Spano M, Evenson DP (1993) Flow cytometric analysis for reproductive biology. Biol Cell 78(1–2):53–62. https://doi.org/10.1016/0248-4900(93)90114-t

Suarez-Trujillo A, Kandula H, Kumar J, Devi A, Shirley L, Thirumalaraju P, Kanakasabapathy MK, Shafiee H, Hart L (2022) Validation of a smartphone-based device to measure concentration, motility, and morphology in swine ejaculates. Transl Anim Sci 6(4):txac119. https://doi.org/10.1093/tas/txac119

Sui H, Sheng M, Luo H, Liu G, Meng F, Cao Z, Zhang Y (2023) Characterization of freezability-associated metabolites in boar semen. Theriogenology 196:88–96. https://doi.org/10.1016/j.theriogenology.2022.11.013

Suo J, Wang J, Zheng Y, Xiao F, Li R, Huang F, Niu P, Zhu W, Du X, He J, Gao Q, Khan A (2024) Recent advances in cryotolerance biomarkers for semen preservation in frozen form-A systematic review. PLoS One 19(5):e0303567. https://doi.org/10.1371/journal.pone.0303567

Sutkeviciene N, Andersson MA, Zilinskas H, Andersson M (2005) Assessment of boar semen quality in relation to fertility with special reference to methanol stress. Theriogenology 63(3):739–747. https://doi.org/10.1016/j.theriogenology.2004.04.006

Szablicka D, Wysokińska A, Pawlak A, Roman K (2022) Morphometry of boar spermatozoa in semen stored at 17°C—the influence of the staining technique. Animals 12(15):1888

Tardif S, Dube C, Chevalier S, Bailey JL (2001) Capacitation is associated with tyrosine phosphorylation and tyrosine kinase-like activity of pig sperm proteins. Biol Reprod 65(3):784–792. https://doi.org/10.1095/biolreprod65.3.784

Thurston LM, Watson PF, Holt WV (1999) Sources of variation in the morphological characteristics of sperm subpopulations assessed objectively by a novel automated sperm morphology analysis system. J Reprod Fertil 117(2):271–280. https://doi.org/10.1530/jrf.0.1170271

Thurston LM, Watson PF, Mileham AJ, Holt WV (2001) Morphologically distinct sperm subpopulations defined by Fourier shape descriptors in fresh ejaculates correlate with variation in boar semen quality following cryopreservation. J Androl 22(3):382–394

Torres MA, Diaz R, Boguen R, Martins SM, Ravagnani GM, Leal DF, Oliveira Mde L, Muro BB, Parra BM, Meirelles FV, Papa FO, Dell'Aqua JA Jr, Alvarenga MA, Moretti Ade S, Sepulveda N, de Andrade AF (2016) Novel flow cytometry analyses of boar sperm viability: can the addition of whole sperm-rich fraction seminal plasma to frozen-thawed boar sperm affect it? PLoS One 11(8):e0160988. https://doi.org/10.1371/journal.pone.0160988

Torres MA, Pedrosa AC, Novais FJ, Alkmin DV, Cooper BR, Yasui GS, Fukumasu H, Machaty Z, de Andrade AFC (2022) Metabolomic signature of spermatozoa established during holding time is responsible for differences in boar sperm freezability dagger. Biol Reprod 106(1):213–226. https://doi.org/10.1093/biolre/ioab200

Tsakmakidis IA, Lymberopoulos AG, Khalifa TA (2010) Relationship between sperm quality traits and field-fertility of porcine semen. J Vet Sci 11(2):151–154. https://doi.org/10.4142/jvs.2010.11.2.151

Valverde A, Madrigal-Valverde M (2019) Evaluación de cámaras de recuento sobre parámetros espermáticos de verracos analizados con un sistema CASA-Mot. Agronomía Mesoamericana 447–458

Valverde A, Madrigal-Valverde M, Lotz J, Bompart D, Soler C (2019a) Effect of video capture time on sperm kinematic parameters in breeding boars. Livest Sci 220:52–56. https://doi.org/10.1016/j.livsci.2018.12.008

Valverde A, Madrigal M, Caldeira C, Bompart D, de Murga JN, Arnau S, Soler C (2019b) Effect of frame rate capture frequency on sperm kinematic parameters and subpopulation structure definition in boars, analysed with a CASA-Mot system. Reprod Domest Anim 54(2):167–175. https://doi.org/10.1111/rda.13320

Valverde A, Barquero V, Soler C (2020) The application of computer-assisted semen analysis (CASA) technology to optimise semen evaluation. A review. J Anim Feed Sci 29(3):189–198. https://doi.org/10.22358/jafs/127691/2020

Valverde A, Calderón Calderón J, Víquez L, Barquero V (2021) Frecuencia de fotogramas óptima para evaluar la cinética espermática de verracos con un sistema CASA-Mot. Agronomía Mesoamericana 32(1):1–18. https://doi.org/10.15517/am.v32i1.41928

Valverde-Abarca A, Madrigal-Valverde M, Solís-Arias J, Paniagua-Madrigal W (2019) Variabilidad en los métodos de estimación de la concentración espermática en verracos. Agron Costarric 43:25–44

van der Horst G, Maree L (2009) SpermBlue: a new universal stain for human and animal sperm which is also amenable to automated sperm morphology analysis. Biotech Histochem 84(6):299–308. https://doi.org/10.3109/10520290902984274

Varea Sanchez M, Bastir M, Roldan ER (2013) Geometric morphometrics of rodent sperm head shape. PLoS One 8(11):e80607. https://doi.org/10.1371/journal.pone.0080607

Verstegen J, Iguer-Ouada M, Onclin K (2002) Computer assisted semen analyzers in andrology research and veterinary practice. Theriogenology 57(1):149–179. https://doi.org/10.1016/s0093-691x(01)00664-1

Vertika S, Singh KK, Rajender S (2020) Mitochondria, spermatogenesis, and male infertility—an update. Mitochondrion 54:26–40. https://doi.org/10.1016/j.mito.2020.06.003

Vianna WL, Bruno DG, Namindome A, Rosseto AC, Rodrigues PHM, Pinese ME, Moretti ASA (2004) Estudo comparativo da eficiência de diferentes técnicas de mensuração da concentração espermática em suínos. Rev Bras Zootec 33:2054–2059

Vicente-Carrillo A, Alvarez-Rodriguez M, Rodriguez-Martinez H (2017) The CatSper channel modulates boar sperm motility during capacitation. Reprod Biol 17(1):69–78. https://doi.org/10.1016/j.repbio.2017.01.001

Vicente-Fiel S, Palacin I, Santolaria P, Hidalgo CO, Silvestre MA, Arrebola F, Yaniz JL (2013) A comparative study of the sperm nuclear morphometry in cattle, goat, sheep, and pigs using a new computer-assisted method (CASMA-F). Theriogenology 79(3):436–442. https://doi.org/10.1016/j.theriogenology.2012.10.015

Visconti PE (2009) Understanding the molecular basis of sperm capacitation through kinase design. Proc Natl Acad Sci USA 106(3):667–668. https://doi.org/10.1073/pnas.0811895106

Visconti PE, Krapf D, de la Vega-Beltran JL, Acevedo JJ, Darszon A (2011) Ion channels, phosphorylation and mammalian sperm capacitation. Asian J Androl 13(3):395–405. https://doi.org/10.1038/aja.2010.69

Vyt P, Maes D, Dejonckheere E, Castryck F, Van Soom A (2004a) Comparative study on five different commercial extenders for boar semen. Reprod Domest Anim 39(1):8–12. https://doi.org/10.1046/j.1439-0531.2003.00468.x

Vyt P, Maes D, Rijsselaere T, Dejonckheere E, Castryck F, Van Soom A (2004b) Motility assessment of porcine spermatozoa: a comparison of methods. Reprod Domest Anim 39(6):447–453. https://doi.org/10.1111/j.1439-0531.2004.00538.x

Waberski D, Meding S, Dirksen G, Weitze K, Leiding C, Hahn R (1994) Fertility of long-term-stored boar semen: Influence of extender (Androhep and Kiev), storage time and plasma droplets in the semen. Anim Reprod Sci 36(1–2):145–151

Wang Z, Gerstein M, Snyder M (2009) RNA-Seq: a revolutionary tool for transcriptomics. Nat Rev Genet 10(1):57–63. https://doi.org/10.1038/nrg2484

Wang C, Guo LL, Wei HK, Zhou YF, Tan JJ, Sun HQ, Jiang SW, Peng J (2019) Logistic regression analysis of the related factors in discarded semen of boars in Southern China. Theriogenology 131:47–51. https://doi.org/10.1016/j.theriogenology.2019.03.012

Xu Z, Xie Y, Zhou C, Hu Q, Gu T, Yang J, Zheng E, Huang S, Xu Z, Cai G, Liu D, Wu Z, Hong L (2020) Expression pattern of seminal plasma extracellular vesicle small RNAs in boar semen. Front Vet Sci 7(929):585276. https://doi.org/10.3389/fvets.2020.585276

Yaniz JL, Soler C, Santolaria P (2015) Computer assisted sperm morphometry in mammals: a review. Anim Reprod Sci 156:1–12. https://doi.org/10.1016/j.anireprosci.2015.03.002

Yaniz JL, Capistros S, Vicente-Fiel S, Hidalgo CO, Santolaria P (2016) A comparative study of the morphometry of sperm head components in cattle, sheep, and pigs with a computer-assisted fluorescence method. Asian J Androl 18(6):840–843. https://doi.org/10.4103/1008-682X.186877

Yeste M, Briz M, Pinart E, Sancho S, Bussalleu E, Bonet S (2010) The osmotic tolerance of boar spermatozoa and its usefulness as sperm quality parameter. Anim Reprod Sci 119(3–4):265–274. https://doi.org/10.1016/j.anireprosci.2010.02.011

Yeste M, Fernandez-Novell JM, Ramio-Lluch L, Estrada E, Rocha LG, Cebrian-Perez JA, Muino-Blanco T, Concha II, Ramirez A, Rodriguez-Gil JE (2015) Intracellular calcium movements of boar spermatozoa during 'in vitro' capacitation and subsequent acrosome exocytosis follow a multiple-storage place, extracellular calcium-dependent model. Andrology 3(4):729–747

Yeste M, Bonet S, Rodriguez-Gil JE, Rivera Del Alamo MM (2018) Evaluation of sperm motility with CASA-Mot: which factors may influence our measurements? Reprod Fertil Dev 30(6):789–798. https://doi.org/10.1071/RD17475

Zhang YT, Liu Y, Liang HL, Xu QQ, Liu ZH, Weng XG (2021) Metabolomic differences of seminal plasma between boars with high and low average conception rates after artificial insemination. Reprod Domest Anim 56(1):161–171. https://doi.org/10.1111/rda.13861

Zhao Y, Gao N, Li X, El-Ashram S, Wang Z, Zhu L, Jiang W, Peng X, Zhang C, Chen Y, Li Z (2020) Identifying candidate genes associated with sperm morphology abnormalities using weighted single-step GWAS in a Duroc boar population. Theriogenology 141:9–15. https://doi.org/10.1016/j.theriogenology.2019.08.031

Zhao G, Zhao X, Bai J, Dilixiati A, Song Y, Haire A, Zhao S, Aihemaiti A, Fu X, Wusiman A (2023) Metabolomic and transcriptomic changes underlying the effects of L-citrulline supplementation on ram semen quality. Animals (Basel) 13(2):217. https://doi.org/10.3390/ani13020217

Zhao Y, Qin J, Sun J, He J, Sun Y, Yuan R, Li Z (2024) Motility-related microRNAs identified in pig seminal plasma exosomes by high-throughput small RNA sequencing. Theriogenology 215:351–360. https://doi.org/10.1016/j.theriogenology.2023.11.028

Zigo M, Kerns K, Sutovsky P (2023) The ubiquitin-proteasome system participates in sperm surface subproteome remodeling during boar sperm capacitation. Biomolecules 13(6):996. https://doi.org/10.3390/biom13060996

Role of Seminal Plasma in the Equine Reproduction

Jordi Miró Roig ⓘ and Jaime Catalán ⓘ

Abstract

Seminal plasma (SP) has a very important role in the equine reproductive strategy. Physiologic endometritis occurs when the semen rises the uterus after a natural mating or an artificial insemination. A large amount of polymorphonuclear neutrophils (PMN) reaches the uterine lumen. These PMN will eliminate many sperm, selecting those that will advance to the fertilization site by means of the neutrophil extracellular traps (NETs) production. SP induces and controls the NETs production. The sperm metabolism produces an important amount of reactive oxygen species (ROS), but the NETosis produces much more, inducing endometrial oxidative stress. SP plasma is an important source of enzymatic and non-enzymatic antioxidants to protect the spermatozoa. SP metabolomics evidenced similar metabolites between horse and donkey semen; however, the amount and the role of each metabolite are different between species. The study of these metabolites is very important to improve sperm metabolism and to design extenders for both species. Recently, SP exosomes and microvesicles have been identified. These vesicles contain proteins and microRNAs whose role is still unknown.

Keywords

Equine reproduction · Seminal plasma · Post-AI endometritis · Antioxidants

J. M. Roig (✉) · J. Catalán
Equine Reproduction Service, Department of Animal Medicine and Surgery, Veterinary Faculty, Universidad Autónoma de Barcelona, Barcelona, Spain
e-mail: jordi.miro@uab.cat

J. C. Gardón, K. Satué Ambrojo (eds.), *Assisted Reproductive Technologies in Animals Volume 1*, https://doi.org/10.1007/978-3-031-73079-5_10

1   Introduction

Routine semen analysis has limited predictive ability of fertility (Colembrander et al. 2003). The increase in analytical capacity through computerized analysis of sperm motility with CASA systems, computerized analysis of morphometry (ASMA), or flow cytometry has slightly improved this predictive capacity. However, this remains low.

This seminal analysis traditionally deals exclusively with the cell, the spermatozoon, but forgets (among others) a fundamental element: seminal plasma (SP) (Fig. 1). SP is the fluid produced in equids by the ampulae of the vas deferens and the sexual glands, prostate, seminal vesicles (or vesicular glands), and bulbourethral glands (Gacem et al. 2020b) (Fig. 2).

The role of SP seems to be fundamental, so the elimination of it in some horses with low fertility and the substitution by SP of other individuals with good fertility significantly increase the fertility of the former.

However, SP has a negative effect on semen preservation, being eliminated, in many cases during semen refrigeration and always in the process of sperm freezing (Miró et al. 2009; Miró and Papas 2018).

Semen from a diluted and refrigerated ejaculate usually shows a significant decrease in viability after 24 h of storage. However, epididymal semen, which has not come into contact with the SP, shows excellent sperm survival up to 5 days after collection or excellent results at freezing–thawing. On the other hand, when epididymal semen is used in artificial insemination, the fertility obtained is poor. This fertility can be significantly increased when SP is added to artificial insemination (Miró et al. 2020).

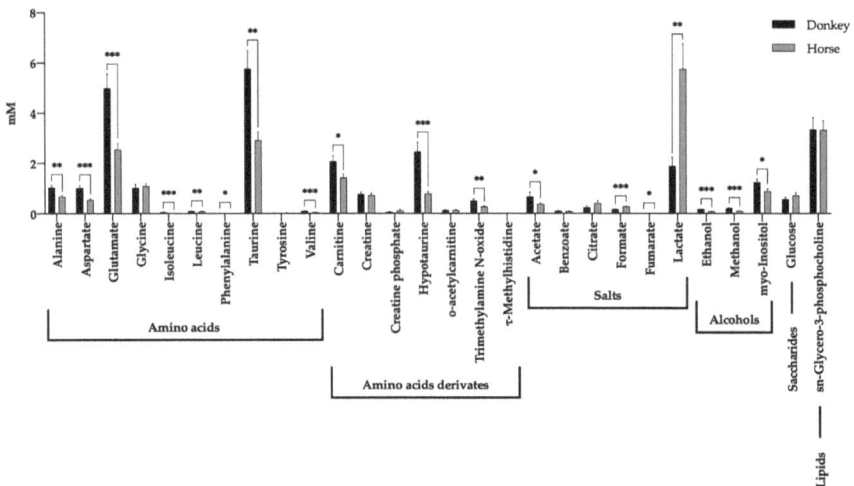

Fig. 1 Concentrations (mean ± SEM; mM) and chemical nature of metabolites in SP samples from donkey ($n = 18$) and horse ($n = 18$). $*P \leq 0.05$; $**P \leq 0.01$; $***P \leq 0.001$. (from Catalán et al. 2023)

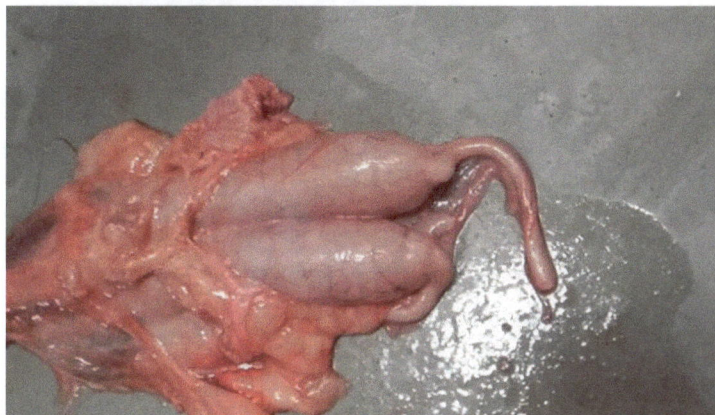

Fig. 2 Donkey accessory seminal glands

It, therefore, seems evident that SP is not good for semen preservation but important for fertility (Serhan et al. 2008; Rota et al. 2012). SP is a complex fluid with a fundamental role in the control of post-insemination inflammatory reaction or oxidative stress, among other functions (Katila 2012, Miró et al. 2013a, b, Miró and Papas 2018; Vilés et al. 2013a), many of which are still unknown.

2 Control of Post-Insemination Uterine Inflammation and Sperm Selection

When a mare is in oestrus, progesterone levels drop, and estrogen levels rise. This situation produces an increase in the vascularization of the uterus and certain extravasation of fluid, edema, which can be observed by ultrasound (Taberner et al. 2008). In this situation, the uterine lumen is aseptic and acellular. But when the semen arrives through copulation or insemination, a great inflammatory response occurs, with the arrival of a large number of polymorphonuclear neutrophils (PMN) (Fig. 3). This response is maximum at 6 h post-insemination and significantly higher on the donkey than on the mare (Troedsson et al. 2001; Alghamdi et al. 2004, 2009; Miró et al. 2013a).

In species of semen deposition at the vaginal level (ruminants, human, bitch ...), thousands of millions of spermatozoa are deposited at the bottom of the vagina. The cervix of the uterus has glands that produce mucus, mucins, and long protein chains. There are two types of mucins. Sulfomucins repel water and form complex buds that act as a stopper. Sulfomucins cover the cervix lumen during diestrus (corpus luteum activity and high progesterone levels). When the female is in oestrus, progesterone goes down, estrogen goes up, and a large number of the other mucins, the sialomucins, are produced, which capture water. They originate a spinning mucus that generates long "lanes." These lanes will act as a filter for sperm. As we have said, billions are deposited at the bottom of the vagina, thousands pass through and

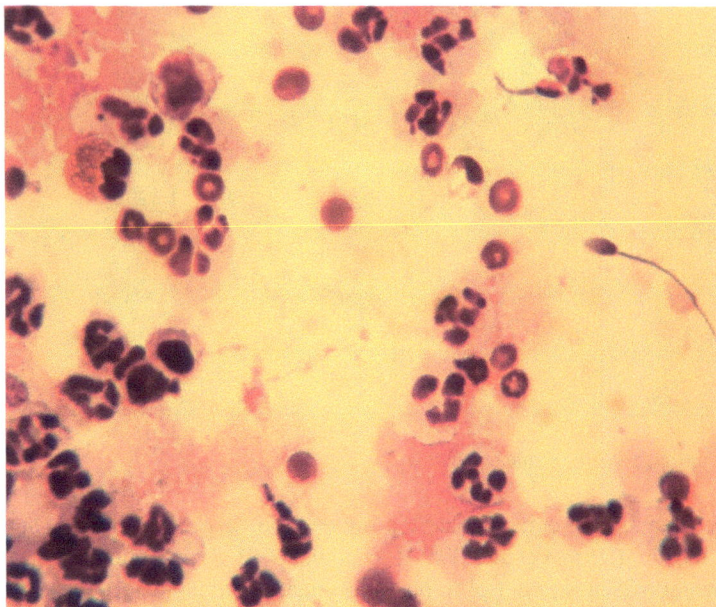

Fig. 3 Post-AI endometrial reaction. Spermatozoa with a lot of PMN

reach the uterus, hundreds reach the oviducts or fallopian tubes, and one will fertilize an oocyte (López Gatius et al. 1996).

But in species of intrauterine deposition (pig, horse, donkey), the cervix is omitted. Who selects sperm? It is this endometrial reaction, these PMNs, that out of billions of spermatozoa deposited will only allow thousands to pass so that hundreds reach the oviducts and one can fertilize.

To try to reduce post-insemination uterine inflammation, a nonsteroidal anti-inflammatory drug, ketoprofen, was used in donkeys. Cytological and humoral inflammation was evaluated by means of cytology, biopsy, and immunohistochemistry of COX-2. Ketoprofen is capable of reducing humoral inflammation, but not the flow of PMNs (cellular inflammation) (Palm et al. 2008; Vilés et al. 2013a).

As with ketoprofen, donkeys were inseminated again with frozen/thawed semen but added SP. It was seen that SP was also capable of controlling the humoral inflammatory response, but neither was the cellular response nor the flow of PMNs (Vilés et al. 2013b).

On the other hand, performing a uterine lavage 6 h post-insemination obtained a fluid rich in PMNs that were quantified with cytometry. Subsequently, an "in vitro" study was carried out; this fluid was incubated with fresh semen, half-diluted semen, or frozen/thawed semen. That means with half or no SP. The PMNs-sperm junction was analyzed at 1, 2, 3, and 4 h of incubation, and it was found that the binding was minimal when all the plasma was present and maximum when not. Therefore, it is clear that SP has a modulatory effect on this PMN-sperm ratio (Miró et al. 2013a, b).

Repeated uterine washes could end up injuring the endometrium. An "in vitro" model by PMNs from the blood isolation was reached with excellent results in horses and donkeys. In several human experiments, to study the relationship between PMNs and microorganisms, PMNs were activated with FMLP (Formyl methyl leucyl phenylalanine). An interesting new discovery was that SP is capable of activating PMNs without the need for FMLP. PMNs quickly captured sperm, but the number of spermatozoa per PMN decreased over time, that is, they were distributed. Another finding, even more interesting, was that phagocytosis by PMNs was scarce (Fig. 4). Most spermatozoa were attached by the head (Fig. 5), and a few by the tail (Fig. 6) to a halo around the PMNs. The amount of sperm was important so that after 3 h spermatozoa were only kept alive when the concentration was greater than 500 million spermatozoa/mL. Under these conditions, most spermatozoa were kept alive together but with energetic tail movements, and some, after 3 h, were released with excellent movement patterns (Miró et al. 2020).

This halo of PMNs is what has been called neutrophil extracellular traps (NETs). PMNs, when activated, degranulate and release the contents of their granules, which are three enzymes (cathepsin, myeloperoxidase, and elastase). This produces lysis of the membranes and exit of the DNA content from the nucleus that forms this halo where sperm will attach themselves like fish that stick to a net around a boat. And it is precisely SP that also induces this response (Mateo-Otero et al. 2021) (Fig. 7).

It seems clear that there must be an important role of SP proteins in controlling all these processes (Portus et al. 2005). In recent studies, not yet published, by proteomics, 2927 different proteins have been isolated in horse SP. Studying the possible role of each is an almost impossible mission. In an attempt to get a closer look at which ones might have an effect, SP was divided into six parts, considering the molecular weight of its proteins, and it was seen that proteins with effect are, above all, between 30 and 100 kDa (Miró et al. 2021). Precisely in this band, there is a DNAse described by Alghamdi and Foster (2004) that can favor the digestion of

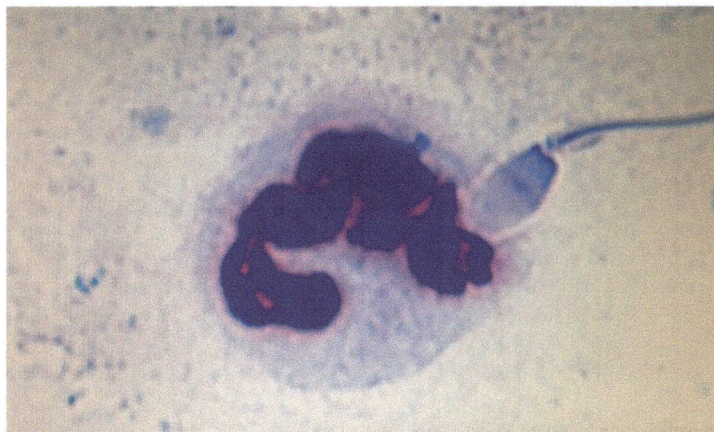

Fig. 4 PMN phagocyting an spermatozoon

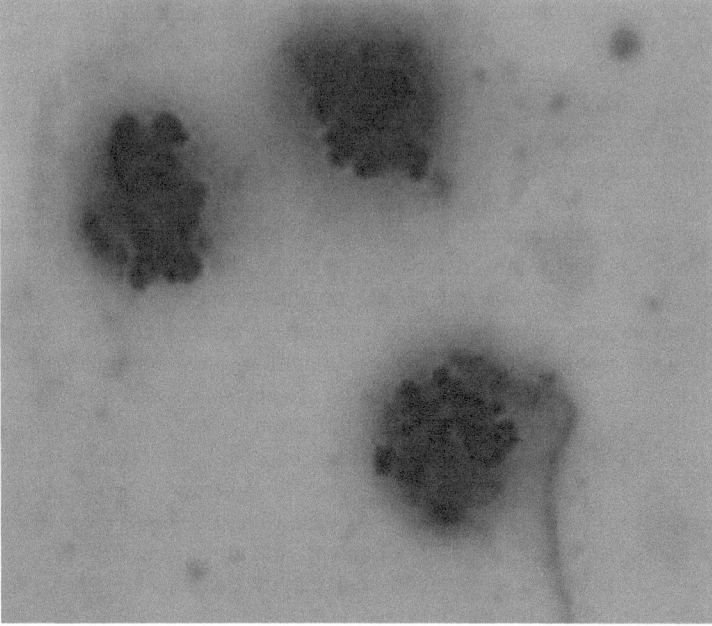

Fig. 5 Spermatozoon attached to PMN by the head

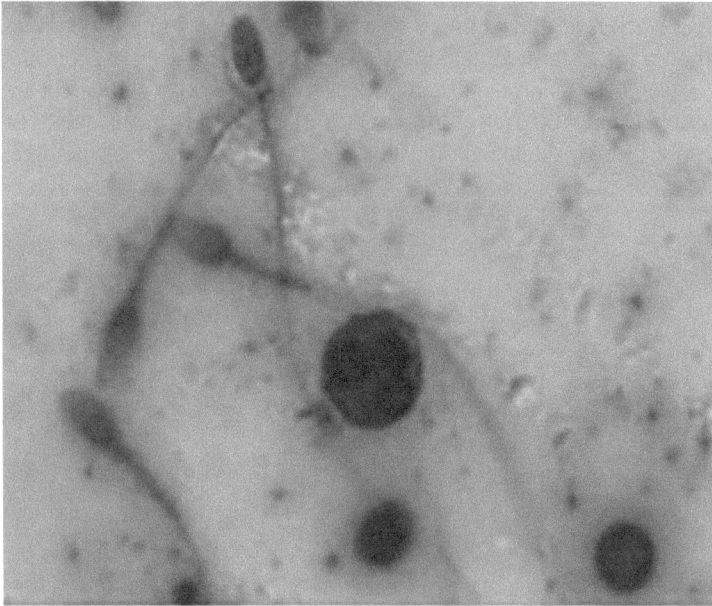

Fig. 6 Spermatozoon attached to the PMN by the tail

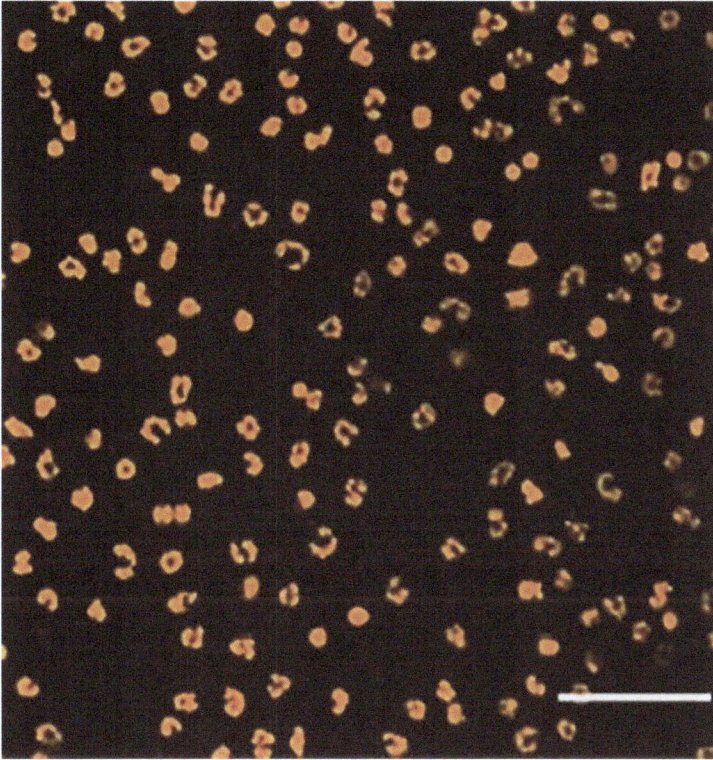

Fig. 7 NETs (PMN extruded DNA) evidenced by confocal imaging and SYTOX orange staining. Scale bar = 20 μm

NETs and some spermatozoa to be released (a possible selection method). Lactate dehydrogenase also can control excess lactic acid. Or enzymes such as SOD (superoxide dismutase), CAT (catalase), GPx (glutathione peroxidase), and GSR (glutathione reductase) are fundamental in the control of ROS (Papas et al. 2019a, b).

Sperm selection, searching for live spermatozoa without anomalies and with good motility for assisted reproduction, has been an interesting field of work in recent years in different species. Silica gel filters are one of the most commonly used methods. However, the use of these filters induces PMNs to perform phagocytosis but not NETosis. That is, they remove sperm but do not keep them alive, with some being released. Something that seems to be part of the reproductive strategy of intrauterine semen deposition species. Possibly, the filters, apart from selecting, wash the selected sperm from proteins attached to the plasma membrane. Proteins may play an important role in controlling the PMN–sperm ratio (Papas et al. 2020).

The testicle is vascularized mainly by the testicular artery, which, when it reaches the testicular cord, begins to curve, a serpentine complex called the Pampiniform plexus. Studying blood flow in the testicular artery by color Doppler ultrasound and pulsar at the level of its origin, the typical arterial pulse is observed, with a systolic peak and diastole. In the center of the cord, the systolic peak decreases significantly and, upon arrival at the testicle, practically disappears. This means that the total flow of blood reaching the testicle is much lower than that to other organs. As a result, the temperature is lower than body temperature, 4–5°C below, and the oxygen pressure is clearly lower (Gacem et al. 2020a). Under these conditions, spermatogenesis occurs, and mature spermatozoa are located immobile at the tail end of the epididymis.

When copulation occurs, sexual stimulation produces the release of oxytocin that causes the emptying of part of the reserves of the tail of the epididymis. The spermatozoa, via vas deferens, enter the abdomen, increase the temperature by 4–5°C and oxygen levels, incorporate fluid from the adjoining glands, with possible sources of energy, and under these conditions, sperm increase their metabolism, especially at the mitochondrial level with energy production at the origin of their tail resulting at the beginning of the sperm movement (Gacem et al. 2020a). The increase in sperm metabolism, however, generates its by-products, on the one hand, lactic acid, which acidifies the environment and ends up killing the sperm themselves. On the other hand, reactive oxygen species (ROS), peroxides, and superoxides generate oxidative stress, peroxidation of the sperm membrane, and also death of sperm. However, SP also contains some proteins that have a negative effect on sperm (Miró et al. 2009). Although, as we will see later, others have a very beneficial effect. We can say then that an ejaculate is a "mega-collective suicide" of more than 15,000 million; with a little luck, only one will be saved, whatever fertilizes.

The spermatozoon, with its metabolism, produces ROS, peroxides, and superoxides, but the amount of ROS produced by NETosis ("the explosion of PMNs") is between three and five times that produced by sperm. As a consequence, intrauterine oxidative stress is important. In this environment, sperm would hardly survive. SP is a great source of antioxidants, enzymatic and non-enzymatic. Among the enzymatic antioxidants, we have already mentioned SOD, CAT, GPx, or GSR before. Donkey sperm runs more than horse sperm and produces more ROS, and the post-insemination inflammatory response is also more exaggerated. Therefore, oxidative stress is higher, but the activity of these enzymes is also significantly higher in donkey SP than in horse plasma (Papas et al. 2019a; Yánez-Ortiz et al. 2021).

Superoxides are highly toxic, so SOD activity is essential to eliminate them. Its activity has a strong correlation with the quality and freezability of semen on both donkeys and horses (Papas et al. 2019b, 2020). On the other hand, a certain amount of peroxides is necessary for sperm training and fertilization so that between CAT, GPx, and GSR, they will maintain a fine balance. On the other hand, a new enzyme, PON (Paroxonase), has been identified, which in both donkeys and horses will act at the level of the sperm itself, preventing lipid peroxidation of the cell membrane. This fine control of sperm membrane oxidation by the PON also has a good correlation with the freezability of the semen of both donkeys and horses (Catalán et al. 2022a, b, 2024).

There are many non-enzymatic antioxidants, and they also play their role, so there is currently a tendency to evaluate the total antioxidant capacity, which also

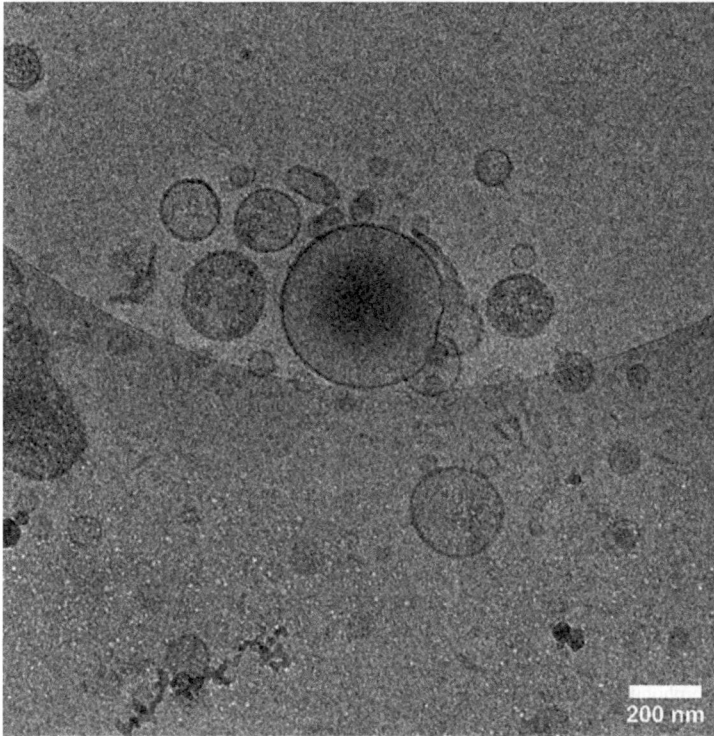

Fig. 8 Horse SP extracellular vesicles

has a strong correlation with the freezing of the semen of both species (González-Arto et al. 2016; Catalán et al., 2022a, b; Catalán et al. 2024).

3   Seminal Plasma Exosomes and Microvesicles

It was long believed that exosomes and microvesicles were cellular debris surrounded by plasma membranes. Donkey or horse SP is rich in exosomes and microvesicles (Fig. 8). The content of these structures is yet to be analyzed, but it seems clear that they contain proteins and RNA fragments (microRNAs) that can also be important signals in the reproductive strategy of each species.

4   Seminal Plasma Metabolomics

The study by nuclear magnetic resonance of the different metabolites of the SP (metabolomics) of donkey and horse showed that, although the metabolites were almost the same, the amounts varied significantly between the two species. Lactate concentrations are much higher in horses, while many amino acids or derivatives are higher in donkeys. When we compare good semen with poor quality semen in both species, we observe that glucose in horses is significantly higher in good ones. On the other hand, glucose does not vary in donkeys, and if some amino acids, amino acids derivatives, salts, or alcohols are higher. It seems evident, then, that the energy sources used by the sperm are different. The horse, for example, seems more dependent on glucose. These findings may be important in the design of extenders for cooled or frozen semen (Catalán et al. 2023).

5 Conclusions

In conclusion, SP has a negative effect on equine semen preservation. However, SP is a complex fluid involved in several steps of equine reproductive strategy. The knowledge of SP functions is very important to increase fertility by artificial insemination, to improve semen preservation, or to improve some fertility problems.

References

Alghamdi AS, Foster DN (2004) Seminal DNase frees spermatozoa entangled in neutrophil extracellular traps. Biol Reprod 73:1174–1181. https://doi.org/10.1095/biolreprod.105.045666

Alghamdi AS, Foster DN, Troedson MH (2004) Equine seminal plasma reduces sperm binding to polymorphonuclear neutrophils (PMNs) and improves fertility of fresh semen inseminated into inflamed uteri. Reproduction 127:593–600. https://doi.org/10.1530/rep.1.00096

Alghamdi A, Lovaas B, Bird S, Lamb G, Rendahl A, Taube P, Foster D (2009) Species-specific interaction of seminal plasma on sperm–neutrophil binding. Anim Reprod Sci 114:331–344. https://doi.org/10.1016/j.anireprosci.2008.10.015

Catalán J, Yánez I, Tvarijonaviciute A, González-Arostegui G, Rubio CP, Yeste M, Miró J, Barranco I (2022b) Impact of seminal plasma antioxidants on donkey sperm cryotolerance. Antioxidants 11:4174. https://doi.org/10.3390/antiox11020417

Catalán J, Yánez I, Martínez Rodero I, Mateo Y, Nolis P, Yeste M, Miró J (2023) Comparison of the metabolite profile of donkey and horse seminal plasma and its relationship with sperm viability and motility. Res Vet Sci 165:105046. https://doi.org/10.1016/j.rvsc.2023.105046

Catalán J, Yánez-Ortiz I, Torres-Garrido M, Ribas-Maynou J, Llavanera M, Barranco I, Yeste M, Miró J (2024) Impact of seminal plasma antioxidants on DNA fragmentation and lípid peroxidation of frozen-thawed horse sperm. Antioxidants 13:322. https://doi.org/10.3390/antiox13030322

Catalán J, Yánez I, Tvarijonaviciute A, González-Arostegui G, Rubio CP, Barranco I, Yeste M, Miró J (2022a) Seminal plasma antioxidants are related to sperm Cryotolerance in the horse. Antioxidants 11:1279. https://doi.org/10.3390/antiox11071279

Colembrander B, Gadella BM, Stout TAE (2003) The predictive value of semen analysis in the evaluation of stallion fertility. Reprod Domest Anim 38:305–311. https://doi.org/10.1046/j.1439-0531.2003.00451.x

Gacem S, Catalán J, Valverde A, Soler C, Miró J (2020a) Optimization of CASA-mot analysis of donkey sperm: optimum frame rate and vàlues of kinematic variables for different counting chamber and fields. Animals 10:1993. https://doi.org/10.3390/ani10111993

Gacem S, Papas M, Catalán J, Miró J (2020b) Examination of jackass (Equus asinus) accessory sex glands by B-mode ultrasound and of testicular artery blood flow by colour pulsed-wave Doppler ultrasound: correlations with semen production. Reprod Domest Anim 00:1–8. https://doi.org/10.1111/rda.13604

González-Arto M, Vicente-Carrillo A, Martínez-Pastor F, Fernández-Alegre E, Roca J, Miró J, Rigau T, Rodríguez-Gil JE, Pérez-Pe R, Muiño-Blanco T, Cebrián-Pérez JA, Casao A (2016) Melatonin receptors MT1 and MT2 are expressed in spermatozoa from several seasonal and nonseasonal breeder species. Theriogenology 86(8):1958–1968. https://doi.org/10.1016/j.theriogenology.2016.06.016

Katila T (2012) Post-mating inflammatory responses of the uterus. Reprod Domest Anim 47:31–41. https://doi.org/10.1111/j.1439-0531.2012.02120.x

López Gatius F, Miró J, Sebastián I, Ibarz A, Labérnia J (1996) Rheological properties of the anterior vaginal fluid from superovulated dairy heifers at estrus. Theriogenology 40:167–180. https://doi.org/10.1016/0093-691X(93)90350-E

Mateo-Otero Y, Zambrano F, Catalán J, Sánchez R, Yeste M, Miró J, Fernández-Fuertes B (2021) Seminal plasma, and not sperm, induces time and concentration-dependent neutrophil extracellular trap release in donkeys. Equine Vet J 54(2):415–426. https://doi.org/10.1111/evj.13457

Miró J, Taberner E, Rivera M, Peña A, Medrano A, Rigau T, Peñalba A (2009) Effects of dilution and centrifugation on the survival of spermatozoa and the structure of motile sperm cell subpopulations in refrigerated Catalonian donkey semen. Theriogenology 72:1017–1022. https://doi.org/10.1016/j.theriogenology.2009.06.012

Miró J, Vilés K, Rabanal R, Marín H (2013b) Influence of seminal plasma on leukocyte migration and COX-2 expression in the jenny endometrium after insemination with frozen-thaled semen. Reprod Domest Anim 48(1):83. https://doi.org/10.1016/j.anireprosci.2013.11.002

Miró J, Vilés K, García W, Jordana J, Yeste M (2013a) Effect of donkey seminal plasma on sperm movement and sperm-polimorphonuclear neutrophils attachment in vitro. Anim Reprod Sci 140:164–172. https://doi.org/10.1016/j.anireprosci.2013.06.007

Miró J, Papas M (2018) Improvement of cryopreservation protocol in both purebred horses including Spanish horses. Span J Agric Res 16(4):e0406. https://doi.org/10.5424/sjar/2018164-13677

Miró J, Morató R, Vilagran I, Taberner E, Bonet S, Yeste M (2020) Preservation of Epididymal stallion sperm in liquid and frozen states. Equine Vet Sci 88:102940. https://doi.org/10.1016/j.jevs.2020.102940

Miró J, Catalán J, Marín H, Yánez-Ortiz I, Yeste M (2021) Specific seminal plasma fractions are responsible for modulation of sperm-PMN binding in the donkey. Animals 11:1388. https://doi.org/10.3390/ani11051388

Palm F, Walter I, Budik S, Kolodziejek J, Nowotny N, Aurich C (2008) Influence of different semen extenders and seminal plasma on PMN migration and on expression of IL-1β, IL-6, TNF-α and COX-2 mRNA in the equine endometrium. Theriogenology 70:843–851. https://doi.org/10.1016/j.theriogenology.2008.04.054

Papas M, Arroyo L, Bassols A, Catalán J, Bonilla S, Gacem S, Yeste M, Miró J (2019a) Activities of antioxidant seminal plasma enzymes (SOD, CAT, GPX and GSR) are higher in jackasses than in stallion and are correlated with sperm motility in jackasses. Theriogenology 140:180–187. https://doi.org/10.1016/j.theriogenology.2019.08.032

Papas M, Catalán J, Fernández-Fuertes B, Arroyo L, Bassols A, Miró J, Yeste M (2019b) Specific activity of superoxide dismutase in stallion seminal plasma is related to sperm cryotolerance. Antioxidants 8:539. https://doi.org/10.3390/antiox8110539

Papas M, Catalán J, Barranca I, Arroyo L, Bassols A, Yeste M, Miró J (2020) Total and specific activities of superoxide dismutase (SOD) in seminal plasma are related with the cryotolerance of jackass spermatozoa. Cryobiology 92:109–116. https://doi.org/10.1016/j.cryobiol.2019.11.043

Portus BJ, Reilas T, Katila T (2005) Effect of seminal plasma on uterine inflammation, contractility and pregnancy rates in mares. Equine Vet J 37:515–519. https://doi.org/10.2746/042516405775314844

Rota A, Panzani D, Sabatini C, Camilo F (2012) Donkey jack (Equus asinus) semen cryopreservation: studies of seminal parameters, postbreeding inflammatory response, and fertility in donkey jennies. Theriogenology 78:1846–1854. https://doi.org/10.1016/j.theriogenology.2012.07.015

Serhan CN, Chiang N, Van Dyke TE (2008) Resolving inflammation: dual anti-inflammatory and pro-resolution lipid mediators. Nat Rev Immunol 8:349–361. https://doi.org/10.1038/nri2294

Taberner E, Medrano A, Peña A, Rigau T, Miró J (2008) Oestrus cycle characteristics and prediction of ovulation in Catalonian jennies. Theriogenology 5(70):1489–1497. https://doi. org/10.1016/j.theriogenology.2008.06.096

Troedsson MH, Loset K, Alghamdi AM, Dahms B, Crabo BG (2001) Interaction between equine semen and the endometrium: the inflammatory response to semen. Anim Reprod Sci 68:273–278. https://doi.org/10.1016/S0378-4320(01)00164-6

Vilés K, Rabanal R, Rodríguez-Prado M, Miró J (2013a) Effect of ketoprofen on the uterine inflammatory response after AI of jennies with frozen semen. Theriogenology 79:1019–1026. https://doi.org/10.1016/j.theriogenology.2013.01.006

Vilés K, Rabanal R, Rodríguez-Prado M, Miró J (2013b) Influence of seminal plasma on leukocyte migration and amount of COX-2 protein in the jenny endometrium after insemination with frozen/thawed semen. Anim Reprod Sci 143:57–63. https://doi.org/10.1016/j. anireprosci.2013.11.002

Yánez-Ortiz I, Catalán J, Mateo-Otero Y, Dordas M, Gacem S, Yeste N, Bassols A, Yeste M, Miró J (2021) Extracellular reactive oxygen species (ROS) production in fresh donkey sperm exposed to reductive stress, oxidative stress and NETosis. Antioxidant 10:1367. https://doi.org/10.3390/ antiox10091367

Male Effect During In Vitro Production of Bovine Embryos

Francisco Báez ⓘ and Nélida Rodríguez-Osorio ⓘ

Abstract

The maternal contribution to fertilization, embryo development, and pregnancy success is undeniable. However, the role of the male in the reproductive outcome, extending beyond genetic material delivery, has only recently gained adequate attention, particularly over the past two decades. This chapter delves into the growing recognition of the male contribution, exploring current trends in bovine in vitro embryo production considering that currently, the number of embryos obtained worldwire by this reproductive biotechnology exceeds those obtained from artificial insemination. This chapter places special emphasis on understanding how sperm origin, quality, and selection methods impact reproductive success. With the rise of in vitro embryo production in the cattle industry, grasping the intricacies of sperm sensitivity to environmental factors and culture conditions becomes paramount. Moreover, the chapter underscores the pivotal role of paternal factors in embryo development, scrutinizing how sperm-borne elements influence gamete interaction, fertilization, and subsequent embryo viability. By delving into these dynamics, this chapter offers insights crucial for optimizing assisted reproductive technologies in bovine breeding programs, with potential implications for other species, including humans.

F. Báez (✉)
Instituto Superior de la Carne, Centro Universitario Regional Noreste, Universidad de la República, Tacuarembó, Uruguay
e-mail: francisco.baez@cut.edu.uy

N. Rodríguez-Osorio
Unidad de Genómica y Bioinformática, Departamento de Ciencias Biológicas, Centro Universitario Regional Litoral Norte, Universidad de la República, Rivera, Salto, Uruguay
e-mail: nelida.rodriguez@unorte.edu.uy

319

J. C. Gardón, K. Satué Ambrojo (eds.), *Assisted Reproductive Technologies in Animals Volume 1*, https://doi.org/10.1007/978-3-031-73079-5_11

Keywords

Sperm quality · Paternal contribution · Embryo · Development

1   Introduction

In cattle, the in vitro production of embryos (IVPE) has gained great importance for the breeding industry and will probably continue to grow in the coming years. Each year, thousands of in vitro-produced cattle embryos from high genetic merit parents are commercialized worldwide. Through bovine IVPE, it is possible to obtain off-spring from elite cows, even after they have been replaced due to age, non-hereditary health reasons, or accidents. The latest report from the International Embryo Technology Society (IETS) highlights that since 2016, the amount of bovine in vitro produced (IVP) embryos exceeds the quantity of embryos derived in vivo. The impact of this reproductive biotechnology is materialized in over half a million calves born annually, thanks to IVPE and embryo transfer (Ferré et al. 2020; Viana 2022).

Besides the commercial applications, there are several advantages of using cattle as a model for the study of mammalian gamete interaction and preimplantation embryo development. First, given the importance of cattle for human food production, the supply of bovine gametes is permanent, with worldwide availability of frozen bovine semen, and cow ovaries from slaughterhouses, as a source of immature oocytes (Isaac et al. 2024). Second, since bovine IVPE has become an efficient method for cattle breeding and selection, highly standardized IVPE protocols have been developed, and countless conditions, media, and supplements have been thoroughly tested throughout the last 40 years (Ferré et al. 2020; Naspinska et al. 2023; Speckhart et al. 2023). Finally, the availability of an annotated bovine reference genome since 2009 (Elsik et al. 2009), with regular updates, has facilitated the development of targeted molecular and genomic tests, allowing to evaluate, at the molecular level, the effect of different culture conditions on embryo gene expression and development.

Historically, great emphasis has been placed on investigating the impact of the maternal contribution to early embryogenesis and how improving oocyte quality reflects on embryo yield. Current research continues to focus on improving conditions during in vitro maturation (IVM) of oocytes to increase their developmental potential and improve the success of reproductive biotechnologies (Roelen 2019; Ferré et al. 2020; Bezerra et al. 2021; Yousefian et al. 2021). However, although the maternal contribution plays a large role in reproductive success, the paternal effect on fertilization, embryo development, blastocyst, and pregnancy rates should also be characterized.

The paternal contribution to embryogenesis goes beyond providing a haploid genome. In addition to the genetic material, there are sperm-borne factors that participate in many events, from meiotic culmination, syngamy, and cleavage to embryo epigenetic regulation (Nanassy and Carrell 2008). Recent research states that, in addition to genetic abnormalities, alterations in sperm epigenetic marks may affect in vitro fertilization (IVF) outcomes (Rahman et al. 2018).

This chapter focuses on the paternal role in embryo development and the relevance of sperm in the production of bovine embryos in vitro. The first part of the chapter will briefly describe the whole process of in vitro production of bovine embryos, to provide a framework for the spermatozoon to meet the oocyte and generate an embryo. Then, it focuses on how sperm origin and quality might determine embryo developmental capacity. Then the chapter delves into the procedures for sperm selection before IVF and their effect on the sperm cell and thus on the outcome of fertilization. The next part of the chapter explores sperm factors crucial for embryo development, from the chromosomes to the ample molecular contribution, besides the genome. Finally, the chapter will focus on sperm epigenetic messages, as a bridge through which the environment might impact embryo development. Figure 1 summarizes all the factors that may affect the sperm cell before and during IVF and then highlights the contributions the sperm makes toward the development of a healthy embryo.

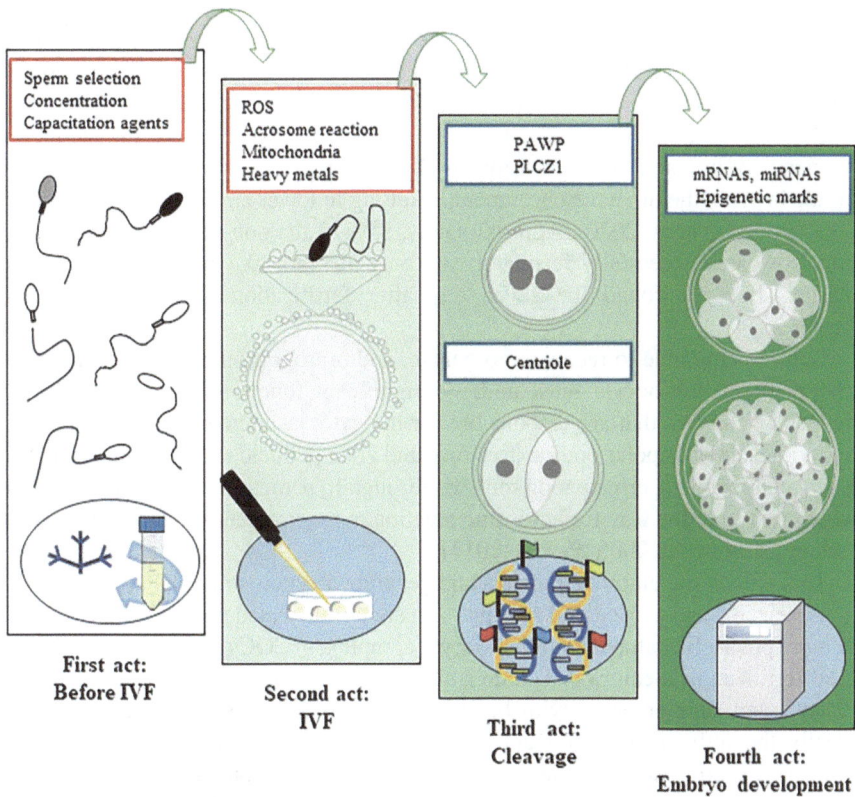

Fig. 1 The spermatozoa in four acts. Four moments related to sperm handling, fertilization, and embryo culture that define success during in vitro production of embryos (IVPE). Before and during in vitro fertilization (IVF), the sperm cell is exposed to stress factors and conditions (listed in the red boxes) that can affect reproductive success. Once fertilization has occurred, several sperm-borne factors (listed in the blue boxes) are indispensable for oocyte activation, blocking polyspermy, and accomplishing the first cell division. The paternal contribution is still needed in the fourth act, for the embryo to go on with development until the blastocyst stage. *ROS* reactive oxygen species, *mRNAs* messenger RNAs, *miRNAs* microRNAs, *PAWP* postacrosomal sheath WW domain-binding protein, *PLCZ1* phospholipase CZ

2   In Vitro Production of Bovine Embryos

Cattle IVF history started in 1977 when bull semen, capacitated in rabbit uterus and estrous cow oviducts, was used to successfully fertilize cow oocytes. The first calf obtained by IVPE was born in 1981, after the transfer of a four-cell embryo into the oviduct of a recipient cow. Two calves were born in 1986 from embryos resulting from IVF using spermatozoa that were capacitated using calcium ionophore. It was not until 1987 that the first calf obtained after IVM of oocytes, followed by IVF, and in vitro culture (IVC) of the resulting embryos was born (Iritani and Niwa 1977; Brackett et al. 1982; Hanada et al. 1986; Fukuda et al. 1990). Since then, bovine IVPE has been refined and adjusted, reaching advances that make it possible to use the whole process to dissect the role and contribution of each gamete on embryo developmental success and understand the biological processes behind embryogenesis.

Currently, the basic stages of IVPE include IVM of immature oocytes, IVF of mature oocytes, and IVC of the presumed zygotes until their cryopreservation or transfer to the uterus. These three stages are generally conducted in sequence to produce bovine embryos. However, despite its widespread use around the world, the efficiency of bovine IVPE is still low, probably in part, due to a failure to properly mimic in vitro, the in vivo environment, resulting in lower embryonic rates (Maruri et al. 2018; He et al. 2020). Approximately, 85% of the oocytes complete maturation, 70% are successfully fertilized, and only 30% to 40% manage to reach the blastocyst stage around the seven day after fertilization (Rizos et al. 2008; Hasler 2014).

Before being able to receive, reorganize, and reprogram the chromosomal load of the sperm, the oocyte must undergo significant functional, biochemical, and molecular changes that are responsible for the cytoplasmic (organelle reorganization), nuclear (first polar body extrusion), and zona pellucida (ZP) maturation. The ejaculated mobile spermatozoon must go through hyperactivation and capacitation, which will pave the way for acrosome reaction, a requirement for successful fertilization (Sirard et al. 2006; Parrish 2014).

To facilitate the union of gametes and generate a zygote under laboratory conditions, selected and capacitated spermatozoa are co-incubated with matured oocytes, or more precisely matured cumulus-oocyte complexes (COCs). The most frequently used fertilization medium is Tyrode's albumin lactate pyruvate, in the absence of glucose and supplemented with heparin (Parrish et al. 1988; Galli et al. 2003). Ideally, the optimal sperm concentration should be determined for each bull in preliminary experiments to achieve the maximum fertilization rate with the lowest level of polyspermy (Yang et al. 1995). For IVF, COCs are placed in contact with sperm for 18 to 20 h. During this time, the sperm penetrate the cumulus cell layers, adhere, and bind to the ZP, and after undergoing the acrosome reaction, they penetrate the ZP. The spermatozoon that manages to reach this point will unite and fuse with the oocyte plasma membrane, causing the activation of the oocyte, which will lead to the formation of male and female pronuclei. After fertilization, cumulus cells progressively lose contact with the oocyte. Presumptive zygotes are cultured for several days until the blastocysts stage, at which embryos are destined for transfer or cryopreservation (Machaty et al. 2012; Parrish 2014; Arshad et al. 2021).

Normal fertilization rates leading to the formation of male and female pronuclei are around 70%, although lower (40–55%) and higher rates (79–81%) have been reported (Hara et al. 2012; Ohlweiler et al. 2013; Sprícigo et al. 2014; Ribas et al. 2018; Báez et al. 2019, 2021; Guibu de Almeida et al. 2022). Fertilization rates exceeding 80% are associated with an increase in polyspermy due to a sperm concentration above 2×10^6 spz/mL, which affects embryonic development rates (Demyda-Peyrás et al. 2015). Maruri et al. (2018) modified the conventional IVF protocol, with a final sperm concentration of 15×10^6 spz/mL for a co-culture period of 5 h, achieving a total blastocyst rate on day 8 of 30%.

In the study by Somfai et al. (2022), sperm concentration was 3×10^6 spz/mL for a co-culture period of 30 min with a blastocyst development rate on day 9 close to 34% and lower rates of polyspermy (16%). Another risk associated with using high sperm concentrations is an increased likelihood of bacterial contamination, which can alter media pH, compromising the fertilizing capacity of sperm (reviewed by Contri et al. 2013). Adjusting seminal evaluation media pH to 7–7.5 allowed for improving viability, mitochondrial activity, and progressive mobility in bull sperm. In this regard, IVPE laboratories should incorporate basic procedures such as material sterilization and asepsis during sperm manipulation. The use of antibiotics, culture media osmolarity, and pH verification are essential, and with these routine practices, repeatable results and high-quality standards could be maintained.

Around half of bovine in vitro-produced embryos do not reach the blastocyst stage. The length of the IVC period could be one of the reasons for the reduced embryo quality in vitro (Lonergan and Fair 2008; Plourde et al. 2012; Ferré et al. 2020; Nix et al. 2023). However, Rizos et al. (2002) showed that for in vivo matured and fertilized oocytes that were subsequently cultured in vitro, blastocyst rates were higher which could suggest that maturation and fertilization also have a deleterious effect. Báez et al. (2024) reported that low oxygen tension during the IVM and IVF phases improves the developmental capacity of bovine embryos, rendering them better prepared for recovering from the vitrification/warming process, which points to the importance of maturation and fertilization phases for blastocyst success.

The rate of embryos that reach the blastocyst stage is a parameter related to two concepts: embryonic quality and developmental competence. Although closely related, these concepts involve different factors. The first one derives from the capacity of the microenvironment to provide adequate support for the development of a healthy embryo capable of managing stress and, more importantly, establishing pregnancy. The concept of competence refers to the intrinsic potential of experiencing development. This potential is highly dependent on uncharacterized factors that may accumulate or may be operative during gametogenesis. The scope and duration of this contribution have not yet been clarified (Plourde et al. 2012). What is widely known is that the origin and quality of the gametes influence the development of competence (Lonergan et al. 2003; Merton et al. 2003; Palma and Sinowatz 2004).

3 Sperm Origin, Variability, and Quality

Gamete quality and culture conditions determine the developmental potential of IVP embryos (Khurana and Niemann 2000; Meirelles et al. 2004). Developmental success also depends on the genetic and epigenetic contributions of both gametes (Chenoweth 2005, 2007). The potential effects of the male gamete on embryo development and embryo quality are probably often underestimated. Sperm quality might be described as the capacity of spermatozoa to move towards and fuse with the oocyte activating it to promote zygote formation and cleavage (Sabés-Alsina et al. 2019; Durairajanayagam et al. 2021).

Variability in IVPE results has been observed within same-breed bulls. There have been differences in early embryo rates (from 56.8% to 90.7%) and blastocyst rates (from 24.7 to 52.1%) for individual Nellore bulls (Leme et al. 2020). These results are similar to reports for Simmental and Holstein bulls (Schneider et al. 1999; Alomar et al. 2008; Llamas-Luceño et al. 2020; Travnickova et al. 2021; Pasquariello et al. 2024). According to Leme et al. (2020), the male contribution does not limit itself to embryo developmental kinetics, yield, or sex rates, but it also involves blastocyst cryotolerance. However, embryo yield and cryotolerance were not related in all cases. Lower cryotolerance could be related to some type of DNA damage or epigenetic alterations.

Chromatin integrity is important not only for fertilization but also for embryo development. Although the oocyte has the capacity to repair sperm DNA (Evenson et al. 2002; Morris et al. 2002), which allows sperm cells with fragmented DNA to achieve fertilization and initiate embryo development, when the extent of sperm DNA fragmentation exceeds the oocyte repair capacity, then embryo development pre and post-implantation can be compromised (Ahmadi and Ng 1999; Ménézo 2006).

As a result of their intense mitochondrial activity, sperm cells produce a variety of molecules that belong to the Reactive Oxygen Species (ROS), including superoxide anion (O_2^-), hydrogen peroxide (H_2O_2), and nitric oxide (NO). These molecules are instrumental in the signal cascade responsible for sperm capacitation and acrosome reaction (Aitken et al. 2010). However, any alteration in ROS levels might induce oxidative stress that can affect sperm function through plasmatic membrane peroxidation, protein damage, and DNA fragmentation (Nichi et al. 2006; Gualtieri et al. 2014; Gürler et al. 2015).

Factors associated with oxidative stress in bovine sperm cells include high environmental temperature, lead and cadmium exposure, lower activity of antioxidant mechanisms, exposure to atmospheric oxygen tension (20% O_2), and cryopreservation (Nichi et al. 2006; Gualtieri et al. 2014; Gürler et al. 2015; Guvvala et al. 2020; He et al. 2020). Freezing and thawing bull semen reduces sperm viability and fertilizing capacity (Peris-Frau et al. 2020). Centrifugation and culture can also increase ROS production affecting sperm cell quality and its reproductive potential. The consequences of oxidative stress include damage to the axoneme and mitochondrial proteins, affecting sperm motility (Chatterjee et al. 2001; Grötter et al. 2019; Peris-Frau et al. 2020; Khan et al. 2021; Sapanidou and Tsantarliotou 2023).

In sexed bovine sperm, these morphological and functional changes appear to be more noticeable, as they are subjected to various adverse conditions during the flow cytometry-based sorting process, compromising their developmental potential (Bermejo-Álvarez et al. 2008; Bermejo-Alvarez et al. 2010; Carvalho et al. 2010; Baldi et al. 2020; Magata et al. 2021). Specifically, in X-sorted sperm, no differences in fertilization rates were observed compared to non-sexed sperm, but a delay in the first mitotic division, early embryonic development, total blastocysts produced, and the number of blastomeres per embryo were noted (Bermejo-Álvarez et al. 2010; Carvalho et al. 2010; Baldi et al. 2020). These findings provide an explanation for the low pregnancy rates after the transfer of embryos derived from sexed semen (Magata et al. 2021).

Several studies suggest that pretreatment of cryopreserved bovine sperm with zinc, D-aspartate, coenzyme Q10, N-acetyl-L-cysteine, and crocin can improve quality and fertilization capacity, enhancing the developmental potential of IVP embryos (Gualtieri et al. 2014; Pérez et al. 2015; Sapanidou et al., 2015). Higher rates of embryonic development were also observed when sexed bovine sperm were pretreated with reduced glutathione during IVF. Therefore, a pretreatment period of sperm with antioxidants could neutralize excessive ROS production while preserving sperm's physiological function (Hu et al. 2016; Sapanidou et al. 2023).

The factor with the highest predictive index for success in IVF is sperm quality, which is reflected in motility, morphology, and the percentage of sperm with intact acrosomes. However, since sperm with pyriform defects apparently have normal acrosomes and motility, one might expect them to penetrate the ZP and enter the cytoplasm under in vitro conditions (Zhang et al. 1998; Walters et al. 2004; Chenoweth 2005). In bovine IVF, the percentage of pyriform sperm penetrating the ZP and reaching the cytoplasm was lower than that of sperm with normal morphology, suggesting that pyriform-shaped sperm are primarily discriminated during ZP binding (Thundathil et al. 1999). Surprisingly, in the study published by Perez Siqueira et al. (2018), sperm with lower values of intact acrosomes and lower mitochondrial membrane potential resulted in higher blastocyst rates. The authors conclude that those parameters, along with the assessment of mobility before the Percoll® gradient, could be key traits for defining the efficiency of bovine IVPE.

For many years, the studies that explored the male role in reproductive failure focused on how male fertility was affected by environment, sanitary conditions, and the nutritional and physiological state (Brito et al. 2003; Chenoweth 2005; Barros et al. 2006; Wrathall et al. 2006; D'Occhio et al. 2007). Today, without disregarding the impact of those aspects, the focus has shifted towards the study of the genome, transcriptome, and epigenome of both gametes, as a tool to understand preimplantation embryonic development in mammals.

Male factor infertility is responsible for approximately 50% of IVF failures. Since male infertility is frequently idiopathic, it is important to consider the potential effects of genetic and epigenetic abnormalities not only on infertility but also on the success of IVF (Nanassy and Carrell 2008). Many studies have shown how the male effect can alter the outcome of assisted reproductive techniques. Several authors indicate that the male effect can be related to sperm chromatin organization

(Ward et al. 2001; Benchaib et al. 2003; Katska-Ksiazkiewicz et al. 2005; Borini et al. 2006; Guibu de Almeida et al. 2022). Salisbury et al. (1977) state that there are differences in sperm DNA, not only between individuals of the same species but even among spermatozoa from the same male.

The male effect during IVF can be observed initially on the cleavage rate, total blastocyst yield, and embryo quality (Ward et al. 2001; Vandaele et al. 2006). The first evidence of the paternal effect on bovine embryo development was detected at the beginning of the synthesis (S) phase during the first mitotic division (Eid et al. 1994; Comizzoli et al. 2000). Comizzoli et al. (2000) showed that the time between the initiation and completion of the first DNA synthesis in bovine embryos is an accurate indicator of the developmental potential of the embryo and that it depends exclusively on the male contribution. Later, Comizzoli's team (2003) reported that paternal pronuclear formation is necessary for the S phase beginning in the maternal pronucleus and that the initiation of an early S phase in both pronuclei is linked to a positive regulation of the pentose phosphate pathway. In high-fertility bulls, the S phase is shorter, which leads to a higher blastocyst yield. These results were confirmed by Ward et al. (2001), Vandaele et al. (2006), and Alomar et al. (2008), who identified higher blastocyst rates for those bulls, producing early dividing zygotes.

When semen from low fertility bulls is used for IVPE, higher polyspermy rates, overexpression of genes related to apoptosis, and cell damage pathways are observed during embryo genome activation affecting embryo rate and viability. In contrast, high-fertility bulls produce a higher quantity of viable blastocysts which is valuable for commercial IVPE laboratories (Guibu de Almeida et al. 2022).

Several studies suggest that alterations in the paternal genome can compromise not only fertilization and embryonic quality but also embryo survival and gestation success, causing in some cases spontaneous abortions (Larson et al. 2000; Evenson et al. 2002; Morris et al. 2002; Benchaib et al. 2003; Borini et al. 2006). To date, studies in humans, cattle, pigs, and birds have highlighted the importance of the male effect, considering origin, age, individual variations, and environmental exposure (Lodge et al. 1971; Ward et al. 2001; Paulson et al. 2001; Rockett et al. 2001; Ward et al. 2003; De Jonge 2005; França et al. 2005; Haugan et al. 2005; Hernández-Ochoa et al. 2006; Chenoweth 2007; Rahman et al. 2011; Guibu de Almeida et al. 2022), which stimulates the refinement of selection and evaluation methods for fresh or frozen sperm (Brito et al. 2003; Cesari et al. 2006; Córdova-Izquierdo et al. 2006; Anguita et al. 2007; Perez Siqueira et al. 2018).

It is evident that certain male factor infertility phenotypes are associated with increased DNA fragmentation and/or aneuploidies that can compromise embryonic development (Lockhart et al. 2023). In general, any alteration in spermatogenesis that results in a high incidence of abnormal sperm has the potential to negatively affect a greater percentage of the sperm than what can be detected with conventional means. In these affected populations, spermatozoa that appear to be morphologically normal may be compromised in terms of their ability to achieve fertilization and sustain embryo development and finally pregnancy (Katska-Ksiazkiewicz et al. 2005). In bulls, differences in reproductive success are often not explained through conventional semen evaluation. Advances in IVPE and embryo transfer may clarify doubts regarding the contribution of the oocyte and sperm genome on embryonic development and survival. The present review aims to address the male factors that limit the survival and normal development of the mammalian embryo, highlighting the most recent lines of work in the study of the male gamete contribution to the embryo.

In cattle, pregnancy rates for IVPE are lower than for those produced in vivo (Ealy et al. 2019; Bouwman and Mullaart 2023). One reason for this may be the increased occurrence of chromosomal abnormalities, both numerical and structural, which could play a role in male factor infertility (Chenoweth 2007). A high proportion of bovine IVP embryos (13.7–80%) exhibit chromosomal imbalance, with mixoploidy being the most frequently observed anomaly (16–72%), negatively affecting embryonic development (Kawarsky et al. 1996; Viuff et al. 1999). In the study by Báez et al. (2014), 88.8% of embryos between 2–8 cells were diploid, and 11.11% were haploid. Tutt et al. (2021) reported a case of haploidy with the absence of paternal genome, suggesting parthenogenesis. The development of polyploid embryos is slow, and they cease development around the third cell cycle, while mixoploid embryos seem to continue development, with polyploid cells mainly located in the blastocyst's trophectoderm (Viuff et al. 2000, 2002). The only classes of aneuploidies whose origins could be most frequently ascribed to the paternal germ line were triploidy/hypotriploidy, possibly resulting from polyspermy or errors in spermatic meiosis. This work also describes a significant effect of the bull's age on the incidence of aneuploidy (Silvestri et al. 2021).

4 Sperm Selection and Preparation: Does It Affect Sperm Quality?

An essential step in the IVF protocol is the selection of motile sperm. During this step, unwanted substances such as extender, cryoprotectants, seminal plasma, detritus, and dead sperm are removed from the culture system. Among the most used sperm selection techniques are swim-up and discontinuous density gradients such as Percoll® and Bovipure® (Cesari et al. 2006; Arias et al. 2017). Although these are based on different principles, both seek the same objective, to select motile sperm. There are other less-used techniques such as migration through hyaluronic acid columns and through glass wool (Samperl and Crabo 1993; Witt et al. 2016).

Swim-up is based on the upward migratory capacity of the sperm due to their own movements and consists of placing the contents of a thawed straw at the bottom of a conical tube, with sperm selection medium for 45 min to 1 h at 38.5 °C. The dense nature of the semen mixed with the extender and cryoprotectants initially keeps sperm at the bottom. Over time, sperm begin to swim out of the extender. This method accumulates the most motile sperm in the upper fraction and leaves those with low or no motility in the lower fraction. The upper fraction of spermatozoa not only has improved motility, compared to those in the lower fraction, but also shows higher metabolic rates and longer flagella. With this method, the isolated sperm portion can be counted, brought down to the desired concentration, and added to the

IVF system. Despite the simplicity of swim-up, the number of sperm cells recovered with this technique is often low (Parrish 2014; Magdanz et al. 2019).

Density gradients allow the separation of motile and viable spermatozoa through sedimentation-centrifugation. Sperm cells are sedimented in a gradient that is in equivalent equilibrium with their own density, which through centrifugation allows the mobile and viable spermatozoa to reach the bottom of the conical tube forming a pellet, acting as a filter for seminal plasma, round cells, debris, non-motile and dead sperm, extenders, and cryoprotectants (Oliveira et al. 2012). Spermatozoa with good nuclear morphology and cellular integrity are denser and are deposited in the higher-density Percoll layer. These gradients can be continuous or discontinuous: Percoll®, Bovipure®, and Puresperm® (Le Lannou and Blanchard 1988; Palomo and Izquierdo 1999; Samardzija et al. 2006). The last two methods are based on colloidal silicon and the concentrations used are combinations of 45% and 90%, 30% and 90%, or 40% and 80%. Dilutions are made in inert or capacitation media, and the centrifugation time varies from 5 to 20 min depending on the protocol. Percoll®, PureSperm®, and Bovipure® are the most frequently used discontinuous gradients sperm separation products (Le Lannou and Blanchard 1988; Malvezzi et al. 2014; Arias et al. 2017; Cajas et al. 2020; Báez et al. 2021; Yousefian et al. 2021).

Although the proportion of motile frozen/thawed sperm is lower for the swim-up technique compared to Percoll®. Some studies have reported higher embryonic division rates for sperm recovered with swim-up (Parrish et al. 1995; Dode et al. 2002). Dode et al. (2002) did not find significant differences in fertilization rates between the two bovine sperm selection methods, while Cesari et al. (2006) reported higher rates of blastocysts in the Percoll® separated sperm, despite not finding differences in hatching rates or embryo sex ratios. For their part, Arias et al. (2017) reported a higher proportion of sperm with intact plasma membrane and acrosome for Percoll® and Bovipure® separation methods compared to swim-up, but no differences in DNA integrity or expression of genes related to metabolism (glyceraldehyde-3-phosphate dehydrogenase *GAPDH*, peptidylprolyl isomerase A *PP1A*, and cytochrome P450 family 19 *CYP19*), inflammation (leptin *LEP*), purine synthesis (hypoxanthine phosphoribosyl transferase 1 *HPRT1*), and DNA compaction (protamine 1 *PRM1*).

After selection, sperm concentration is measured, to adjust the amount of sperm to the quantity of matured oocytes. Conventional semen straws are packaged at a concentration of 12–20 million sperm per milliliter (Maicas et al. 2020). For IVF, a concentration of $1–2 \times 10^6$ spermatozoa per milliliter is normally preferred. When IVF is performed with sexed semen, for which the usual concentration is 2×10^6 spermatozoa per straw, 3–4 straws are normally used (Holm et al. 1999; Ferré et al. 2020; Maicas et al. 2020). Sperm recovery is also performed with Percoll®, and the final concentration is adjusted to 2.5×10^6 spz/mL or 5000 sperm cells per fertilization drop of 40–50 µL (Trigal et al. 2012; Magata et al. 2021).

Conventional or sexed sperm cells must be treated with enabling factors, and added to the IVF medium, to promote structural and biochemical changes in the spermatozoa that allow them to penetrate the ZP and fuse with the oocyte. Among the capacitating factors are fatty acid-free bovine serum albumin, which removes cholesterol from the plasma membrane; β-amino acids and catecholamines (penicillamine, hypotaurine, and epinephrine—PHE), which maintain motility and increase penetration rates; caffeine, a known phosphodiesterase inhibitor that enhances and prolongs motility increasing sperm penetration rates; and heparin, which binds to the plasma membrane allowing the entry of extracellular calcium and inducing spermatic capacitation (Parrish et al. 1986; Miller et al. 1990; Coscioni et al. 2001; Xia and Ren 2009). The use of this glycosaminoglycan has become widespread in IVF protocols because it improves fertilization and total blastocyst rates when cryopreserved bovine sperm are used. The most used concentration of this compound in the IVF medium is 10 µg/mL in frozen/thawed sperm, while for sexed semen the concentrations range from 5 µg/mL to 40 µg/mL (Saeki et al. 1995; Dode et al. 2002; Blondin et al. 2009; An et al. 2017). The response to heparin concentration for in vitro capacitation in sexed or non-sexed sperm is variable between bulls (Lu et al. 2003; Lu and Seidel 2004; An et al. 2017).

A PubMed® biomedical literature search on the effects of bovine sperm preparation methods rendered approximately 90 results, mostly published in the last 20 years. In this regard, further research is needed to improve the mobility and fertilization potential of fresh, cryopreserved, conventional, or sexed spermatozoa. The conjunction of sperm quality assays with IVF is a valuable tool to study the male effect during fertilization and embryonic development, and appropriate preventive, corrective, or predictive measures could be developed.

5   Paternal Factors Crucial for Embryo Development

In mammals, the entire sperm, including the tail, enters the oocyte during fertilization (Fukuda et al. 2011). Hence, besides the essential haploid paternal genome, the spermatozoon delivers to the oocyte several factors that determine the success of fertilization, cleavage, and the development of the embryo and extraembryonic tissues. The most important sperm-borne oocyte activating factors are spermatic postacrosomal sheath WW domain-binding protein (PAWP), and phospholipase CZ (PLCZ1), which are involved in the calcium signaling cascade needed for oocyte activation and meiotic culmination (Krawetz 2005; Wu et al. 2007; Lockhart et al. 2023).

During fertilization, the sperm also delivers to the oocytes the typical proximal and the atypical distal centriole (Uzbekov et al. 2023). Based on the bovine IVF model, Avidor-Reis and Fishman have proposed that during fertilization, the atypical centriole is released in the zygote, nucleates a new centriole, and acts as a second centriole (Avidor-Reiss 2018; Fishman E 2019). After fertilization, typical and atypical centrioles sequester maternal pericentriolar material and reconstitute the centrosome of the zygote. Subsequently, both centrioles separate and each one forms a centrosome, an aster, a new centriole, and a spindle pole at different stages of zygote development. In bovine zygotes and embryos resulting from IVF, the distal centriole of the sperm disappears at the end of the first mitotic division.

Centrioles appear in two- to four-cell embryos (Uzbekov et al. 2023). Centriole formation includes the appearance of atypical centrioles with microtubule triplets of various lengths.

The sperm centrosome is primarily responsible for nucleating and organizing the sperm aster, which pushes the sperm head toward the oocyte center and guides the migration of the female pronucleus for union with the male pronucleus, completing the fertilization process. In the midpiece of the spermatozoon, the proximal and distal centrioles are surrounded by mitochondria (Uzbekov et al., 2023). Mitochondrial functionality and the integrity of their membrane potential are requirements for sperm motility, hyperactivation, capacitation, acrosin activity, acrosomal reaction, and DNA integrity. It is known that sperm mitochondrial DNA (mtDNA) is susceptible to oxidative damage and mutations that could compromise function and lead to fertility issues (Durairajanayagam et al. 2021).

However, the paternal contribution does not end here: as early as 1989, it was established that the spermatozoon delivers to the oocyte/zygote a complex cargo of transcripts, which amount to 10–100 fg of RNA (Pessot et al. 1989; Li and Zhou 2012). Despite their low cytoplasmic content, sperm-borne RNAs include messenger RNAs (mRNAs), long noncoding RNAs (lncRNAs), micro RNAs (miRNAs), small interfering RNAs (siRNAs), Piwi-interacting RNAs (piRNAs), and transfer RNA-derived small RNAs (tsRNAs) (Ostermeier et al. 2004; Martins and Krawetz 2005; Lalancette et al. 2008; Dadoune 2009; Hamatani 2012; Kawano et al. 2012; Gòdia et al. 2018; Zhang et al. 2024).

Despite being smaller than the payload of RNAs contributed by the oocyte, these rich repertoires of sperm RNAs seem to be not just a remnant from spermatogenesis but a coherent collection of additional paternal information, beyond DNA (Das et al. 2013). A functional role was proposed for sperm-borne transcripts since they are delivered to oocytes and detected in zygotes; hence sperm transcript cargo could be markers of male factor contribution (Li and Zhou 2012; Jodar et al. 2013). Two recent reviews that summarize the current knowledge of the origin, types, and functional roles of human sperm RNAs, and their potential to be used as biomarkers of fertility and reproductive outcomes, might also serve as a source for bovine sperm-borne RNA research given the conserved nature of several transcripts (Santiago et al. 2022; Hernández-Silva et al. 2022).

The advent of transcriptomic technologies, expression microarrays, and RNA sequencing has made it easier to dissect the sperm transcriptome. Studies in bovine spermatozoa transcriptome profiling have revealed that bull spermatozoa contain transcripts for about 13,000 genes, similar to what has been reported for other mammals (Feugang et al. 2010; Selvaraju et al. 2017). Most spermatozoa transcripts seem to be fragmented in nature; however, some paternal transcripts seem to play a crucial role in the initiation of zygotic transcription and the early stages of embryogenesis, influencing processes such as cell division, differentiation, and development of various tissues. Among the most abundant bovine transcripts are protamine 1 *PRM1*, fatty acid binding protein 1 *FABP1*, sterol-binding domain containing 1 *SCP2D1*, nuclear receptor subfamily 2 group E member 3 *NR2E3*, BCL2 like 11,

BCL2L11, and BRCA1 DNA repair-associated *BRCA1*. The last two transcripts seem to be involved in embryo development.

The presence of specific mRNAs in sperm cells suggests that they may participate in shaping the early molecular landscape of the developing embryo (Bonache et al. 2012; Dhawan et al. 2019). Several studies have even identified different mRNA profiles in sperm cells from males with varying fertility statuses, or under different diets or environmental conditions (Stringer et al. 2006; Garrido et al. 2009; Feugang et al. 2010; Bansal et al. 2015; Nätt et al. 2019; Urena et al. 2022). The sperm transcriptome cargo can be so specific that it has been established that X- and Y-bearing sperm cells have distinctive RNA signatures. This indicates that not all genes expressed in spermatids are completely shared among them through cytoplasmic bridges. Instead, a particular collection of transcripts is retained based on the sex chromosome present in the spermatid, resulting in a haploid sex-specific transcriptome in sperm cells (Bhutani et al. 2021).

In addition to mRNAs, several non-coding RNAs have been found in sperm and are believed to be transferred to the oocyte upon fertilization. These RNA molecules may have different epigenetic functions regulating maternal mRNA degradation, embryo genome activation, and gene expression. The most studied among them are miRNAs, small non-coding RNAs that can posttranscriptionally regulate gene expression. The delivery of paternal miRNAs to the oocyte may contribute to the modulation of gene expression patterns during early embryonic development. These small RNA molecules have been implicated in processes such as embryonic stem cell differentiation, cell fate determination, and the establishment of cell lineages.

Sperm miRNA profiles are affected by the male's fertility status and the content of miRNA in fertile and infertile men, and bulls is markedly different, particularly for miRNA molecules involved in chromatin modification and embryogenesis (Salas-Huetos et al. 2016; Ying Zhang et al. 2024). Different miRNA profiles have also been detected in sperm depending on the bull's age, which may mediate the effects of paternal age on early embryonic development (Wu et al. 2020). Several sperm miRNAS have been singled out in mice, humans, and bovine, as biomarkers of first cleavage, cell proliferation, and embryo development including mmu-miR-34c, has-miR-34c, hsa-miR-191, has-miR-149, and bta-miR-202, and bta-miR-216b (Liu et al. 2012; Alves et al. 2019; Xu et al. 2020; Li et al. 2021; Wang et al. 2021; Cui et al. 2023).

It seems that paternal miRNAs, siRNAs, and tsRNAs are important for the preimplantation developmental program and play a critical role in the transcriptomic homeostasis of zygotes and two-cell embryos since sperm with aberrant miRNA profiles produced embryos with reduced preimplantation developmental potential (Yuan et al. 2016). As reviewed by Yang (2023), embryonic development and offspring phenotype might be modulated by sperm sncRNAs. Exposure of male mice to stress alters the sperm miRNA content, which elicits changes in the offspring regulation of the hypothalamic-pituitary-adrenal stress axis, which might indicate that miRNAs could allow transgenerational transfer of information in response to environmental changes (Rodgers et al. 2013; Dahlen et al. 2023).

Besides the transcriptomic cargo, which can also be considered a type of epigenetic information, sperm DNA methylation has been explored as the better-understood mechanism for epigenetic regulation. Distinctive patterns of DNA methylation in specific sperm genomic *loci* have been associated with decreased fecundity in humans and bovines (Jenkins et al. 2016; Kropp et al. 2017). They have even been linked with male-caused recurrent spontaneous abortion (Ma et al. 2023). Most DNA methylation marks are erased from the paternal genome during the first embryonic cell cycles. However, several genomic regions maintain their methylation status, which has been widely studied for imprinted genes. It is important to point out that sperm DNA methylation can be affected by DNA fragmentation, which can interfere with the active demethylation process and disrupt the insertion of histones into the male chromatin in the male pronucleus, after fertilization (Rajabi et al. 2018).

Histone content and histone modifications in sperm are valuable indicators of sperm quality since histones and their posttranslational modifications play a significant role in sperm DNA packaging and function (Miller et al. 2010). During spermatogenesis, most histones are removed from spermatid chromatin in a sequential process: core histones are replaced by transition proteins 1 and 2 (TP1 and TP2); then transition proteins are replaced by testis-specific protamines that have a lower molecular weight and allow the sperm DNA to be tightly coiled into the donut shape protamine toroid (Balhorn 2007). During this process, histone 4 H4 hyperacetylation is critical for histone removal (Sonnack et al. 2002). H4 acetylation changes histone tail charge from positive to neutral, thus reducing the interaction between the negatively charged DNA phosphate group and histone N termini. Alterations in H4 hyperacetylation have been found in low-fertility bulls and have been linked with alterations in chromatin structure and abnormalities in sperm morphology and function (Ugur et al. 2019).

However, not all histones are removed from sperm chromatin, and nearly 15% of DNA in mature sperm cells remains coiled around histones (Oliva 2006). Alterations in the modifications in the tails of the retained sperm histones and the methylation status of the DNA coiled around those histones have been associated with unexplained male factor infertility and poor blastocyst development (Denomme et al. 2017). For these men, methylation for DNA coiled around retained histones was lower than in men with normal fertility, showing that, as it has been previously proven for other cell types, there is a link between histone modifications and DNA methylation patterns.

6   Environmental and Microbiological Factors

Heat stress caused by rising temperatures and global warming has led to adverse effects on the physiology, well-being, health, and reproduction of cattle. Heat stress results in cellular function decrease in various reproductive tissues in cattle, negatively impacting hormonal secretion, gametes—the most sensitive structures—and embryonic development (Llamas Luceño et al. 2020; Morrell 2020; Roth 2020;

Báez et al. 2022). The intensity of heat stress can be assessed by the temperature-humidity index (THI) (Roth 2021). In a study conducted by Seifi-Jamadi et al. (2020), exposure of Belgian Blue bulls to elevated ambient temperatures (maximum THI of 83.7) during the last stages of spermatogenesis, corresponding to 14–28 days before semen collection, negatively affected sperm quality. They observed a higher percentage of sperm with abnormal morphologies, increased ROS generation after thawing, decreased chromatin condensation, motility, and higher H_2O_2 generation. Sperm collected after exposure to high THI generated lower rates of embryonic cleavage and total blastocysts obtained in vitro. Similarly, Llamas Luceño et al. (2020) demonstrated that sperm from Holstein-Friesian dairy bulls collected during higher THI (77.9–80.5) exhibited reduced viability, leading to decreased blastocyst development and delayed hatching compared to semen collected during lower THI (51.8–55.0).

Scrotal insulation has become the method of choice to mimic the effects of heat stress on sperm cells (Rahman et al. 2011; Sabés-Alsina et al. 2019). In a study by Fernandes et al. (2008), Nellore bulls were subjected to scrotal insulation for 5 days. After 14 days post-insulation, sperm characterized by defects in the head, nuclear vacuoles, abnormal chromatin, and lower embryonic rates were observed. In another study, bovine sperm were subjected to thermal shock (41 °C) for 4 h before IVF. The results showed low blastocyst rates and a high rate of apoptosis in the blastocysts (Rahman et al. 2014). Sperm chromatin is highly vulnerable to heat, leading to a tendency to decrease the size of maternal and paternal pronuclei, reflected in low IVF rates (Rahman et al. 2018).

Environmental contaminants are gaining increasing attention due to their harmful effects on livestock productivity and fertility. The presence of heavy metals in water and grass is potentially dangerous for humans and animals due to their bioaccumulation (Guvvala et al. 2020; Wrzecińska et al. 2021). Among toxic elements, lead (Pb) and cadmium (Cd) have been reported in human and bovine sperm, altering morphology and fertilization potential (Méndez et al. 2011, 2012; Chand et al. 2019). According to Wrzecińska et al. (2021), the origin of both metals in the environment is attributed to forest fires, erosion, and pesticides, accumulating in the liver, kidneys, lungs, brain, bones, testicles, epididymis, and semen. In the study by Chand et al. (2019), 0.23 µg/mL of Pb and 0.11 µg/mL of Cd were detected in fresh semen samples from bulls. An in vitro study found that exposure to 750 µgmol/L of Cd in bovine sperm led to increased lipid peroxidation, membrane damage, low motility, and infertility (Arabi and Mohammadpour 2006).

In males, Cd can inhibit DNA repair and decrease antioxidant activity. Pb can cuase Leydig cells atrophy and is associated with sperm abnormalities such as azoospermia, asthenozoospermia, and teratozoospermia. Overall, exposure of livestock to toxic metals can cause embryotoxicity and alterations in spermatogenesis (Verma et al. 2018; Guvvala et al. 2020). Future research appears justified to characterize these two elements and explore others (such as arsenic and mercury) that may be masking the cause of low fertilization and embryonic development rates in bovine sperm.

Sperm contamination with mycotoxins is another aspect to consider. The exposure of sperm to single mycotoxins or binary mixtures did not affect their cellular characteristics. However, exposure to the ternary mixtures reduced mitochondrial membrane potential and increased the proportion of spermatozoa with ROS (Kalo et al., 2023).

From a microbiological perspective, laboratories dedicated to IVPE that incorporate strict cleaning and sterilization practices are still prone to bacterial contamination. Non-sterilized air, personnel, and materials, as well as follicular fluid and semen, are common sources for a diverse variety of bacteria and fungi in embryo culture (Borges et al. 2020; Li et al. 2022). Bacterial contamination of semen is not uncommon and has been reported in sperm used for IVF. Kim and collaborators (1998) isolated bacteria from frozen-thawed semen from four bulls and identified *Pseudomonas aeruginosa*, *Staphylococcus sciuri*, *Enterobacter cloacae*, *Acinetobacter calcoaceticus*, and *Flavobacterium* spp. The presence of these bacteria during IVF negatively affects viability and subsequent embryonic development. Bacterial growth leads to an increase in seminal plasma pH and a significant reduction in sperm mobility, resulting in low IVF rates (Contri et al. 2013). The removal of cumulus cells from presumptive zygotes, which is routinely carried out through mechanical agitation after fertilization, might reduce bacterial contamination and allow embryonic development.

Mycoplasma bovis has also been found in commercial semen, negatively affecting fertilization (Eaglesome, 1990). According to Pohjanvirta et al. (2023), semen naturally contaminated with *Mycoplasma bovis* cannot transmit infection to embryos produced by IVP.

Wrathall et al. (2006) reported the presence in bull semen of four viruses: enzootic bovine leukosis virus (EBLV), bovine herpesvirus-1 (BoHV-1), bovine viral diarrhea virus (BVDV), and bluetongue virus (BTV), posing the risk of producing infected embryos. Particularly, the risk of infection in embryos with EBLV and BTV is very low. Embryos infected with BoHV-1 can be treated with trypsin, while for BVDV, there is insufficient information on how the virus is carried in semen or how it can interact with sperm, oocytes, and embryos.

The development and application of assisted reproductive techniques introduced environmental changes that might increase the distribution of these pathogens in cattle (Gard et al. 2007). According to the last edition of the IETS manual (Stringfellow and Seidel 2009), a way to minimize it is to use IVF semen from certified breeders that regularly evaluate the health status of their semen donor. The manual also states that the use of semen from BVDV or BHV-1 positive males is strictly forbidden, since these viruses can persist in the IVPE systems, independently of the semen processing method.

7   Conclusions

The success of IVPE in cattle is intricately tied to various paternal factors, including sperm origin, quality, and the contribution of genetic and epigenetic elements. Sperm quality significantly influences embryo development, with abnormalities in

DNA fragmentation, chromatin integrity, and mitochondrial function impacting fertilization rates and embryonic viability. Moreover, the paternal genome delivers essential factors, including transcripts and epigenetic information, influencing early embryonic processes. Environmental factors, such as heat stress and exposure to contaminants, pose significant challenges to sperm quality and embryonic development, necessitating careful management practices. The utilization of antioxidants or protective molecules to supplement sperm media presents promising potential, counteracting the redox imbalance induced during in vitro manipulation, which deprives spermatozoa of their natural antioxidant protection, leading to an increase in ROS levels. Additionally, microbiological contamination, though controllable, requires strict hygiene measures to mitigate adverse effects on IVF outcomes. Understanding and addressing these paternal and environmental factors are crucial for improving the efficiency and reliability of IVPE in cattle, ultimately enhancing reproductive success and productivity in the livestock industry.

References

Ahmadi A, Ng SC (1999) Fertilizing ability of DNA-damaged spermatozoa. J Exp Zool 284:696–704. https://doi.org/10.1002/(SICI)1097-010X(19991101)284:6<696::AID-JEZ11> 3.0.CO;2-E

Aitken RJ, Baker MA, De Iuliis GN, Nixon B (2010) New insights into sperm physiology and pathology. Handb Exp Pharmacol 198:99–115. https://doi.org/10.1007/978-3-642-02062-9_7

Alomar M, Tasiaux H, Remacle S, George F, Paul D, Donnay I (2008) Kinetics of fertilization and development, and sex ratio of bovine embryos produced using the semen of different bulls. Anim Reprod Sci 107:48–61. https://doi.org/10.1016/j.anireprosci.2007.06.009

Alves MBR, de Arruda RP, De Bem THC, Florez-Rodriguez SA, de Sá Filho MF, Belleannée C, Meirelles FV, da Silveira JC, Perecin F, Celeghini ECC (2019) Sperm-borne miR-216b modulates cell proliferation during early embryo development via K-RAS. Sci Rep 9:1–14. https://doi.org/10.1038/s41598-019-46775-8

An LY, Chaubal SA, Liu Y, Chen Y, Nedambale TL, Xu J, Xue F, Moreno JF, Tao S, Presicce GA, Du F (2017) Significant heparin effect on bovine embryo development during sexed in vitro fertilization. J Reprod Dev 63:175–183. https://doi.org/10.1262/jrd.2016-142

Anguita B, Vandaele L, Mateusen B, Maes D, Van Soom A (2007) Developmental competence of bovine oocytes is not related to apoptosis incidence in oocytes, cumulus cells and blastocysts. Theriogenology 67:537–549. https://doi.org/10.1016/j.theriogenology.2006.09.004

Arabi M, Mohammadpour AA (2006) Adverse effects of cadmium on bull spermatozoa. Vet Res Commun 30:943–951. https://doi.org/10.1007/s11259-006-3384-3

Arias ME, Andara K, Briones E, Felmer R (2017) Bovine sperm separation by swim-up and density gradients (Percoll and BoviPure): effect on sperm quality, function and gene expression. Reprod Biol 17:126–132. https://doi.org/10.1016/j.repbio.2017.03.002

Arshad U, Sagheer M, González-Silvestry FB, Hassan M, Sosa F (2021) Vitrification improves in-vitro embryonic survival in Bos taurus embryos without increasing pregnancy rate post embryo transfer when compared to slow-freezing: A systematic meta-analysis. Cryobiology 101:1–11. https://doi.org/10.1016/j.cryobiol.2021.06.007

Avidor-Reiss T (2018) Rapid evolution of sperm produces diverse centriole structures that reveal the most rudimentary structure needed for function. Cells 7:1–16. https://doi.org/10.3390/cells7070067

Báez F, Camargo Á, Reyes AL, Márquez A, Paula-Lopes F, Viñoles C (2019) Time-dependent effects of heat shock on the zona pellucida ultrastructure and in vitro developmental competence of bovine oocytes. Reprod Biol 19:195–203. https://doi.org/10.1016/j.repbio.2019.06.002

Báez F, de Brun V, Rodríguez-Osorio N, Viñoles C (2024) Low oxygen tension during in vitro embryo production improves the yield, quality, and cryotolerance of bovine blastocysts. Anim Sci J 95:e13941. https://doi.org/10.1111/asj.13941

Báez F, Gómez B, de Brun V, Rodríguez-Osorio N, Viñoles C (2021) Effect of ethanol on parthenogenetic activation and α-tocopherol supplementation during in vitro maturation on developmental competence of summer-collected bovine oocytes. Curr Issues Mol Biol 43:2253–2265. https://doi.org/10.3390/cimb43030158

Báez F, López Darriulat R, Rodríguez-Osorio N, Viñoles C (2022) Effect of season on germinal vesicle stage, quality, and subsequent in vitro developmental competence in bovine cumulus-oocyte complexes. J Therm Biol 103:1–8. https://doi.org/10.1016/j.jtherbio.2021.103171

Báez F, Méndez-López Y, Villamediana-Monreal P (2014) Complemento cromosómico de embriones bovinos (Bos taurus indicus) producidos in vitro. Revista Ciencia 22:14–20

Baldi E, Tamburrino L, Muratori M, Degl'Innocenti S, Marchiani S (2020) Adverse effects of in vitro manipulation of spermatozoa. Anim Reprod Sci 220:106314. https://doi.org/10.1016/j.anireprosci.2020.106314

Balhorn R (2007) The protamine family of sperm nuclear proteins. Genome Biol 8:1–8. https://doi.org/10.1186/gb-2007-8-9-227

Bansal SK, Gupta N, Sankhwar SN, Rajender S (2015) Differential genes expression between fertile and infertile spermatozoa revealed by transcriptome analysis. PLoS One 10:1–21. https://doi.org/10.1371/journal.pone.0127007

Barros CM, Pegorer MF, Vasconcelos JLM, Eberhardt BG, Monteiro FM (2006) Importance of sperm genotype (indicus versus taurus) for fertility and embryonic development at elevated temperatures. Theriogenology 65:210–218. https://doi.org/10.1016/j.theriogenology.2005.09.024

Benchaib M, Braun V, Lornage J, Hadj S, Salle B, Lejeune H, Guérin JF (2003) Sperm DNA fragmentation decreases the pregnancy rate in an assisted reproductive technique. Hum Reprod 18:1023–1028. https://doi.org/10.1093/humrep/deg228

Bermejo-Álvarez P, Lonergan P, Rath D, Gutiírrez-Adan A, Rizos D (2010) Developmental kinetics and gene expression in male and female bovine embryos produced in vitro with sex-sorted spermatozoa. Reprod Fertil Dev 22:426–436. https://doi.org/10.1071/RD09142

Bermejo-Álvarez P, Rizos D, Rath D, Lonergan P, Gutiérrez-Adán A (2008) Can bovine in vitro-matured oocytes selectively process X- or Y-sorted sperm differentially? Biol Reprod 79:594–597. https://doi.org/10.1095/biolreprod.108.070169

Bermejo-Alvarez P, Rizos D, Rath D, Lonergan P, Gutierrez-Adan A (2010) Sex determines the expression level of one third of the actively expressed genes in bovine blastocysts. Proc Natl Acad Sci 107:3394. https://doi.org/10.1073/pnas.0913843107

Bezerra FTG, Dau AMP, Van Den Hurk R, Silva JRV (2021) Molecular characteristics of oocytes and somatic cells of follicles at different sizes that influence in vitro oocyte maturation and embryo production. Domest Anim Endocrinol 74:106485. https://doi.org/10.1016/j.domaniend.2020.106485

Bhutani K, Stansifer K, Ticau S, Bojic L, Villani AC, Slisz J, Cremers CM, Roy C, Donovan J, Fiske B, Friedman RC (2021) Widespread haploid-biased gene expression enables sperm-level natural selection. Science 371:1–9. https://doi.org/10.1126/science.abb1723

Blondin P, Beaulieu M, Fournier V, Morin N, Crawford L, Madan P, King WA (2009) Analysis of bovine sexed sperm for IVF from sorting to the embryo. Theriogenology 71:30–38. https://doi.org/10.1016/j.theriogenology.2008.09.017

Bonache S, Mata A, Ramos MD, Bassas L, Larriba S (2012) Sperm gene expression profile is related to pregnancy rate after insemination and is predictive of low fecundity in normozoospermic men. Hum Reprod 27:1556–1567. https://doi.org/10.1093/humrep/des074

Borges ED, Berteli TS, Reis TF, Silva AS, Vireque AA (2020) Microbial contamination in assisted reproductive technology: source, prevalence, and cost. J Assist Reprod Genet 37:53–61. https://doi.org/10.1007/s10815-019-01640-5

Borini A, Tarozzi N, Bizzaro D, Bonu MA, Fava L, Flamigni C, Coticchio G (2006) Sperm DNA fragmentation: paternal effect on early post-implantation embryo development in ART. Hum Reprod 21:2876–2881. https://doi.org/10.1093/humrep/del251

Bouwman AC, Mullaart E (2023) Screening of in vitro-produced cattle embryos to assess incidence and characteristics of unbalanced chromosomal aberrations. JDS Comm 4:101–105. https://doi.org/10.3168/jdsc.2022-0275

Brackett B, Bousquet D, Boice M, Donawick W, Evans JDM (1982) Normal development following in vitro fertilization in the cow. Biol Reprod 27:147–158

Brito LFC, Barth AD, Bilodeau-Goeseels S, Panich PL, Kastelic JP (2003) Comparison of methods to evaluate the plasmalemma of bovine sperm and their relationship with in vitro fertilization rate. Theriogenology 60:1539–1551. https://doi.org/10.1016/S0093-691X(03)00174-2

Cajas YN, Cañón-Beltrán K, de Guevara ML, Millán de la Blanca MG, Ramos-Ibeas P, Gutiérrez-Adán A, Rizos D, González EM (2020) Antioxidant nobiletin enhances oocyte maturation and subsequent embryo development and quality. Int J Mol Sci 21:1–18. https://doi.org/10.3390/ijms21155340

Carvalho JO, Sartori R, Machado GM, Mourão GB, Dode MAN, Agca Y, Monson RL, Northey DL, Peschel DE, Schaefer DM, Rutledge JJ (2010) Quality assessment of bovine cryopreserved sperm after sexing by flow cytometry and their use in in vitro embryo production. Theriogenology 50:1521–1530. https://doi.org/10.1016/j.theriogenology.2010.06.030

Cesari A, Kaiser GG, Mucci N, Mutto A, Vincenti A, Fornés MW, Alberio RH (2006) Integrated morphophysiological assessment of two methods for sperm selection in bovine embryo production in vitro. Theriogenology 66:1185–1193. https://doi.org/10.1016/j.theriogenology.2006.03.029

Chand N, Tyagi S, Prasad R, Dutta D, Sirohi AS, Sharma A, Tyagi R (2019) Effect of heavy metals on oxidative markers and semen quality parameters in HF crossbred bulls. Ind J Anim Sci 89:632–636. https://doi.org/10.56093/ijans.v89i6.91114

Chatterjee S, De Lamirande E, Gagnon C (2001) Cryopreservation alters membrane sulfhydryl status of bull spermatozoa: protection by oxidized glutathione. Mol Reprod Dev 60:498–506. https://doi.org/10.1002/mrd.1115

Chenoweth PJ (2005) Genetic sperm defects. Theriogenology 64:457–468. https://doi.org/10.1016/j.theriogenology.2005.05.005

Chenoweth PJ (2007) Influence of the male on embryo quality. Theriogenology 68:308–315. https://doi.org/10.1016/j.theriogenology.2007.04.002

Comizzoli P, Guienne BML, Heyman Y, Renard JP (2000) Onset of the first S-phase is determined by a paternal effect during the G1-phase in bovine zygotes. Biol Reprod 62:1677–1684. https://doi.org/10.1095/biolreprod62.6.1677

Comizzoli P, Urner F, Sakkas D, Renard JP (2003) Up-regulation of glucose metabolism during male pronucleus formation determines the early onset of the S phase in bovine zygotes. Biol Reprod 68:1934–1940. https://doi.org/10.1095/biolreprod.102.011452

Contri A, Gloria A, Robbe D, Valorz C, Wegher L, Carluccio A (2013) Kinematic study on the effect of pH on bull sperm function. Anim Reprod Sci 136:252–259. https://doi.org/10.1016/j.anireprosci.2012.11.008

Córdova-Izquierdo A, Oliva JH, Lleó B, García-Artiga C, Corcuera BD, Pérez-Gutiérrez JF (2006) Effect of different thawing temperatures on the viability, in vitro fertilizing capacity and chromatin condensation of frozen boar semen packaged in 5 ml straws. Anim Reprod Sci 92:145–154. https://doi.org/10.1016/j.anireprosci.2005.05.011

Coscioni AC, Reichenbach HD, Schwartz J, LaFalci VSN, Rodrigues JL, Brandelli A (2001) Sperm function and production of bovine embryos in vitro after swim-up with different calcium and caffeine concentration. Anim Reprod Sci 67:59–67. https://doi.org/10.1016/S0378-4320(01)00116-6

Cui L, Fang L, Zhuang L, Shi B, Lin CP, Ye Y (2023) Sperm-borne microRNA-34c regulates maternal mRNA degradation and preimplantation embryonic development in mice. Reprod Biol Endocrinol 21:1–11. https://doi.org/10.1186/s12958-023-01089-3

D'Occhio MJ, Hengstberger KJ, Johnston SD (2007) Biology of sperm chromatin structure and relationship to male fertility and embryonic survival. Anim Reprod Sci 101:1–17. https://doi.org/10.1016/j.anireprosci.2007.01.005

Dadoune JP (2009) Spermatozoal RNAs: what about their functions? Microsc Res Tech 72:536–551

Dahlen CR, Amat S, Caton JS, Crouse MS, Da Silva Diniz WJ, Reynolds LP (2023) Paternal effects on fetal programming. Anim Reprod 20:e20230076. https://doi.org/10.1590/1984-3143-AR2023-0076

Das PJ, McCarthy F, Vishnoi M, Paria N, Gresham C, Li G, Kachroo P, Sudderth AK, Teague S, Love CC, Varner DD, Chowdhary BP, Raudsepp T (2013) Stallion sperm transcriptome comprises functionally coherent coding and regulatory RNAs as revealed by microarray analysis and RNA-seq. PLoS One 8:e56535. https://doi.org/10.1371/journal.pone.0056535

De Jonge C (2005) Biological basis for human capacitation. Hum Reprod Update 11:205–214. https://doi.org/10.1093/humupd/dmi010

Demyda-Peyrás S, Dorado J, Hidalgo M, Moreno-Millán M (2015) Influence of sperm fertilising concentration, sperm selection method and sperm capacitation procedure on the incidence of numerical chromosomal abnormalities in IVF early bovine embryos. Reprod Fertil Dev 27:351–359. https://doi.org/10.1071/RD13285

Denomme MM, McCallie BR, Parks JC, Schoolcraft WB, Katz-Jaffe MG (2017) Alterations in the sperm histone-retained epigenome are associated with unexplained male factor infertility and poor blastocyst development in donor oocyte IVF cycles. Hum Reprod 32:2443–2455. https://doi.org/10.1093/humrep/dex317

Dhawan V, Kumar M, Malhotra N, Singh N, Dadhwal V, Dada R (2019) Paternal contributions in early embryonic gene expression: role in early pregnancy loss. Fertil Steril 112:e49–e50. https://doi.org/10.1016/j.fertnstert.2019.07.257

Dode MAN, Rodovalho NC, Ueno VG, Fernandes CE (2002) The effect of sperm preparation and co-incubation time on in vitro fertilization of Bos indicus oocytes. Anim Reprod Sci 69:15–23. https://doi.org/10.1016/S0378-4320(01)00148-8

Durairajanayagam D, Singh D, Agarwal A, Henkel R (2021) Causes and consequences of sperm mitochondrial dysfunction. Andrologia 53:1–15. https://doi.org/10.1111/and.13666

Eaglesome MGM (1990) The effect of mycoplasma bovis on fertilization processes in vitro with bull spermatozoa and zona-free hamster oocytes. Vet Microbiol 21:329–337

Ealy AD, Wooldridge LK, McCoski SR (2019) Board invited review: post-transfer consequences of in vitro-produced embryos in cattle. J Anim Sci 97:2555–2568. https://doi.org/10.1093/jas/skz116

Eid LN, Lorton SF, Parrish JJ (1994) Paternal influence on S-phase in the first cell cycle of the bovine embryo. Biol Reprod 51:1232–1237. https://doi.org/10.1095/biolreprod51.6.1232

Elsik CG, Tellam RL, Worley KC, Gibbs RA, Muzny DM, Weinstock GM, Zhao F-Q et al (2009) The genome sequence of taurine cattle: A window to ruminant biology and evolution. Science 324:522–528. https://doi.org/10.1126/science.1169588

Evenson DP, Larson KL, Jost LK (2002) Sperm chromatin structure assay: its clinical use for detecting sperm DNA fragmentation in male infertility and comparisons with other techniques. J Androl 23:25–43. https://doi.org/10.1002/j.1939-4640.2002.tb02599.x

Fernandes CE, Dode MAN, Pereira D, Silva AEDF (2008) Effects of scrotal insulation in Nellore bulls (Bos taurus indicus) on seminal quality and its relationship with in vitro fertilizing ability. Theriogenology 70:1560–1568. https://doi.org/10.1016/j.theriogenology.2008.07.005

Ferré LB, Kjelland ME, Strøbech LB, Hyttel P, Mermillod P, Ross PJ (2020) Review: recent advances in bovine in vitro embryo production: reproductive biotechnology history and methods. Animal 14:991–1004. https://doi.org/10.1017/S1751731119002775

Feugang JM, Rodriguez-Osorio N, Kaya A, Wang H, Page G, Ostermeier GC, Topper EK, Memili E (2010) Transcriptome analysis of bull spermatozoa: implications for male fertility. Reprod Biomed Online 21:312–324. https://doi.org/10.1016/j.rbmo.2010.06.022

Fishman EA-RT (2019) It takes two to tango. Reproduction 1:34. https://doi.org/10.1038/s41559-016-0034

França LR, Avelar GF, Almeida FFL (2005) Spermatogenesis and sperm transit through the epididymis in mammals with emphasis on pigs. Theriogenology 63:300–318. https://doi.org/10.1016/j.theriogenology.2004.09.014

Fukuda M, Fukuda K, Shimizu T, Nobunaga M, Byskov AG, Yding Andersen C (2011) Reply: the sex ratio of offspring is associated with the mothers age at menarche. Hum Reprod 26:2589–2590. https://doi.org/10.1093/humrep/der206

Fukuda Y, Ichikawa M, Naito K, Toyoda Y (1990) Birth of normal calves resulting from bovine oocytes matured, fertilized, and cultured with cumulus cells in vitro up to the blastocyst stage. Biol Reprod 42:114–119. https://doi.org/10.1095/biolreprod42.1.114

Galli C, Duchi R, Crotti G, Turini P, Ponderato N, Colleoni S, Lagutina I, Lazzari G (2003) Bovine embryo technologies. Theriogenology 59:599–616. https://doi.org/10.1016/S0093-691X(02)01243-8

Gard JA, Givens MD, Stringfellow DA (2007) Bovine viral diarrhea virus (BVDV): epidemiologic concerns relative to semen and embryos. Theriogenology 68:434–442. https://doi.org/10.1016/j.theriogenology.2007.04.003

Garrido N, Martínez-Conejero JA, Jauregui J, Horcajadas JA, Simón C, Remohí J, Meseguer M (2009) Microarray analysis in sperm from fertile and infertile men without basic sperm analysis abnormalities reveals a significantly different transcriptome. Fertil Steril 91:1307–1310. https://doi.org/10.1016/j.fertnstert.2008.01.078

Gòdia M, Mayer FQ, Nafissi J, Castelló A, Rodríguez-Gil JE, Sánchez A, Clop A (2018) A technical assessment of the porcine ejaculated spermatozoa for a sperm-specific RNA-seq analysis. Syst Biol Reprod Med 64:291–303. https://doi.org/10.1080/19396368.2018.1464610

Grötter LG, Cattaneo L, Marini PE, Kjelland ME, Ferré LB (2019) Recent advances in bovine sperm cryopreservation techniques with a focus on sperm post-thaw quality optimization. Reprod Domest Anim 54:655–665. https://doi.org/10.1111/rda.13409

Gualtieri R, Barbato V, Fiorentino I, Braun S, Rizos D (2014) Theriogenology treatment with zinc, D -aspartate, and coenzyme Q10 protects bull sperm against damage and improves their ability to support embryo development. Theriogenology 82:592–598. https://doi.org/10.1016/j.theriogenology.2014.05.028

Guibu de Almeida T, Mingoti RD, Signori de Castro L, Perez Siqueira AF, dos Santos R, Hamilton T, Kubo Fontes P, Gouveia Nogueira MF, Alves MF, Basso AC, Pecora Milazzotto M, Assumpção OD'A, M. E. (2022) Paternal effect does not affect in vitro embryo morphokinetics but modulates molecular profile. Theriogenology 178:30–39. https://doi.org/10.1016/j.theriogenology.2021.10.027

Gürler H, Malama E, Heppelmann M, Calisici O, Leiding C, Kastelic JP, Bollwein H (2015) Effects of cryopreservation on sperm viability, synthesis of reactive oxygen species, and DNA damage of bovine sperm. Theriogenology 86:562–571. https://doi.org/10.1016/j.theriogenology.2016.02.007

Guvvala PR, Ravindra JP, Selvaraju S (2020) Impact of environmental contaminants on reproductive health of male domestic ruminants: a review. Environ Sci Pollut Res 27:3819–3836. https://doi.org/10.1007/s11356-019-06980-4

Hamatani T (2012) Human spermatozoal RNAs. Fertil Steril 97:275–281. https://doi.org/10.1016/j.fertnstert.2011.12.035

Hanada A, Enya Y, Suzuki T (1986) Birth of calves by non-surgical transfer of in vitro fertilized embryos obtained from oocytes matured in vitro. Japanese J Anim Reprod 32:208

Hara H, Hwang IS, Kagawa N, Kuwayama M, Hirabayashi M, Hochi S (2012) High incidence of multiple aster formation in vitrified-warmed bovine oocytes after in vitro fertilization. Theriogenology 77:908–915. https://doi.org/10.1016/j.theriogenology.2011.09.018

Hasler JF (2014) Forty years of embryo transfer in cattle : A review focusing on the journal *Theriogenology*, the growth of the industry in North America, and personal reminisces. Theriogenology 81:152–169. https://doi.org/10.1016/j.theriogenology.2013.09.010

Haugan T, Reksen O, Gröhn YT, Kommisrud E, Ropstad E, Sehested E (2005) Seasonal effects of semen collection and artificial insemination on dairy cow conception. Anim Reprod Sci 90:57–71. https://doi.org/10.1016/j.anireprosci.2005.02.002

He H, Zhang H, Li Q, Fan J, Pan Y, Zhang T, Robert N, Zhao L, Hu X, Han X, Yang S, Cui Y, Yu S (2020) Low oxygen concentrations improve yak oocyte maturation and enhance the developmental competence of preimplantation embryos. Theriogenology 156:46–58. https://doi.org/10.1016/j.theriogenology.2020.06.022

Hernández-Ochoa I, Sánchez-Gutiérrez M, Solís-Heredia MJ, Quintanilla-Vega B (2006) Spermatozoa nucleus takes up lead during the epididymal maturation altering chromatin condensation. Reprod Toxicol 21:171–178. https://doi.org/10.1016/j.reprotox.2005.07.015

Hernández-Silva G, Caballero-Campo P, Chirinos M (2022) Sperm mRNAs as potential markers of male fertility. Reprod Biol 22:100636. https://doi.org/10.1016/J.REPBIO.2022.100636

Holm P, Booth PJ, Schmidt MH, Greve T, Callesen H (1999) High bovine blastocyst in a static in vitro production system SOFaa medium supplemented with sodium citrate and myo-inositol with or without serum-proteins. Theriogenology:684–700

Hu T, Zhu H, Sun W, Hao H, Zhao X, Du W, Wang Z (2016) Sperm pretreatment with glutathione improves IVF embryos development through increasing the viability and antioxidative capacity of sex-sorted and unsorted bull semen. J Integr Agric 15:2326–2335. https://doi.org/10.1016/S2095-3119(16)61402-8

Iritani A, Niwa K (1977) Capacitation of bull spermatozoa and fertilization in vitro of cattle follicular oocytes matured in culture. J Reprod Fertil 50:119–121. https://doi.org/10.1530/jrf.0.0500119

Isaac E, Berg DK, Pfeffer PL (2024) Using extended growth of cattle embryos in culture to gain insights into bovine developmental events on embryonic days 8 to 10. Theriogenology 214:10–20. https://doi.org/10.1016/j.theriogenology.2023.10.004

Jenkins TG, Aston KI, Meyer TD, Hotaling JM, Shamsi MB, Johnstone EB, Cox KJ, Stanford JB, Porucznik CA, Carrell DT (2016) Decreased fecundity and sperm DNA methylation patterns. Fertil Steril 105:51–57.e3. https://doi.org/10.1016/j.fertnstert.2015.09.013

Jodar M, Selvaraju S, Sendler E, Diamond MP, Krawetz SA (2013) The presence, role and clinical use of spermatozoal RNAs. Hum Reprod Update 19:604–624. https://doi.org/10.1093/humupd/dmt031

Kalo D, Mendelson P, Komsky-Elbaz A, Voet H (2023) The effect of mycotoxins and their mixtures on bovine spermatozoa characteristics. Toxins 15:1–22

Katska-Ksiazkiewicz L, Bochenek M, Rynska B (2005) Effect of quality of sperm chromatin structure on in-vitro production of cattle embryos. Arch Anim Breed 48:32–39. https://doi.org/10.5194/aab-48-32-2005

Kawano M, Kawaji H, Grandjean V, Kiani J, Rassoulzadegan M (2012) Novel small noncoding RNAs in mouse spermatozoa, Zygotes and Early Embryos. PLoS ONE 7:e44542. https://doi.org/10.1371/journal.pone.0044542

Kawarsky SJ, Basrur PK, Stubbings RB, Hansen PJ, King WA (1996) Chromosomal abnormalities in bovine embryos and their influence on development. Biol Reprod 54:53–59. https://doi.org/10.1095/biolreprod54.1.53

Khan IM, Cao Z, Liu H, Khan A, Rahman SU, Khan MZ, Sathanawongs A, Zhang Y (2021) Impact of cryopreservation on spermatozoa freeze-thawed traits and relevance OMICS to assess sperm Cryo-tolerance in farm animals. Front Vet Sci 8:1–14. https://doi.org/10.3389/fvets.2021.609180

Khurana NK, Niemann H (2000) Effects of oocyte quality, oxygen tension, embryo density, cumulus cells and energy substrates on cleavage and morula/blastocyst formation of bovine embryos. Theriogenology 54:741–756. https://doi.org/10.1016/S0093-691X(00)00387-3

Kim I, Son D, Lee H, Yang B, Lee D, Suh G, Lee K (1998) Bacterial in semen used for IVF affect embryo viability but can be removed by stripping cumulus cells by vortexing. Theriogenology 50:293–300

Krawetz SA (2005) Paternal contribution: new insights and future challenges. Nat Rev Genet 6:633–642

Kropp J, Carrillo JA, Namous H, Daniels A, Salih SM, Song J, Khatib H (2017) Male fertility status is associated with DNA methylation signatures in sperm and transcriptomic profiles of bovine preimplantation embryos. BMC Genomics 18:280. https://doi.org/10.1186/s12864-017-3673-y

Lalancette C, Miller D, Li Y, Krawetz SA (2008) Paternal contributions: new functional insights for spermatozoal RNA. J Cell Biochem 104:1570–1579. https://doi.org/10.1002/jcb.21756

Larson KL, DeJonge CJ, Barnes AM, Jost LK, Evenson DP (2000) Sperm chromatin structure assay parameters as predictors of failed pregnancy following assisted reproductive techniques. Hum Reprod 15:1717–1722. https://doi.org/10.1093/humrep/15.8.1717

Le Lannou D, Blanchard Y (1988) Nuclear maturity and morphology of human spermatozoa selected by Percoll density gradient centrifugation or swim-up procedure. J Reprod Fertil 84:551–556. https://doi.org/10.1530/jrf.0.0840551

Leme LO, Carvalho JO, Franco MM, Dode MAN (2020) Effect of sex on cryotolerance of bovine embryos produced in vitro. Theriogenology 141:219–227. https://doi.org/10.1016/j.theriogenology.2019.05.002

Li R, Du F, Ou S, Ouyang N, Wang W (2022) A new method to rescue embryos contaminated by bacteria. F and S Reports 3:168–171. https://doi.org/10.1016/j.xfre.2022.05.002

Li H, Li L, Lin C, Hu M, Liu X, Wang L, Le F, Jin F (2021) Decreased miR-149 expression in sperm is correlated with the quality of early embryonic development in conventional in vitro fertilization. Reprod Toxicol 101:28–32. https://doi.org/10.1016/j.reprotox.2021.02.005

Li C, Zhou X (2012) Gene transcripts in spermatozoa: markers of male infertility. Clin Chim Acta 413:1035–1038. https://doi.org/10.1016/j.cca.2012.03.002

Liu WM, Pang RTK, Chiu PCN, Wong BPC, Lao K, Lee KF, Yeung WSB (2012) Sperm-borne microRNA-34c is required for the first cleavage division in mouse. Proc Natl Acad Sci USA 109:490–494. https://doi.org/10.1073/pnas.1110368109

Llamas Luceño N, de Souza Ramos Angrimani D, de Cássia Bicudo L, Szymańska KJ, Van Poucke M, Demeyere K, Meyer E, Peelman L, Mullaart E, Broekhuijse MLWJ, Van Soom A (2020) Exposing dairy bulls to high temperature-humidity index during spermatogenesis compromises subsequent embryo development in vitro. Theriogenology 141:16–25. https://doi.org/10.1016/j.theriogenology.2019.08.034

Llamas-Luceño N, Hostens M, Mullaart E, Broekhuijse M, Lonergan P, Van Soom A (2020) High temperature-humidity index compromises sperm quality and fertility of Holstein bulls in temperate climates. J Dairy Sci 103:9502–9514. https://doi.org/10.3168/jds.2019-18089

Lockhart K, Fallon L, Ortega M (2023) Paternal determinants of early embryo development. Reprod Fertil Dev, A-H 36:43. https://doi.org/10.1071/RD23172

Lodge JR, Fechheimer NS, Jaap RG (1971) The relationship of in vivo sperm storage interval to fertility and embryonic survival in the chicken. Biol Reprod 5:252–257. https://doi.org/10.1093/biolreprod/5.3.252

Lonergan P, Fair T (2008) In vitro-produced bovine embryos—dealing with the warts. Theriogenology 69:17–22. https://doi.org/10.1016/j.theriogenology.2007.09.007

Lonergan P, Rizos D, Gutierrez-Adan A, Fair T, Boland MP (2003) Oocyte and embryo quality: effect of origin, culture conditions and gene expression patterns. Reprod Domest Anim 38:259–267. https://doi.org/10.1046/j.1439-0531.2003.00437.x

Lu KH, Seidel GE (2004) Effects of heparin and sperm concentration on cleavage and blastocyst development rates of bovine oocytes inseminated with flow cytometrically-sorted sperm. Theriogenology 62:819–830. https://doi.org/10.1016/j.theriogenology.2003.12.029

Lu KH, Seidel GE, Mendes JOB, Burns PD, De La Torre-Sanchez JF, Seidel GE (2003) Effect of heparin on cleavage rates and embryo production with four bovine sperm preparation protocols. Theriogenology 60:331–340. https://doi.org/10.1016/S0093-691X(03)00029-3

Ma R-H, Zhang Z-G, Zhang Y-T, Jian S-Y, Li B-Y (2023) Detection of aberrant DNA methylation patterns in sperm of male recurrent spontaneous abortion patients. Zygote 31:163–172. https://doi.org/10.1017/S0967199422000648

Machaty Z, Peippo J, Peter A (2012) Production and manipulation of bovine embryos: techniques and terminology. Theriogenology 78:937–950. https://doi.org/10.1016/j.theriogenology.2012.04.003

Magata F, Urakawa M, Matsuda F, Oono Y (2021) Developmental kinetics and viability of bovine embryos produced in vitro with sex-sorted semen. Theriogenology 161:243–251. https://doi.org/10.1016/j.theriogenology.2020.12.001

Magdanz V, Boryshpolets S, Ridzewski C, Eckel B, Reinhardt K (2019) The motility-based swim-up technique separates bull sperm based on differences in metabolic rates and tail length. PLoS One 14:1–16. https://doi.org/10.1371/journal.pone.0223576

Maicas C, Holden SA, Drake E, Cromie AR, Lonergan P, Butler ST (2020) Fertility of frozen sex-sorted sperm at 4×106 sperm per dose in lactating dairy cows in seasonal-calving pasture-based herds. J Dairy Sci 103:929–939. https://doi.org/10.3168/jds.2019-17131

Malvezzi H, Sharma R, Agarwal A, Abuzenadah AM, Abu-Elmagd M (2014) Sperm quality after density gradient centrifugation with three commercially available media: A controlled trial. Reprod Biol Endocrinol 12:1–7. https://doi.org/10.1186/1477-7827-12-121

Martins RP, Krawetz SA (2005) RNA in human sperm. Asian J Androl 7:115–120. https://doi.org/10.1111/j.1745-7262.2005.00048.x

Maruri A, Cruzans PR, Lorenzo MS, Tello MF, Teplitz GM, Carou MC, Lombardo DM (2018) Embryotrophic effect of a short-term embryo coculture with bovine luteal cells. Theriogenology 119:143–149. https://doi.org/10.1016/j.theriogenology.2018.06.032

Meirelles FV, Caetano AR, Watanabe YF, Ripamonte P, Carambula SF, Merighe GK, Garcia SM (2004) Genome activation and developmental block in bovine embryos. Anim Reprod Sci 82–83:13–20. https://doi.org/10.1016/j.anireprosci.2004.05.012

Méndez Y, Báez F, Quintero-Moreno AVP (2012) Efecto de la exposición in vitro de espermatozoides humanos a plomo. Ciencias 20:5–11

Méndez Y, Báez F, Villamediana P (2011) Efecto de la exposición in vitro de espermatozoides humanos a cadmio (CdCl2). Perinatologia y Reproducción Humana:198–204

Ménézo YJR (2006) Paternal and maternal factors in preimplantation embryogenesis: interaction with the biochemical environment. Reprod Biomed Online 12:616–621. https://doi.org/10.1016/S1472-6483(10)61188-1

Merton JS, De Roos APW, Mullaart E, De Ruigh L, Kaal L, Vos PLAM, Dieleman SJ (2003) Factors affecting oocyte quality and quantity in commercial application of embryo technologies in the cattle breeding industry. Theriogenology 59:651–674. https://doi.org/10.1016/S0093-691X(02)01246-3

Miller D, Brinkworth M, Iles D (2010) Paternal DNA packaging in spermatozoa: more than the sum of its parts? DNA, histones, protamines and epigenetics. Reproduction 139:287–301. https://doi.org/10.1530/REP-09-0281

Miller DJ, Winer MA, Ax RL (1990) Heparin-binding proteins from seminal plasma bind to bovine spermatozoa and modulate capacitation by heparin. Biol Reprod 42:899–915. https://doi.org/10.1095/biolreprod42.6.899

Morrell JM (2020) Theriogenology heat stress and bull fertility. Theriogenology 153:62–67. https://doi.org/10.1016/j.theriogenology.2020.05.014

Morris ID, Ilott S, Dixon L, Brison DR (2002) The spectrum of DNA damage in human sperm assessed by single cell gel electrophoresis (comet assay) and its relationship to fertilization and embryo development. Hum Reprod 17:990–998. https://doi.org/10.1093/humrep/17.4.990

Nanassy L, Carrell DT (2008) Paternal effects on early embryogenesis. J Exp Clin Assist Reprod 5:1–9. https://doi.org/10.1186/1743-1050-5-2

Naspinska R, Moreira da Silva MH, Moreira da Silva F (2023) Current advances in bovine in vitro maturation and embryo production using different antioxidants: A review. J Dev Biol 11:1–9. https://doi.org/10.3390/jdb11030036

Nätt D, Kugelberg U, Casas E, Nedstrand E, Zalavary S, Henriksson P, Nijm C, Jäderquist J, Sandborg J, Flinke E, Ramesh R, Örkenby L, Appelkvist F, Lingg T, Guzzi N, Bellodi C, Löf M, Vavouri T, Öst A (2019) Human sperm displays rapid responses to diet. PLoS Biol 17:1–25. https://doi.org/10.1371/journal.pbio.3000559

Nichi M, Bols PEJ, Züge RM, Barnabe VH, Goovaerts IGF, Barnabe RC, Cortada CNM (2006) Seasonal variation in semen quality in Bos indicus and Bos taurus bulls raised under tropical conditions. Theriogenology 66:822–828. https://doi.org/10.1016/j.theriogenology.2006.01.056

Nix J, Marrella M, Ali O, Rhoads M, Ealy A, Biase F (2023) Cleavage kinetics is a better indicator of embryonic developmental competency than brilliant cresyl blue staining of oocytes. Anim Reprod Sci 248:107174. https://doi.org/10.1016/j.anireprosci.2022.107174

Ohlweiler LU, Brum DS, Leivas FG, Moyses AB, Ramos RS, Klein N, Mezzalira JC, Mezzalira A (2013) Intracytoplasmic sperm injection improves in vitro embryo production from poor quality bovine oocytes. Theriogenology 79:778–783. https://doi.org/10.1016/j.theriogenology.2012.12.002

Oliva R (2006) Protamines and male infertility. Hum Reprod Update 12:417–435. https://doi.org/10.1093/humupd/dml009

Oliveira LZ, Arruda RP, Celeghini ECC, de Andrade AFC, Perini AP, Resende MV, Miguel MCV, Lucio AC, Hossepian de Lima VFM (2012) Effects of discontinuous Percoll gradient centrifugation on the quality of bovine spermatozoa evaluated with computer-assisted semen analysis and fluorescent probes association. Andrologia 44:9–15. https://doi.org/10.1111/j.1439-0272.2010.01096.x

Ostermeier GC, Miller D, Huntriss JD, Diamond MP, Krawetz SA (2004) Delivering spermatozoan RNA to the oocyte. Nature 429:154. https://doi.org/10.1038/429154a

Palma GA, Sinowatz F (2004) Male and female effects on the in vitro production of bovine embryos. Anat Histol Embryol 33:257–262. https://doi.org/10.1111/j.1439-0264.2004.00543.x

Palomo MJ, Izquierdo D, Mogas T, Paramio MT (1999) Effect of semen preparation on IVF of prepubertal goat oocytes. Theriogenology 51:927–940

Parrish JJ (2014) Bovine in vitro fertilization: in vitro oocyte maturation and sperm capacitation with heparin. Theriogenology 81:67–73. https://doi.org/10.1016/j.theriogenology.2013.08.005

Parrish JJ, Krogenaes A, Susko-Parrish J (1995) Effect of bovine sperm separation by either swim-up or percoll method on success of in vitro fertilization and early embryonic development. Thermal Biol 44:859–869

Parrish JJ, Susko-Parrish JL, Leibfried-Rutledge ML, Critser ES, Eyestone WH, First NL (1986) Bovine in vitro fertilization with frozen-thawed semen. Theriogenology 25:591–600. https://doi.org/10.1016/0093-691X(86)90143-3

Parrish JJ, Susko-Parrish J, Winer MA, First NL (1988) Capacitation of bovine sperm by heparin. Biol Reprod 38:1171–1180. https://doi.org/10.1095/biolreprod38.5.1171

Pasquariello R, Pennarossa G, Arcuri S, Fernandez-Fuertes B, Lonergan P, Brevini TAL, Gandolfi F (2024) Sperm fertilizing ability in vitro influences bovine blastocyst miRNA content. Theriogenology 222:1–9. https://doi.org/10.1016/j.theriogenology.2024.03.016

Paulson RJ, Milligan RC, Sokol RZ (2001) The lack of influence of age on male fertility. Am J Obstet Gynecol 184:818–824. https://doi.org/10.1067/mob.2001.113852

Pérez L, Arias ME, Sánchez R, Felmer R, Sapanidou V, Taitzoglou I, Tsakmakidis I, Kourtzelis I, Fletouris D, Theodoridis A, Zervos I, Tsantarliotou M (2015) N-acetyl-L-cysteine pretreatment protects cryopreserved bovine spermatozoa from reactive oxygen species without compromising the in vitro developmental potential of intracytoplasmic sperm injection embryos. Andrologia 84:1273–1282. https://doi.org/10.1111/and.12412

Perez Siqueira AF, De Castro LS, De Assis PM, De Cássia Bicudo L, Mendes CM, Nichi M, Visintin JA, Ortiz D'Ávila Assumpção ME (2018) Sperm traits on in vitro production (IVP) of bovine embryos: too much of anything is good for nothing. PLoS One 13:1–16. https://doi.org/10.1371/journal.pone.0200273

Peris-Frau P, Soler AJ, Iniesta-Cuerda M, Martín-Maestro A, Sánchez-Ajofrín I, Medina-Chávez DA, Fernández-Santos MR, García-álvarez O, Maroto-Morales A, Montoro V, Garde JJ (2020) Sperm cryodamage in ruminants: understanding the molecular changes induced by the cryopreservation process to optimize sperm quality. Int J Mol Sci 21:2781. https://doi.org/10.3390/ijms21082781

Pessot CA, Brito M, Figueroa J, Concha II, Yañez A, Burzio LO (1989) Presence of RNA in the sperm nucleus. Biochem Biophys Res Commun 158:272–278. https://doi.org/10.1016/S0006-291X(89)80208-6

Plourde D, Vigneault C, Lemay A, Breton L, Gagné D, Laflamme I, Blondin P, Robert C (2012) Contribution of oocyte source and culture conditions to phenotypic and transcriptomic variation in commercially produced bovine blastocysts. Theriogenology 78:116–131. https://doi.org/10.1016/j.theriogenology.2012.01.027

Pohjanvirta T, Vähänikkilä N, Mutikainen M, Lindeberg H, Pelkonen S, Peippo J, Autio T (2023) Transmission of mycoplasma bovis infection in bovine in vitro embryo production. Theriogenology 199:43–49. https://doi.org/10.1016/j.theriogenology.2023.01.011

Rahman MB, Schellander K, Luceño NL, Van Soom A (2018) Heat stress responses in spermatozoa: mechanisms and consequences for cattle fertility. Theriogenology 113:102–112. https://doi.org/10.1016/j.theriogenology.2018.02.012

Rahman MB, Vandaele L, Rijsselaere T, El-Deen MS, Maes D, Shamsuddin M, Van Soom A (2014) Bovine spermatozoa react to in vitro heat stress by activating the mitogen-activated protein kinase 14 signalling pathway. Reprod Fertil Dev 26:245–257. https://doi.org/10.1071/RD12198

Rahman MB, Vandaele L, Rijsselaere T, Maes D, Hoogewijs M, Frijters A, Noordman J, Granados A, Dernelle E, Shamsuddin M, Parrish JJ, Van Soom A (2011) Scrotal insulation and its relationship to abnormal morphology, chromatin protamination and nuclear shape of spermatozoa in Holstein-Friesian and Belgian blue bulls. Theriogenology 76:1246–1257. https://doi.org/10.1016/j.theriogenology.2011.05.031

Rajabi H, Mohseni-kouchesfehani H, Eslami-Arshaghi T, Salehi M (2018) Sperm DNA fragmentation affects epigenetic feature in human male pronucleus. Andrologia 50:1–7. https://doi.org/10.1111/and.12800

Ribas BN, Missio D, Junior Roman I, Neto NA, Claro I, dos Santos Brum D, Leivas FG (2018) Superstimulation with eCG prior to ovum pick-up improves follicular development and fertilization rate of cattle oocytes. Anim Reprod Sci 195:284–290. https://doi.org/10.1016/j.anireprosci.2018.06.006

Rizos D, Clemente M, Bermejo-Alvarez P, De La Fuente J, Lonergan P, Gutiérrez-Adán A (2008) Consequences of in vitro culture conditions on embryo development and quality. Reprod Domest Anim 43:44–50. https://doi.org/10.1111/j.1439-0531.2008.01230.x

Rizos D, Ward F, Duffy P, Boland MP, Lonergan P (2002) Consequences of bovine oocyte maturation, fertilization or early embryo development in vitro versus in vivo: implications for blastocyst yield and blastocyst quality. Mol Reprod Dev 61:234–248. https://doi.org/10.1002/mrd.1153

Rockett JC, Mapp FL, Garges JB, Luft JC, Mori C, Dix DJ (2001) Effects of hyperthermia on spermatogenesis, apoptosis, gene expression, and fertility in adult male mice. Biol Reprod 65:229–239. https://doi.org/10.1095/biolreprod65.1.229

Rodgers AB, Morgan CP, Bronson SL, Revello S, Bale TL (2013) Paternal stress exposure alters sperm MicroRNA content and reprograms offspring HPA stress axis regulation. J Neurosci 33:9003–9012. https://doi.org/10.1523/JNEUROSCI.0914-13.2013

Roelen BAJ (2019) Bovine oocyte maturation: acquisition of developmental competence. Reprod Fertil Dev 32:98–103. https://doi.org/10.1071/RD19255

Roth Z (2020) Reproductive physiology and endocrinology responses of cows exposed to environmental heat stress–experiences from the past and lessons for the present. Theriogenology 155:150–156. https://doi.org/10.1016/j.theriogenology.2020.05.040

Roth Z (2021) Heat stress reduces maturation and developmental capacity in bovine oocytes. Reprod Fertil Dev 33:66–75. https://doi.org/10.1071/RD20213

Sabés-Alsina M, Lundeheim N, Johannisson A, López-Béjar M, Morrell JM (2019) Relationships between climate and sperm quality in dairy bull semen: A retrospective analysis. J Dairy Sci 102:5623–5633. https://doi.org/10.3168/jds.2018-15837

Saeki K, Nagao Y, Hoshi M, Nagai M (1995) Effects of heparin, sperm concentration and bull variation on in vitro fertilization of bovine oocytes in protein-free medium. Theriogenology 43:751–759

Salas-Huetos A, Blanco J, Vidal F, Grossmann M, Pons MC, Garrido N, Anton E (2016) Spermatozoa from normozoospermic fertile and infertile individuals convey a distinct miRNA cargo. Andrology 4:1028–1036. https://doi.org/10.1111/andr.12276

Salisbury GW, Hart RG, Lodge JR (1977) The spermatozoan genome and fertility. Am J Obstet Gynecol 128:342–350. https://doi.org/10.1016/0002-9378(77)90635-4

Samardzija M, Karadjole M, Matkovic M, Cergolj M, Getz I, Dobranic T, Tomaskovic A, Petric J, Surina J, Grizelj J, Karadjole T (2006) A comparison of BoviPure® and Percoll® on bull sperm separation protocols for IVF. Anim Reprod Sci 91:237–247. https://doi.org/10.1016/j.anireprosci.2005.04.005

Samper JC, Crabo BG (1993) Assay of capacitated, freeze-damaged and extended stallion spermatozoa by filtration. Theriogenology 39:1209–1220

Santiago J, Silva JV, Howl J, Santos MAS, Fardilha M (2022) All you need to know about sperm RNAs. Hum Reprod Update 28:67–91. https://doi.org/10.1093/humupd/dmab034

Sapanidou V, Taitzoglou I, Tsakmakidis I, Kourtzelis I, Fletouris D, Theodoridis A, Zervos I, Tsantarliotou M (2015) Antioxidant effect of crocin on bovine sperm quality and in vitro fertilization. Theriogenology 84:1273–1282. https://doi.org/10.1016/j.theriogenology.2015.07.005

Sapanidou V, Tsantarliotou MP, Lavrentiadou SN (2023) A review of the use of antioxidants in bovine sperm preparation protocols. Anim Reprod Sci 251:107215. https://doi.org/10.1016/j.anireprosci.2023.107215

Schneider C, Ellington J, Wright R (1999) Relationship between bull field fertility and in vitro embryo production using sperm preparation methods with and without somatic cell co-culture. Theriogenology 51:1085–1098. https://doi.org/10.1016/S0093-691X(99)80013-2

Seifi-Jamadi A, Zhandi M, Kohram H, Llamas N, Leemans B, Henrotte E, Latour C, Demeyere K, Meyer E, Van Soom A (2020) Influence of seasonal differences on semen quality and subsequent embryo development of Belgian blue bulls. Theriogenology 158:8–17. https://doi.org/10.1016/j.theriogenology.2020.08.037

Selvaraju S, Parthipan S, Somashekar L, Kolte AP, Krishnan Binsila B, Arangasamy A, Ravindra JP (2017) Occurrence and functional significance of the transcriptome in bovine (Bos taurus) spermatozoa. Sci Rep 7:1–14. https://doi.org/10.1038/srep42392

Silvestri G, Canedo-Ribeiro C, Serrano-Albal M, Labrecque R, Blondin P, Larmer SG, Marras G, Tutt DAR, Handyside AH, Farré M, Sinclair KD, Griffin DK (2021) Preimplantation genetic testing for aneuploidy improves live birth rates with in vitro produced bovine embryos: A blind retrospective study. Cells 10:1–16. https://doi.org/10.3390/cells10092284

Sirard M-A, Blondin P, Robert C (2006) Contribution of the oocyte to embryo quality. Theriogenology 65:126–136. https://doi.org/10.1016/j.theriogenology.2005.09.020

Somfai T, Ogata K, Takeda K, H. Y. (2022) Bulk vitrification of in vitro produced bovine zygotes without reducing developmental competence to the blastocyst stage. Cryobiology 106:32–38

Sonnack V, Failing K, Bergmann M, Steger K (2002) Expression of hyperacetylated histone H4 during normal and impaired human spermatogenesis. Andrologia 34:384–390. https://doi.org/10.1046/j.1439-0272.2002.00524.x

Speckhart SL, Wooldridge LK, Ealy AD (2023) An updated protocol for in vitro bovine embryo production. STAR Protocols 4:101924. https://doi.org/10.1016/j.xpro.2022.101924

Sprícigo JFW, Morais K, Ferreira AR, Machado GM, Gomes ACM, Rumpf R, Franco MM, Dode MAN (2014) Vitrification of bovine oocytes at different meiotic stages using the Cryotop method: assessment of morphological, molecular and functional patterns. Cryobiology 69:256–265. https://doi.org/10.1016/j.cryobiol.2014.07.015

Stringer KA, Tobias M, Neill HCO, Franklin CC, Physiol AJ, Cell L, Manual U et al (2006) A low temperature, ultrahigh vacuum, microwave-frequency-compatible scanning tunneling microscope. J Androl 27:205–211. https://doi.org/10.1016/j.theriogenology.2008.12.006

Stringfellow D, Seidel SM (2009) Manual of international embryo transfer society (IETS), 4th edn. IETS, Maryland Heights, MO

Thundathil J, Palasz AT, Mapletoft RJ, Barth AD (1999) An investigation of the fertilizing characteristics of pyriform-shaped bovine spermatozoa. Anim Reprod Sci 57:35–50. https://doi.org/10.1016/S0378-4320(99)00058-5

Travnickova I, Hulinska P, Kubickova S, Hanzalova K, Kempisty B, Nemcova L, Machatkova M (2021) Production of sexed bovine embryos in vitro can be improved by selection of sperm treatment and co-culture system. Reprod Domest Anim 56:864–871. https://doi.org/10.1111/rda.13926

Trigal B, Gómez E, Caamaño JN, Muñoz M, Moreno J, Carrocera S, Martín D, Diez C (2012) In vitro and in vivo quality of bovine embryos in vitro produced with sex-sorted sperm. Theriogenology 78:1465–1475. https://doi.org/10.1016/j.theriogenology.2012.06.018

Tutt DAR, Silvestri G, Serrano-Albal M, Simmons RJ, Kwong WY, Guven-Ates G, Canedo-Ribeiro C, Labrecque R, Blondin P, Handyside AH, Griffin DK, Sinclair KD (2021) Analysis of bovine blastocysts indicates ovarian stimulation does not induce chromosome errors, nor discordance between inner-cell mass and trophectoderm lineages. Theriogenology 161:108–119. https://doi.org/10.1016/j.theriogenology.2020.11.021

Ugur MR, Kutchy NA, Menezes EB, Ul-Husna A, Haynes BP, Uzun A, Kaya A, Topper E, Moura AA, Memili E (2019) Retained acetylated histone four in bull sperm associated with fertility. Front Vet Sci 6:1–10. https://doi.org/10.3389/fvets.2019.00223

Urena I, Gonzalez C, Ramon M, Godia M, Clop A, Calvo JH, Carabano MJ, Serrano M (2022) Exploring the ovine sperm transcriptome by RNAseq techniques. I effect of seasonal conditions on transcripts abundance. PLoS One 17:1–27. https://doi.org/10.1371/journal.pone.0264978

Uzbekov R, Singina GN, Shedova EN, Banliat C, Avidor-Reiss T, Uzbekova S (2023) Centrosome formation in the bovine early embryo, vol 12

Vandaele L, Mateusen B, Maes D, de Kruif A, Van Soom A (2006) Is apoptosis in bovine in vitro produced embryos related to early developmental kinetics and in vivo bull fertility? Theriogenology 65:1691–1703. https://doi.org/10.1016/j.theriogenology.2005.09.014

Verma R, Vijayalakshmy K, Chaudhiry V (2018) Detrimental impacts of heavy metals on animal reproduction: A review. ~ 27 ~. J Entomol Zool Stud 6:27–30

Viana JHM (2022) 2021 Statistics of embryo production and transfer in domestic farm animals. Embryo Technology Newsletter 40:22–40

Viuff D, Greve T, Avery B, Hyttel P, Brockhoff PB, Thomsen PD (2000) Chromosome aberrations in in vitro-produced bovine embryos at days 2-5 post-insemination. Biol Reprod 63:1143–1148. https://doi.org/10.1095/biolreprod63.4.1143

Viuff D, Palsgaard A, Rickords L, Lawson LG, Greve T, Schmidt M, Avery B, Hyttel P, Thomsen PD (2002) Bovine embryos contain a higher proportion of polyploid cells in the trophectoderm than in the embryonic disc. Mol Reprod Dev 62:483–488. https://doi.org/10.1002/mrd.90004

Viuff D, Rickords L, Offenberg H, Hyttel P, Avery B, Greve T, Olsaker I, Williams JL, Callesen H, Thomsen PD (1999) A high proportion of bovine blastocysts produced in vitro are mixoploid. Biol Reprod 60:1273–1278. https://doi.org/10.1095/biolreprod60.6.1273

Walters AH, Eyestone WE, Saacke RG, Pearson RE, Gwazdauskas FC (2004) Sperm morphology and preparation method affect bovine embryonic development. J Androl 25:554–563. https://doi.org/10.1002/j.1939-4640.2004.tb02826.x

Wang M, Du Y, Gao S, Wang Z, Qu P, Gao Y, Wang J, Liu Z, Zhang J, Zhang Y, Qing S, Wang Y (2021) Sperm-borne miR-202 targets SEPT7 and regulates first cleavage of bovine embryos via cytoskeletal remodeling. Development (Cambridge) 148. https://doi.org/10.1242/dev.189670

Ward F, Dimitrios R, Doreen C, Quinn KBM, Lonergan P (2001) Paternal influence on the time of First embryonic cleavage post insemination and the implications for subsequent bovine embryo development in vitro and fertility in vivo. Mol Reprod Dev 60:47–55. https://doi.org/10.1046/j.1439-0531.2003.00437.x

Ward F, Rizos D, Boland MP, Lonergan P (2003) Effect of reducing sperm concentration during IVF on the ability to distinguish between bulls of high and low field fertility: work in progress. Theriogenology 59:1575–1584. https://doi.org/10.1016/S0093-691X(02)01202-5

Witt KD, Beresford L, Bhattacharya S, Brian K, Coomarasamy A, Hooper R, Kirkman-Brown J, Khalaf Y, Lewis SE, Pacey A, Pavitt S, West R, Miller D (2016) Hyaluronic acid binding sperm selection for assisted reproduction treatment (HABSelect): study protocol for a multicentre randomised controlled trial. BMJ Open 6:1–10. https://doi.org/10.1136/bmjopen-2016-012609

Wrathall AE, Simmons HA, Van Soom A (2006) Evaluation of risks of viral transmission to recipients of bovine embryos arising from fertilisation with virus-infected semen. Theriogenology 65:247–274. https://doi.org/10.1016/j.theriogenology.2005.05.043

Wrzecińska M, Kowalczyk A, Cwynar P, Czerniawska-Piątkowska E (2021) Disorders of the reproductive health of cattle as a response to exposure to toxic metals. Biology 10:1–16. https://doi.org/10.3390/biology10090882

Wu C, Blondin P, Vigneault C, Labrecque R, Sirard MA (2020) Sperm miRNAs— potential mediators of bull age and early embryo development. BMC Genomics 21:798. https://doi.org/10.1186/s12864-020-07206-5

Wu ATH, Sutovsky P, Manandhar G, Xu W, Katayama M, Day BN, Park KW, Yi YJ, Yan WX, Prather RS, Oko R (2007) PAWP, a sperm-specific WW domain-binding protein, promotes meiotic resumption and pronuclear development during fertilization. J Biol Chem 282:21164–21175. https://doi.org/10.1074/jbc.M609132200

Xia J, Ren D (2009) The BSA-induced ca(2+) influx during sperm capacitation is CATSPER channel-dependent. Reprod Biol Endocrinol 7:1–9. https://doi.org/10.1186/1477-7827-7-119

Xu H, Wang X, Wang Z, Li J, Xu Z, Miao M, Chen G, Lei X, Wu J, Shi H, Wang K, Zhang T, Sun X (2020) MicroRNA expression profile analysis in sperm reveals hsa-mir-191 as an auspicious omen of in vitro fertilization. BMC Genomics 21:165. https://doi.org/10.1186/s12864-020-6570-8

Yang X, Giles J, Liu Z, Jiang S, Presicce G, Foote R, W. C. (1995) In vitro fertilization and embryo development are influenced by IVF technicians and bulls. Theriogenology 360:360

Yang C, Zeng QX, Liu JC, Yeung WSB, Zhang JV, Duan YG (2023) Role of small RNAs harbored by sperm in embryonic development and offspring phenotype. Andrology 11:770–782

Ying Z, Labrecque R, Tremblay P, Plessis C, Dufour P, Hélène Martin MAS (2024) Sperm-borne tsRNAs and miRNAs analysis in relation to dairy cattle fertility. Theriogenology 215:241–248

Yousefian I, Zare-Shahneh A, Goodarzi A, Baghshahi H, Fouladi-Nashta AA (2021) The effect of tempo and MitoTEMPO on oocyte maturation and subsequent embryo development in bovine model. Theriogenology 176:128–136. https://doi.org/10.1016/j.theriogenology.2021.09.016

Yuan S, Schuster A, Tang C, Yu T, Ortogero N, Bao J, Zheng H, Yan W (2016) Sperm-borne miRNAs and endo-siRNAs are important for fertilization and preimplantation embryonic development. Development (Cambridge) 143:635. https://doi.org/10.1242/dev.131755

Zhang BR, Larsson B, Lundeheim N, Rodriguez-Martinez H (1998) Sperm characteristics and zona pellucida binding in relation to field fertility of frozen-thawed semen from dairy AI bulls. Int J Androl 21:207–216. https://doi.org/10.1046/j.1365-2605.1998.00114.x

Zhang Ying LR, Plessis Clément DPTP, Martin Hélène SMA (2024) Sperm-borne tsRNAs and miRNAs analysis in relation to dairy cattle fertility. Theriogenology 215:241–248. https://doi.org/10.1016/J.THERIOGENOLOGY.2023.11.029

Sperm Mitochondria: Quantitative Regulation and Its Impact on Sperm Quality

Hiroaki Funahashi [ID], Hai Thanh Nguyen [ID], and Takuya Wakai [ID]

Abstract

Although spermatozoa produce the energy necessary for their proper functions through glycolysis and oxidative phosphorylation to maintain the energy, the sperm mitochondria are thought to be active in producing energy to deliver the ejected paternal gametes in the female reproductive tract to the site of fertilization and then into the maternal gamete, oocyte, since they are degraded by the oocyte after its spermatozoon penetrated. Mitochondrial content in spermatozoa is varied among males, due to various factors through the process of spermatogenesis, consequently making to affect sperm penetrability. Therefore, the content may be a potential sperm biomarker for the quality. Since mitochondrial DNA copy number/mitochondria content in spermatozoa is determined mainly through spermatogenesis, especially from the secondary spermatocyte to fully differentiated spermatozoa, some abnormalities and oxidative stress during spermatogenesis may be the main factors associated with the quantitative change of mitochondrial contents in spermatozoa, affecting their reproductive efficiency. Here, insight into the mechanisms of how sperm mitochondrial contents interact with these factors to make the quantitative alterations will be reviewed.

Keywords

Spermatozoa · Mitochondria · Mitochondrial DNA · Motility

H. Funahashi (✉) · T. Wakai
Faculty of Environmental, Life, Natural Science and Technology, Okayama University, Okayama, Japan
e-mail: hirofun@okayama-u.ac.jp; t2wakai@okayama-u.ac.jp

H. T. Nguyen
Faculty of Animal Science and Veterinary Medicine, Nong Lam University, Ho Chi Minh City, Vietnam
e-mail: hai.nguyenthanh@hcmuaf.edu.vn

© The Author(s), under exclusive license to Springer Nature Switzerland AG 2024
J. C. Gardón, K. Satué Ambrojo (eds.), *Assisted Reproductive Technologies in Animals Volume 1*, https://doi.org/10.1007/978-3-031-73079-5_12

349

1   Introduction

Spermatozoa are extremely specialized cells in that they have haploid amounts of paternal genes, are smaller than other somatic cells, and are motile through the active movement of their elongated flagella. During differentiation and meiotic division from spermatogonial stem cells to round spermatids in the seminiferous tubules of the testis, a large number of mitochondria are lost with the cytoplasm, in which major antioxidants exist. Furthermore, almost all of the cytoplasm is removed from spermatozoa during spermiation from round spermatids, consequently remaining a small number of mitochondria at the midpiece region. Finally, spermatozoa are formed from three structure parts: head, midpiece, and tail. The sperm head contains the nucleus, which has haploid amounts of paternal DNA, and the acrosome, which includes various enzymes necessary for penetration into the oocyte and for activation. The midpiece region, in which mitochondrial sheaths coil around a very long flagellum and a characteristic structure is also found in the villi, is connected to the head at the neck, which contains the proximal centriole, which plays an important role in organizing the spindle for cleavage and cell division in embryos. Spermatozoa lose the acrosome at binding to the zona pellucida, and all components without the paternal genome and centrosome are destroyed in the zygote after penetration; the last two remaining components will contribute significantly to embryonic development.

Although reducing NADH and some ATP are provided by the cytosolic glycolytic system, spermatozoa produce energy, ATP, at mitochondria not only for viability and motility but also for various functions during penetration, such as capacitation and acrosome reaction (Boguenet et al. 2021; Kumar 2023). In fact, spermatozoa consume oxygen the most during capacitation and acrosome reactions (Ramio-Lluch et al. 2011). Since mitochondrial activities in spermatozoa, reflected as mitochondrial membrane potential (MMP), are positively correlated with the motility, capacitation, or acrosome reaction and consequently the penetrability (Agnihotri et al. 2016; Athurupana et al. 2015; Park and Gye 2024; Zhang et al. 2019), the quality and status of mitochondria should be very important for sperm functions. Recently, mitochondrial DNA (mtDNA) copy number has been recognized to be significantly correlated with semen quality and fertility of spermatozoa (Nguyen et al. 2023; Pi et al. 2024; Popova et al. 2022; Sun et al. 2022). Any problems in regulating mtDNA copy number and the transcription levels contribute to dysfunction in oxidative phosphorylation, reduced mitochondrial activity, and consequently poor seminal quality with abnormal sperm parameters (Guo et al. 2017; Nakada et al. 2006). Mitochondrial DNA in spermatozoa may be affected by damage factors as compared with the nuclear genome due to no protective DNA-binding proteins and no DNA repair capacity, as well as its close location to the inner mitochondrial membrane which produces endogenous ROS (Boguenet et al. 2021).

A review of the mechanisms regulating sperm mtDNA copy number/mitochondria content and their relationships with the potential affecting factors may improve the understanding of the problems associated with paternal gamete affecting reproductive efficiency.

2   Sperm Motility

Spermatozoa obtain the motility for fertilization during the maturation process through the epididymis. In general, sperm motility is divided into two patterns, one that ejaculated cells have for normal forward movement and the other is the hyper-activated, jumpy motility that capacitated sperm exhibit near the site of fertilization. The flagellar segments, which make sperm motility, have a common cytoskeletal structure consisting of nine outer doublet microtubules, each with two dynein arms, surrounding and separated from two central singlet microtubules by radial spokes. Flagellar movement is induced by the ATP-dependent, periodic contact of the tip of the dynein arm with the adjacent microtubule and movement over it. Therefore, in both sperm motility patterns, ATP consumed by the dynein-ATPase in the sperm flagella must be supplied as an energy source.

On the motility of spermatozoa, not only anaerobic glycolysis in the cytoplasm (Mukai and Okuno 2004; Takei et al. 2014) but also oxidative phosphorylation in mitochondria (Marchetti et al. 2002) has been known to be importantly contributing to produce energy, although it is possible that the energy site utilized by spermatozoa is species-specific. Spermatozoa in mice, pigs, and humans appear to use the glycolytic system as their main source of energy (Miki et al. 2004), while oxidative phosphorylation is sourced mainly by horses (Moraes and Meyers 2018), and both are used by cattle (Losano et al. 2017). Despite the extremely short length of sperm tails in pigs and humans compared to rodents, including mice, it is surprising that sperm from all of these animals use the glycolytic system primarily in the cytoplasm to generate active motility. Since the completion of glycolysis by transition from pyruvate to lactate by the action of lactic dehydrogenase or alternatively transporting pyruvate into the mitochondria matrix through a specific pyruvate transporter for oxidative phosphorylation could be a critical step for glycolysis and oxidative phosphorylation, respectively (Boguenet et al. 2021), the balance between pyruvate and lactate may be species-specific different in spermatozoa in these species and consequently may make the difference in energy sources. In addition, since glucose concentrations in the oviduct are kept very low and glucose inhibits the acquisition of fertilization potential in bull spermatozoa (Piomboni et al. 2012), the oxidative phosphorylation system in mitochondria utilizing various substrates, as well as the glycolytic system, may also be utilized as a response to the high glucose sensitivity of the bovine gametes.

Additionally, L-arginine increases the generation of nitric oxide (NO), a short-lived free radical, which seems to reduce lipid peroxidation of spermatozoa by inactivating superoxide, although a low level of NO is known to induce sperm capacitation, through the enhancement of tyrosine phosphorylation in sperm proteins. Since phosphodiesterase-5, which specifically degrades cGMP produced by NO-mediated activation of soluble guanylyl cyclase, is expressed in spermatozoa, and the use of its specific inhibitor, sildenafil, significantly induces capacitation (Ioki et al. 2016), this NO-mediated reaction pathway appears to be largely responsible for changes in sperm motility. Furthermore, L-carnitine detected in the epididymis, seminal plasma, and spermatozoa transport acetyl and acyl groups, which are essential for mitochondrial metabolism, across the mitochondrial inner membrane, likely accelerating the metabolism of long-chain fatty acids in mitochondria (De Luca et al. 2021). L-carnitine indirectly also seems to improve the motility of spermatozoa by participating in the energetic metabolism.

3   Mitochondria in the Midpiece Region of Spermatozoa

As reviewed well by Boguenet et al. (Boguenet et al. 2021), mammalian spermatozoa have 22–80 mitochondria, each of which contains ~11 copies of maternally derived mitochondrial DNA (mtDNA) (Darr et al. 2017), forming 10 to 12 mitochondrial coils in the mid-piece (Gu et al. 2019). Human mtDNA is a double-stranded circular structure with 16,569 base-pairs (Anderson et al. 1981), which code for a part (13 subunits) of the electron transport chain complexes I, III, IV, and V (ATP synthase), except complex II encoded by the nuclear DNA, as well as 22 tRNAs and the rRNA subunits (Amaral et al. 2013; St John et al. 2000). Mitochondria DNA has few non-coding bases, except for the displacement loop (D loop) (Anderson et al. 1981). All the other proteins in mitochondria are encoded by nuclear genes (Amaral et al. 2013). Sperm mitochondria provide reducing equivalents, such as NADH and $FADH_2$, through different reactions at shuttles, such as malate/aspartate (including the aspartate/glutamate carrier and the malate/2-oxoglutarate carrier), glycerol-3-phosphate and lactate/pyruvate shuttles, beta-oxidation, and Krebs cycle to produce ATP at the electron transport chain, including ATP synthase, on the inner mitochondrial membrane (Boguenet et al. 2021).

Whereas functions of the mitochondria also include the production of reactive oxygen species (ROS) during hyperpolarization of MMP, apoptosis following the failures due to impaired oxidative phosphorylation or excess ROS production, and calcium homeostasis (Meyers et al. 2019; Ramalho-Santos and Amaral 2013), ROS may be essential to transduce signals and maintain homeostasis (Zhang et al. 2021), associated with sperm motility, capacitation/acrosome reaction, and penetrability (Amaral et al. 2013; Gibb et al. 2014). Therefore, suitable ROS levels should be critical for the intact functions of spermatozoa. In spermatozoa, mitochondria, especially the electron transport chain complexes I and III (Koppers et al. 2008), are the main place where ROS is converted from 0.2 to 2% of the intracellular oxygen (Piomboni et al. 2012). Polyunsaturated fatty acids, which are contained in large quantities in the sperm plasma membrane, undergo lipid peroxidation by ROS, and this reduces the integrity, consequently increasing the fluidity of the membrane (Sanocka and Kurpisz 2004). Excess ROS also damages the nuclear DNA with double- or single-strand breaks and alterations. Since sperm mtDNA copy number is quite lower than somatic cells (10^2–10^4 copies) and since spermatozoa had lost most of the cytoplasm where the major place existing antioxidants, a mutation in sperm mtDNA due to oxidative stress induced by a large amount of ROS appears to easily result in impaired motility and consequently reduced penetrability (Piomboni et al. 2012). In general, spermatozoa can be protected from overabundant ROS by various scavengers, such as glutathione peroxidase and peroxiredoxin, at epididymal epithelium during maturation (Boguenet et al. 2021). After ejaculation, the seminal plasma contains an antioxidant system including various factors, such as superoxide dismutase (which catalyzes the superoxide anion dismutation reaction, protects polyunsaturated fatty acids from oxidation and DNA from fragmentation), catalase (which catalyzes the transformation of hydrogen peroxide into oxygen and water), and glutathione peroxidase (which contains three isoforms, including mitochondrial one associated with sperm motility and catalyzes the reduction of hydrogen peroxide and organic peroxides) to maintain an optimal condition for sperm mitochondria against ROS (De Luca et al. 2021).

Furthermore, since blocking the mitochondrial calcium transporter, which is a main channel for mediating the absorption of calcium ion through the inner membrane (Patergnani et al. 2011), induces reduction in ATP levels in spermatozoa (Bravo et al. 2015), sperm mitochondria regulate calcium homeostasis and consequently affecting intracellular Ca^{2+} stores (Costello et al. 2009), whereas the excessive mitochondrial calcium concentration is likely to open the permeability transition pores, to release cytochrome c, and to consequently stimulate the intrinsic apoptotic pathway in spermatozoa (Boguenet et al. 2021). Also, since Ca^{2+} signaling in spermatozoa plays critically important roles in regulating processes of capacitation, hyperactivation, and acrosome reaction (Harayama 2013; Nowicka-Bauer and Szymczak-Cendlak 2021; Suarez 2008) and, in fact, Ca^{2+} seems to induce hyperactivation by working directly to the flagellar axoneme (Ho and Suarez 2001), mitochondria could contribute indirectly to these important functional processes of spermatozoa.

Mitochondria are highly dynamic organelles making constant fusion and fission (Al Ojaimi et al. 2022). During the process of biogenesis, mitochondria make fusion together by the action of mainly mitofusion-1/2 (MFN1/2; MFN1 is a GTPase existing on the outer membrane for mitochondrial fusion (Alghamdi 2024), although MFN2 is ~80% identical to MFN1 in sequence, has a proline-rich site accounting for specific protein-protein interactions (Filadi et al. 2018)) and optic atrophy 1 (OPA1; the main regulator of mitochondrial inner membrane fusion (Belenguer and Pellegrini 2013)) expressed on the surface (Fig. 1). Both these GTPases, MFN1/MFN2, which primarily regulate mitofusion, are upregulated during spermatogonial differentiation (Zhang et al. 2022). Deletion of Mfn1 and Mfn2 from mouse pro-spermatogonia causes mitochondrial dysfunction in spermatogonia and spermatocytes (Chen et al. 2020; Zhang et al. 2022). Since MFN2 has been detected in human spermatozoa and the expression level was higher in normal than asthenospermic patients (Fang et al. 2018), this fusion mechanism seems to be active in the dynamics of spermatozoa, whereas levels of mitofusion and fission may be kept minimal to make sure normal sperm mitochondrial function (Zhang et al. 2022).

On the other hand, mitochondria with low activity or damaged by oxidative stress, such as ROS, make fission by the action of a cytosolic GTPase, dynamin-1-like protein (DNM1L), and the receptors on the mitochondrial outer membrane, such as mitochondrial fission factor (Mff) (Rios et al. 2023). The large cytosolic GTPase, DNM1L, is also expressed in human spermatozoa and could be the target of posttranslational modification processes (sumoylation) to control and regulate

BIOGENESIS

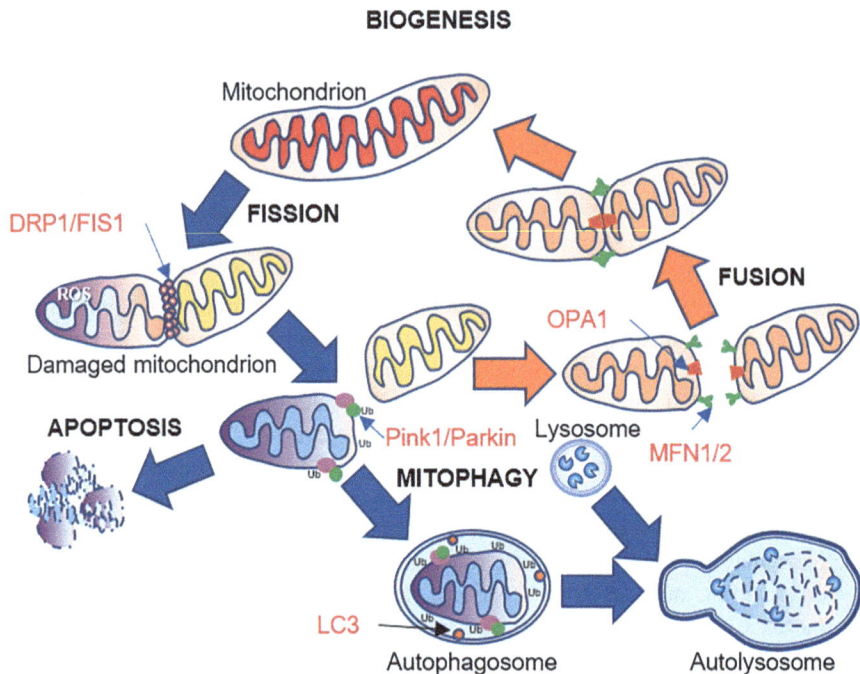

Fig. 1 Schematic dynamics of mitochondrial fusion/fission and the biogenesis. During the biogenesis process, active mitochondria fuse together by the action of MFN1/2 and OPA1 expressed on the surface. On the other hand, mitochondria with low activity or damaged by oxidative stress, such as ROS, make fission by the action of DRP1/FIS1. Damaged mitochondria are detached from the working parts of Pink1/Parkin. Damaged mitochondria are tagged by ubiquitin on the surface by the action of Pink1/Parkin, taken up within autophagosomes, and degraded within lysosome-bound autolysosomes (mitophagy)

mitochondrial dynamics, occurring in mature spermatozoa (Marchiani et al. 2011; Marchiani et al. 2014). These damaged mitochondria are detached from the working parts of Pink1/Parkin (Fig. 1). The detached mitochondrial membranes and the matrix protein, such as mitochondrial transcriptional factor A (TFAM) and prohibitin, are tagged by ubiquitin by the action of Pink1/Parkin (Antelman et al. 2008; Song et al. 2016; Thompson et al. 2003), taken up within autophagosomes and degraded within lysosome-bound autolysosomes (mitophagy). Ubiquitination of mitochondria has been detected during spermatogenesis from the secondary spermatocyte to fully differentiated spermatozoa (Sutovsky et al. 2000). Also, specific endonucleases seem to be active in removing damaged mitochondria during spermatogenesis (Nishimura et al. 2006).

The mtDNA transcription and replication during mitochondrial biogenesis appear to be regulated by some factors, including TFAM (Amaral et al. 2013; Faja et al. 2019; Guo et al. 2017). TFAM is a transcriptional element responsible for the formation of primers to initiate mtDNA replication (Amaral et al. 2007). The transcription/replication deeply involves TFAM, of which transcription and translation are promoted by combining the PGC-1α with the target, nuclear respiratory factor 1/2 (NRF1/2; transcription factors of nuclear genes coding for mitochondrial import proteins and also factors responsible for mtDNA translation), consequently resulting in mitochondrial biogenesis and functions by the role of POLG which is the mitochondrial polymerase essential for mtDNA replication (Fig. 2).

During spermatogenesis, the TFAM expression is physiologically downregulated, probably associated with the control of mtDNA in spermatozoa (Rantanen et al. 2001). Interestingly, it was recently found that TFAM redistributed phosphorylation dependently from the mitochondria to the nucleus of sperm during spermatogenesis, resulting in devoid of intact mtDNA and discontinuation of the replication (Fig. 3) (Lee et al. 2023).

4   Mitochondria during Spermatogenesis

During spermatogenesis from spermatogonial stem cells, mitochondrial morphology, as well as the size, drastically changes from conventional type at spermatogonia and early spermatocyte stages to condensed type at late spermatocyte and early spermatid stages, through intermediate type at zygotene and pachytene spermatocyte stages (Fig. 3) (Zhang et al. 2022). Differentiation of spermatogonial stem cells to spermatogonia appears to require mitochondrial fusion (Varuzhanyan et al. 2019) and a metabolic shift from glycolysis to oxidative phosphorylation in mitochondria, because of higher expression of glycolytic enzymes and hypoxia-responsive factors in spermatogonial stem cells and upregulation of factors associated with mitochondrial biogenesis and oxidative phosphorylation in differentiating progenitors (Zhang et al. 2022). The requirement of lactate and pyruvate for the survival of spermatocytes (Courtens and Ploen 1999) and spermatids and higher expression of the pyruvate carrier MPC1 in spermatocyte mitochondria suggest a dependence of spermatocytes on oxidative phosphorylation in mitochondria (Varuzhanyan et al. 2019). Between the round spermatid and mature sperm stages (during spermiation with a condensed nucleus, acrosome, and tail formations), mammalian spermatozoa drastically reduced mitochondrial contents (Boguenet et al. 2021; Gu et al. 2019) by discarding their cytoplasm, which decreases approximately one-fifth to sixth in humans (Diez-Sanchez et al. 2003) and one-eighth to tenth in mice (Vertika et al. 2020). During this time, the morphology of mitochondria changes again to intermediate type, finally forming the midpiece region by coiling along the flagellum (Zhang et al. 2022). Also, autophagy is vitally included in the remodeling of

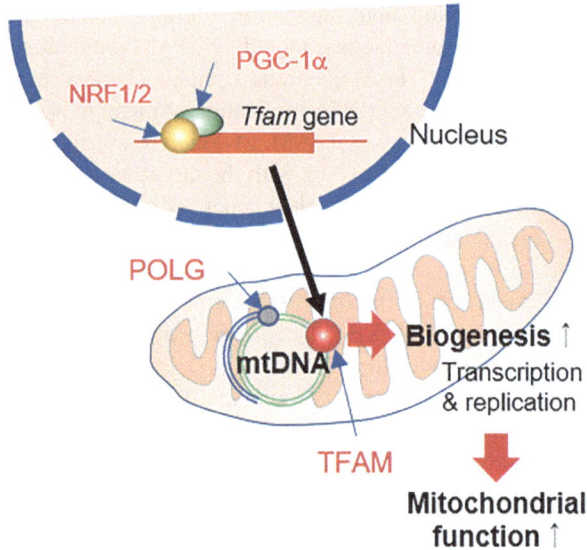

Fig. 2 Transcription and replication of mtDNA are deeply involved in TFAM, of which transcription and translation are promoted by the PGC-1α/NRF1/2 complex, consequently resulting in enhanced mitochondrial biogenesis and functions

spermatids to spermatozoa (Varuzhanyan and Chan 2020), especially in spermatid polarization, acrosome biogenesis, and cytoplasmic removal during spermatogenesis (Shang et al. 2016); due to a core autophagy gene, Atg7 (Komatsu et al. 2005) play critical roles in those. Furthermore, during the formation of the acrosome, a key feature during spermiation, mitochondria seem to provide membranes to the acrosome because mitochondrial cardiolipin exists in the acrosome (Ren et al. 2019).

Molecular mechanisms on the midpiece formation during spermatogenesis (Miyata et al. 2021; Shimada et al. 2021; Varuzhanyan et al. 2021; Zhang et al. 2012) and cytoplasmic deletion from elongating spermatids during spermiation (Shimada and Ikawa 2023; Shimada et al. 2023) have recently been clarified by CRISPR/Cas9 gene editing researches (Fig. 4). Kinesin light chain 3 (KLC3) is a kinesin light chain expressed in post-meiotic male germ cells, causes mitochondrial aggregation in a microtubule-dependent manner, and forms the mitochondrial sheath structure in the midpiece during spermiogenesis (Zhang et al. 2012). Armadillo repeat containing 12 (ARMC12), which is a mitochondrial peripheral membrane protein and works as an adherence factor, is associated with elongating properly at the interlocking step and causes normal coiling along the flagellum (Shimada et al. 2021). A calcineurin-binding protein SPATA33 binds sperm calcineurin, which is a calcium-dependent phosphatase, to the mitochondria and regulates midpiece flexibility and motility (Miyata et al. 2021). Mitochondrial fission 1 protein (FIS1) controls mitochondrial morphology and promotes spermatid maturation (Varuzhanyan et al. 2021). A testis-specific protein, coiled-coil domain containing 183 (CCDC183), expressing in round spermatids, has also been demonstrated to

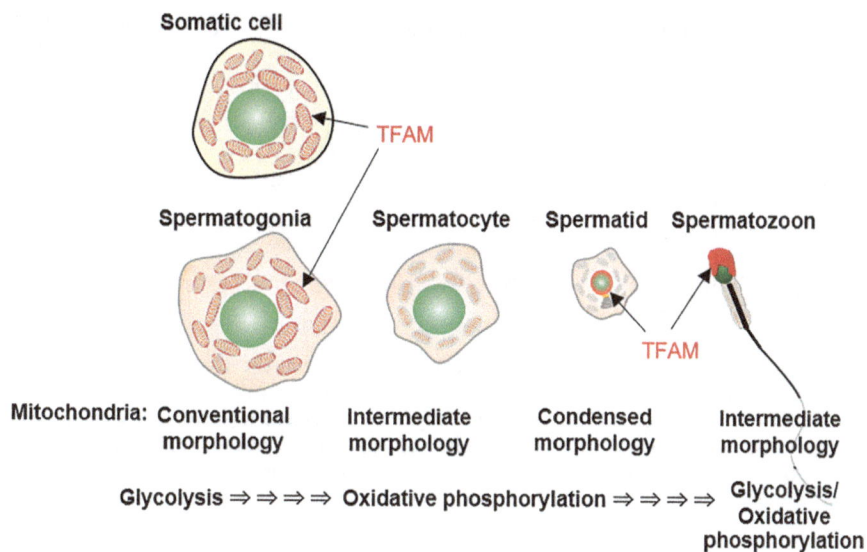

Fig. 3 TFAM localizes exclusively to the mitochondria in somatic cells and spermatogonia, whereas the transcriptional factor is speculated to move to the nucleus of spermatocyte (Lee et al. 2023). In contrast, TFAM appears to redistribute to the nucleus of spermatid and spermatozoon, but not in the mitochondria at the midpiece region of spermatozoa. During spermatogenesis, the morphology of mitochondria and energy source (glycolysis/oxidative phosphorylation) also change drastically

associate with the connection of centrioles with the cell surface and nucleus after elongation of axonemal microtubules during spermiogenesis, enabling a proper flagellar compartment (Shimada and Ikawa 2023). Furthermore, testis-specific serine kinase substrate (TSKS) and TSKS-derived nuage (TDN) play an important role in spermiation by eliminating cytoplasmic contents from the spermatid cytoplasm (Shimada et al. 2023). The identification of factors affecting the expression levels of these proteins and consequently how they affect mitochondrial content in spermatozoa is still unknown, but it is expected that they will be elucidated in the future.

In spermatozoa, energy production goes back to glycolysis in many species, including mice, pigs, and humans (Miki et al. 2004), but except horses which utilize oxidative phosphorylation (Moraes and Meyers 2018) and cattle which use both (Losano et al. 2017). In the former species, spermatozoa seem to survive by using glycolytic energy and utilizing oxidative phosphorylation for motility and penetration into oocytes (Zhang et al. 2022). Germ cell-specific glycolytic enzymes, such as spermatogenic cell-specific type 1 hexokinase (Nakamura et al. 2008), phosphoglycerate kinase 2 (Danshina et al. 2010), and glycerol-3-phosphate acyltransferase 2 (Garcia-Fabiani et al. 2017), seem to support metabolic requirement during spermatogenesis (Zhang et al. 2022).

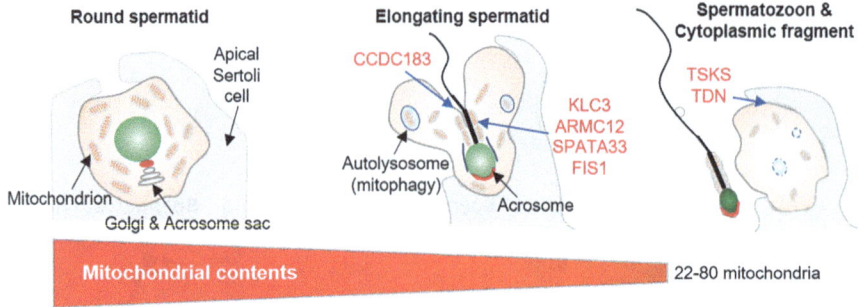

Fig. 4 Schematic mitochondrial dynamics during spermatogenesis. During spermatogenesis, especially during the elongation of spermatid, as the flagellum and acrosome are formed, mature mitochondria are relocated around the midpiece region of the flagellum (sperm tail), and most part of the cytoplasm is separated from the spermatozoa. During these processes, mitochondrial contents relocated in the sperm midpiece are reduced up to 22–80. During spermiation, the majority of the cytoplasm is detached from the sperm by the retention of its association with the apical Sertoli cell

5 Quantitative Regulation of Mitochondrial DNA Copy Number in Paternal Gametes during Spermatogenesis and Fertilization

Since mtDNA contains only exons (no introns), exhibits a homoplasmy state, and is not packaged as nucleosomes (lacking protective histones), as well as existing close to the site of ROS production through the mitochondrial electron transport chain (Boguenet et al. 2021), these appear to be easily having mutations and deletion due to oxidative damage than the nuclear ones. Although some DNA repair enzymes have been found in mitochondria, further information may be required to make clear the repair system of the mtDNA. The mutation rate of mtDNA in spermatozoa is estimated about 10 to 100 times higher than that of nuclear one (Richter et al. 1988, 1998), probably due to no transmission through fertilization (Durairajanayagam et al. 2021).

Mitochondrial DNA copy number may be elevated by any abnormalities in the physiological downregulation of TFAM during spermatogenesis (Rantanen et al. 2001). Since it has been recently found that TFAM is redistributed from the mitochondria to the nucleus during spermatogenesis, resulting in devoid of intact mtDNA and discontinuation of the replication (Lee et al. 2023), abnormalities in TFAM relocalization during spermatogenesis may also affect the mtDNA copy number. Furthermore, a negative correlation of mtDNA copy number has been observed, not only with TFAM (Guo et al. 2017) but also with POLG expression levels (Amaral et al. 2007). Spermatozoa with lower mtDNA copy number also contain higher levels of cytochrome c oxidase subunits 1 and 4 (COX-1 and COX-4) and PGC-1α (Guo et al. 2017). Since COX-1 expression level is negatively correlated with mtDNA copy number, but positively with both TFAM and PGC-1α (Amaral et al. 2007), therefore, some transcription and replication factors (such as COX-1, COX-4, PGC-1α, POLG), as well as TFAM, may contribute to the control of normal mtDNA copy number/mitochondria content during spermatogenesis. However, the mtDNA copy number in spermatozoa is relatively stable throughout their life (Darr et al. 2017; Nguyen et al. 2023; Orsztynowicz et al. 2016; Song and Lewis 2008) and does not appear to be affected by aging.

Because the DNA encodes 13 subunits of the electron transport chain complexes I, III, IV, and V (ATP synthase), 22 tRNAs, and the rRNA subunits (St John et al. 2000), damages in mtDNA, including point mutations and multiple deletions, have been known to make a reduction in sperm motility and functions, if the damage is severe, developing asthenozoospermia and oligoasthenozoospermia (Durairajanayagam et al. 2021). Deletions in sperm mtDNA seem to be related to poor scores in the parameters for sperm quality (Song and Lewis 2008), as well as a large disruption to the electron transport chain complexes to produce ATP (O'Connell et al. 2002). Although the common 4977-bp mtDNA deletion, which is positively correlated with the intracellular level of 8-hydroxy-2′ deoxyguanosine, a biomarker of oxidative DNA damage, is well-known to accumulate in various tissues during aging, males with asthenozoospermia and oligoasthenozoospermia also have a higher incidence of mutation in this deletion (Durairajanayagam et al. 2021). Not only the degree of mtDNA deletion but also mtDNA copy number in spermatozoa seem to be negatively correlated with the motility and successful fertilization rates following IVF but, of course, not with those after ICSI.

During fertilization, since mitochondria and the mtDNA of penetrated spermatozoon are made degradation in the zygote and are dissolved in the embryo by a ubiquitin-mediated mechanism before embryonic genome activation (St John et al. 2010), protecting from mitochondrial heteroplasmy (the coexistence of multiply-derived mtDNA haplotypes) and realizing uniparental mtDNA transmission, paternal mtDNA is never transmitted to the embryo of the next generation in mammals. A target (degradation signal) of ubiquitination of sperm mitochondria could be prohibitin, the major protein of the inner mitochondrial membrane of sperm mitochondria (Sutovsky et al. 2000; Sutovsky et al. 2004). Although the ubiquitin labeling of paternal mitochondria takes place during spermatogenesis, the signals on the sperm mitochondrial membrane seem to be detected in the male reproductive tract before fertilization (Sutovsky et al. 1999). The recent progress in more sophisticated assisted reproductive techniques, such as cytoplasmic transfer and incorporation of extraneous mtDNA, has indicated a need for further experiments on the effects of the transmission of heterogeneous mtDNA to offspring (St John et al. 2004; St John and St John 2002). Indeed, modification of the mtDNA copy number in oocytes by mitochondrial supplementation changes the gene expression profiles during embryonic development (St John et al. 2023).

6   Sperm Functions Affected by the mtDNA Copy Number/ Mitochondria Content

Because mtDNA encodes 13 subunits of the electron transport chain complexes (St John et al. 2000), damages in the DNA, including point mutations and multiple deletions, have been known to have negative effect on energy production in mitochondria of paternal gametes, reducing the sperm motility and functions, if the damage is severe, developing oligoasthenozoospermia and asthenozoospermia (Durairajanayagam et al. 2021). Deletions in sperm mtDNA seem to be related to poor scores in the parameters for sperm quality (Song and Lewis 2008), as well as a large disruption to the electron transport chain complexes to produce ATP (O'Connell et al. 2002). Although the common 4977-bp mtDNA deletion, which is positively correlated with the intracellular level of 8-hydroxy-2′ deoxyguanosine, a biomarker of oxidative DNA damage, is well-known to accumulate in various tissues during aging, males with oligoasthenozoospermia and asthenozoospermia also have a higher incidence of mutation in this deletion (Durairajanayagam et al. 2021). Therefore, the degree of deletion in mtDNA, as well as nuclear DNA encoding mitochondrial proteins (Amaral et al. 2013), seems to be negatively correlated with the motility and successful fertilization rates following IVF but maybe not with those after ICSI.

Mitochondrial DNA copy number in actively motile spermatozoa is significantly lower than less motile cells (May-Panloup et al. 2003). Currently, there are numerous studies showing negative correlations between mtDNA copy number/mitochondria content and the quality of spermatozoa in humans (Boguenet et al. 2022; Diez-Sanchez et al. 2003; May-Panloup et al. 2003; Song and Lewis 2008; Vozdova et al. 2022; Yao et al. 1996), mouse (Luo et al. 2011), cattle (Nguyen et al. 2023), stallion (Darr et al. 2017; Orsztynowicz et al. 2016), and pig (Guo et al. 2017), dog (Hesser et al. 2017). Abnormally high copy number of mtDNA in spermatozoa may result from dysregulation of mtDNA replication, severe oxidative stress, and/or energy metabolism problems, during spermatogenesis (Gabriel et al. 2012; Song and Lewis 2008). Although low mtDNA copy number/mitochondrial content does not affect sperm motility and MMP (Boguenet et al. 2022; Darr et al. 2016; Guo et al. 2017), higher content and copy number decrease the motility and potential due to extreme oxidative stress under ROS overproduction, consequently decreasing sperm motility (Nguyen et al. 2023). Excessive ROS levels in spermatozoa will cause reductions in the quality parameters, including decreases in MMP, ATP production, and consequent motility (Marchetti et al. 2004; Wang et al. 2003), due to oxidative stress (Guo et al. 2017). Or conversely, excessive ROS production in low-quality spermatozoa may cause stimulated mitochondrial biogenesis and consequently increase mtDNA copy number through the overexpression of TFAM, a main regulator for mtDNA replication (Faja et al. 2019; Jiang et al. 2017; Luo et al. 2011).

Also, oxidative stress by excessive ROS production may induce the opening of the mitochondrial permeability transition pore and activation of the caspase cascade by extrusion of cytochrome c through those, consequently triggering apoptosis.

Apoptosis is known to be a programmed cell death that may take place through either the intrinsic (mitochondrial) pathway or extrinsic (death receptor) pathway, which are regulated by BCL-2/BCL-XL (anti-apoptosis effectors) and BAX-BAK (apoptotic effectors). The activation of caspases induced in the midpiece region of spermatozoa makes reductions in the MMP and motility and promotion of DNA disintegration (Grunewald et al. 2008; Paasch et al. 2004).

On the other hand, heat stress will increase ROS and mtDNA deletion rates in spermatozoa, inducing lipid peroxidation and motility impairment (Capela et al. 2022), through membrane fluidity alteration (Aitken 2017). Indeed, exposure of boar spermatozoa to heat stress (42 °C) induces a significant reduction in intracellular ATP content and motility (Gong et al. 2017). However, mtDNA copy number seems to be stable in spermatozoa through seasons during the year (Vozdova et al. 2022).

7 Conclusion

In the current review, we have attempted to review about potential quantitative regulation of mtDNA copy number/mitochondria content of spermatozoa and the factors associated with the alteration. It is quite unique that sperm with relatively low mtDNA copy number mitochondria perform important functions in the critical fertilization event that links life to the next generation. It is not yet clear why this downregulation of mtDNA copy number during spermatogenesis is required for normal spermatozoa, but it will need further discussions, in relation to the mechanism that prevents normal fertilization from passing on paternal mitochondria to the next generation (Birky Jr. 2001).

References

Agnihotri SK, Agrawal AK, Hakim BA, Vishwakarma AL, Narender T, Sachan R, Sachdev M (2016) Mitochondrial membrane potential (MMP) regulates sperm motility. In Vitro Cell Dev Biol Anim 52:953–960. https://doi.org/10.1007/s11626-016-0061-x

Aitken RJ (2017) Reactive oxygen species as mediators of sperm capacitation and pathological damage. Mol Reprod Devel 84:1039–1052. https://doi.org/10.1002/mrd.22871

Al Ojaimi M, Salah A, El-Hattab AW (2022) Mitochondrial fission and fusion: molecular mechanisms, biological functions, and related disorders. Membranes 12:16. https://doi.org/10.3390/membranes12090893

Alghamdi A (2024) A detailed review of pharmacology of MFN1 (mitofusion-1)-mediated mitochondrial dynamics: implications for cellular health and diseases. Saudi Pharmaceutical J 32:102012. https://doi.org/10.1016/j.jsps.2024.102012

Amaral A, Lourenco B, Marques M, Ramalho-Santos J (2013) Mitochondria functionality and sperm quality. Reproduction 146:R163–R174. https://doi.org/10.1530/REP-13-0178

Amaral A, Ramalho-Santos J, St John JC (2007) The expression of polymerase gamma and mitochondrial transcription factor a and the regulation of mitochondrial DNA content in mature human sperm. Hum Reprod 22:1585–1596. https://doi.org/10.1093/humrep/dem030

Anderson S, Bankier AT, Barrell BG, de Bruijn MH, Coulson AR, Drouin J, Eperon IC, Nierlich DP, Roe BA, Sanger F, Schreier PH, Smith AJ, Staden R, Young IG (1981) Sequence and

organization of the human mitochondrial genome. Nature 290:457–465. https://doi.
org/10.1038/290457a0

Antelman J, Manandhar G, Yi YJ, Li R, Whitworth KM, Sutovsky M, Agca C, Prather RS,
Sutovsky P (2008) Expression of mitochondrial transcription factor a (TFAM) during porcine
gametogenesis and preimplantation embryo development. J Cell Physiol 217:529–543. https://
doi.org/10.1002/jcp.21528

Athurupana R, Takahashi D, Ikoki S, Funahashi H (2015) Trehalose in glycerol-free extender
enhances post-thaw survival of boar spermatozoa. J Reprod Dev 61:205–210. https://doi.
org/10.1262/jrd.2014-152

Belenguer P, Pellegrini L (2013) The dynamin GTPase OPA1: more than mitochondria? Biochim
Biophys Acta 1833:176–183. https://doi.org/10.1016/j.bbamcr.2012.08.004

Birky CW Jr (2001) The inheritance of genes in mitochondria and chloroplasts: laws, mechanisms, and models. Annu Rev Genet 35:125–148. https://doi.org/10.1146/annurev.
genet.35.102401.090231

Boguenet M, Bouet PE, Spiers A, Reynier P, May-Panloup P (2021) Mitochondria: their role in
spermatozoa and in male infertility. Hum Reprod Update 27:697–719. https://doi.org/10.1093/
humupd/dmab001

Boguenet M, Desquiret-Dumas V, Goudenege D, Bris C, Boucret L, Blanchet O, Procaccio V,
Bouet PE, Reynier P, May-Panloup P (2022) Mitochondrial DNA content reduction in the most
fertile spermatozoa is accompanied by increased mitochondrial DNA rearrangement. Hum
Reprod 37:669–679. https://doi.org/10.1093/humrep/deac024

Bravo A, Treulen F, Uribe P, Boguen R, Felmer R, Villegas JV (2015) Effect of mitochondrial
calcium uniporter blocking on human spermatozoa. Andrologia 47:662–668. https://doi.
org/10.1111/and.12314

Capela L, Leites I, Romao R, Lopes-da-Costa L, Pereira R (2022) Impact of heat stress on bovine
sperm quality and competence. Animals [Electronic Resource] 12:09. https://doi.org/10.3390/
ani12080975

Chen W, Sun Y, Sun Q, Zhang J, Jiang M, Chang C, Huang X, Wang C, Wang P, Zhang Z, Chen X,
Wang Y (2020) MFN2 plays a distinct role from MFN1 in regulating Spermatogonial differentiation. Stem Cell Rep 14:803–817. https://doi.org/10.1016/j.stemcr.2020.03.024

Costello S, Michelangeli F, Nash K, Lefievre L, Morris J, Machado-Oliveira G, Barratt C,
Kirkman-Brown J, Publicover S (2009) Ca2+-stores in sperm: their identities and functions.
Reproduction 138:425–437. https://doi.org/10.1530/REP-09-0134

Courtens JL, Ploen L (1999) Improvement of spermatogenesis in adult cryptorchid rat testis by intratesticular infusion of lactate. Biol Reprod 61:154–161. https://doi.org/10.1095/
biolreprod61.1.154

Danshina PV, Geyer CB, Dai Q, Goulding EH, Willis WD, Kitto GB, McCarrey JR, Eddy EM,
O'Brien DA (2010) Phosphoglycerate kinase 2 (PGK2) is essential for sperm function and
male fertility in mice. Biol Reprod 82:136–145. https://doi.org/10.1095/biolreprod.109.079699

Darr CR, Cortopassi GA, Datta S, Varner DD, Meyers SA (2016) Mitochondrial oxygen consumption is a unique indicator of stallion spermatozoal health and varies with cryopreservation
media. Theriogenology 86:1382–1392. https://doi.org/10.1016/j.theriogenology.2016.04.082

Darr CR, Moraes LE, Connon RE, Love CC, Teague S, Varner DD, Meyers SA (2017) The relationship between mitochondrial DNA copy number and stallion sperm function. Theriogenology
94:94–99. https://doi.org/10.1016/j.theriogenology.2017.02.015

De Luca MN, Colone M, Gambioli R, Stringaro A, Unfer V (2021) Oxidative stress and male
fertility: role of antioxidants and Inositols. Antioxidants 10:13. https://doi.org/10.3390/
antiox10081283

Diez-Sanchez C, Ruiz-Pesini E, Lapena AC, Montoya J, Perez-Martos A, Enriquez JA, Lopez-
Perez MJ (2003) Mitochondrial DNA content of human spermatozoa. Biol Reprod 68:180–185.
https://doi.org/10.1095/biolreprod.102.005140

Durairajanayagam D, Singh D, Agarwal A, Henkel R (2021) Causes and consequences of sperm
mitochondrial dysfunction. Andrologia 53:e13666. https://doi.org/10.1111/and.13666

Faja F, Carlini T, Coltrinari G, Finocchi F, Nespoli M, Pallotti F, Lenzi A, Lombardo F, Paoli D (2019) Human sperm motility: a molecular study of mitochondrial DNA, mitochondrial transcription factor a gene and DNA fragmentation. Mol Biol Rep 46:4113–4121. https://doi.org/10.1007/s11033-019-04861-0

Fang F, Ni K, Shang J, Zhang X, Xiong C, Meng T (2018) Expression of mitofusin 2 in human sperm and its relationship to sperm motility and cryoprotective potentials. Exp Biol Med 243:963–969. https://doi.org/10.1177/1535370218790919

Filadi R, Greotti E, Pizzo P (2018) Highlighting the endoplasmic reticulum-mitochondria connection: focus on Mitofusin 2. Pharmacol Res 128:42–51. https://doi.org/10.1016/j.phrs.2018.01.003

Gabriel MS, Chan SW, Alhathal N, Chen JZ, Zini A (2012) Influence of microsurgical varicocelectomy on human sperm mitochondrial DNA copy number: a pilot study. J Assist Reprod Genet 29:759–764. https://doi.org/10.1007/s10815-012-9785-z

Garcia-Fabiani MB, Montanaro MA, Stringa P, Lacunza E, Cattaneo ER, Santana M, Pellon-Maison M, Gonzalez-Baro MR (2017) Glycerol-3-phosphate acyltransferase 2 is essential for normal spermatogenesis. Biochem J 474:3093–3107. https://doi.org/10.1042/BCJ20161018

Gibb Z, Lambourne SR, Aitken RJ (2014) The paradoxical relationship between stallion fertility and oxidative stress. Biol Reprod 91:77. https://doi.org/10.1095/biolreprod.114.118539

Gong Y, Guo H, Zhang Z, Zhou H, Zhao R, He B (2017) Heat stress reduces sperm motility via activation of glycogen synthase kinase-3alpha and inhibition of mitochondrial protein import. Front Physiol 8:718. https://doi.org/10.3389/fphys.2017.00718

Grunewald S, Said TM, Paasch U, Glander HJ, Agarwal A (2008) Relationship between sperm apoptosis signalling and oocyte penetration capacity. Int J Androl 31:325–330. https://doi.org/10.1111/j.1365-2605.2007.00768.x

Gu NH, Zhao WL, Wang GS, Sun F (2019) Comparative analysis of mammalian sperm ultrastructure reveals relationships between sperm morphology, mitochondrial functions and motility. Reprod Biol Endocrinol 17:66. https://doi.org/10.1186/s12958-019-0510-y

Guo H, Gong Y, He B, Zhao R (2017) Relationships between mitochondrial DNA content, mitochondrial activity, and boar sperm motility. Theriogenology 87:276–283. https://doi.org/10.1016/j.theriogenology.2016.09.005

Harayama H (2013) Roles of intracellular cyclic AMP signal transduction in the capacitation and subsequent hyperactivation of mouse and boar spermatozoa. J Reprod Dev 59:421–430. https://doi.org/10.1262/jrd.2013-056

Hesser A, Darr C, Gonzales K, Power H, Scanlan T, Thompson J, Love C, Christensen B, Meyers S (2017) Semen evaluation and fertility assessment in a purebred dog breeding facility. Theriogenology 87:115–123. https://doi.org/10.1016/j.theriogenology.2016.08.012

Ho HC, Suarez SS (2001) Hyperactivation of mammalian spermatozoa: function and regulation. [review] [61 refs]. Reproduction 122:519–526. https://doi.org/10.1530/rep.0.1220519

Ioki S, Wu QS, Takayama O, Motohashi HH, Wakai T, Funahashi H (2016) A phosphodiesterase type-5 inhibitor, sildenafil, induces sperm capacitation and penetration into porcine oocytes in a chemically defined medium. Theriogenology 85:428–433. https://doi.org/10.1016/j.theriogenology.2015.09.013

Jiang M, Kauppila TES, Motori E, Li X, Atanassov I, Folz-Donahue K, Bonekamp NA, Albarran-Gutierrez S, Stewart JB, Larsson NG (2017) Increased Total mtDNA copy number cures male infertility despite unaltered mtDNA mutation load. Cell Metab 26:429–36.e4. https://doi.org/10.1016/j.cmet.2017.07.003

Komatsu M, Waguri S, Ueno T, Iwata J, Murata S, Tanida I, Ezaki J, Mizushima N, Ohsumi Y, Uchiyama Y, Kominami E, Tanaka K, Chiba T (2005) Impairment of starvation-induced and constitutive autophagy in Atg7-deficient mice. J Cell Biol 169:425–434. https://doi.org/10.1083/jcb.200412022

Koppers AJ, De Iuliis GN, Finnie JM, McLaughlin EA, Aitken RJ (2008) Significance of mitochondrial reactive oxygen species in the generation of oxidative stress in spermatozoa. J Clin Endocrinol Metab 93:3199–3207. https://doi.org/10.1210/jc.2007-2616

Kumar N (2023) Sperm mitochondria, the driving force behind human spermatozoa activities: its functions and dysfunctions–a narrative review. Curr Mol Med 23:332–340. https://doi.org/10.2174/1566524022666220408104047

Lee W, Zamudio-Ochoa A, Buchel G, Podlesniy P, Marti Gutierrez N, Puigros M, Calderon A, Tang HY, Li L, Mikhalchenko A, Koski A, Trullas R, Mitalipov S, Temiakov D (2023) Molecular basis for maternal inheritance of human mitochondrial DNA. Nat Genet 55:1632–1639. https://doi.org/10.1038/s41588-023-01505-9

Losano JDA, Padin JF, Mendez-Lopez I, Angrimani DSR, Garcia AG, Barnabe VH, Nichi M (2017) The stimulated glycolytic pathway is able to maintain ATP levels and kinetic patterns of bovine Epididymal sperm subjected to mitochondrial uncoupling. Oxid Med Cell Longev 2017:1682393. https://doi.org/10.1155/2017/1682393

Luo Y, Liao W, Chen Y, Cui J, Liu F, Jiang C, Gao W, Gao Y (2011) Altitude can alter the mtDNA copy number and nDNA integrity in sperm. J Assist Reprod Genet 28:951–956. https://doi.org/10.1007/s10815-011-9620-y

Marchetti C, Jouy N, Leroy-Martin B, Defossez A, Formstecher P, Marchetti P (2004) Comparison of four fluorochromes for the detection of the inner mitochondrial membrane potential in human spermatozoa and their correlation with sperm motility. Hum Reprod 19:2267–2276. https://doi.org/10.1093/humrep/deh416

Marchetti C, Obert G, Deffosez A, Formstecher P, Marchetti P (2002) Study of mitochondrial membrane potential, reactive oxygen species, DNA fragmentation and cell viability by flow cytometry in human sperm. Hum Reprod 17:1257–1265. https://doi.org/10.1093/humrep/17.5.1257

Marchiani S, Tamburrino L, Giuliano L, Nosi D, Sarli V, Gandini L, Piomboni P, Belmonte G, Forti G, Baldi E, Muratori M (2011) Sumo1-ylation of human spermatozoa and its relationship with semen quality. Int J Androl 34:581–593. https://doi.org/10.1111/j.1365-2605.2010.01118.x

Marchiani S, Tamburrino L, Ricci B, Nosi D, Cambi M, Piomboni P, Belmonte G, Forti G, Muratori M, Baldi E (2014) SUMO1 in human sperm: new targets, role in motility and morphology and relationship with DNA damage. Reproduction 148:453–467. https://doi.org/10.1530/REP-14-0173

May-Panloup P, Chretien MF, Savagner F, Vasseur C, Jean M, Malthiery Y, Reynier P (2003) Increased sperm mitochondrial DNA content in male infertility. Hum Reprod 18:550–556. https://doi.org/10.1093/humrep/deg096

Meyers S, Bulkeley E, Foutouhi A (2019) Sperm mitochondrial regulation in motility and fertility in horses. Reprod Dom Anim 54(Suppl 3):22–28. https://doi.org/10.1111/rda.13461

Miki K, Qu W, Goulding EH, Willis WD, Bunch DO, Strader LF, Perreault SD, Eddy EM, O'Brien DA (2004) Glyceraldehyde 3-phosphate dehydrogenase-S, a sperm-specific glycolytic enzyme, is required for sperm motility and male fertility. Proc Natl Acad Sci USA 101:16501–16506. https://doi.org/10.1073/pnas.0407708101

Miyata H, Oura S, Morohoshi A, Shimada K, Mashiko D, Oyama Y, Kaneda Y, Matsumura T, Abbasi F, Ikawa M (2021) SPATA33 localizes calcineurin to the mitochondria and regulates sperm motility in mice. Proc Natl Acad Sci USA 118:31. https://doi.org/10.1073/pnas.2106673118

Moraes CR, Meyers S (2018) The sperm mitochondrion: organelle of many functions. Anim Reprod Sci 194:71–80. https://doi.org/10.1016/j.anireprosci.2018.03.024

Mukai C, Okuno M (2004) Glycolysis plays a major role for adenosine triphosphate supplementation in mouse sperm flagellar movement. Biol Reprod 71:540–547. https://doi.org/10.1095/biolreprod.103.026054

Nakada K, Sato A, Yoshida K, Morita T, Tanaka H, Inoue S, Yonekawa H, Hayashi J (2006) Mitochondria-related male infertility. Proc Natl Acad Sci USA 103:15148–15153. https://doi.org/10.1073/pnas.0604641103

Nakamura N, Miranda-Vizuete A, Miki K, Mori C, Eddy EM (2008) Cleavage of disulfide bonds in mouse spermatogenic cell-specific type 1 hexokinase isozyme is associated with increased hexokinase activity and initiation of sperm motility. Biol Reprod 79:537–545. https://doi.org/10.1095/biolreprod.108.067561

Nguyen HT, Do SQ, Kobayashi H, Wakai T, Funahashi H (2023) Negative correlations of mitochondrial DNA copy number in commercial frozen bull spermatozoa with the motil-

ity parameters after thawing. Theriogenology 210:154–161. https://doi.org/10.1016/j.
theriogenology.2023.07.027

Nishimura Y, Yoshinari T, Naruse K, Yamada T, Sumi K, Mitani H, Higashiyama T, Kuroiwa T
(2006) Active digestion of sperm mitochondrial DNA in single living sperm revealed by optical
tweezers. Proc Natl Acad Sci USA 103:1382–1387. https://doi.org/10.1073/pnas.0506911103

Nowicka-Bauer K, Szymczak-Cendlak M (2021) Structure and function of ion channels regulating
sperm motility-an overview. Int J Mol Sci 22:23. https://doi.org/10.3390/ijms22063259

O'Connell M, McClure N, Lewis SE (2002) Mitochondrial DNA deletions and nuclear DNA fragmentation in testicular and epididymal human sperm. Hum Reprod 17:1565–1570. https://doi.
org/10.1093/humrep/17.6.1565

Orsztynowicz M, Pawlak P, Podstawski Z, Nizanski W, Partyka A, Gotowiecka M, Kosiniak-
Kamysz K, Lechniak D (2016) Mitochondrial DNA copy number in spermatozoa of fertile
stallions. Reprod Dom Anim 51:378–385. https://doi.org/10.1111/rda.12689

Paasch U, Grunewald S, Dathe S, Glander HJ (2004) Mitochondria of human spermatozoa are pref-
erentially susceptible to apoptosis. Ann N Y Acad Sci 1030:403–409. https://doi.org/10.1196/
annals.1329.050

Park SH, Gye MC (2024) Inhibition of mitochondrial cyclophilin D, a downstream target of glycogen synthase kinase 3alpha, improves sperm motility. Reprod Biol Endocrinol 22:15. https://
doi.org/10.1186/s12958-024-01186-x

Patergnani S, Suski JM, Agnoletto C, Bononi A, Bonora M, De Marchi E, Giorgi C, Marchi S,
Missiroli S, Poletti F, Rimessi A, Duszynski J, Wieckowski MR, Pinton P (2011) Calcium
signaling around mitochondria associated membranes (MAMs). Cell Commun Signal 9:19.
https://doi.org/10.1186/1478-811X-9-19

Pi Y, Huang Z, Xu X, Zhang H, Jin M, Zhang S, Lin G, Hu L (2024) Increases in computationally
predicted deleterious variants of unknown significance and sperm mtDNA copy numbers may
be associated with semen quality. Andrology 12:585–598. https://doi.org/10.1111/andr.13521

Piomboni P, Focarelli R, Stendardi A, Ferramosca A, Zara V (2012) The role of mitochondria in energy production for human sperm motility. Int J Androl 35:109–124. https://doi.
org/10.1111/j.1365-2605.2011.01218.x

Popova D, Bhide P, D'Antonio F, Basnet P, Acharya G (2022) Sperm mitochondrial DNA copy
numbers in normal and abnormal semen analysis: a systematic review and meta-analysis.
BJOG Int J Obstet Gynaecol 129:1434–1446. https://doi.org/10.1111/1471-0528.17078

Ramalho-Santos J, Amaral S (2013) Mitochondria and mammalian reproduction. Mol Cell
Endocrinol 379:74–84. https://doi.org/10.1016/j.mce.2013.06.005

Ramio-Lluch L, Fernandez-Novell JM, Pena A, Colas C, Cebrian-Perez JA, Muino-Blanco
T, Ramirez A, Concha II, Rigau T, Rodriguez-Gil JE (2011) 'In vitro' capacitation and
acrosome reaction are concomitant with specific changes in mitochondrial activity in
boar sperm: evidence for a nucleated mitochondrial activation and for the existence of a
capacitation-sensitive subpopulational structure. Reprod Dom Anim 46:664–673. https://doi.
org/10.1111/j.1439-0531.2010.01725.x

Rantanen A, Jansson M, Oldfors A, Larsson NG (2001) Downregulation of Tfam and mtDNA
copy number during mammalian spermatogenesis. Mamm Genome 12:787–792. https://doi.
org/10.1007/s00335-001-2052-8

Ren M, Xu Y, Erdjument-Bromage H, Donelian A, Phoon CKL, Terada N, Strathdee D, Neubert
TA, Schlame M (2019) Extramitochondrial cardiolipin suggests a novel function of mitochondria in spermatogenesis. J Cell Biol 218:1491–1502. https://doi.org/10.1083/jcb.201808131

Richter C, Park JW, Ames BN (1988) Normal oxidative damage to mitochondrial and nuclear DNA
is extensive. Proc Natl Acad Sci USA 85:6465–6467. https://doi.org/10.1073/pnas.85.17.6465

Richter C, Suter M, Walter PB (1998) Mitochondrial free radical damage and DNA repair.
Biofactors 7:207–208. https://doi.org/10.1002/biof.5520070308

Rios L, Pokhrel S, Li SJ, Heo G, Haileselassie B, Mochly-Rosen D (2023) Targeting an alloste-
ric site in dynamin-related protein 1 to inhibit Fis1-mediated mitochondrial dysfunction. Nat
Commun 14:4356. https://doi.org/10.1038/s41467-023-40043-0

Sanocka D, Kurpisz M (2004) Reactive oxygen species and sperm cells. Reprod Biol Endocrinol 2:12. https://doi.org/10.1186/1477-7827-2-12

Shang Y, Wang H, Jia P, Zhao H, Liu C, Liu W, Song Z, Xu Z, Yang L, Wang Y, Li W (2016) Autophagy regulates spermatid differentiation via degradation of PDLIM1. Autophagy 12:1575–1592. https://doi.org/10.1080/15548627.2016.1192750

Shimada K, Ikawa M (2023) CCDC183 is essential for cytoplasmic invagination around the flagellum during spermiogenesis and male fertility. Development 150:01. https://doi.org/10.1242/dev.201724

Shimada K, Park S, Miyata H, Yu Z, Morohoshi A, Oura S, Matzuk MM, Ikawa M (2021) ARMC12 regulates spatiotemporal mitochondrial dynamics during spermiogenesis and is required for male fertility. Proc Natl Acad Sci USA 118:09. https://doi.org/10.1073/pnas.2018355118

Shimada K, Park S, Oura S, Noda T, Morohoshi A, Matzuk MM, Ikawa M (2023) TSKS localizes to nuage in spermatids and regulates cytoplasmic elimination during spermiation. Proc Natl Acad Sci USA 120:e2221762120. https://doi.org/10.1073/pnas.2221762120

Song GJ, Lewis V (2008) Mitochondrial DNA integrity and copy number in sperm from infertile men. Fertil Steril 90:2238–2244. https://doi.org/10.1016/j.fertnstert.2007.10.059

Song WH, Yi YJ, Sutovsky M, Meyers S, Sutovsky P (2016) Autophagy and ubiquitin-proteasome system contribute to sperm mitophagy after mammalian fertilization. Proc Natl Acad Sci USA 113:E5261–E5270. https://doi.org/10.1073/pnas.1605844113

St John JC, Facucho-Oliveira J, Jiang Y, Kelly R, Salah R (2010) Mitochondrial DNA transmission, replication and inheritance: a journey from the gamete through the embryo and into offspring and embryonic stem cells. Hum Reprod Update 16:488–509. https://doi.org/10.1093/humupd/dmq002

St John JC, Lloyd R, El Shourbagy S (2004) The potential risks of abnormal transmission of mtDNA through assisted reproductive technologies. Reprod Biomed Online 8:34–44. https://doi.org/10.1016/s1472-6483(10)60496-8

St John JC, Okada T, Andreas E, Penn A (2023) The role of mtDNA in oocyte quality and embryo development. Mol Reprod Devel 90:621–633. https://doi.org/10.1002/mrd.23640

St John JC, Sakkas D, Barratt CL (2000) A role for mitochondrial DNA and sperm survival. J Androl 21:189–199. https://doi.org/10.1002/j.1939-4640.2000.tb02093.x

St John JC, St John JC (2002) The transmission of mitochondrial DNA following assisted reproductive techniques. Theriogenology 57:109–123

Suarez SS (2008) Control of hyperactivation in sperm. Hum Reprod Update 14:647–657. https://doi.org/10.1093/humupd/dmn029

Sun B, Hou J, Ye YX, Chen HG, Duan P, Chen YJ, Xiong CL, Wang YX, Pan A (2022) Sperm mitochondrial DNA copy number in relation to semen quality: a cross-sectional study of 1164 potential sperm donors. BJOG Int J Obstet Gynaecol 129:2098–2106. https://doi.org/10.1111/1471-0528.17139

Sutovsky P, Moreno RD, Ramalho-Santos J, Dominko T, Simerly C, Schatten G (2000) Ubiquitinated sperm mitochondria, selective proteolysis, and the regulation of mitochondrial inheritance in mammalian embryos. Biol Reprod 63:582–590. https://doi.org/10.1095/biolreprod63.2.582

Sutovsky P, Moreno RD, Ramalho-Santos J, Dominko T, Simerly C, Schatten G (1999) Ubiquitin tag for sperm mitochondria. Nature 402:371–372. https://doi.org/10.1038/46466

Sutovsky P, Van Leyen K, McCauley T, Day BN, Sutovsky M (2004) Degradation of paternal mitochondria after fertilization: implications for heteroplasmy, assisted reproductive technologies and mtDNA inheritance. Reprod Biomed Online 8:24–33. https://doi.org/10.1016/s1472-6483(10)60495-6

Takei GL, Miyashiro D, Mukai C, Okuno M (2014) Glycolysis plays an important role in energy transfer from the base to the distal end of the flagellum in mouse sperm. J Exp Biol 217:1876–1886. https://doi.org/10.1242/jeb.090985

Thompson WE, Ramalho-Santos J, Sutovsky P (2003) Ubiquitination of prohibitin in mammalian sperm mitochondria: possible roles in the regulation of mitochondrial inheritance and sperm quality control. Biol Reprod 69:254–260. https://doi.org/10.1095/biolreprod.102.010975

Varuzhanyan G, Chan DC (2020) Mitochondrial dynamics during spermatogenesis. J Cell Sci 133:16. https://doi.org/10.1242/jcs.235937

Varuzhanyan G, Ladinsky MS, Yamashita SI, Abe M, Sakimura K, Kanki T, Chan DC (2021) Fis1 ablation in the male germline disrupts mitochondrial morphology and mitophagy, and arrests spermatid maturation. Development 148:15. https://doi.org/10.1242/dev.199686

Varuzhanyan G, Rojansky R, Sweredoski MJ, Graham RLJ, Hess S, Ladinsky MS, Chan DC (2019) Mitochondrial fusion is required for spermatogonial differentiation and meiosis. eLife 8:09. https://doi.org/10.7554/eLife.51601

Vertika S, Singh KK, Rajender S (2020) Mitochondria, spermatogenesis, and male infertility-an update. Mitochondrion 54:26–40. https://doi.org/10.1016/j.mito.2020.06.003

Vozdova M, Kubickova S, Kopecka V, Sipek J, Rubes J (2022) Association between sperm mitochondrial DNA copy number and deletion rate and industrial air pollution dynamics. Sci Rep 12:8324. https://doi.org/10.1038/s41598-022-12328-9

Wang X, Sharma RK, Gupta A, George V, Thomas AJ, Falcone T, Agarwal A (2003) Alterations in mitochondria membrane potential and oxidative stress in infertile men: a prospective observational study. Fertil Steril 80(Suppl 2):844–850. https://doi.org/10.1016/s0015-0282(03)00983-x

Yao YQ, Ng V, Yeung WS, Ho PC (1996) Profiles of sperm morphology and motility after discontinuous multiple-step Percoll density gradient centrifugation. Andrologia 28:127–131. https://doi.org/10.1111/j.1439-0272.1996.tb02768.x

Zhang B, Wang Y, Wu C, Qiu S, Chen X, Cai B, Xie H (2021) Freeze-thawing impairs the motility, plasma membrane integrity and mitochondria function of boar spermatozoa through generating excessive ROS. BMC Vet Res 17:127. https://doi.org/10.1186/s12917-021-02804-1

Zhang G, Yang W, Zou P, Jiang F, Zeng Y, Chen Q, Sun L, Yang H, Zhou N, Wang X, Liu J, Cao J, Zhou Z, Ao L (2019) Mitochondrial functionality modifies human sperm acrosin activity, acrosome reaction capability and chromatin integrity. Hum Reprod 34:3–11. https://doi.org/10.1093/humrep/dey335

Zhang Y, Ou Y, Cheng M, Saadi HS, Thundathil JC, van der Hoorn FA (2012) KLC3 is involved in sperm tail midpiece formation and sperm function. Dev Biol 366:101–110. https://doi.org/10.1016/j.ydbio.2012.04.026

Zhang Z, Miao J, Wang Y (2022) Mitochondrial regulation in spermatogenesis. Reproduction 163:R55–R69. https://doi.org/10.1530/REP-21-0431

Part V

Gamete-Maternal Interactions in the Establishment of Pregnancy

The Dialogue into the Sow Genital Tract: An Essential Process for Fertility

Santa María Toledo Guardiola ⓘ, Carmen Matás Parra ⓘ, and Cristina Soriano Úbeda ⓘ

Abstract

The pork meat industry is one of the most significant in the world, and artificial insemination (AI) is one of the most widely employed techniques. The increased utilization of the complete ejaculate, which contains the highest concentration of seminal plasma (SP) for inseminating sows, provides insight into the advancement of this field, with a focus on improving the health of the offspring. It has been demonstrated that SP contains immunosuppressive and immunomodulatory factors that support maternal immune tolerance and pro-angiogenic factors that support blood vessel formation and, among other functions, promote embryonic development through reproductive pathways and specific signaling to the female genital tract. These pathways involve extracellular vesicles (EVs), which are double-membraned nanovesicles secreted by most cells and represent a mode of intercellular communication through the transfer of bioactive molecules (DNAs, RNAs, miRNAs, proteins, and lipids) in their contents to target cells. An understanding of the pathways that gametes take and the interaction between them and the vesicles is crucial for elucidating the molecular processes occurring within the female reproductive tract during insemination and fertilization. Moreover, the characterization of the proteins and molecules involved is essential for advancing our knowledge of the intercellular communication between gametes and EVs.

S. M. T. Guardiola · C. M. Parra (✉)
Department of Physiology, Faculty of Veterinary Medicine, University of Murcia Campus Espinardo, Murcia, Spain
e-mail: smaria.toledo@um.es; cmatas@um.es

C. S. Úbeda
Animal Reproduction and Obstetrics, Department of Veterinary Medicine, Surgery, and Anatomy. University of León. Campus Vegazana, León, Spain
e-mail: c.soriano.ubeda@unileon.es

J. C. Gardón, K. Satué Ambrojo (eds.), *Assisted Reproductive Technologies in Animals Volume 1*, https://doi.org/10.1007/978-3-031-73079-5_13

Keywords

Artificial insemination · Assisted reproductive techniques · Extracellular vesicles · Female genital tract · Fertilization

1   Introduction

1.1   Significance of Artificial Insemination (AI) in Swine, Both at the Agri-food Industry and at the Level of the Artificial Reproductive Techniques (ARTs)

The Organization for Economic Cooperation and Development and the Food and Agriculture Organization of the United Nations (OCDE-FAO) predicts a 14% increase in global meat production, with demand growing in proportion to the world's population, increasing to reach 374 million metric tons by the year 2030 (OCDE-FAO 2023). Pork meat is of great importance in the human diet, accounting for about 33% of the total increase in meat consumption. Furthermore, its consumption is expected to continue to increase over the next 10 years, reaching 127 metric tons. Given these factors, global pork consumption is projected to increase by 0.3% per year to reach 35.4 kg/per capita in 2030 (OCDE-FAO 2023).

Spain is the main European producer (followed by Germany) with almost 34 million pigs registered in 2023. In addition, Germany and Spain are the third and fourth nations, both countries together contributing about 10% of the total pig production obtained in the world Interporc. This increase is largely dependent on the successful reproductive management of herds through the correct implementation of reproductive biotechnologies available today (Choudhary et al. 2016). By the beginning of the new millennium, AI rates had increased to ~60% in the United States, from 12 to 60% in South America, and from 40 to 85% in Europe (Waberski et al. 2019).

AI was the first and the oldest major biotechnology introduced to improve reproduction in domestic animals (Foote 2002). AI involves the collection of semen from the male and its dilution in an appropriate medium to produce multiple doses of semen. Subsequently, semen doses are placed through different methods in the female's reproductive tract (reviewed by Abad et al. (2007). AI is used in the swine industry worldwide and is considered the best breeding tool applied to domestic animals under commercial conditions (Waberski et al. 2019). Compared to natural mating, AI reduces the risk of disease transmission, allows the introduction of superior genes into sow herds and, consequently, efficiently improves pig production and pork quality meat and leads to better profitability of each boar ejaculate (Riesenbeck 2011; Riesenbeck et al. 2015). As a result, AI in pigs has reached high levels of efficiency in most major pork-producing countries, contributing significantly to herd health and genetic progress (Waberski et al. 2019). Currently, AI in pigs is used in approximately 90% of farms in several European countries and is considered one of the most successful animal reproduction techniques (Roca et al. 2006).

2 The Pig as a Research Model in ARTs

The economic importance of pigs has led to the creation of various systems to produce pigs as a source of human food, with ongoing efforts to optimize these systems by reducing costs and increasing production efficiency (Day 2000). In this context, advances in reproductive technologies have been aimed at improving the production of pigs both for consumption and for research purposes.

In addition to their importance in agriculture, the growing interest in biotechnological applications in pigs, especially in the generation of transgenic animals for biomedical research, is driven by their closer physiological similarity to humans than other large domestic animals. This progress has coincided with improvements in ARTs tailored to this species (Romar et al. 2019).

Although not traditionally considered an obvious model organism, the pig has recently gained relevant in research due to its anatomy, genetics, and physiology, which are more similar to humans, than those of other classic animal models, such as mice. Significant advances have been made in using this species as a biomedical model in genetic engineering, fueling interest in the potential use of pigs as donors of specific proteins and even organs to improve human health (Mourad & Gianello 2017; Watson et al. 2016).

3 The Ejaculate and the Importance of Seminal Plasma

3.1 The Ejaculate

Ejaculation is the release of semen from the male reproductive system. The ejaculate is a complex suspension of spermatozoa in a heterogeneous composite fluid, known as seminal plasma (SP). The SP is composed of contributions from the testis, epididymis, and/or the accessory sex glands. The volume of the ejaculate and SP composition vary among animal classes and species which is directly related to differences in presence and the size of the glands (Rodriguez-Martinez et al. 2021). Ejaculation is a highly coordinated physiological process involving neurological and muscular events that occur in two distinct phases: (i) emission, in which semen travels from the epididymis through the vas deferens mixing with fluids from the male accessory glands and deposits in the urethra, and (ii) ejection of the semen through rhythmic contractions of the pelvic floor and bulbospongiosus muscles pushing the semen towards the penis urethra and expelling it in several spurts (this process is known as the expulsion or ejaculation proper) (Rodríguez-Martínez et al. 2009). In boars, emission and ejaculation repeat as waves for 5–10 min, during which the complete ejaculate (250–300 mL in a mature boar) is sequentially verted into the female cervix lumen, or it is manually collected *ex corpore* into a recipient (Wallgren et al. 2010).

The ejaculation in boars is characterized by the release of semen in three visually differentiated fractions, each with specific color and consistency, associated with different cellular concentrations and biochemical compositions

(Rodríguez-Martínez et al. 2009). The previous, pre-spermatic fraction (PSF) (~25 mL) is composed of urethral content and secretions from bulbourethral glands and prostate, but usually without sperm cells. PSF is characterized by containing clear SP, some gel, and a significant amount of contamination with cell debris, urine, and smegma from the preputium. It is usually discarded due to its high bacterial count. The next fraction emitted is the sperm-rich fraction (SRF), which accounts for 10–30% of the total ejaculate volume and is easily recognized by its dense white color. This fraction contains the highest concentration of spermatozoa (80–90% of all sperm cells) and SP (derived from testes, epididymis, seminal vesicles, and prostate). The following fraction, the post-SRF, represents the largest volume of ejaculate (80%–90% of the total volume ejaculated) but contains a low concentration of spermatozoa and a large volume of SP (derived from seminal vesicles, prostate, and bulbourethral glands), giving it a watery appearance. The transition fraction between the SRF and the post-SRF, called the intermediate fraction, is considered a transition phase which is constituted by a higher volume of SP than SRF, low concentration of spermatozoa, and grayish color (Fig. 1). Finally, there is a latter emission that provides a tapioca-like flocculent secretion that coagulates on contact with the vesicular fluid of the SP, as seen when the entire ejaculate is collected in an open recipient. The role of this process in vivo is to retain the ejaculate in the uterus and minimize the transcervical reflux commonly seen with liquid semen where the gel component is consistently filtered away. Whose increasing amounts of "tapioca-like flocula" signal the end of the ejaculation process (reviewed by Mann and Mann 1981; Rodríguez-Martínez et al. 2009).

The SRF is the specific portion of the semen that is typically collected manually using the gloved-hand method. This method was originally adopted because it simplifies the handling of a significant number of spermatozoa in relatively small volumes, typically around 50–70 mL during the preparation of AI doses. In addition, it is widely recognized that spermatozoa from the two different fractions of a boar's ejaculate respond differently to various sperm biotechnologies, such as cryopreservation, as demonstrated in studies by Saravia et al. (2009). Recent research has demonstrated that spermatozoa from the SRF have superior cryosurvival rates compared to those from the entirety of an ejaculate, which includes both the SRF and post-SRF portions, as described in the study carried out by Li et al. (2018).

3.2   Seminal Plasma (SP) Composition

The SP is a complex fluid produced by the mixture of the epididymal fluid and the secretion of the sexual accessory glands, in the boar the vesicular, prostate, and bulbourethral glands. In boars, the SP represents the major portion of the ejaculate volume (approximately 95%) (Rodríguez-Martínez et al. 2009) and acts primarily as a transport medium for sperm on its arduous journey through the male and then female reproductive tract following ejaculation (Bromfield 2016). Sperm are produced in the testes and need to travel through the female reproductive system to

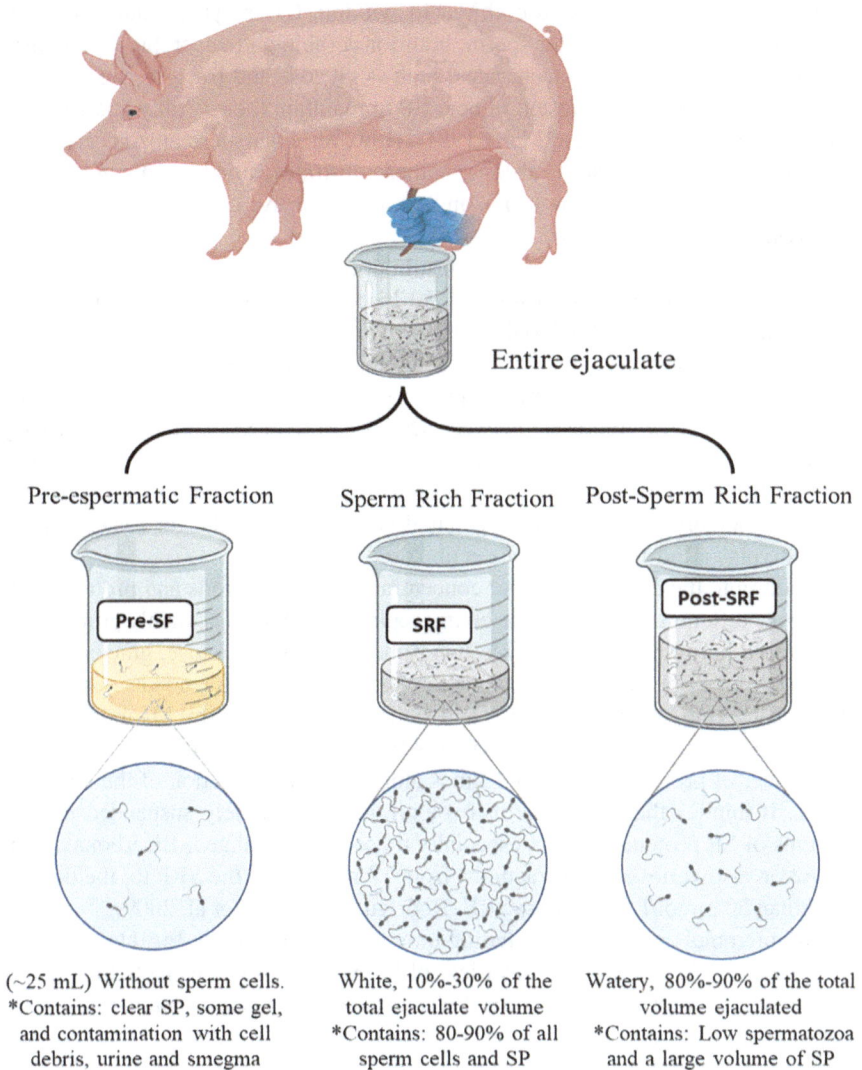

Fig. 1 Sperm fractions of the boar ejaculate. The composition of boar ejaculate is heterogeneous, comprising distinct fractions with varying contents and volumes. The initial fraction expelled is the pre-sperm fraction, which is devoid of sperm and contains a substantial amount of bacteria, urine, and smegma. It is usually discarded due to its high bacterial count. The subsequent fraction is the sperm-rich fraction, comprising numerous sperm and seminal plasma, which is utilized in artificial insemination. The final fraction expelled is the post-sperm fraction, which contains a greater proportion of seminal plasma and a smaller number of sperm. Image of own origin made with bio-render.com

reach the egg, and SP provides the fluid environment that allows sperm to swim and move effectively (Garner and Hafez 2016).

SP is also characterized by a high content of water, ions, sugars, organic acids, hormones, and a variety of amino acids and proteins. SP also contains various nutrients, including fructose, ascorbic acid (vitamin C), enzymes, and other substances that provide energy for sperm, both for aerobic and anaerobic metabolism (Parrilla et al. 2020). Monosaccharides, such as glucose and fructose, are the main energy sources for sperm present in boar SP to maintain their vitality and function during the journey (Mann and Mann 1981). However, boar spermatozoa can, in the absence of monosaccharides, also use other substrates, breaking down glucose, fructose, mannose, lactate, pyruvate, and citrate to survive under anaerobic conditions (Garner and Hafez 2016).

Moreover, proteins are one of the most important components of boar SP, with concentrations ranging from a total of 30 to 60 g/L (Rodríguez-Martínez et al. 2009). Many SP proteins bind to the surface of spermatozoa to modulate their functional capacity (Parrilla et al. 2019). These proteins also protect spermatozoa during their transit through the sow genital tract (Bromfield 2016). In boar, SP proteins are spermadhesins, which represent between 75–90% of total SP protein content. They attach to the sperm plasma membrane to varying degrees from the testis to the ejaculate. Collectively, they have been associated with multiple effects on spermatozoa including membrane stabilization, capacitation, and spermoviduct or zona pellucida (ZP) interplay (Rodríguez-Martínez et al. 2009).

In addition, the composition and concentrations of these ions and proteins vary among the different fractions of the boar ejaculate. The SRF has mainly proteins but also steroid hormones such as glycerophosphorylcholine, fructose, glucose, inositol, citrate, bicarbonate, and zinc, while the post-SRF has high amounts of proteins, bicarbonate, zinc, sodium, chlorine, and sialic acid (reviewed by Mann and Mann 1981). The protein concentration increases from SRF towards the post-SRF, being the number of proteins increased 4 times along with the secretion of the vesicular glands. It implies that most spermatozoa are not immediately suspended in high amounts of SP proteins in vivo, especially those first ejaculated. Bicarbonate concentration also varies among fractions, from 17 mmol/L in the SRF to, the double, more than 30 mmol/L in the post-SRF (Rodríguez-Martínez et al. 2009).

Antimicrobial constituents including immunoglobulins of the IgA class are also constituents of SP. In addition, a variety of hormonal substances including androgens, estrogens, prostaglandins, FSH, LH, and chorionic gonadotrophin, like growth hormone, insulin, glucagon, prolactin, relaxin, thyroid releasing hormone, and enkephalins have been detected in SP as well (Garner and Hafez 2016).

4   Seminal Plasma (SP) Function

4.1   Sperm Functionality Modulation

It is well established that SP modulates sperm viability, function, and the ability to interact with the uterine epithelium and oocyte for successful fertilization (Rodriguez-Martinez et al. 2021). Moreover, the influence of SP extends beyond fertility, as its infusion into the uterus during the estrous persists throughout the preimplantation period, leading to modifications in the endometrial and embryonic transcriptome by upregulating genes and pathways related to maternal immune tolerance, embryonic development, implantation, and pregnancy progress (Martinez et al. 2019; Parrilla et al. 2020).

Some authors like Martinez et al. (2019) found that SP infusions prior to conventional AI indirectly influence preimplantation embryo development in pigs showing a higher percentage of embryos in advanced stages of the preimplantation development 6 days post-AI relative to the control group (buffer infusions before AI). This provokes an increase in the proportion of blastocyst obtained, an increased preimplantation embryo development, and an altered expression of the endometrial genes and pathways potentially involved in embryo development. This is of particular importance for pig AI programs to decrease the immune response of the recipient females against the heterospermic seminal doses (i.e., different parents) that are commonly inseminated (Martinez et al. 2019).

SP triggers genetic and epigenetic pathways in spermatozoa that produce lasting changes in the female immune response with significant implications for fertility since SP increases the chances of the sperm reaching and fertilizing the egg as well as it has the potential to influence the development of the embryo (Martínez et al. 2020). Moreover, the effects of SP on the female environment thus have far-reaching consequences for offspring, independent of the genetic and epigenetic pathways controlled by sperm (Baxter and Drake 2019).

The pH of the male and female reproductive tracts can vary, and it is essential for sperm to be in an environment that supports their survival. SP helps regulate and buffer the pH to ensure sperm can survive in the acidic conditions of the female reproductive tract. The pH of boar SP ranges between 7.3–7.8, so it is considered alkaline (Garner and Hafez 2016). This slight alkalinity favors sperm when they are in the vagina, where the pH is acidic (Schjenken and Robertson 2020). It has been shown that the ion bicarbonate plays important roles in sperm physiology, both by maintaining intracellular pH (ipH), the homeostasis of the cell, and by modulating sperm motility and membrane stability (Soriano-Úbeda et al. 2017, 2019).

Moreover, it is known that boar sperm are highly susceptible to oxidative stress due to the high polyunsaturated fatty acids (PUFAs) content in the plasma membrane. Oxidative stress is caused by redox unbalance surpassing a physiological range could undermine the structural and functional integrity of sperm. The over-accumulation of reactive oxygen species (ROS) in sperm produces oxidation of lipids, proteins, and DNA that leads to lipid peroxidation, oxidation of essential structural proteins and enzymes, and mutations of DNA, which would impair sperm motility, the integrity of plasma membrane and acrosome, and mitochondrial function and eventually lead to infertility (Li et al. 2022). For all of these, to minimize these harmful effects, SP contains antioxidants (enzymatic and nonenzymatic) (Strzeżek and Lecewicz 1999) secreted by the prostate that are a potent natural free radical scavenger that play an essential role in extracting excessive ROS and protecting sperm cells against redox disorder in spermatozoa (Barranco et al. 2021).

5   Preparing the Uterus for Implantation

The SP plays several roles including modulation of sperm function, interaction with the epithelia, and secretions of the female genital tract and also as a carrier of immunomodulatory effects, potentially helping to modulate the female immune response to sperm, making it more favorable for fertilization. It is known that exposure of the female genital tract to SP triggers ovulation both at coitus in natural mating or using seminal doses in AI (Chen et al. 1985). This effect is produced by altering the endocrine-immune-cytokine network in preovulatory follicles, resulting in a shortening of the interval between LH peak and ovulation time (Waberski et al. 1997). Moreover, it has been shown that porcine SP administered just before ovulation modulates *corpus luteum* (CL) development via recruitment and activation of ovarian macrophages (Gangnuss et al. 2004). The physiological consequences of the induction of ovulation and accelerated CL development with increased progesterone (P_4) production (O'Leary et al. 2006) contribute to postovulatory tissue remodeling and steroidogenic function of luteal cells (Care et al. 2013).

The SP also promotes subtle changes in uterine gene expression associated with sperm transport and protection (Chen et al. 2014). It appears to induce changes in the expression of specific genes in the oviductal sperm reservoir (the utero-tubal junction) (Atikuzzaman et al. 2017) and induces an altered transcription in several tissues in the female tract that modulates embryo development (Bromfield 2016). Moreover, SP initiates the immune adaptation to promote endometrial receptivity to implantation, placental development, and uterine vascular adaptation to placentation (Adams et al. 2016).

6   Offspring Health

In a study published in 2014, Bromfield et al. (2014) evaluated the consequences of ablating the SP semen in mice by surgical excision of the seminal vesicle gland. Their findings demonstrated that SP deficiency resulted in reduced fecundity, disrupted blastocyst development, and altered placental development. Conception was substantially impaired, and when pregnancy did occur, placental hypertrophy was evident in late gestation. Following birth, the growth trajectory and metabolic parameters of the progeny were found to be altered, with the greatest impact observed in males, who exhibited obesity, distorted metabolic hormones, reduced glucose tolerance and hypertension.

These findings demonstrated that the composition of SP has an influx on the progeny and demonstrated the paternal role of SP in the growth and health of male offspring, which aligns with the conclusions of other authors in the field. Watkins et al. (2018) demonstrated that the semen from males fed with a low protein diet significantly reduces levels of TNF, IL-1β, IFN-γ, MIP-1α, and CSF3 (formerly G-CSF) in the uteri of females mated with them at day 3.5 of gestation in mice. Furthermore, they observed a reduction in the expression of several prostaglandin synthesis genes and genes associated with regulatory T-cell (Tregs)-mediated responses, including Cd3e, Foxp3, and H2-Ab1. Moreover, in adulthood, offspring of low-protein diet-fed males have been shown to display increased adiposity, hypotension, cardiovascular dysfunction, and poor glucose tolerance responses (Watkins et al. 2014). These data suggest that SP from sub-optimally fed males causes a blunting of the normal uterine inflammatory and cell signaling environment during the development of the preimplantation embryo (Morgan and Watkins 2019).

In contemporary AI centers, the ejaculate collection method is undergoing a transition from manual to semi-automated techniques. This shift is driven by considerations related to hygiene standards and labor cost efficiency, moving away from collecting only the SRF to collecting the entire ejaculate (Aneas et al. 2008). This change increases the relevance of SP, as the large SP volume of the post-SRF fraction will now become a part of the collected ejaculate. In this context, researchers such as Luongo et al. (2022) evaluated the effect of cumulative fractions of the ejaculate in seminal doses on reproductive performance after AI (fertility and prolificacy) in field conditions and offspring analysis (growth and blood assay). Their findings indicated that the presence of all ejaculate fractions within the seminal doses does not impair nor improve the reproductive performance in swine production (Luongo et al. 2022).

7 What Happens in the Female Tract Once the Ejaculate Has Been Deposited?

7.1 Into the Uterus

Boars deposit approximately 250 mL of ejaculate, containing 50 to 60 billion sperm and SP, into the proximal part of the cervix and the distal part of the uterine body of sows. However, only a small subpopulation of spermatozoa reaches the oviduct and the site of fertilization (Soriano-Úbeda et al. 2017). The sperm is transported through the uterus with the aid of coordinated contractions of the myometrium (Rath et al. 2016). P_4 and estrogens regulate the myometrium's activity whereby the frequency and amplitude of contractions reach their peak during estrus. Daily variations are dependent on varying hormone levels including estrogens, oxytocin, and prostaglandins (Langendijk et al. 2005), such as PGF2α released immediately after insemination (Claus 1990). The interaction between SP and the uterus is complex and involves several physiological processes:

1. *Microbial clearance:* The cervix plays a key role in microbial clearance and defense against infection from bacterial and viral pathogens by eliminating the excess sperm and microorganisms introduced into the uterus during mating (Robertson and Sharkey 2016). After insemination, an acute inflammatory response is observed in the endometrium and cervix and is accompanied by the recruitment of neutrophils from the peripheral blood that cross the epithelial surface into the luminal cavity to maintain microbial homeostasis by removing debris and pathogens introduced during coitus. This rapid neutrophilic leukocy-

tosis induced immediately by SP contact is common to all mammals and is thought to maintain microbial homeostasis of the female tract (Schjenken and Robertson 2020). The female fights the invasion of pathogens immunologically foreign cells and other antigenic material present in the ejaculate (proteins/peptides) by inducing a transient genital inflammation (Schjenken and Robertson 2014). However, considering that spermatozoa are also foreign cells, such transient inflammation is followed by the establishment of a longer-lasting state of immunological tolerance (Rodriguez-Martinez et al. 2021) to those spermatozoa that randomly colonize the oviductal sperm reservoirs and at the very earliest time of the embryo's first contact with maternal tissues (Álvarez-Rodríguez et al. 2020).

2. *Immune tolerance:* In the cervix, it has been shown that SP provokes a cascade of cellular and molecular events characterized by an inflammatory response, inducing pro-inflammatory cytokines followed by a post-mating leukocytosis response and a rapid recruitment of neutrophils into the uterine lumen. These neutrophils bind to viable but not non-viable spermatozoa (Taylor et al. 2008) regulated partly by porcine SP proteins (PSP-I and PSP-II), which are major proteins in boar SP (Rodríguez-Martínez et al. 2011). It may engage in the selection of sperm most competent for fertilization and induce immune tolerance to paternal antigens potentially expressed on the resulting semi-allogeneic conceptus (reviewed by Schjenken and Robertson 2014).

Sperm and neutrophil interactions may contribute to the activation of the adaptive immune response to antigens in SP. Neutrophils have the capacity to act as antigen-presenting cells and prime T cells and can modulate the antigen-presenting environment through the induction of Treg cells (Nadkarni et al. 2016). In addition to neutrophils, SP induces the accumulation of macrophages, dendritic cells, granulocytes, and lymphocytes within the endometrial stroma (Schjenken and Robertson 2014) after insemination in pigs (Robertson 2007) followed by concomitant endometrial hypertrophy, which is consistent with the observations of uterine vascularity, luminal fluid content, and weight by O'Leary et al. (2004). While the luminal neutrophil response resolves within 24 h (Rozeboom et al. 1998), macrophages and dendritic cells are professional antigen-processing and antigen-presenting cells that infiltrate the endometrium and persist and differentiate in the tissue, expanding the paternal major histocompatibility antigen (MHC class II+; Swine Leukocyte Antigen complex, SLA II in pigs), and initiate T-cell activation that mediates immune tolerance to the same paternal alloantigens expressed by the implanting embryo during the preimplantation period (Moldenhauer et al. 2017).

Upon deposition in female reproductive tract tissues, seminal transforming growth factor beta (TGF-β) which is a very potent, multifunctional cytokine present in SP (Robertson et al. 2002) interacts with uterine and/or cervical epithelial cells, triggering the female tract response to SP and initiating a cascade of downstream effects. TGF-β has significant immune-regulatory and pro-tolerogenic effects on lymphocytes, macrophages, and dendritic cells (Gorelik and Flavell 2002). TGF-β also regulates T-cell differentiation in immunosuppressive regulatory T lymphocytes which play a critical role in immune tolerance in pigs during implantation and placental development by an antigen-specific action (O'Leary et al. 2006). This supports the creation of a favorable immune environment that aids implantation and subsequent fetal development.

3. *Sperm sequestration:* After coitus, neutrophils have a second important function of removing excess sperm by phagocytosis or sequestration of the sperm into large aggregates associated with NETs. In humans, there is evidence for selective sperm phagocytosis, which filters out morphologically abnormal spermatozoa (Tomlinson et al. 1992). Selection may occur based on morphological or antigenic parameters and differential resistance to phagocytosis rather than fertilization ability. Even apparently viable and morphologically normal spermatozoa are targeted (Robertson 2007). In sows, semen backflow causes sperm loss, with up to 70% of the volume and 40% of the inseminated sperm being lost through the vestibule to the outside in almost all inseminated females (Steverink et al. 1998). However, after natural mating, a bulbourethral plug is deposited at the end of ejaculation into the sow's vagina to prevent semen backflow. One mechanism to prevent the selective action of backflow and neutrophil attack is sperm adhesion to the apical membrane of uterine epithelial cells (UECs). Communication with the UECs is mainly maintained by interactions mediated by spermadhesin, lectin, or lectin-like proteins (Taylor et al. 2008). Research suggests that sperm selection may depend on a specific morphology as only viable, membrane-intact, and motile sperm attach to the UECs (reviewed by (García-Vázquez et al. 2016).

4. *Preparation for endometrial receptivity:* In pigs, SP infusion induces the expression of the cytokines CSF2, IL6, and CCL2. This is accompanied by the recruitment of macrophages, dendritic cells, and T cells into the endometrial stroma (Bischoft et al. 1994; O'Leary et al. 2004). These white blood cells, especially macrophages, are involved in postovulatory tissue remodeling by restructuring the endometrial environment after ovulation to aid in implantation and placental development (Hunt and Robertson 1996). They also directly affect endometrial receptivity by secreting angiogenic factors and inducing the expression of embryo attachment ligands in epithelial cells in mice (Robertson and Sharkey 2016). Macrophages also contribute to ovarian vascular homeostasis, corpora lutea formation, and steroidogenic function of luteal cells (Care et al. 2013).

5. *Uterine decidual response:* SP may influence angiogenesis, as it can cause an increase in vasodilation and edema which has been observed with a macroscopic study of the endometrium after coitus in pigs (O'Leary et al. 2004). To reinforce this idea, authors such as Toledo-Guardiola et al. (2024a, b) demonstrate that uterine vascularization was significantly lower in non-inseminated sows compared to those inseminated with different fractions of the boar ejaculate. These results indicate that the interactions between the SP and the uterus could stimulate the formation of new vessels in the endometrium.

Uterine macrophages recruited by SP participate in the vascular response to implantation by synthesizing vascular endothelial growth factor (VEGF) in a tightly regulated pattern throughout the estrous cycle and early pregnancy (Schjenken and Robertson 2020). In mice, it has been demonstrated that pro-angiogenic macrophages play a critical role in the rapid neovascularization that is essential for luteal development and P_4 synthesis (Kaczmarek et al. 2013). Angiogenesis is necessary for decidualization, placentation, yolk sac formation, and embryonic vascular development (Klauber et al. 1997). The regulation of angiogenesis is the most important potential pathway for macrophage effects on implantation. Activated macrophages can influence each phase of the angiogenic process, including altering the local extracellular matrix, inducing endothelial cell migration and proliferation, and forming capillaries (Robertson 2007).

Furthermore, TGF-β, present in SP, plays a pivotal role in the promotion of new blood vessel formation and prostaglandins (PGs) (Kaczmarek et al. 2013). It has been shown that uterine dendritic cells express TGF-β which affects angiogenesis as some studies in horses and pigs have shown that TGF-β promotes coordinated blood vessel maturation, leading to an increase in vasodilation and edema following coitus (Schjenken and Robertson 2020). In addition, supplementing the semen extender used for pig AI with TGF-β has been shown to increase placental efficiency (Rhodes et al. 2006). These factors are paracrine regulators of endothelial cell function during endometrial angiogenesis, leading to increased density of the vascular network exhibited 10 days after SP infusion in the pig.

7.2   In the Oviduct

In the oviduct, those spermatozoa with normal morphology and motility bind to the oviductal epithelial cells (OECs) and form the sperm reservoir (SR), where they will remain in a state of low-activity or uncapacitated status (Rodríguez-Martínez et al. 2009).

At a physiological level, it has been shown that sperm can be arrested in the porcine SR by the interaction of several mechanisms: The most obvious is the presence of a thick mucus, high in hyaluronan (HA) (Tienthai et al. 2000) during most of the preovulatory period (Johansson et al. 2000), and a lower pH in the SR, supported by the low bicarbonate levels registered in vivo in utero-tubal junction (UTJ) which decrease sperm motility (Rodríguez-Martínez et al. 2005). In any case, the HA-rich fluid that embeds spermatozoa would help resident spermatozoa evade recognition by the female immune system and increase their survival by delaying capacitation (Rodriguez-Martinez et al. 2001). The SR selects the spermatozoa, extends their lifespan, preserves their functionality to allow spermatozoa quiescence due to their slightly acid pH (pH 6.5), and storages them up until ovulation (reviewed by Soriano-Úbeda et al. (2017).

When ovulation occurs, spermatozoa separate from the oviductal epithelium and swim toward the ampullary region, where fertilization takes place. Changes in the viscosity of the intraluminal mucus (Johansson et al. 2000), the flow of intraluminal fluid, and the coordinated contractions of the myometrium and the myosalpinx towards the UTJ (Álvarez-Rodríguez et al. 2020), perhaps influenced by the peri-ovulatory surge of P_4 from the ovary (Hunter and Rodriguez-Martinez 2004), could all contribute to detachment of epithelium-bound spermatozoa (Fazeli 2011). In porcine, fertilization takes place in the ampullary-isthmic junction (Hunter and Rodriguez-Martinez 2002) and begins when the capacitated spermatozoa in immediate proximity to the oocyte pass through the cumulus-oophorous and interact with the glycoproteins of the ZP of the oocyte (E. Kim et al. 2008).

Collectively, these semen-induced changes in female tissues are intended to improve the transportation of the most competent male spermatozoa to the oocyte, maximizing the chances of successful fertilization, embryo development, and implantation.

8   Extracellular Vesicles (EVs)

8.1   What Are Extracellular Vesicles?

Extracellular vesicles (EVs) are membrane-bound nanovesicles (30–1000 nm) of endosomal and plasma membrane origin, composed of a lipid bilayer, and are found in all cells that are capable of secreting different types of membrane vesicles, a process conserved throughout evolution from bacteria (Deatherage and Cookson 2012) to humans (Schorey et al. 2015) and plants (Robinson et al. 2016). EVs have been isolated from most cell types and biological fluids, such as urine (Pisitkun et al. 2004), saliva (Ogawa et al. 2011), blood (Caby et al. 2005), breast milk (Admyre et al. 2007), amniotic fluid (Asea et al. 2008), and others including reproductive fluids: SP (Piehl et al. 2013; Sullivan and Saez 2013), follicular fluid (FF) (da Silveira et al. 2012), uterine fluid (UF) (Burns et al. 2014), and oviductal fluid (OF) (Alcântara-Neto et al. 2019; Almiñana et al. 2018). EVs play an important role in intercellular communication and the regulation of physiological and pathological processes. They contain and can transfer different bioactive molecules (lipids, proteins, and nucleic acids such as DNA, RNAs, mRNAs, and especially miRNAs) (reviewed by Raposo and Stoorvogel 2013) and have surface receptors/ligands from the cell of origin that can selectively interact with specific target cells (Rana et al. 2012).

EVs serve as an umbrella term for two distinct types known as exosomes and microvesicles (MVs). These EVs are categorized based on their origin and size (Van Niel et al. 2018). Exosomes, which range in size from 30 to 200 nm, originate from endocytic pathways and are intraluminal vesicles formed by the inward budding of the endosomal membrane during the maturation of multivesicular bodies (MVBs) and are released from cells upon fusion of these MVBs to the plasma membrane (Harris et al. 2020). In contrast, MVs are larger in diameter, typically in the range of 200 to 1000 nm and are formed by outward budding and cleavage of the plasma membrane of the cell, resulting in their release into the extracellular environment (Lopera-Vasquez et al. 2017).

8.2   Where Do EVs Come from?

In the mid-1990s, EVs were reported to be secreted by B lymphocytes and dendritic cells with potential functions related to immune regulation and were considered for use as vehicles in antitumor immune responses (Raposo et al. 1996; Zitvogel et al. 1998). However, MVs, formerly referred to as "platelet dust," were first described as subcellular material derived from platelets in normal plasma and serum (Wolf 1967). Although MVs have been studied primarily for their role in blood coagulation (Sims et al. 1988), more recently they have been reported to play a role in cell-to-cell communication in various cell types, including cancer cells, where they are commonly referred to as "oncosomes" (Al-Nedawi et al. 2008).

EVs have been found in all biological fluids and are known to be secreted by specialized cells (Van Niel et al. 2018). Waqas et al. (2017) demonstrated that ciliated cells produce and secrete EVs into the oviduct of turtles. Because ciliated cells are concentrated in the ampulla, whereas secretory cells are mainly abundant in the isthmus (Winuthayanon and Li 2018), it is possible that secretory cells in the ampulla secrete EVs differently from secretory cells in the isthmus region. Therefore, the spatiotemporal regulation of EVs from different oviductal regions may provide unique biological actions tailored for sperm migration/function, fertilization, preimplantation embryo development, and embryo transport (Harris et al. 2020). Similarly, Burns et al. (2018) used electron microscopy to detect the presence of multivesicular bodies (MVBs) in the apical region of both luminal and glandular epithelial cells in the endometrium of cyclic ewes.

8.3   How EVs Secretion Is Regulated?

Traditionally, intercellular communication has involved three mechanisms: contact-dependent signaling via membrane-bound signaling molecules (receptors) or gap junctions, short-range paracrine signaling via secreted soluble molecules such as cytokines and chemokines, and long-range endocrine signaling via secreted hormones. Recent studies have revealed the existence of EVs that are released by cells into the extracellular environment and can serve as vehicles for the transfer of proteins, lipids, and RNA between cells both locally (autocrine and paracrine) and remotely (Machtinger et al. 2016; Raposo and Stoorvogel 2013).

Because exosomes are formed by MVBs and MVs are formed by direct budding from the plasma membrane, the cellular machinery involved in their formation and release is likely to be different, although mechanistic elements may be shared (Möbius et al. 2002). The molecular machinery involved in the biogenesis of MVBs en route to degradation has been elucidated since the initial discovery of yeast mutants defective in transport to the vacuole, the yeast analog of the mammalian lysosome. These evolutionarily conserved proteins assemble into four multiprotein complexes: the Endosomal Sorting Complex Responsible for Transport (ESCRT)

-0, -I, -II, and -III, which associate with accessory proteins (e.g., Alix and VPS4) (Raposo and Stoorvogel 2013).

The discovery of the ESCRT machinery as a driver of membrane shaping and scission was the first breakthrough in uncovering the mechanisms involved in the formation of MVBs and intraluminal vesicles. The ESCRT machinery acts in a step-wise manner, with ESCRT-0, ESCRT-I, and ESCRT-II complexes recognizing and sequestering ubiquitylated membrane proteins at the endosomal limiting membrane of MVBs, and recruiting, via the ESCRT-III complex, is responsible for membrane budding and actual scission of intraluminal vesicles. Accordingly, ESCRT-0 appears to be required for exosome formation and/or secretion by dendritic cells (Tamai et al. 2010), and ESCRT-II has also been shown to be an RNA binding complex (Irion and St Johnston 2007), raising the possibility that it may also function to select RNA for incorporation into EVs.

The machinery involved in the cleavage/release of MVs from the plasma membrane and those involved in the mobilization of secretory MVBs to the cell periphery, their docking, and fusion with the cell surface is still at an early stage of understanding. These processes require the cytoskeleton (actin and microtubules), associated molecular motors (kinesins and myosins), molecular switches (small GTPases), and the fusion machinery (SNAREs and tethering factors; Cai et al. 2007). The release of EVs has been found to be regulated in several cellular model systems.

It is well established that steroid hormones induce significant changes in the oviductal transcriptome (Acuña et al. 2017) and OF composition throughout the estrous cycle (Lamy et al. 2016). These changes are essential to provide an optimal environment for sperm storage, capacitation, gamete transport, fertilization, and early embryo development (Hunter and Rodriguez-Martinez 2004). Thus, it is very likely that the secretion and composition of EVs are also under the regulation of ovarian hormones in the oviduct and in the uterus. For example, Greening et al. (2016) showed that the protein content of uterine extracellular EVs (uEVs) released by human endometrial epithelial cells (EECs) during in vitro culture was altered by treatment with estrogen and P_4, suggesting hormone-specific changes in EVs cargo. This is similar to Burns et al. (2018) who demonstrated that the amount of EVs in the lumen of the ovine uterus increased from days 10 to 14 of the estrous cycle and also, after 14 days of P_4 treatment, suggesting that P_4 may regulate endometrial epithelial production of EVs and their release into the uterine lumen. In addition, a recent publication on bovine oviductal extracellular vesicles (oEVs) has provided the first evidence for the variation of oEVs cargo by hormonal influence showing that OECs can release oEVs into the oviductal environment at different stages of the estrous cycle and that steroid hormones significantly regulate their molecular cargo (Almiñana et al. 2018). Taken together, the dynamic EVs protein profile and the rapid adaptation of EVs size and quantity suggest that reproductive EVs secretion is also under the hormonal influence during the estrous cycle (Laezer et al. 2020).

8.4   Binding and Endocytosis Mechanism

Once released into the extracellular space, EVs can reach recipient cells and deliver their contents to elicit functional responses and promote phenotypic changes that affect their physiological or pathological state (Van Niel et al. 2018). Intercellular communication mediated by EVs requires membrane docking, activation of surface receptors and signaling, and either vesicle internalization (endocytosis) or fusion with target cells (Théry et al. 2009). The specificity of exosome (or other EVs) binding to target cells is likely determined by specific interactions between proteins enriched on the surface of EVs and receptors on the plasma membrane of the recipient cells. Several mediators of these interactions are known, including tetraspanins, integrins, lipids, lectins, heparan sulfate proteoglycans, and extracellular matrix (EMC) components (Harris et al. 2020).

EV membranes commonly contain proteins such as tetraspanins (CD63, CD9, CD81, heat shock protein 70; HSP70) and glycophosphatidylinositol-anchored proteins (Machtinger et al. 2016). These molecules facilitate interaction between EVs and recipient cells by adhering to lipids and ligands on the cell surface. For example, integrins on EVs can interact with adhesion molecules such as intercellular adhesion molecules (ICAM), on the surface of recipient cells (Morelli et al. 2004). Additionally, the interaction of integrins with EMC, specifically fibronectin and laminin, has been shown to play a crucial role in the binding of exosomes and MVs to recipient cells (Leiss et al. 2008). Proteoglycans and lectins, which are present in EVs at the plasma membrane, contribute to the docking and/or attachment of these vesicles to recipient cells (Van Niel et al. 2018). Exosomal tetraspanins, a family of over 30 proteins that are composed of four transmembrane domains (Hemler 2003), may also regulate cell targeting as they have been shown to interact with integrins and to promote exosome docking and uptake by selected recipient cells (Rana et al. 2012).

The lipid composition of EVs may affect their targeting of recipient cells. Phosphatidylserine, for instance, can attract specific lipid-binding proteins like galectin 5 or annexin 5 (Barrès et al. 2010; Frey and Gaipl 2011), which in turn facilitate vesicle docking to the target cell membrane. After binding to recipient cells, EVs can either remain stably associated with the plasma membrane or dissociate, and fuse directly with it, or be internalized through different endocytic pathways (Raposo and Stoorvogel 2013).

8.5   EVs Content and Function

Due to their origin, EVs from different cell types contained in the extracellular membrane (Fig. 2): endosome-associated proteins that are involved in intracellular trafficking (e.g., Rab GTPases, SNAREs, and annexins), membrane proteins including tetraspanins (e.g., CD9, CD63, CD81, CD82, and flotillin 1 and 2), as well as lipids (e.g., phosphatidylserine, cholesterol, ceramide, and sphingolipids), matrix glycoproteins involved in cell adhesion (e.g., integrins, lactadherins, fibronectins

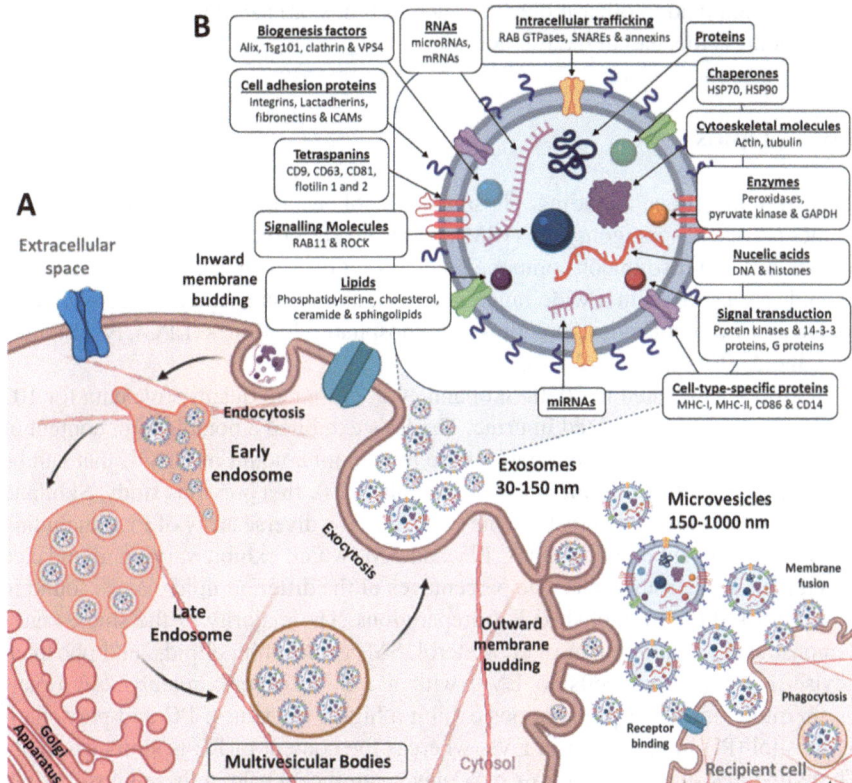

Fig. 2 The origin and secretion of EVs. (**a**) Extracellular vesicles (EVs) are formed by outward budding of the cell plasma membrane, which results in the formation of microvesicles. Alternatively, EVs are generated by the inward budding of the limiting membrane, which leads to the formation of early endosomes. These structures mature into late endosomes and, finally, multivesicular bodies (MVBs). The intraluminal vesicles (ILVs) within the lumen of the MVBs fuse with the plasma membrane to release ILVs, which are then referred to as exosomes. (**b**) A study of extracellular vesicle composition revealed that they can carry a variety of cargoes, including proteins, lipids, and nucleic acids. The content of these vesicles can vary significantly between cells and conditions. The specific composition of extracellular vesicles has a direct impact on their fate and function, underscoring the importance of selective cargo sorting mechanisms. Image of own origin made with biorender.com

and ICAMs), and cell type-specific proteins such as MHC-I, MHC-II, CD86, and CD14. The lumen of the EVs also contains cytoplasmic enzymes (e.g., peroxidases, pyruvate kinase, and GAPDH), as well as signaling molecules (e.g, RAB11 and ROCK) and signal transduction molecules (protein kinases and 14–3-3 proteins). Additionally, it discussed proteins involved in MVB biogenesis (e.g., Alix, Tsg101, and VPS4), chaperones (HSP70 and HSP90), cytoskeletal molecules (actin and tubulin), and nucleic acids (microRNA, mRNA, DNA, and histones) (Van Niel et al. 2018). Despite the paucity of literature on the subject, the existing research on EVs has focused on the proteomics and miRNAs of vesicular content, with relatively little attention paid to other molecules such as lipids or enzymes. In this article, we aim to address this gap in the literature.

8.6   Lipids

Phospholipids (PL), including phosphatidylcholines (PC), phosphatidylethanolamines (PE), and sphingomyelins (SM), represent key structural components of cell membranes and also major components of EVs. Glycerophospholipids and sphingolipids are involved in a wide range of cell signaling pathways and act as precursors to many biomolecules, including lysophosphatidylcholines (LPC), lysoPE, and eicosanoids (Banliat et al. 2020).

In a study published in 2020, Skotland et al. presented quantitative data for 107 lipid species in EVs excreted in urine. The EVs exhibited a notable high content of cholesterol (CHOL), reaching 63% of the maximum amount of CHOL that can be included in model membranes (Skotland et al. 2020). In a previous study, Skotland et al. (2019) demonstrated that bovine OF contains a diverse array of PLs, including PC, PE, lysophosphatidylcholines, PE, and SM, which exhibit varying abundance throughout the estrous cycle. The percentages of the different lipid classes found in EVs are based on 10 published EV preparations. The majority of the studies have demonstrated an enrichment of cholesterol, SM, glycosphingolipids, and phosphatidylserine (PS) from cells to EVs, with a ratio of approximately 2–3 times. Furthermore, most EV preparations exhibit a higher content of PC and phosphatidylinositol (PI) in cells than in EVs, whereas the content of PE is similar in both. These findings suggest that oEVs may play a significant role in these phospholipid profiles. Moreover, significant differences in various PC and SM were reported between in vitro-produced cattle blastocysts and their in vivo-conceived counterparts, indicating that the maternal environment modulates the phospholipid composition of oocytes and developing embryos (Skotland et al. 2019).

8.7   Proteins

Previous studies have investigated the protein content of vesicles in different species, including cattle (Almiñana et al. 2018; Piibor et al. 2023), sheep (Burns et al. 2016; Soleilhavoup et al. 2016), and pigs (Laezer et al. 2020); (Toledo-Guardiola et al. 2023). As previously stated, the content of EVs is highly regulated by hormonal changes and the cellular origin of the vesicles. Therefore, it is crucial to differentiate between vesicles from the same species based on their origin (uterus or oviduct) and their reproductive status (estrous cycle or pregnancy). The complete proteome of uterine and oviductal tissues and vesicles in pigs is well documented. Seytanoglu et al. (2008) studied the proteome of uterine and oviductal tissues, respectively, while Bidarimath et al. (2017) and Laezer et al. (2020) focused on the proteome of uterine and oviductal vesicles. In addition, previous studies have separately examined the proteome of EVs during pregnancy (Burns et al. 2016), and the estrous cycle (Kim et al. 2018), however (Toledo-Guardiola et al. 2023), was the first to compare the protein cargo of oEVs and uEVs of nonpregnant and pregnant sows with different ejaculate fractions of the boar ejaculate. Notably, this study identified four common proteins in the oEVs and uEVs proteome of inseminated and cyclic sows: 15SMg (2+)-ATPase p97 subunit, annexin (ANXA), glyceraldehyde-3-phosphate dehydrogenase (GAPDH), and tubulin alpha 1 chain (TUBA1C) (Toledo-Guardiola et al. 2023).

This study also identifies proteins that are exclusively present in oEVs from cyclic sows such as 14–3-3 domain-containing protein (YWHA), which is known to regulate normal oogenesis and oocyte maturation (Eisa et al. 2019) and galectin-1 (GAL1), which promotes the proliferation of porcine ovarian granulosa cells (Walzel et al. 2004). Some of the exclusive proteins in oEVs from pregnant sows were the heat shock protein 70 kDa (HSP70) which has been shown to enhance the in vitro survival of boar and bull spermatozoa when they are briefly exposed to the protein before interacting with oocytes (Elliott et al. 2009). Glyceraldehyde-3-Phosphate Dehydrogenase (GAPDH) is another protein found in oEVs from pregnant sows which increases the number of ciliated cells in the murine oviduct (Nakano et al. 2017).

Proteins found exclusively in the uEVs of cyclic sows were Clathrin heavy chain (CLTC), which is known to be involved in the development of the placenta and the associate vasculature during the early stages of pregnancy (Demari et al. 2016) and Beta-2 microglobulin (β-2 m), which is involved in immune recognition of foreign pathogens and transplanted tissues as well as in the discrimination of self from non-self (Joyce et al. 2008). Exclusive proteins in uEVs from pregnant sows were ezrin (EZR) which plays a key role in ZP pre-fertilization, hardening, and subsequent fertilization (García-Martínez et al. 2020), and heat shock protein HSP90-alpha (HSPAA901) which plays an important role in the controlled inflammatory response required for conceptus implantation and trophoblast growth (Jee et al. 2021) (Table 1).

8.8   miRNAs

The integration of EVs into the information transmission and control process introduces a further layer of complexity, as they carry a diverse array of nucleic acids, including small non-coding RNAs (sncRNAs), tRNA-derived sncRNAs (tRNA fragments), microRNAs (miRNAs), small nuclear RNAs, and mitochondrial-derived small RNAs (Schuster et al. 2016). Each subtype of small non-coding RNAs (sncRNAs) exhibits differences in their biogenesis, length, role, and mechanisms for accomplishing their likely biological functions (Jodar et al. 2013).

miRNAs are single-stranded sncRNA molecules with a length of 16–28 nucleotides. They function to regulate gene expression by interacting with the 3′ untranslated regions (3' UTRs) of mRNA target transcripts (forming semi-complementary structures between mRNA and miRNAs) and promoting degradation or inhibiting

Table 1 Some proteins present in the extracellular vesicles (EVs) of the sow reproductive tract with a potential role in reproductive processes

Protein	Function	Specie	Reference
Glyceraldehyde-3-phosphate dehydrogenase (GAPDH)	Increase the number of ciliated cells	Mouse	(Nakano et al. 2017)
Tubulin alpha 1 chain (TUBA1C)	Remodeling of the apical surface during uterine receptivity	Rat	(Kalam et al. 2018)
Annexin 2 (ANXA)	Membrane trafficking and fusion events	Bovine	(Almiñana et al. 2018)
14–3-3 domain-containing protein (YWHA)	Oogenesis / oocyte maturation	Mouse	(Eisa et al. 2019)
Galectin-1 (GAL1)	Proliferation of ovarian granulosa cells	Porcine	(Walzel et al. 2004)
Tubulin beta chain (TUBB1C)	Remodeling of the apical surface during uterine receptivity	Rat	(Kalam et al. 2018)
Heat shock protein 70 kDa (HSP70)	In vitro survival of boar and bull spermatozoa	Boar Bull	(Elliott et al. 2009)
Clathrin heavy chain (CLTC)	Development of the placenta and vasculature during pregnancy	Human	(Demari et al. 2016)
Beta-2 microglobulin	Immune recognition of foreign pathogens Discrimination of self from non-self	Porcine	(Joyce et al. 2008)
EZR	ZP hardening	Porcine	(García-Martínez et al. 2020)
HSP90-alpha	Implantation/trophoblast growth	Human	(Jee et al. 2021)

translation. The transfer of EV-derived unique miRNAs to recipient cells represents a previously undescribed mechanism of gene-based communication between mammalian cells, which ultimately regulates cell phenotype and function (Mathivanan et al. 2010).

A plethora of studies have demonstrated the pivotal role of microRNAs in the development and maturation of germ cells. It is evident that microRNAs are necessary for meiosis in germ cells, as demonstrated by Luense et al. (2009). Moreover, the absence of microRNAs in oocytes has been demonstrated to result in meiotic spindle defects, particularly during the I and II division. This suggests a potential role for these molecules in regulating chromosome segregation. In contrast to oocytes, the effects of microRNA absence on male germ cell maturation appear to be more widespread, including apoptosis, chromosomal misalignment, acrosome defects, morphological defects, and impaired motility (reviewed in Hilz et al. 2017). It has been demonstrated that miRNAs are predicted to target key elements in pathways related to follicular growth and oocyte maturation in mammals, including the wingless signaling pathway (WNT), transforming growth factor beta (TGF-β), mitogen-activated protein kinase (MAPK), neurotrophin, epidermal growth factor receptor (ErbB) pathways, and ubiquitin-mediated pathways (Santonocito et al. 2014).

In recent years, several studies have focused on miRNAs in FF or culture media as possible biomarkers for increasing IVF success rates (Rosenbluth et al. 2014). EV-encapsulated miRNAs, in particular, are protected from degradation and exhibit remarkable stability in biological fluids. This property may greatly facilitate the translation of the growing understanding of miRNA biology into clinical applications (Machtinger et al. 2016). Moreover, the investigation of EV-encapsulated miRNAs offers a number of key biological advantages over the analysis of total miRNAs. While total miRNAs in human biofluids or supernatants from cell cultures may be released from apoptotic cells or cell debris, EV-derived miRNAs are actively released by viable cells and are expected to represent an active means of communication between cells and tissues locally or systemically. In particular, miRNAs that are encapsulated by EVs may have a distinct role compared to miRNAs in biofluids, as they transfer biological information to recipient cells (Machtinger et al. 2016).

Research conducted by da Silveira et al. (2012) demonstrated that FF exosomes contain miRNAs which may contribute to follicular growth and oocyte maturation in the mare identifying changes in potential targets belonging to the TGF-β signaling member family in follicle growth and development in young and old mares. Moreover, subsequent studies have demonstrated that EV miR-125 and miR-199 are elevated in FF from preovulatory follicles and are present in equine follicular somatic cells such as granulosa cells. They target *LIF* and *PTGS2* genes, respectively, which are pathways known to be involved in oocyte maturation and cumulus expansion. Similarly, miR-21, miR-132, and miR-212 are examples of microRNAs that are regulated by hCG/LH in FF and periovulatory granulosa cells of mares. Predicted targets include *PTEN* and *BMPR2* genes which are involved in cell survival and final differentiation (Schauer et al. 2013).

In addition to proteins, epididymosomes carry microRNAs within the epididymal fluid. This suggests that microRNAs may be involved in epididymal function and gamete maturation. Similar to the ovarian follicle, regional-specific microRNAs are present in epididymal epithelial cells. Furthermore, the microRNA content of sperm changes as they traverse the epididymis, and these alterations are also associated with the presence of cell-secreted vesicles. Belleannée et al. (2013) demonstrated that bovine epididymosomes collected from the epididymal caput (head) and cauda (tail) regions contain different microRNA profiles and do not always reflect the repertoire present in epididymal cells. The let-7 and miR-200 families as well as miR-26a, miR-103, and miR-191 were found to be present in high abundance in both epididymosomal populations (Belleannée et al. 2013). In addition, miR-145, miR-143, miR-214, and miR-199 were found to be present in higher quantities in epididymosomes from the caput, while miR-654, miR-1224, and miR-395 were present in higher quantities in epididymosomes collected from the cauda region (Belleannée et al. 2013). Although the capacity of exosomal miRNAs to regulate gene expression in distal tissues has been demonstrated in other systems (Castaño et al. 2018), the regulatory effects of specific paternal exosomal RNAs on the female tract cells require experimental validation (Table 2).

Table 2 Some miRNAs present in the extracellular vesicles (EVs) of reproductive fluids and the associated pathways that they regulate

microRNAs	Source	Target genes	Signaling pathways and processes	Effect	Species	Reference
miR-125 **miR-199**	ffEVs	*LIF* *PTGS2*	Cell cycle Oocyte maturation Cumulus expansion	Differed between small and large follicles	Equine	(da Silveira et al. 2012)
miR-21 **miR-132** **miR-212**	ffEVs	*PTEN* *BTG2* *PTEN*	Cell signaling, survival and proliferation/differentiation	Regulated by hCG/LH in periovulatory granulosa cells Differed between small and large follicles	Equine	(da Silveira et al. 2012) (Schauer et al. 2013)
miR-145 **miR-143** **miR-214** **miR-199**	spEVs	*HERC1* *RASSF2*	Proteasomal degradation/ Ubiquitination pathway	Sperm maturation in the epididymis	Human	(Belleannée et al. 2013).
miR-34c-5p	oEVs	*BCL2*	Centromere regulation	Zygote development	Mouse	(Fereshteh et al. 2018)
miR-126–5p **miR- 296–5p** **miR-16** **miR-17-5p**	eEVs	*WNT* *Integrin*	VEGF signaling Inflammation by chemokines T and B cell activation	Cell proliferation, apoptosis, and angiogenesis	Porcine	(Bidarimath et al. 2017)
Bta-miR-98 **miRNA-499**	pEVs	*Unkown*	NF-kB signaling pathway	Regulate local inflammation	Bovine	(Zhao et al. 2018)

9 Reproductive Fluids EVs

9.1 EVs from the Male Reproductive Tract

9.1.1 Epididymosomes

Epididymosomes are EVs produced through apocrine secretion by epididymal epithelial cells. They consist of blebs that form at the apical pole of the epididymal secretory cells, containing selected organelles (including vesicles of various sizes), detaching from the cell and releasing their encapsulated cargo into the extracellular compartment upon disintegration (Sullivan 2015). This type of secretion does not involve the Golgi apparatus or the fusion of secretory vesicles with the plasma membrane before protein secretion and is probable that proteins of the cytoskeleton, mainly actin, are involved in this mechanism of apocrine secretion at the step of bleb release (Hermo and Robaire 2002).

The first studies of this field such as Yanagimachi et al. (1985) identified a population of small membranous vesicles, residing near the surface of epididymal spermatozoa in the Chinese hamster. These vesicles are found in the distal caput of the epididymis, decreasing in number in the cauda, and bind exclusively to the sperm region covering the acrosome (Yanagimachi et al. 1985). The existence of epididymosomes has been confirmed in the epididymal fluid of various mammalian species, including mice (Rejraji et al. 2006), bulls (Frenette et al. 2002; Frenette and Sullivan 2001), boars (Álvarez-Rodríguez et al. 2019), and humans (Thimon et al. 2008). These small structures ranging in size from 20–100 nm are characterized by their heterogeneous cargo of macromolecules and a membrane that is highly enriched in cholesterol (Sullivan 2015). In mice, epididymosomes contain high levels of sphingomyelin and polyunsaturated membranous fatty acids, particularly arachidonic acids. The concentration of sphingomyelin in epididymosomes increases during epididymal transit and represents half of the phospholipids in the caudal epididymosomes in murine cauda (Rejraji et al. 2006). It was hypothesized that these vesicles, which are rich in cholesterol, also could transfer cholesterol to the sperm membrane to stabilize it (Davis 1980).

9.1.2 Prostasomes

At ejaculation, cauda epididymal spermatozoa are mixed with secretions of the reproductive tract accessory glands (Ronquist and Hedstrom 1977). EVs result from the secretion of the prostate, and these accessory glands are called prostasomes, which are EVs secreted by exocytosis when the MVB fuse with the plasma membrane of prostate acinar cells (reviewed by Sullivan and Saez (2013). Prostasomes secreted by the human prostate were the first described and, to date, the most extensively studied prostasomes (Ronquist 2012).

As epididymosomes, prostasomes form a heterologous population of multilayered lipid membranes with regard to size and appearance at the electron microscopic level (Aalberts et al. 2012). Prostasomes have a high phospholipid ratio of approximately 2:1, with sphingomyelin accounting for half of their phospholipids (Arienti et al. 1999). This unique lipid composition gives these vesicles particular biophysical characteristics and contributes to their stability in acidic vaginal conditions (Arienti et al., 1998a, b). Apart from all these processes, it has been discovered that spEVs may transfer cholesterol, sphingomyelin, and saturated glycophospholipids to sperm after fusion, leading to a decrease in the fluidity of the sperm membrane and preventing early capacitation and a premature acrosome reaction (Baskaran et al. 2020).

9.2 EVs from the Female Reproductive Tract

9.2.1 Vaginosomes

Vaginosomes which are EVs released from the vagina (vEVs) were identified in the vaginal luminal epithelium and vaginal luminal fluid (VLF) (Al-Dossary et al. 2013). These EVs were observed in the lumen and also embedded between squamous epithelial and keratinized cells of vaginal cross-sections in mice, consistent with their endogenous origin (Fereshteh et al. 2019).

vEVs are characterized by their biochemical markers (CD9 and HSC70) and are known to contain PMCA4 protein, which is transferred to spermatozoa. However, this protein is present in a significantly lower abundance than its uterine and oviductal counterparts (Al-Dossary et al. 2013). This is in contrast to SPAM1, which has not been studied in the VLF, but has been detected in the vaginal luminal epithelium (which releases proteins into the luminal fluid) at significantly higher levels than in oviductal tissues (Zhang and Martin-DeLeon 2003).

These proteins are known to prevent premature sperm capacitation and acrosome reaction, thereby increasing the likelihood of P_4-induced sperm acrosome reaction proteins (Fereshteh et al. 2019). In addition, vEVs also contain PMCA4a, which sperm can utilize (Al-Dossary et al. 2013).

9.2.2 Uterosomes

EVs released from the uterus are known as uterosomes (uEVs) and appear to be generated by the apocrine pathway, similar to other reproductive EVs. In the mouse system, the protein cargo carried by uterosomes is hormonally regulated, appearing at high levels during proestrus/estrus and only marginally present during diestrus/metestrus (Martin-DeLeon 2016). The presence of exosomal biochemical markers such as CD9 and CD63 on the apical surface of human endometrial epithelium suggests the involvement of the apocrine pathway in the formation of uterosomes, which have been found to be positive for tetraspanins CD9- and CD63- (Ng et al. 2013). Uterosomes deliver proteins from their cargo to spermatozoa, affecting their function and promoting sperm capacitation during their transit through the female genital tract (Martin-DeLeon 2016). They modulate reproductive processes such as follicular development in the ovary and oocyte maturation (Machtinger et al. 2016), maternal-embryonic communication (Almiñana et al. 2017), and the pregnancy establishment in mammals. All of this may indicate a specific mechanism utilized by the uterine microenvironment to aid in the fertilization process (Burns et al. 2014) and early embryonic development (Burns et al. 2016).

9.2.3 Oviductosomes

The oviductosomes (oEVs) are derived from the endocytic secretion of the OECs, which are part of the OF. These vesicles are known to release their bioactive contents into the extracellular environment via an apocrine pathway (reviewed by Harris et al. 2020). In recent years, oEVs have received considerable attention due to their potential to serve as natural nano-shuttles transporting essential components from the oviduct to gametes and embryos.

They play important roles in sperm capacitation, fertilization, and early embryo development (Almiñana et al. 2018). Research has demonstrated that oEVs bind to and fuse with the head and the intermediate portion of spermatozoa when they detach from the SR (Caballero et al. 2013). This transfer of protein content aids in the acquisition of a specialized flagellar movement known as hyperactivation (Soriano-Úbeda et al. 2017). According to Visconti (2009), hyperactivation increases the bending amplitude of the sperm tail which can overcome the binding between the sperm and the epithelium, acquiring the ability to fertilize the oocyte. In fact, authors such as Toledo-Guardiola et al. (2024a, b) have shown that the incubation of sperm with porcine oEVs from different stages of the estrous cycle plays a modulatory role in sperm functionality by interacting with spermatozoa, affecting motility and capacitation, and participating in sperm-oocyte interaction.

Furthermore, research has demonstrated that supplementing embryo development media with EVs derived from murine OF can enhance blastocyst rate and quality (Qu et al. 2017). This technique has also been shown to improve embryo cryosurvival in bovine (Lopera-Vasquez et al. 2016) and porcine (Fu et al. 2022). It is important to note that the size of oEVs can vary between different species (Alcântara-Neto et al. 2019). Additionally, the secretion of these EVs is under the hormonal influence during the estrous cycle, which can cause changes in both size and protein cargo (Laezer et al. 2020). Alcântara-Neto et al. (2019) were the first to confirm the existence of porcine oEVs and identified two subpopulations within them differentiated by their diameter. The first group, corresponding to exosomes, has a diameter of 30–150 nm. The second group, classified as MVs, has a diameter between 150–1000 nm.

Once spermatozoa enter the female genital tract, they encounter uEVs and oEVs, respectively. These EVs are transferred to the male gamete to enhance hyperactive sperm motility and fertilization potential (Nguyen et al. 2016). Spermatozoa can take up EVs from different parts of the female tract which can affect their competition (Bridi et al. 2020). It has been shown that EV exchange may be proposed as an emerging pathway through which female reproductive tract cells interact with spermatozoa (Murdica et al. 2020).

10 Interactions of EVs with Gametes

10.1 Effects of Seminal Plasma EVs (spEVs) on Sperm

The occurrence of membranous vesicles along the male reproductive tract and in the ejaculated semen appears to be a common feature among different species, including humans (Sullivan and Saez 2013). The primary function of these vesicles in sperm physiology is post-testicular sperm processing and protection. While epididymosomes transfer proteins and other molecules to sperm cells, thereby facilitating their development of the requisite functions for fertilization and maturation which occurs within the epididymis (Sullivan 2015), prostasomes provide a protective barrier that enhances sperm survival and functionality within the female reproductive tract, thereby increasing their chances of successfully reaching and fertilizing the egg (Aalberts et al. 2014) (Fig. 3).

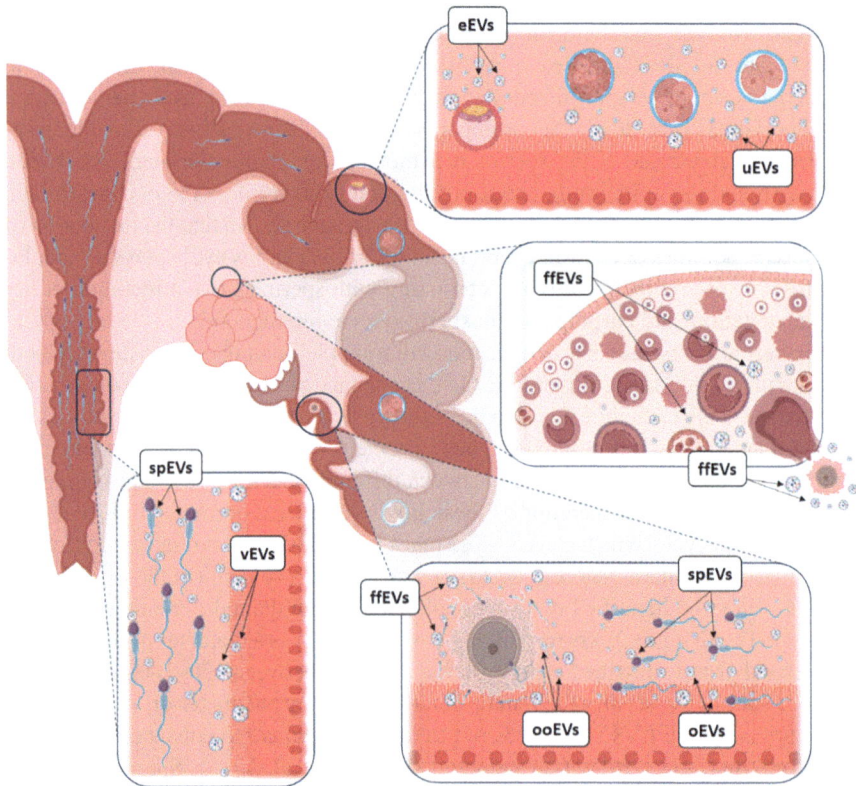

Fig. 3 The interaction of extracellular vesicles with sperm in the reproductive tract of the sow. Upon entering the female reproductive tract, sperm are mixed with SP-containing extracellular vesicles (EVs) from the seminal plasma (spEVs; epididysomes, and prostasomes). Subsequently, these extracellular vesicles interact with the extracellular vesicles of the vagina (vEVs). Upon passing the cervix and entering the uterus, the sperm maintain rapid contact with the uterine extracellular cells (uEVs) and continue towards the oviductal space, where they mix with the extracellular vesicles of the oviduct (oEVs). At this juncture, the sperm reach the ampullary isthmic junction, the site of fertilization. Upon contact with the extracellular vesicles of the oocyte (ffEVs and ooEVs), fertilization occurs. Once fertilization has occurred, the resulting embryo begins to release embryonic extracellular vesicles (eEVs). These vesicles come into contact with the oEVs and uEVs on the journey of the embryo to the implantation site in the female reproductive tract. Image of own origin made with biorender.com

10.1.1 Epididymosomes

It is known that epididymosomes, delivered to the intraluminal compartment of the epididymis, are involved in transferring proteins to the maturing spermatozoa. When epididymosomes collected from a proximal epididymal segment are incubated in vitro with spermatozoa collected distally, only a subset of epididymosome-associated proteins is transferred to the male gamete. Additionally, it has also been shown that this transfer is saturable over time, and it is temperature and pH dependent (Sullivan 2015). Some researchers have demonstrated that the transfer of biotinylated proteins from caput epididymosomes to the acrosomal cap and the midpiece of mouse spermatozoa is temperature dependent, with maximum transfer occurring between 32 °C and 37 °C (Aalberts et al. 2014). Zhou et al. (2018) suggested that the binding mechanism of epididymosomes could be related to key elements including SNARE proteins. These proteins have well-described roles in regulating membrane fusion activity in alternative tissue models. This activity requires the complementary action of different SNARE proteins contributed by vesicles (v-SNARE proteins) and target (t-SNARE proteins) membranes.

Research has demonstrated that bovine epididymosomes collected from different segments of the epididymis can transfer distinct protein repertoires to spermatozoa. Additionally, an overabundance of one population of epididymosomes (collected from the caput segment) does not significantly affect the transfer efficiency of the other population (collected from the cauda segment) during simultaneous co-incubation with spermatozoa (Frenette et al. 2006). In bull, epididymosome-associated proteins that are known to be transferred to sperm surface during epididymal maturation include enzymes of the polyol pathway, aldose reductase, sorbitol dehydrogenase, and macrophage migration inhibitory factor (MIF) cytokine, which are associated with epididymosomes. This association suggests that these vesicles may play a role in modulating sperm motility during epididymal maturation (Frenette et al. 2004). In addition to what was described above, epididymosomes release proteins such as glutathione peroxidase 5 (GPX5), which protects the transiting epididymal spermatozoa from oxidative stress and preserves their DNA integrity (Chabory et al. 2009). Furthermore, epididymosomes have a dual role in protecting sperm and eliminating defective sperm cells through association with ubiquitin, which is involved in enzymatic degradation of proteins by proteasome (Sutovsky et al. 2001), and epididymal sperm binding protein 1 (ELSPBP1) with these small membrane vesicles (D'Amours et al. 2012) suggests a potential role in eliminating defective spermatozoa during the epididymal transit.

Taken together, epididymosomes have multiple functions aimed at the maturation of mammalian spermatozoa, ultimately generating fully functional male gametes through the acquisition of new proteins during epididymal transit (Saez et al. 2003). The biochemical mechanisms underlying the transfer of macromolecules from epididymosomes to maturing spermatozoa are elusive and the subject of ongoing research.

10.1.2 Prostasomes

Prostasomes, secreted by the prostate during ejaculation, appear to play a protective role for ejaculated spermatozoa and modulate some sperm physiological parameters such as motility and capacitation through membrane fusion with the sperm (Sullivan and Saez 2013). The mechanism responsible for the protein transfer across prostasomes has not yet been elucidated. However, several hypotheses have been proposed for the cell-to-cell transfer of proteins from prostasomes. Briefly, the acquisition could be mediated in one of three ways: (1) through the intervention of plasma lipid carrier proteins, (2) by interactions between a donor and an acceptor membrane via a "flip-over" phenomenon, or (3) through vesicles that could be processed by endocytosis and deliver the carried proteins back to the plasma membrane of the acceptor cell (Ilanguramaran et al. 1996). It is known that the protein transfer mechanism is pH and zinc concentration dependent (Frenette et al. 2002). This is consistent with the fact that it has been demonstrated that the fusion between human prostasomes and ejaculated sperm is favored at a slightly acidic pH (Arienti et al. 1997). Although prostasomes fused with spermatozoa at pH 4–5, a non-physiological condition (Arienti et al. 1997), the interaction of these membrane vesicles with spermatozoa was thought to be hydrophobic in nature (Ronquist et al. 1990). Therefore, it is possible that the pH of the vaginal contents is low enough to allow fusion to occur, even considering the alkaline pH of prostatic secretions, which prevents early EV sperm fusion. From this perspective, fusion with prostasomes could be a mechanism to protect spermatozoa (Arienti et al. 1997).

Prostasomes are rich in divalent cations, such as calcium, zinc, and magnesium. They are believed to regulate the concentration of these cations near spermatozoa, which helps to modulate flagellar motility (Fabiani et al. 1995). The mechanism by which prostasomes inhibit or prevent the progression of capacitation is not yet known. However, the transfer of cholesterol to the sperm cell's plasma membrane, independent of membrane fusion, is a potential candidate (Cross and Mahasreshti 1997). Cholesterol is a known inhibitory factor of capacitation, as determined by sperm response to P_4. Prostasomes which represent approximately 40% of the total cholesterol present in SP were found to inhibit the P_4-stimulated acrosome reaction of human spermatozoa in vitro (Cross 1996). Notably, however, while prostasomes required capacitating conditions to bind to live sperm cells, they affect sperm function by regulating sperm membrane fluidity, subsequently reducing the level of Tyr-P-preventing premature capacitation, and delaying acrosomal reaction (Aalberts et al. 2013; Pons-Rejraji et al. 2011).

Prostasomes are also rich in GDP, ADP, and ATP, and many proteins on their surface possess a catalytic activity (Fabiani et al. 1994). According to Saez et al. (2000), prostasomes interfere with the production of reactive oxygen species (ROS) by polymorphonuclear neutrophils in blood or semen. This suggests that prostasomes may inhibit the depletion of cholesterol from the sperm plasma membrane by interfering with sperm cell ROS production. Prostasomes may also inhibit progression to late capacitation events, thus preventing premature induction of the acrosome reaction (Samanta et al. 2018).

Furthermore, prostasomes are proposed to protect ejaculated spermatozoa against immune responses in the female reproductive tract. The immunosuppressive activity of prostasomes is due to the presence of several complement inhibitory molecules such as CD55 and CD46, both of which inhibit the C3-convertase, and CD59 (protectin), which inhibits the formation of the membrane attack complex (Saez et al. 2003). This is achieved by modulating the complement pathway, inhibiting monocyte and neutrophil phagocytosis, and lymphocyte proliferation (Aalberts et al. 2014). Additionally, they possess antibacterial properties, providing supplemental protection to ejaculated spermatozoa (Saez et al. 2003).

Taken together, the diverse functions of prostasomes aim to protect spermatozoa after ejaculation, preserving their full functional capacities before encountering the oocyte (Saez et al. 2003).

10.2   Effect of Vaginal EVs (vEVs) on Sperm

Upon entering the female reproductive system, sperm encounter the vaginal environment and vaginal EVs (vEVs), which impact sperm function (Fig. 3). It has been suggested that vaginosomes transfer proteins that are already phosphorylated in tyrosine residues to spermatozoa during a 15 to 30 min incubation period (Franchi et al. 2020). Additionally, a similar pattern of estrogen-regulated expression of uterine sperm adhesion molecule 1 (SPAM1) was also detected. These proteins were identified not only in uterosomes but also in EVs in the oviduct, as well as in the oviductal and vaginal epithelia and/or fluid, where they are likely to be present in EVs (Martin-DeLeon 2016).

Interactions between sperm and vEVs have been shown to be pH dependent. This fits well with the known alkaline pH of prostatic secretions, which prevents early EV/sperm fusion, and the acidic pH of the vagina, which acts as a trigger for fusion (Arienti et al. 1997; Frenette et al. 2002).

10.3   Effects of Uterine EVs (uEVs) on Sperm

The environment in the uterus and fallopian tube plays a critical role in sperm survival, transport, and capacitation. This decisive effect is most likely carried out via the oviduct and the EVs present in the uterus (Fan et al. 2023). In recent years, some studies have reported the presence of EVs in UF and OF (named uterosomes and oviductosomes, respectively) associated with diverse physiological effects, for instance, embryo-endometrial interaction and sperm motility (Al-Dossary et al. 2013; Burns et al. 2014; Griffiths et al. 2008; Machtinger et al. 2016; Ng et al. 2013) (Fig. 3).

Previously, it was thought that the uterine environment did not affect spermatozoa due to their rapid transit through this region of the reproductive tract. However, Franchi et al. (2016) demonstrated that a brief co-incubation of uEVs secreted by EECs with human spermatozoa rapidly interacted with the sperm plasma membrane, stimulating sperm capacitation. The study found that the level of induced acrosome reaction and the induction of Tyr-P increased, undergoing the acrosome reaction and enhancing the fertilizing capability.

Furthermore, uterosomes can deliver hyaluronidase PH-20 (SPAM1) to the sperm membrane. The acquisition of SPAM1, by mature caudal epididymal sperm from uEVs in vitro, appears to be involved in the maturational process of capacitation as it enhances hyaluronic-acid-binding ability and increases oocyte cumulus penetration efficiency (Zhang and Martin-DeLeon 2003). Additionally, it induces acrosome reaction related to calcium signaling to promote fertilization (Griffiths et al. 2008). This suggests that in vivo transfer may occur during sperm transit in the female tract and that this transfer may be a component of the capacitation process (Martin-DeLeon 2016). Griffiths et al. (2008) argued that acquiring maximal levels of SPAM1 on the surface of sperm only after arrival in the ampulla is advantageous, as fertilization occurs at the ampullary-isthmic junction and SPAM1 is implicated in the acrosome reaction. Therefore, the incremental acquisition of SPAM1 on sperm before arriving in the ampulla serves to prevent premature acrosome reaction before arriving at the fertilization site.

In addition, several authors have demonstrated the role of EVs from the female reproductive tract origin to modulate the sperm functionality (Almiñana et al. 2018; Laezer et al. 2020), and as it is plausible to think, the function of these uEVs on the sperm functionality varies at different stages of the cycle. A recent study by Piibor et al. (2023) has demonstrated that the protein content of uEVs varies during different phases of the estrous cycle in bovine. This is consistent with findings by Murdica et al. (2020) which showed that the uptake of uEVs by spermatozoa is significantly higher when the EVs are derived from endometrial cells in the proliferative stage in women. They found that uEVs derived from the proliferative stage can improve sperm capacitation and fertilization ability more than those uEVs derived from the secretory stage, resulting in a significantly higher response (Murdica et al. 2020).

10.4 Effects of Oviductal EVs (oEVs) on Sperm

Numerous studies have been conducted on the OF and its regulation of sperm function and polyspermy, particularly in farm animals. In the past decade, studies have shifted their focus to the effects of oEVs on sperm function (Harris et al. 2020). EVs from the OF, also known as oviductosomes, were identified for the first time in the murine oviduct (Al-Dossary et al. 2013) (Fig. 3).

As the mechanism of EVs cargo delivery to spermatozoa has been shown to be fusogenic, during estrous, oviductosomes were reported to carry transmembrane proteins that can be transferred to sperm in vitro (Al-Dossary et al. 2013), highlighting SPAM1, which plays multiple roles in sperm function (Martin-Deleon 2011). This protein has a relevant role in sperm capacitation, and their levels depend on hormonal status (Al-Dossary et al. 2013; Griffiths et al. 2008). Al-Dossary et al. (2013) also described the implications of oEVs in sperm storage, promotion of capacitation, regulation of the acrosome reaction, and the hyperactivated motility in mice. They have demonstrated that plasma membrane calcium ATPase 4 (PMCA4), a Ca^{2+} ion discharge pump, is secreted by murine oviduct epithelial cells in oEVs and interacts with sperm. Mice oEVs can transfer PMCA4 to spermatozoa, promoting Ca^{2+} outflow and inducing sperm capacitation, sperm hyperactivation, and the acrosomal reaction. The researchers have also shown some years later that the mechanism by which EVs transfer PMCA4 and other transmembrane proteins to the sperm plasma membrane at various parts, including the acrosome, head, midpiece, and principal piece, may be related to fusion proteins, $\alpha v\beta 3$ integrin (αv subunit), and CD9 (Al-Dossary et al. 2015).

Other authors like Franchi et al. (2020) found that oEVs collected from the isthmus and ampulla incorporate into sperm cells within a few minutes, reaching a maximum after 2 h. This process maintains cell survival and stimulates protein Tyr-P and intracellular Ca^{2+} increase in bovine spermatozoa in vitro, suggesting that they trigger signals involved in capacitation. It has been shown that cat oEVs can also bind to spermatozoa at the acrosome and middle part, increasing sperm motility and maintaining acrosome integrity, as well as improving sperm fertilization ability. This could be due to the many proteins found in oEVs that promote sperm motility and fertilization, such as oviduct-specific glycoprotein (OVGP1), T-complex protein 1 subunit gamma, and T-complex protein 1 subunit alpha (de Ferraz et al. 2019).

In addition, the proteasome-mediated ubiquitin-dependent protein degradation process and positive regulation of RNA polymerase pathways were found to be among the top five most enriched biological processes in oEVs. Proteins belonging to the ubiquitin-proteasome system (UPS) have been shown to play a key role in sperm surface remodeling during capacitation. These proteins are involved in the formation of the oviductal SR, which helps to reduce polyspermy and supports and establishes a sequential release of freshly capacitated spermatozoa (Zigo et al. 2019). The removal of some proteins (i.e., spermadhesins) covering the surface of the spermatozoa appears to be necessary for the disruption of the sperm-oviductal epithelium interaction, with some studies suggesting that this event is regulated by UPS-dependent proteolysis (Zigo et al. 2023).

Finally, authors (Alcântara-Neto et al. 2019) demonstrated that adding oEVs to the IVF culture medium reduces polyspermic penetrations in vitro in porcine species. They showed that supplementation with oEVs during IVF increases monospermy, improving the efficiency of IVF compared with the control. The hypothesis is that proteins such as OVGP1 and myosin heavy chain 9 (MYH9) from the oEVs bind to the ZP, modifying its carbohydrate and protein composition causing subsequent ZP hardening and preventing polyspermy. Some years before, further experiments have shown that oEVs can pass through the ZP and are known to be internalized by blastomeres (Saadeldin et al. 2014). Later studies have found that annexin proteins from porcine oEVs bind to the sperm, preventing many of them from reaching the oocyte, thus improving IVF efficiency. Additionally, porcine oEVs were found to reduce sperm motility while maintaining their viability, which is associated with the oviductal SR (Alcântara-Neto et al. 2020).

10.5 Effects of Oocyte EVs (ooEVs) on the Sperm during Fertilization

Although the molecular basis of fusion in this context is not yet fully understood, research studies have shown the ability of oocyte EVs (ooEVs) to fuse with sperm and induce an acrosome reaction in an in vitro porcine model (Siciliano et al. 2008). The most exciting part is that recent research studies have shown that oocytes can communicate with sperm through ooEVs before fertilization (Fig. 3). For instance, Miyado et al. (2008) have demonstrated that mature eggs also deliver ooEVs to sperm, providing a new and exciting insight into a previously unknown aspect of fertilization. Additionally, Barraud-Lange et al. (2007) demonstrated in a mouse model that oocyte-derived EVs from the plasma membrane are capable of transferring proteins to the fertilizing sperm in the perivitelline space (PVS) before direct interaction between both gametes. The transfer of these ooEVs is crucial for the reorganization of the sperm membrane and its fusion with the oocyte. These findings suggest a potential mechanism for EV-mediated fertilization and warrant further investigation (Barraud-Lange et al. 2012).

The transfer of oocyte membrane fragments containing CD9 tetraspanin onto the sperm head, which does not express CD9 otherwise, before gamete fusion was observed in mice. The fusion between the sperm and CD9 null oocytes can be rescued through co-incubation with wild-type oocytes (Barraud-Lange et al. 2007). Another tetraspanin, CD81, produced primarily by cumulus cells and localized to the inner region of the ZP, plays a crucial role in fertilization, specifically in the acrosome reaction and fusion-related events before membrane fusion. Both CD81 and CD9 are transferred to the sperm via EVs when the sperm penetrates the PVS (Miyado et al. 2008). Further supporting these oEVs-sperm interactions, it has been shown that miRNA-34c-5p can be transferred from oEVs to the sperm head. Since the first cleavage of the zygote requires miRNA-34c-5p to act on the centromere, and miRNA-34c-5p can only be obtained from sperm, oEVs can indirectly affect zygote development via sperm (Fereshteh et al. 2018).

11 Interactions of EVs with the Embryo

11.1 Effects of Oviductal EVs (oEVs) on the Embryo

The initial stages of embryo development in eutherian mammals occur in the oviduct, where the embryo spends approximately 4–8 days, depending on the species. The OF contains oEVs, which are released from the oviductal epithelium and then internalized by preimplantation embryos, ultimately improving embryo development and quality through the bioactive molecules contained in these vesicles (Fu et al. 2020) (Fig. 3). This growing evidence is supported by several studies, such as Almiñana et al. (2017) who isolated oEVs from bovine OF and identified the presence of OVGP1, which can bind to and then interact with the ZP of the oocyte or early embryo, causing the ZP to block polyspermy fertilization and ensuring normal embryo development (Coy et al. 2008). Additionally, it is also known that during ciliary beating and oviductal muscle contractions, OVGP1 can prevent the spread of nutrients and ions, thereby stabilizing the microenvironment of early embryos (Hunter 1994).

EVs produced by healthy oviducts are beneficial for embryo development. For example, in pigs, oEVs can promote embryonic development by increasing blastocyst formation rate and attachment ability (Fang et al. 2022). The content of oEVs may be the cause of the beneficial effect, not only by promoting early embryo development but alleviating the damage to the embryo from adverse factors. In fact, it has been shown that supplementing embryo development media with oEVs can improve blastocyst rate and quality in rodents (Qu et al. 2019) and enhance embryo cryosurvival in bovine (Lopera-Vasquez et al. 2016). Authors such as Harris et al. (2020) suggested some years ago that adding oEVs to the in vitro culture conditions of spermatozoa could improve the quality of pig embryo production by better mimicking the oviductal environment.

The secretion of oEVs and uEVs has been demonstrated to be time dependent and specific to the physiological status of the female genital tract (reviewed by Bidarimath et al. 2021). Toledo-Guardiola et al. (2023) examined the proteome of oEVs from post-fertilized sows, we observed that some of the most enriched biological processes were the vesicle mediated-transport pathway, which contains proteins previously reported in bovine oEVs, such as heat shock protein family A member 8 (HSPA8) and heat-shock protein family A (HSPA70) (Almiñana et al. 2017), which play important roles in the gamete/embryo-oviduct interactions. HSPA8 and HSP70 are induced chaperones released by the porcine oviductal epithelium upon environmental stress to maintain protein structural homeostasis (Liman 2017; Nollen and Morimoto 2002). In addition, HSP70 has been shown to increase in vitro survival of boar and bull spermatozoa when spermatozoa are briefly exposed to HSP70 in vitro prior to interacting with oocytes (Elliott et al. 2009).

Another of the most enriched biological processes found in oEVs was the innate immune response pathway, where we found proteins such as glyceraldehyde-3-phosphate dehydrogenase (GAPDH). Researchers, such as Almiñana et al. (2018) and Nakano et al. (2017), found that GAPDH mRNA is also a component of mouse oEVs isolated from oviductal mesenchymal cell lines, where they demonstrated that these specific mRNAs exert a functional effect by increasing the number of ciliated cells and improve embryo quality (Lopera-Vasquez et al. 2017).

11.2   Effects of Uterine EVs (uEVs) on the Embryo

During pregnancy, uEVs consist primarily of EVs produced by uterine gland epithelial cells (Hu et al. 2022; O'Neil et al. 2020) (Fig. 3). The embryo may respond differently to endometrial EVs at different times to prepare for implantation. The different effects of uEVs on embryos at different stages of the pregnancy may be due to different hormone levels acting on the cargo carried by endometrial EVs. In addition, to increase the number of uEVs produced by the endometrium, P_4 can also regulate the miRNA content (Fan et al. 2023). For instance, EVs secreted at the ninth day of gestation can promote embryonic trophoblast cell proliferation and migration, whereas uEVs at the 12th or 15th day of gestation can inhibit this embryonic process (Hu et al. 2022).

Researchers such as Zhou et al. (2020) profiled EVs obtained from serum samples of pregnant pigs on days 9, 12, and 15 after insemination. They concluded that miR-92b-3p and miR-17-5p could be identified in the serum exosomes as early as the ninth day of pregnancy. Therefore, these miRNAs could serve as biomarkers for early pregnancy. It has been shown that during embryonic development, EECs cultured with estrogen and P_4 produce uEVs with higher levels of laminin subunit alpha-5 (LAMA5), which mediates cell attachment, migration, and tissue invasion, compared to EECs cultured with estrogen alone. These cells, via the order of uEVs, activate the adhesive FAK signal and upregulate the expression of FAKTYR397, which can enhance embryonic trophoblast cell adhesion (Greening et al. 2016).

It has been suggested that the endometrial epithelium releases EVs that are involved in the transfer of signaling miRNAs and adhesion molecules either to the blastocyst or to the adjacent endometrium in the uterine cavity, which, in turn, may affect endometrial receptivity and implantation (Machtinger et al. 2016). Analysis of these uEVs from cyclic and pregnant sheep revealed miRNAs and proteins expressed by both the conceptus trophectoderm and the endometrial epithelium, such as cathepsin L1 and prostaglandin synthase 2. In sheep, endogenous beta-retroviruses (enJSRVs) play a critical role in regulating conceptus trophectoderm development and placental growth (G. Burns et al. 2014). P_4-induced miRNA can promote embryonic adhesion by regulating PI3K/AKT, BMP, and small RNA post-transcriptional silencing (G. W. Burns et al. 2018).

Interestingly, GO analysis identified the protein stabilization pathway as one of the most enriched pathways in inseminated sows. Within this pathway, there are proteins such as HSP90 Alpha Family Class B Member 1 (HSP90AB1) and HSP90 Alpha Family Class A Member 1 (HSP90AA1) (Toledo-Guardiola et al. 2023). HSP90 is a cofactor for steroid hormone receptors and is released from the receptor complex upon hormone-receptor ligand interaction produced by uterine cells (Neuer et al. 2000). HSP90 also plays an essential role in the controlled inflammatory response required for conceptus implantation and trophoblast growth (Jee et al. 2021).

In the uEVs of pregnant sows, it has also been found that the protein localization to the plasma membrane pathway has some proteins such as Rap1, a member of the GTPase family. It is normally in the active GTP-bound form, which is the major regulator of cell-cell junction (Gaonac'h-Lovejoy et al. 2020) and cell adhesion (Boettner and Van Aelst 2009). Furthermore, Rap1 has been implicated in embryo implantation, as Rap1 can induce cell-cell junction stabilization (Gaonac'h-Lovejoy et al. 2020). Activation of Rap1 has been shown to be involved in the establishment of cell polarity (Freeman et al. 2017), suggesting that Rap1a is involved in the repair of endometrial injury (T. Zhang et al. 2023), including physiological processes that occur during the establishment of pregnancy, such as embryo implantation and uterine fibroid development (Toledo-Guardiola et al. 2023).

During the peri-implantation period, there is communication between the mother and the conceptus, and the EVs produced both in the oviduct (Mazzarella et al.

2021) and the uterus (Kusama et al. 2018) are involved in this communication. It has also been suggested that the endometrial epithelium releases EVs that are involved in the transfer of signaling miRNAs and adhesion molecules either to the blastocyst or to the adjacent endometrium, affecting endometrial receptivity and implantation (Machtinger et al. 2016).

11.3   Interactions of Embryonic EVs (eEVs)

Some authors have identified proteins secreted by EVs released by the conceptus trophectoderm and the endometrial epithelium, such as cathepsin L1 and PG synthase 2 in sheep, which play a critical role in regulating conceptus trophectoderm development and placental growth (Burns et al. 2014) (Fig. 3). The first evidence of embryo-derived EVs (eEVs) was demonstrated by the co-culture of porcine parthenogenetic and somatic cell nuclear transference embryos (Saadeldin et al. 2014). EVs have also been found in the chorioallantoic membrane and porcine trophectoderm on day 20 of pregnancy. Porcine trophectoderm-derived EVs have been shown to promote the proliferation of aortic endothelial cells, which have miRNAs predicted to modulate angiogenesis and placental developmental pathways, playing an important role in the communication between the conceptus and maternal endometrium affecting the establishment of pregnancy (Bidarimath et al. 2017). The authors showed that EVs derived from porcine trophectoderm cells (eEVs) contain several crucial proteins and miRNAs including miR-126–5P, miR- 296–5P, miR-16, and miR-17-5p, which play a major role in cell proliferation, apoptosis, and, most importantly, angiogenesis. One of the miRNAs packaged in exosomes derived from porcine umbilical cord blood, miR-150, stimulated the proliferation, cell migration, and tube formation of umbilical vein endothelial cells (Bidarimath et al. 2017).

Studies suggest that the miRNA cargo present in the eEVs can modulate the immune response, ultimately helping in achieving immune tolerance during early pregnancy. Researchers such as Zhao et al. (2018) reported that bta-miR-98 is a likely regulator of the maternal immune system during bovine pregnancy based on the miRNA profiles of EVs and bioinformatics analysis. Similarly, miRNA-499 was found to regulate local inflammation at the bovine maternal-fetal interface by inhibiting NF-kB signaling. Disruption of miR-499 can increase the risk of pregnancy failure due to severe local inflammatory processes, placental resorption in early pregnancy, or even fetal growth restriction.

12   ARTs with EVs

12.1   Future Approaches with Reproductive EVs

12.1.1   EVs as Predictive Biomarkers

The study of EVs in reproduction has the potential to expand our current understanding of the normal physiology of the reproductive process, including the identification of high-quality sperm and oocytes as well as pathological conditions such as implantation failure. EVs hold promise for identifying noninvasive biomarkers and for developing novel therapies to increase reproductive success (Machtinger et al. 2016). Several studies have focused on miRNAs in FF or culture media as potential biomarkers for increasing IVF success rates (Rosenbluth et al. 2014). In particular, miRNAs encapsulated in EVs are shielded from degradation and are remarkably stable in biological fluids. This property may greatly facilitate the translation of the growing understanding of miRNA biology into clinical applications (Machtinger et al. 2016).

Biomarkers for Female Reproductive Cancer

Although tissue biopsies are considered the gold standard for cancer evaluation, they are invasive and provide only a small sample of the whole tumor. It has been shown that EVs can eliminate these limitations (Capra and Lange-Consiglio 2020). Collecting body fluids for screening certain diseases could avoid the need for invasive examinations (Lin et al. 2015). The molecular cargo inside the EVs is protected from degradation and can serve as biomarkers for noninvasive cancer diagnosis (Chen et al. 2018). Furthermore, miRNA analysis from certain EVs may be used to distinguish between benign and malignant diseases. Indeed, it has been shown that the expression levels of miR-200 family members (miR-200a, miR-200b, and miR-200c) could differentiate between malignant and benign ovarian tumors with a sensitivity of 88% and specificity of 90% (Meng et al. 2016).

Biomarkers for Female Fertility

Phthalates and phenols are classes of potential endocrine-disrupting chemicals found in the present mainly in the environment and in food. These components have been shown to influence fertility in women (Machtinger et al. 2018). A correlation has been identified between the concentration of phthalates and phenols and their impact on the EV-miRNA profile in FF, with a negative impact on female fertility. Increased urinary concentration of phenols has been associated with altered expression of miR-125b, miR-24, and miR-375, which have been shown to play an important role in animal oocyte maturation and fertilization (Martinez et al. 2019). In rodents, overexpression of a miR-125b mimic blocks the expression of specific genes required for embryos to progress beyond the two-cell stage (Kim et al. 2016). Elevated levels of miR-24 in bovine culture media have been linked to failed embryo differentiation (Kropp and Khatib 2015). Meanwhile, overexpression of miR-375 in granulosa cells and oocytes has been found to block proliferation, increase the apoptosis rate of cumulus cells in cows (Chen et al. 2017), and suppress estradiol production and follicular growth proliferation in porcine granulosa cells (Yu et al. 2017).

Biomarkers for Embryo Quality

An important technique used in human medicine to detect and prevent transmission of single-gene disorders is the removal of cells from the trophectoderm by biopsy (Gutiérrez-Mateo et al. 2009). Identifying specific markers for embryo quality can eliminate the need for embryo biopsy and its inherent risks. For instance, the levels of miRNA-20a and miRNA-30c levels in eEVs were significantly higher than those in unimplanted embryos, and both had some predictive value for embryo implantation (Capalbo et al. 2016). In recent years, studies have shown that miRNAs found in EVs in plasma and serum can be used as a diagnostic tool for embryo implantation. In addition, uEVs can also indicate the readiness of the endometrium for embryo implantation, as evidenced by the upregulation of miRNA-34c-5p (Tan et al. 2020). Moreover, miRNA-362-3p expression in uEVs was found to be significantly higher in nonpregnant patients following controlled ovarian stimulation-embryo transfer than in pregnant patients (Li et al. 2020).

Biomarker for Early Abortion

Placental EVs (pEVs) were isolated from maternal blood using a chromatographic/immunosorbent procedure with antibody agarose beads against placental alkaline phosphatase protein (PLAP) which is present exclusively on EVs secreted from placentae. pEVs can be distinguished from other EVs by the presence of placenta-specific proteins (e.g., PLAP4) and miRNAs such as the chromosome 19 miRNA cluster that are exclusively expressed in the placenta (Sabapatha et al. 2006).

Pohler et al. (2017) discovered 194 and 211 circulating EVs-derived miRNAs on days 17 and 24 of gestation, respectively, confirming their increase during gestation. However, three miRNAs (miR-25, $-16a/b$, and -3596) were present in greater abundance either in control animals or in cows experiencing embryonic-mortality compared to pregnant animals on day 17. The increase in miR-25 abundance in cows with embryonic mortality might indicate early embryonic death or a systemic response to pregnancy loss. The increase in miR-25 abundance in cows experiencing embryonic mortality might indicate early embryonic death or a systemic response to pregnancy loss. According to pathway analysis, changes in specific miRNAs in cows with embryonic mortality compared to pregnant animals could impact the upregulation of pathways involved in prostaglandin production (Bazer 2013).

EVs as a Diagnostic Tool for Reproductive Diseases

In recent years, noninvasive prenatal testing has revolutionized prenatal screening for common fetal aneuploidies and Y-chromosome abnormalities using cell-free DNA (cfDNA) in maternal plasma (Bianchi et al. 2015). If embryonic and/or fetal DNA can be identified either in the culture media or maternal circulation via cfDNA in EVs, they could potentially serve as biomarkers for Y-chromosome-specific DNA3. Further, EVs could potentially be used for noninvasive prenatal genetic diagnostic techniques to detect aneuploidy before embryo transfer or in the very early stages of pregnancy, instead of embryo biopsy as is currently done (Saadeldin et al. 2015).

During the first trimester of pregnancy, pEVs are released into the maternal blood by extravillous trophoblast and/or syncytiotrophoblast cells. These EVs promote immunosuppression and maternal immune tolerance to the fetus by suppressing maternal T-cell signaling components. In humans, the concentration of EVs in plasma increases during pregnancy and with gestational age. In fact, the concentration of EVs in maternal peripheral blood is 20 times greater than that found in nonpregnant women (Parks et al. 2018). This increase in EVs may be clinically useful in diagnosing placental dysfunction. It has been demonstrated that certain proteins such as the syncytiotrophoblast protein syncytin-2 are significantly downregulated in EVs derived from the placentae of pregnant women with preeclampsia compared to healthy control pregnancies (Vargas et al. 2014). Therefore, placental EVs and specific miRNAs have been further examined as potential biomarkers for diagnosing preeclampsia (Li et al. 2020). EVs isolated from plasma preeclamptic pregnancies were found to impair angiogenesis in human umbilical vein endothelial cells and to express abundant sFlt-1 (soluble fms-like tyrosine kinase-1) and sEng (soluble endoglin) (Chang et al. 2018).

13   Other Future Uses of EVs

13.1   EVs as a Drug Delivery Vector

EVs are currently being investigated for their potential clinical applications, particularly for targeted drug delivery as an alternative to nanoparticle-mediated delivery. In the field of reproduction, nanoparticles have been experimentally used to load sperm with exogenous genetic material that is transferred to the oocyte during fertilization (Kim et al. 2010). However, the long-term consequences of using exogenous nanoparticles during conception are unknown. In this respect, EVs, as endogenous carriers of biomolecules, may be particularly advantageous. They are naturally present in mammal biofluids and can also be engineered for tissue-specific transfers, such as the transfer of selected compounds into gametes and embryos to increase reproductive success (Barkalina et al. 2015).

13.2   EVs as a Treatment of Reproductive Disease

Amniotic-derived EVs (amEVs) were recently used to treat endometritis in a mare to achieve a successful pregnancy (Lange-Consiglio et al. 2020). The mare, an 11-year-old Holstein Friesian, had a history of failed pregnancies despite numerous insemination attempts over many years. After a final insemination attempt using a stallion of proven fertility, the collection of an 8-day-old embryo suggested that the mare was affected by implantation failure related to endometritis. In their 2020 study, Lange-Consiglio et al. treated the mare with two cycles of intrauterine administration of amEVs that induced an improvement in the classification of endometritis and resulted in a successful pregnancy that ended with the birth of a foal (Lange-Consiglio et al. 2020). This is the only known paper that has used EVs in the treatment of reproductive diseases. Probably, amEVs were able to restore the injured endometrium and reestablish the proper communication for successful embryo implantation due to their anti-inflammatory and regenerative effects (Capra and Lange-Consiglio 2020).

13.3   EVs as Reproductive Regenerative Therapy

Stem cells have been shown to secrete EVs, which is not surprising given their undifferentiated nature and the potential that stem cells carry. Stem cell therapy has been heralded as the ultimate regenerative therapy since stem cells are the mother of all cells and can potentially produce any cell type (Kalra et al. 2016). The pathways of stem cell action include homing, self-differentiation, paracrine signaling, and immune regulation (Naji et al. 2017). EVs play a crucial role in various processes that can be applied in regenerative therapy. For instance, they can recruit stem cells and transmit stem cell information (Qu et al. 2022). Cells at the lesion site release EVs and recruitment factors to recruit surrounding tissues or peripheral stem cells (Chen et al. 2022). Stem cells can either release EVs to play a distant role or return to the lesion site to self-differentiate, replace damaged cells, and release EVs or other therapeutic factors (Fu et al. 2022). Studies have shown that EVs released by stem cells can restore ovarian function (Liu et al. 2020) and repair endometrial cells (Yao et al. 2019).

13.4   EVs as Semen Cryoprotectors

Moreover, studying the content of EVs could lead to the use of specific donor cells or to the development of synthetic or semisynthetic vesicles to transport specific components of interest, mimicking the maternal environment (de ávila et al. 2018). For instance, liposomes are artificial lipid vesicles that can optimize cryopreservation of semen and modulate different aspects of male reproductive functions, especially in preserving the motility and acrosomal integrity of spermatozoa when subjected to a cold shock (5 °C) using phosphatidylcholine: cholesterol liposomes (Holt et al. 1988). Similarly, other authors discovered that boar spermatozoa incubated with liposomes made from a complex mixture of selected phospholipids, before the freezing-thawing process, exhibited better motility and viability than sperm incubated with liposomes made from fresh sperm head plasma membranes (He et al. 2001).

14   Conclusions

It has become clear that the prostate, embryo, oviduct, uterus, and vagina all produce messengers that transport information to the semen, OF, UF, and vaginal secretions, which collectively form the environment for gametes and embryos during different stages of the reproductive process. This information exchange within the environment in which gametes interact and form an embryo is crucial for successful reproduction and EVs play a key role in that exchange of information. EVs are highly specific and versatile vehicles that are involved in intercellular communication. One major advantage over secreted signaling molecules is that EVs deliver signals over long distances without dilution or degradation, securely transferring biomolecules within their capsule. This review summarizes the role of EVs in cell-to-cell communication in reproduction, with a focus on pig reproduction. Although there is clear evidence that EVs play crucial roles in oocyte maturation, fertilization, implantation, and embryo-maternal crosstalk, challenges remain in a partial understanding of which signaling biomolecules carried by EVs mediate cell-to-cell communication. Further investigations are necessary to better understand the physiological processes of spermatozoa and oocyte interaction, cell development, embryo formation and development, embryo implantation, and other processes related to pregnancy maintenance and progress up to birth.

Declaration of Funding This research was supported by the Spanish Ministry of Science and Innovation (PID2019-106380RBI00/ AEI/10.13039/501100011033) and the European Union 'Next Generation EU/PRTR' (PDC2022–133589-I00/ AEI/10.13039/501100011033/Unión Europea NextGenerationEU/PRTR).

Conflicts of Interest The authors declare no conflicts of interest.

References

Aalberts M, Sostaric E, Wubbolts R, Wauben MWM, Nolte-T Hoen ENM, Gadella BM, Stout TAE, Stoorvogel W (2013) Spermatozoa recruit prostasomes in response to capacitation induction. Biochim Biophys Acta 1834(11):2326–2335. https://doi.org/10.1016/j.bbapap.2012.08.008

Aalberts M, Stout TAE, Stoorvogel W (2014) Prostasomes: extracellular vesicles from the prostate. Reproduction 147(1):1–14. https://doi.org/10.1530/REP-13-0358

Aalberts M, van Dissel-Emiliani FMF, van Adrichem NPH, van Wijnen M, Wauben MHM, Stout TAE, Stoorvogel W (2012) Identification of distinct populations of prostasomes that differentially express prostate stem cell antigen, annexin A1, and GLIPR2 in humans. Biol Reprod 86(3):1–8. https://doi.org/10.1095/biolreprod.111.095760

Abad M, Garcia JC, Sprecher DJ, Cassar G, Friendship RM, Buhr MM, Kirkwood RN (2007) Effect of insemination-ovulation interval and addition of seminal plasma on sow fertility to insemination of cryopreserved sperm. Reprod Domest Anim 42(4):418–422. https://doi.org/10.1111/j.1439-0531.2006.00801.x

Acuña OS, Avilés M, López-Úbeda R, Guillén-Martínez A, Soriano-Úbeda C, Torrecillas A, Coy P, Izquierdo-Rico MJ (2017) Differential gene expression in porcine oviduct during the oestrous cycle. Reprod Fertil Dev 29(12):2387–2399. https://doi.org/10.1071/RD16457

Adams GP, Ratto MH, Silva ME, Carrasco RA (2016) Ovulation-inducing factor (OIF/NGF) in seminal plasma: a review and update. Reprod Domest Anim 51:4–17. https://doi.org/10.1111/rda.12795

Admyre C, Johansson SM, Rahman Qazi K, Filén J-J, Lahesmaa R, Norman MA, Neve EP, Scheynius A, Gabrielsson S (2007) Exosomes with immune modulatory features are present in human breast Milk. J Immunol 179:1969–1978. http://www.expasy.org/sprot/

Alcântara-Neto AS, Fernandez-Rufete M, Corbin E, Tsikis G, Uzbekov R, Garanina AS, Coy P, Almiñana C, Mermillod P (2019) Oviduct fluid extracellular vesicles regulate polyspermy during porcine in vitro fertilisation. Reprod Fertil Dev 1–10:409. https://doi.org/10.1071/RD19058

Alcântara-Neto AS, Schmaltz L, Caldas E, Blache MC, Mermillod P, Almiñana C (2020) Porcine oviductal extracellular vesicles interact with gametes and regulate sperm motility and survival. Theriogenology 155:240–255. https://doi.org/10.1016/j.theriogenology.2020.05.043

Al-Dossary AA, Bathala P, Caplan JL, Martin-DeLeon PA (2015) Oviductosome-sperm membrane interaction in cargo delivery: detection of fusion and underlying molecular players using three-dimensional super-resolution structured illumination microscopy (SR-SIM). J Biol Chem 290(29):17710–17723. https://doi.org/10.1074/jbc.M114.633156

Al-Dossary AA, Strehler EE, Martin-DeLeon PA (2013) Expression and secretion of plasma membrane Ca2+-ATPase 4a (PMCA4a) during murine estrus: association with oviductal exosomes and uptake in sperm. PLoS One 8(11):e80181. https://doi.org/10.1371/journal.pone.0080181

Almiñana C, Corbin E, Tsikis G, Soares A, Neto A, Labas V, Reynaud K, Galio L, Uzbekov R, Garanina AS, Druart X (2017) Oviduct extracellular vesicles protein content and their role during oviduct-embryo cross-talk. Reproduction 154(4):253–268. https://doi.org/10.1530/REP-17-0054

Almiñana C, Tsikis G, Labas V, Uzbekov R, da Silveira JC, Bauersachs S, Mermillod P (2018) Deciphering the oviductal extracellular vesicles content across the estrous cycle: implications for the gametes-oviduct interactions and the environment of the potential embryo. BMC Genomics 19(1):1–27. https://doi.org/10.1186/s12864-018-4982-5

Al-Nedawi K, Meehan B, Micallef J, Lhotak V, May L, Guha A, Rak J (2008) Intercellular transfer of the oncogenic receptor EGFRvIII by microvesicles derived from tumour cells. Nat Cell Biol 10(5):619–624. https://doi.org/10.1038/ncb1725

Álvarez-Rodríguez M, Ljunggren SA, Karlsson H, Rodriguez-Martínez H (2019) Exosomes in specific fractions of the boar ejaculate contain CD44: A marker for epididymosomes? Theriogenology 140:143–152. https://doi.org/10.1016/j.theriogenology.2019.08.023

Álvarez-Rodríguez M, Martinez CA, Wright D, Rodríguez-Martinez H (2020) The role of semen and seminal plasma in inducing large-scale genomic changes in the female porcine periovulatory tract. Sci Rep 10(1):1–16. https://doi.org/10.1038/s41598-020-60810-z

Aneas SB, Gary BG, Bouvier BP (2008) Collectis® automated boar collection technology. Theriogenology 70(8):1368–1373. https://doi.org/10.1016/j.theriogenology.2008.07.011

Arienti G, Carlini E, De Cosmo AM, Di Profio P, Palmerini CA (1998a) Prostasome-like particles in stallion semen 1. Biol Reprod 59:309–313. https://academic.oup.com/biolreprod/article/59/2/309/2740820

Arienti G, Carlini E, Nicolucci A, Cosmi EV, Santi F, Palmerini CA (1999) The motility of human spermatozoa as influenced by prostasomes at various pH levels. Biol Cell 91(1):51–54. https://doi.org/10.1111/j.1768-322x.1999.tb01083.x

Arienti G, Carlini E, Palmerini CA (1997) Fusion of human sperm to Prostasomes at acidic pH. J Membr Biol 155:89–94

Arienti G, Carlini E, Saccardi C, Palmerini CA (1998b) Interactions between prostasomes and leukocytes. Biochim Biophys Acta 1425:36–40

Asea A, Jean-Pierre C, Kaur P, Rao P, Linhares IM, Skupski D, Witkin SS (2008) Heat shock protein-containing exosomes in mid-trimester amniotic fluids. J Reprod Immunol 79(1):12–17. https://doi.org/10.1016/j.jri.2008.06.001

Atikuzzaman M, Álvarez-Rodríguez M, Carrillo AV, Johnsson M, Wright D, Rodríguez-Martínez H (2017) Conserved gene expression in sperm reservoirs between birds and mammals in response to mating. BMC Genomics 18(98):1–18. https://doi.org/10.1186/s12864-017-3488-x

Banliat C, Le Bourhis D, Bernardi O, Tomas D, Labas V, Salvetti P, Guyonnet B, Mermillod P, Saint-Dizier M (2020) Oviduct fluid extracellular vesicles change the phospholipid composition of bovine embryos developed in vitro. Int J Mol Sci 21(15):1–13. https://doi.org/10.3390/ijms21155326

Barkalina N, Jones C, Wood MJA, Coward K (2015) Extracellular vesicle-mediated delivery of molecular compounds into gametes and embryos: learning from nature. Hum Reprod Update 21(5):627–639. https://doi.org/10.1093/humupd/dmv027

Barranco I, Rubio CP, Tvarijonaviciute A, Rodriguez-Martinez H, Roca J (2021) Measurement of oxidative stress index in seminal plasma can predict in vivo fertility of liquid-stored porcine artificial insemination semen doses. Antioxidants 10(8):1–12. https://doi.org/10.3390/antiox10081203

Barraud-Lange V, Boissonnas CC, Serres C, Auer J, Schmitt A, Lefèvre B, Wolf JP, Ziyyat A (2012) Membrane transfer from oocyte to sperm occurs in two CD9-independent ways that do not supply the fertilising ability of Cd9-deleted oocytes. Reproduction 144(1):53–66. https://doi.org/10.1530/REP-12-0040

Barraud-Lange V, Naud-Barriant N, Bomsel M, Wolf J, Ziyyat A (2007) Transfer of oocyte membrane fragments to fertilizing spermatozoa. FASEB J 21(13):3446–3449. https://doi.org/10.1096/fj.06-8035hyp

Barrès C, Blanc L, Bette-Bobillo P, André S, Mamoun R, Gabius HJ, Vidal M (2010) Galectin-5 is bound onto the surface of rat reticulocyte exosomes and modulates vesicle uptake by macrophages. Blood 115(3):696–705. https://doi.org/10.1182/blood-2009-07-231449

Baskaran S, Panner Selvam MK, Agarwal A (2020) Exosomes of male reproduction. In: Advances in clinical chemistry, vol 95. Academic Press Inc., pp 149–163. https://doi.org/10.1016/bs.acc.2019.08.004

Baxter FA, Drake AJ (2019) Non-genetic inheritance via the male germline in mammals. Philos Trans R Soc Lond B Biol Sci 374(1770):20180118. https://doi.org/10.1098/rstb.2018.0118

Bazer FW (2013) Pregnancy recognition signaling mechanisms in ruminants and pigs. J Anim Sci Biotechnol 4(23):1–10. https://doi.org/10.1186/2049-1891-4-23

Belleannée C, Légaré C, Calvo É, Thimon V, Sullivan R (2013) MicroRNA signature is altered in both human epididymis and seminal microvesicles following vasectomy. Hum Reprod 28(6):1455–1467. https://doi.org/10.1093/humrep/det088

Bianchi DW, Parsa S, Bhatt S, Halks-Miller M, Kurtzman K, Sehnert AJ, Swanson A (2015) Fetal sex chromosome testing by maternal plasma DNA sequencing: clinical laboratory experience and biology. Obstet Gynecol 125(2):375–382. https://doi.org/10.1097/AOG.0000000000000637

Bidarimath M, Khalaj K, Kridli RT, Kan FWK, Koti M, Tayade C (2017) Extracellular vesicle mediated intercellular communication at the porcine maternal-fetal interface: A new paradigm for conceptus-endometrial cross-talk. Sci Rep 7:1–14. https://doi.org/10.1038/srep40476

Bidarimath M, Lingegowda H, Miller JE, Koti M, Tayade C (2021) Insights into extracellular vesicle/exosome and miRNA mediated bi-directional communication during porcine pregnancy. Front Vet Sci 8:1–13. https://doi.org/10.3389/fvets.2021.654064

Bischoft RJ, Lee C-S, Brandon MR, Meeusen E (1994) Inflammatory response in the pig uterus induced by seminal plasma. J Reprod Immunol 26:131–146

Boettner B, Van Aelst L (2009) Control of cell adhesion dynamics by Rap1 signaling. Current Opinion in Cell Biology 21:684. https://doi.org/10.1016/j.ceb.2009.06.004

Bridi A, Perecin F, da Silveira JC (2020) Extracellular vesicles mediated early embryo–maternal interactions. Int J Mol Sci 21(3):1–15. https://doi.org/10.3390/ijms21031163

Bromfield JJ (2016) A role for seminal plasma in modulating pregnancy outcomes in domestic species. Reproduction 152(6):R223–R232. https://doi.org/10.1530/REP-16-0313

Bromfield JJ, Schjenken JE, Chin PY, Care AS, Jasper MJ, Robertson SA (2014) Maternal tract factors contribute to paternal seminal fluid impact on metabolic phenotype in offspring. Proc Natl Acad Sci USA 111(6):2200–2205. https://doi.org/10.1073/pnas.1305609111

Burns G, Brooks K, Wildung M, Navakanitworakul R, Christenson LK, Spencer TE (2014) Extracellular vesicles in luminal fluid of the ovine uterus. PLoS One 9(3):1–11. https://doi.org/10.1371/journal.pone.0090913

Burns GW, Brooks KE, O'neil EV, Hagen DE, Behura SK, Spencer TE (2018) Progesterone effects on extracellular vesicles in the sheep uterus. Biol Reprod 1–26. https://doi.org/10.1093/biolre/ioy011/4810746

Burns GW, Brooks KE, Spencer TE (2016) Extracellular vesicles originate from the conceptus and uterus during early pregnancy in sheep. Biol Reprod 94(3):1–11. https://doi.org/10.1095/biolreprod.115.134973

Caballero JN, Frenette G, Belleannée C, Sullivan R (2013) CD9-positive microvesicles mediate the transfer of molecules to bovine spermatozoa during Epididymal maturation. PLoS One 8(6):1–12. https://doi.org/10.1371/journal.pone.0065364

Caby MP, Lankar D, Vincendeau-Scherrer C, Raposo G, Bonnerot C (2005) Exosomal-like vesicles are present in human blood plasma. Int Immunol 17(7):879–887. https://doi.org/10.1093/intimm/dxh267

Cai H, Reinisch K, Ferro-Novick S (2007) Coats, tethers, Rabs, and SNAREs work together to mediate the intracellular destination of a transport vesicle. Dev Cell 12(5):671–682. https://doi.org/10.1016/j.devcel.2007.04.005

Capalbo A, Ubaldi FM, Cimadomo D, Noli L, Khalaf Y, Farcomeni A, Ilic D, Rienzi L (2016) MicroRNAs in spent blastocyst culture medium are derived from trophectoderm cells and can be explored for human embryo reproductive competence assessment. Fertil Steril 105(1):225–235. https://doi.org/10.1016/j.fertnstert.2015.09.014

Capra E, Lange-Consiglio A (2020) The biological function of extracellular vesicles during fertilization, early embryo—maternal crosstalk and their involvement in reproduction: review and overview. Biomol Ther 10(11):1–26. https://doi.org/10.3390/biom10111510

Care AS, Diener KR, Jasper MJ, Brown HM, Ingman WV, Robertson SA (2013) Macrophages regulate corpus luteum development during embryo implantation in mice. J Clin Invest 123(8):3472–3487. https://doi.org/10.1172/JCI60561

Castaño C, Kalko S, Novials A, Párrizas M (2018) Obesity-associated exosomal miRNAs modulate glucose and lipid metabolism in mice. Proc Natl Acad Sci USA 115(48):12158–12163. https://doi.org/10.1073/pnas.1808855115

Chabory E, Damon C, Lenoir A, Kauselmann G, Kern H, Zevnik B, Garrel C, Saez F, Cadet R, Henry-Berger J, Schoor M, Gottwald U, Habenicht U, Drevet JR, Vernet P (2009) Epididymis seleno-independent glutathione peroxidase 5 maintains sperm DNA integrity in mice. J Clin Invest 119(7):2074–2085. https://doi.org/10.1172/jci38940

Chang X, Yao J, He Q, Liu M, Duan T, Wang K (2018) Exosomes from women with preeclampsia induced vascular dysfunction by delivering sFLT (soluble fms-like tyrosine kinase)-1 and SENG (soluble endoglin) to endothelial cells. Hypertension 72(6):1381–1390. https://doi.org/10.1161/HYPERTENSIONAHA.118.11706

Chen BX, Yuen ZX, Pan GW (1985) Semen-induced ovulation in the bactrian camel (Camelus bactrianus). J Reprod Fertil 73:335–339

Chen H, Liu C, Jiang H, Gao Y, Xu M, Wang J, Liu S, Fu Y, Sun X, Xu J, Zhang J, Dai L (2017) Regulatory role of miRNA-375 in expression of BMP15/GDF9 receptors and its effect on proliferation and apoptosis of bovine cumulus cells. Cell Physiol Biochem 41(2):439–450. https://doi.org/10.1159/000456597

Chen IH, Aguilar HA, Paez Paez JS, Wu X, Pan L, Wendt MK, Iliuk AB, Zhang Y, Tao WA (2018) Analytical pipeline for discovery and verification of glycoproteins from plasma-derived extracellular vesicles as breast cancer biomarkers. Anal Chem 90(10):6307–6313. https://doi.org/10.1021/acs.analchem.8b01090

Chen K, Li Y, Xu L, Qian Y, Liu N, Zhou C, Liu J, Zhou L, Xu Z, Jia R, Ge YZ (2022) Comprehensive insight into endothelial progenitor cell-derived extracellular vesicles as a promising candidate for disease treatment. Stem Cell Res Ther 13(1):1–16. https://doi.org/10.1186/s13287-022-02921-0

Chen X, Zhu H, Hu C, Hao H, Zhang J, Li K, Zhao X, Qin T, Zhao K, Zhu H, Wang D (2014) Identification of differentially expressed proteins in fresh and frozen-thawed boar spermatozoa by iTRAQ-coupled 2D LC-MS/MS. Reproduction 147(3):321–330. https://doi.org/10.1530/REP-13-0313

Choudhary KK, Kavya KM, Jerome A, Sharma RK (2016) Advances in reproductive biotechnologies. Veterinary. WORLD 9(4):388–395. https://doi.org/10.14202/vetworld.2016.388-395

Claus R (1990) Physiological role of seminal components in the reproductive tract of the female pig. J Reprod Fertil Suppl 40:117–131

Coy P, Cánovas S, Mondéjar I, Dolores Saavedra M, Romar R, Grulló L, Matás C, Avilés M (2008) Oviduct-specific glycoprotein and heparin modulate sperm-zona pellucida interaction during fertilization and contribute to the control of polyspermy. PNAS 105(41):15809–15814. https://doi.org/10.1073/pnas.0804422105

Cross NL (1996) Human seminal plasma prevents sperm from becoming Acrosomally responsive to the agonist, progesterone: cholesterol is the major inhibitor. Biol Reprod 54:138–145. https://academic.oup.com/biolreprod/article/54/1/138/2761855

Cross NL, Mahasreshti P (1997) Prostasome fraction of human seminal plasma prevents sperm from becoming acrosomally responsive to the agonist progesterone. Arch Androl 39:39–44

da Silveira JC, Veeramachaneni DNR, Winger QA, Carnevale EM, Bouma GJ (2012) Cell-secreted vesicles in equine ovarian follicular fluid contain mirnas and proteins: A possible new form of cell communication within the ovarian follicle. Biol Reprod 86(3):1–10. https://doi.org/10.1095/biolreprod.111.093252

D'Amours O, Bordeleau LJ, Frenette G, Blondin P, Leclerc P, Sullivan R (2012) Binder of sperm 1 and epididymal sperm binding protein 1 are associated with different bull sperm subpopulations. Reproduction 143(6):759–771. https://doi.org/10.1530/REP-11-0392

Davis BK (1980) Interaction of lipids with the plasma membrane of sperm cells. I. The Antifertilization action of cholesterol. Arch Androl 5:249–254

Day BN (2000) Reproductive biotechnologies: current status in porcine reproduction. Anim Reprod Sci 60:161–172. www.elsevier.comrlocateranireprosci

de ávila ACFCM, Andrade GM, Bridi A, Gimenes LU, Meirelles FV, Perecin F, da Silveira JC (2018) Extracellular vesicles and its advances in female reproduction. Anim Reprod 16(1):31–38. https://doi.org/10.21451/1984-3143-AR2018-0101

Deatherage BL, Cookson BT (2012) Membrane vesicle release in bacteria, eukaryotes, and archaea: A conserved yet underappreciated aspect of microbial life. Infect Immun 80(6):1948–1957. https://doi.org/10.1128/IAI.06014-11

Demari J, Mroske C, Tang S, Nimeh J, Miller R, Lebel RR (2016) CLTC as a clinically novel gene associated with multiple malformations and developmental delay. Am J Med Genet A 170(4):958–966. https://doi.org/10.1002/ajmg.a.37506

Eisa AA, De S, Detwiler A, Gilker E, Ignatious AC, Vijayaraghavan S, Kline D (2019) YWHA (14-3-3) protein isoforms and their interactions with CDC25B phosphatase in mouse oogenesis and oocyte maturation. BMC Dev Biol 19(1):1–22. https://doi.org/10.1186/s12861-019-0200-1

Elliott RMA, Lloyd RE, Fazeli A, Sostaric E, Georgiou AS, Satake N, Watson PF, Holt WV (2009) Effects of HSPA8, an evolutionarily conserved oviductal protein, on boar and bull spermatozoa. Reproduction 137(2):191–203. https://doi.org/10.1530/REP-08-0298

Fabiani R, Johansson L, Lundkvist Ö, Ulmsten U, Ronquist G (1994) Promotive effect by prostasomes on normal human spermatozoa exhibiting no forward motility due to buffer washings. Eur J Obstet Gynecol Reprod Biol 57:181–188

Fabiani R, Johanssonb L, Lundkvistb D, Ronquist G (1995) Prolongation and improvement of prostasome promotive sperm forward motility effect on. Eur J Obstet Gynecol Reprod Biol 58:191–198

Fan W, Qi Y, Wang Y, Yan H, Li X, Zhang Y (2023) Messenger roles of extracellular vesicles during fertilization of gametes, development and implantation: recent advances. Front Cell Dev Biol 10:1–19. https://doi.org/10.3389/fcell.2022.1079387

Fang X, Tanga BM, Bang S, Seo C, Kim H, Saadeldin IM, Lee S, Cho J (2022) Oviduct epithelial cell-derived extracellular vesicles improve porcine trophoblast outgrowth. Vet Sci 9(609):1–13. https://doi.org/10.3390/vetsci9110609

Fazeli A (2011) Maternal communication with gametes and embryo: A personal opinion. Reprod Domest Anim 46(SUPPL. 2):75–78. https://doi.org/10.1111/j.1439-0531.2011.01870.x

Fereshteh Z, Bathala P, Galileo DS, Martin-DeLeon PA (2019) Detection of extracellular vesicles in the mouse vaginal fluid: their delivery of sperm proteins that stimulate capacitation and modulate fertility. J Cell Physiol 234(8):12745–12756. https://doi.org/10.1002/jcp.27894

Fereshteh Z, Schmidt SA, Al-Dossary AA, Accerbi M, Arighi C, Cowart J, Song JL, Green PJ, Choi K, Yoo S, Martin-DeLeon PA (2018) Murine Oviductosomes (OVS) microRNA profiling during the estrous cycle: delivery of OVS-borne microRNAs to sperm where miR-34c-5p localizes at the centrosome. Sci Rep 8(1):1–18. https://doi.org/10.1038/s41598-018-34409-4

de Ferraz AMM, Carothers A, Dahal R, Noonan MJ, Songsasen N (2019) Oviductal extracellular vesicles interact with the spermatozoon's head and mid-piece and improves its motility and fertilizing ability in the domestic cat. Sci Rep 9(1):1–12. https://doi.org/10.1038/s41598-019-45857-x

Foote RH (2002) The history of artificial insemination: selected notes and notables 1. Am Soc Anim Sci 80:1–10

Franchi A, Cubilla M, Guidobaldi HA, Bravo AA, Giojalas LC (2016) Uterosome-like vesicles prompt human sperm fertilizing capability. Mol Hum Reprod 22(12):833–841. https://doi.org/10.1093/molehr/gaw066

Franchi A, Moreno-Irusta A, Domínguez EM, Adre AJ, Giojalas LC (2020) Extracellular vesicles from oviductal isthmus and ampulla stimulate the induced acrosome reaction and signaling events associated with capacitation in bovine spermatozoa. J Cell Biochem 121(4):2877–2888. https://doi.org/10.1002/jcb.29522

Freeman SA, Christian S, Austin P, Iu I, Graves ML, Huang L, Tang S, Coombs D, Gold MR, Roskelley CD (2017) Applied stretch initiates directional invasion through the action of Rap1 GTPase as a tension sensor. J Cell Sci 130(1):152–163. https://doi.org/10.1242/jcs.180612

Frenette G, Girouard J, Sullivan R (2006) Comparison between epididymosomes collected in the intraluminal compartment of the bovine caput and cauda epididymidis. Biol Reprod 75(6):885–890. https://doi.org/10.1095/biolreprod.106.054692

Frenette G, Lessard C, Sullivan R (2002) Selected proteins of "'Prostasome-like particles'" from Epididymal cauda fluid are transferred to Epididymal caput spermatozoa in bull. Biol Reprod 67:308–313. http://www.biolreprod.org

Frenette G, Lessard C, Sullivan R (2004) Polyol pathway along the bovine epididymis. Mol Reprod Dev 69(4):448–456. https://doi.org/10.1002/mrd.20170

Frenette G, Sullivan R (2001) Prostasome-like particles are involved in the transfer of P25b from the bovine Epididymal fluid to the sperm surface. Mol Reprod Dev 59:115–121

Frey B, Gaipl US (2011) The immune functions of phosphatidylserine in membranes of dying cells and microvesicles. Semin Immunopathol 33(5):497–516. https://doi.org/10.1007/s00281-010-0228-6

Fu B, Ma H, Liu D (2020) Extracellular vesicles function as bioactive molecular transmitters in the mammalian oviduct: an inspiration for optimizing in vitro culture systems and improving delivery of exogenous nucleic acids during preimplantation embryonic development. Int J Mol Sci 21(6):1–16. https://doi.org/10.3390/ijms21062189

Fu B, Ma H, Zhang D, Wang L, Li Z, Guo Z, Liu Z, Wu S, Meng X, Wang F, Chen W, Liu D (2022) Porcine oviductal extracellular vesicles facilitate early embryonic development via relief of endoplasmic reticulum stress. Cell Biol Int 46(2):300–310. https://doi.org/10.1002/cbin.11730

Gangnuss S, Sutton-McDowall ML, Robertson SA, Armstrong DT (2004) Seminal plasma regulates corpora lutea macrophage populations during early pregnancy in mice. Biol Reprod 71(4):1135–1141. https://doi.org/10.1095/biolreprod.104.027425

Gaonac'h-Lovejoy V, Boscher C, Delisle C, Gratton JP (2020) Rap1 is involved in Angiopoietin-1-induced cell-cell junction stabilization and endothelial cell sprouting. Cells 9(1):1–15. https://doi.org/10.3390/cells9010155

García-Martínez S, Gadea J, Coy P, Romar R (2020) Addition of exogenous proteins detected in oviductal secretions to in vitro culture medium does not improve the efficiency of in vitro fertilization in pigs. Theriogenology 157:490–497. https://doi.org/10.1016/j.theriogenology.2020.08.017

García-Vázquez F, Gadea J, Matás C, Holt W (2016) Importance of sperm morphology during sperm transport and fertilization in mammals. Asian J Androl 18(6):844–850. https://doi.org/10.4103/1008-682X.186880

Garner DL, Hafez ESE (2016) Spermatozoa and seminal plasma. In: Physiology of reproduction, pp 96–109

Gorelik L, Flavell RA (2002) Transforming growth factor-β in T-cell biology. Nat Rev Immunol 2(1):46–53. https://doi.org/10.1038/nri704

Greening DW, Nguyen HPT, Elgass K, Simpson RJ, Salamonsen LA (2016) Human endometrial exosomes contain hormone-specific cargo modulating trophoblast adhesive capacity: insights into endometrial-embryo interactions. Biol Reprod 94(2):1–15. https://doi.org/10.1095/biolreprod.115.134890

Griffiths GS, Miller KA, Galileo DS, Martin-DeLeon PA (2008) Murine SPAM1 is secreted by the estrous uterus and oviduct in a form that can bind to sperm during capacitation: acquisition enhances hyaluronic acid-binding ability and cumulus dispersal efficiency. Reproduction 135(3):293–301. https://doi.org/10.1530/REP-07-0340

Gutiérrez-Mateo C, Sánchez-García JF, Fischer J, Tormasi S, Cohen J, Munné S, Wells D (2009) Preimplantation genetic diagnosis of single-gene disorders: experience with more than 200 cycles conducted by a reference laboratory in the United States. Fertil Steril 92(5):1544–1556. https://doi.org/10.1016/j.fertnstert.2008.08.111

Harris EA, Stephens KK, Winuthayanon W (2020) Extracellular vesicles and the oviduct function. Int J Mol Sci 21(21):1–20. https://doi.org/10.3390/ijms21218280

He L, Bailey JL, Buhr MM (2001) Incorporating lipids into boar sperm decreases chilling sensitivity but not capacitation potential 1. Biol Reprod 64:69–79. http://www.biolreprod.org

Hemler ME (2003) Tetraspanin proteins mediate cellular penetration, invasion, and fusion events and define a novel type of membrane microdomain. Annu Rev Cell Dev Biol 19:397–422. https://doi.org/10.1146/annurev.cellbio.19.111301.153609

Hermo L, Robaire B (2002) Epididymal cell types and their functions. In: Robaire, Hinton (eds) The epididymis: from molecules to clinical praactice. Kluwer Academic/Plenum Publishers, pp 81–102

Hilz S, Fogarty EA, Modzelewski AJ, Cohen PE, Grimson A (2017) Transcriptome profiling of the developing male germ line identifies the miR-29 family as a global regulator during meiosis. RNA Biol 14(2):219–235. https://doi.org/10.1080/15476286.2016.1270002

Holt WV, Morris GJ, Coulson G, North RD (1988) Direct observation of cold-shock effects in ram spermatozoa with the use of a programmable Cryomicroscope. J Exp Zool 246:305–314

Hu Q, Zang X, Ding Y, Gu T, Shi J, Li Z, Cai G, Liu D, Wu Z, Hong L (2022) Porcine uterine luminal fluid-derived extracellular vesicles improve conceptus-endometrial interaction during implantation. Theriogenology 178:8–17. https://doi.org/10.1016/j.theriogenology.2021.10.021

Hunt JS, Robertson SA (1996) Uterine macrophages and environmental programming for pregnancy success. J Reprod Immunol 32:1–25

Hunter RHF (1994) Modulation of gamete and embryonic microenvironments by oviduct glycoproteins. Mol Reprod Dev 39:176–181

Hunter RHF, Rodriguez-Martinez H (2002) Analysing mammalian fertilisation: reservations and potential pitfalls with an in vitro approach. Zygote 10:11–15

Hunter RHF, Rodriguez-Martinez H (2004) Capacitation of mammalian spermatozoa in vivo, with a specific focus on events in the fallopian tubes. Mol Reprod Dev 67(2):243–250. https://doi.org/10.1002/mrd.10390

Ilanguramaran S, Robinson PJ, Hoessli DC (1996) Transfer of exogenous glycosylphos- phatidylinositol (GPI) -linked molecules to plasma membranes. Cell Biol 6:163–167

Irion U, St Johnston D (2007) Bicoid RNA localization requires specific binding of an endosomal sorting complex. Nature 445(7127):554–558. https://doi.org/10.1038/nature05503

Jee B, Dhar R, Singh S, Karmakar S (2021) Heat shock proteins and their role in pregnancy: redefining the function of "old rum in a new bottle". Frontiers in Cell and Developmental Biology 9:1–18. https://doi.org/10.3389/fcell.2021.648463

Jodar M, Selvaraju S, Sendler E, Diamond MP, Krawetz SA (2013) The presence, role and clinical use of spermatozoal RNAs. Hum Reprod Update 19(6):604–624. https://doi.org/10.1093/humupd/dmt031

Johansson M, Tienthai P, Rodríguez Martínez H (2000) Histochemistry and ultrastructure of the intraluminal mucus in the sperm reservoir of the pig oviduct. J Reprod Dev 46(3):183–192

Joyce MM, Burghardt JR, Burghardt RC, Hooper, Neil R, Bazer FW, Johnson GA (2008) Uterine MHC class I molecules and 2-microglobulin are regulated by progesterone and conceptus interferons during pig pregnancy 1. J Immunol 181:2494–2505. http://journals.aai.org/jimmunol/article-pdf/181/4/2494/1261524/zim01608002494.pdf

Kaczmarek MM, Krawczynski K, Filant J (2013) Seminal plasma affects prostaglandin synthesis and angiogenesis in the porcine uterus. Biol Reprod 88(3):1–11. https://doi.org/10.1095/biolreprod.112.103564

Kalam SN, Dowland S, Lindsay L, Murphy CR (2018) Microtubules are reorganised and fragmented for uterine receptivity. Cell Tissue Res 374(3):667–677. https://doi.org/10.1007/s00441-018-2887-x

Kalra H, Drummen GPC, Mathivanan S (2016) Focus on extracellular vesicles: introducing the next small big thing. Int J Mol Sci 17(2):1–30. https://doi.org/10.3390/ijms17020170

Kim E, Yamashita M, Kimura M, Honda A, Kashiwabara SI, Baba T (2008) Sperm penetration through cumulus mass and zona pellucida. Int J Dev Biol 52(5–6):677–682. https://doi.org/10.1387/ijdb.072528ek

Kim JM, Park JE, Yoo I, Han J, Kim N, Lim WJ, Cho ES, Choi B, Choi S, Kim TH, Te Pas MFW, Ka H, Lee KT (2018) Integrated transcriptomes throughout swine oestrous cycle reveal dynamic changes in reproductive tissues interacting networks. Sci Rep 8(1):1–14. https://doi.org/10.1038/s41598-018-23655-1

Kim KH, Seo YM, Kim EY, Lee SY, Kwon J, Ko JJ, Lee KA (2016) The miR-125 family is an important regulator of the expression and maintenance of maternal effect genes during preimplantational embryo development. Open Biol 6(11):1–15. https://doi.org/10.1098/rsob.160181

Kim TS, Lee SH, Gang GT, Lee YS, Kim SU, Koo DB, Shin MY, Park CK, Lee DS (2010) Exogenous DNA uptake of boar spermatozoa by a magnetic nanoparticle vector system. Reprod Domest Anim 45(5):e201–e206. https://doi.org/10.1111/j.1439-0531.2009.01516.x

Klauber N, Rohan M, Flynn E, D'Amato RJ (1997) Critical components of the female reproductive pathway are suppressed by the angiogenesis inhibitor AGM-1470. Nat Med 3(4):443–446

Kropp J, Khatib H (2015) Characterization of microRNA in bovine in vitro culture media associated with embryo quality and development. J Dairy Sci 98(9):6552–6563. https://doi.org/10.3168/jds.2015-9510

Kusama K, Nakamura K, Bai R, Nagaoka K, Sakurai T, Imakawa K (2018) Intrauterine exosomes are required for bovine conceptus implantation. Biochem Biophys Res Commun 495(1):1370–1375. https://doi.org/10.1016/j.bbrc.2017.11.176

Laezer I, Palma-Vera SE, Liu F, Frank M, Trakooljul N, Vernunft A, Schoen J, Chen S (2020) Dynamic profile of EVs in porcine oviductal fluid during the periovulatory period. Reproduction 159:371–382. https://doi.org/10.1530/REP

Lamy J, Labas V, Harichaux G, Tsikis G, Mermillod P, Saint-Dizier M (2016) Regulation of the bovine oviductal fluid proteome. Reproduction 152(6):629–644. https://doi.org/10.1530/REP-16-0397

Lange-Consiglio A, Funghi F, Cantile C, Idda A, Cremonesi F, Riccaboni P (2020) Case report: use of amniotic microvesicles for regenerative medicine treatment of a Mare with chronic endometritis. Front Vet Sci 7(347):1–6. https://doi.org/10.3389/fvets.2020.00347

Langendijk P, Soede NM, Kemp B (2005) Uterine activity, sperm transport, and the role of boar stimuli around insemination in sows. Theriogenology 63(2 SPEC):500–513. https://doi.org/10.1016/j.theriogenology.2004.09.027

Leiss M, Beckmann K, Girós A, Costell M, Fässler R (2008) The role of integrin binding sites in fibronectin matrix assembly in vivo. Curr Opin Cell Biol 20(5):502–507. https://doi.org/10.1016/j.ceb.2008.06.001

Li H, Ouyang Y, Sadovsky E, Parks WT, Chu T, Sadovsky Y (2020) Unique microRNA signals in plasma exosomes from pregnancies complicated by preeclampsia. Hypertension 75(3):762–771. https://doi.org/10.1161/HYPERTENSIONAHA.119.14081

Li J, Parrilla I, Ortega MD, Martínez EA, Rodríguez-Martínez H, Roca J (2018) Post-thaw boar sperm motility is affected by prolonged storage of sperm in liquid nitrogen. A retrospective study. Cryobiology 80:119–125. https://doi.org/10.1016/j.cryobiol.2017.11.004

Li R, Wu X, Zhu Z, Lv Y, Zheng Y, Lu H, Zhou K, Wu D, Zeng W, Dong W, Zhang T (2022) Polyamines protect boar sperm from oxidative stress in vitro. J Anim Sci 100(4):1–15. https://doi.org/10.1093/jas/skac069

Liman N (2017) Heat shock proteins (HSP)-60, −70, −90, and 105 display variable spatial and temporal immunolocalization patterns in the involuting rat uterus. Anim Reprod 14(4):1072–1086. https://doi.org/10.21451/1984-3143-AR917

Lin J, Li J, Huang B, Liu J, Chen X, Chen XM, Xu YM, Huang LF, Wang XZ (2015) Exosomes: novel biomarkers for clinical diagnosis. Sci World J 2015:1–8. https://doi.org/10.1155/2015/657086

Liu C, Yin H, Jiang H, Du X, Wang C, Liu Y, Li Y, Yang Z (2020) Extracellular vesicles derived from mesenchymal stem cells recover fertility of premature ovarian insufficiency mice and the effects on their offspring. Cell Transplant 29:1–11. https://doi.org/10.1177/0963689720923575

Lopera-Vasquez R, Hamdi M, Fernandez-Fuertes B, Maillo V, Beltran-Brena P, Calle A, Redruello A, Lopez-Martin S, Gutierrez-Adan A, Yanez-Mo M, Ramirez MA, Rizos D (2016) Extracellular vesicles from BOEC in in vitro embryo development and quality. PLoS One 11(2):1–23. https://doi.org/10.1371/journal.pone.0148083

Lopera-Vasquez R, Hamdi M, Maillo V, Gutierrez-Adan A, Bermejo-Alvarez P, Angel Ramirez M, Yanez-Mo M, Rizos D (2017) Effect of bovine oviductal extracellular vesicles on embryo development and quality in vitro. Reproduction 153(4):461–470. https://doi.org/10.1530/REP-16-0384

Luense LJ, Carletti MZ, Christenson LK (2009) Role of dicer in female fertility. Trends Endocrinol Metab 20(6):265–272. https://doi.org/10.1016/j.tem.2009.05.001

Luongo C, Llamas-López PJ, Hernández-Caravaca I, Matás C, García-Vázquez FA (2022) Should all fractions of the boar ejaculate be prepared for insemination rather than using the sperm rich only? Biology 11(2):1–15. https://doi.org/10.3390/biology11020210

Machtinger R, Gaskins AJ, Racowsky C, Mansur A, Adir M, Baccarelli AA, Calafat AM, Hauser R (2018) Urinary concentrations of biomarkers of phthalates and phthalate alternatives and IVF outcomes. Environ Int 111:23–31. https://doi.org/10.1016/j.envint.2017.11.011

Machtinger R, Laurent LC, Baccarelli AA (2016) Extracellular vesicles: roles in gamete maturation, fertilization and embryo implantation. In: Human reproduction update, vol 22. Oxford University Press, pp 182–193. https://doi.org/10.1093/humupd/dmv055

Mann T, Mann L (1981) Male reproductive function and the composition of semen: general considerations. In: Male reproductive function and semen. Springer-Verlag, Berling Heidelberg, pp 1–2

Martin-DeLeon P (2016) Uterosomes: Exosomal cargo during the estrus cycle and interaction with sperm. Front Biosci 8:115–122

Martin-Deleon PA (2011) Germ-cell hyaluronidases: their roles in sperm function. In. Int J Androl 34:e306. https://doi.org/10.1111/j.1365-2605.2010.01138.x

Martínez CA, Cambra JM, Gil MA, Parrilla I, Álvarez-Rodríguez M, Rodríguez-Martínez H, Cuello C, Martínez EA (2020) Seminal plasma induces overexpression of genes associated with embryo development and implantation in day-6 porcine blastocysts. Int J Mol Sci 21(10). https://doi.org/10.3390/ijms21103662

Martinez CA, Cambra JM, Parrilla I, Roca J, Ferreira-Dias G, Pallares FJ, Lucas X, Vazquez JM, Martinez EA, Gil MA, Rodriguez-Martinez H, Cuello C, Álvarez-Rodriguez M (2019) Seminal plasma modifies the transcriptional pattern of the endometrium and advances embryo development in pigs. Front Vet Sci 6(465):1–16. https://doi.org/10.3389/fvets.2019.00465

Mathivanan S, Ji H, Simpson RJ (2010) Exosomes: Extracellular organelles important in intercellular communication. J Proteome 73(10):1907–1920. https://doi.org/10.1016/j.jprot.2010.06.006

Mazzarella R, Bastos NM, Bridi A, del Collado M, Andrade GM, Pinzon J, Prado CM, Silva LA, Meirelles FV, Pugliesi G, Perecin F, da Silveira JC (2021) Changes in Oviductal cells and small extracellular vesicles miRNAs in pregnant cows. Front Vet Sci 8:1–14. https://doi.org/10.3389/fvets.2021.639752

Meng X, Müller V, Milde-Langosch K, Trillsch F, Pantel K, Schwarzenbach H (2016) Diagnostic and prognostic relevance of circulating exosomal miR-373, miR-200a, miR-200b and miR-200c in patients with epithelial ovarian cancer. Oncotarger 7(13):16923–16935. www.impactjournals.com/oncotarget/

Miyado K, Yoshida K, Yamagata K, Sakakibara K, Okabe M, Wang X, Miyamoto K, Akutsu H, Kondo T, Takahashi Y, Ban T, Ito C, Toshimori K, Nakamura A, Ito M, Miyado M, Mekada E, Umezawa A (2008) The fusing ability of sperm is bestowed by CD9-containing vesicles released from eggs in mice. PNAS 105(35):12921–12926. www.pnas.org/cgi/content/full/

Möbius W, Ohno-Iwashita Y, Van Donselaar EG, Oorschot VMJ, Shimada Y, Fujimoto T, Heijnen HFG, Geuze HJ, Slot JW (2002) Immunoelectron microscopic localization of cholesterol using biotinylated and non-cytolytic Perfringolysin O. J Histochem Cytochem 50(1):43–55. http://www.jhc.org

Moldenhauer LM, DIener KR, Hayball JD, Robertson SA (2017) An immunogenic phenotype in paternal antigen-specific CD8 + T cells at embryo implantation elicits later fetal loss in mice. Immunol Cell Biol 95(8):705–715. https://doi.org/10.1038/icb.2017.41

Morelli AE, Larregina AT, Shufesky WJ, Sullivan MLG, Stolz DB, Papworth GD, Zahorchak AF, Logar AJ, Wang Z, Watkins SC, Falo LD, Thomson AW (2004) Endocytosis, intracellular sorting, and processing of exosomes by dendritic cells. Blood 104(10):3257–3266. https://doi.org/10.1182/blood-2004-03-0824

Morgan HL, Watkins AJ (2019) The influence of seminal plasma on offspring development and health. Semin Cell Dev Biol 97:131–137. https://doi.org/10.1016/j.semcdb.2019.06.008

Mourad NI, Gianello P (2017) Gene Editing, Gene Therapy, and Cell Xenotransplantation: cell Transplantation Across Species. Curr Transplant Rep 4(3):193–200. https://doi.org/10.1007/s40472-017-0157-6

Murdica V, Giacomini E, Makieva S, Zarovni N, Candiani M, Salonia A, Vago R, Viganò P (2020) In vitro cultured human endometrial cells release extracellular vesicles that can be uptaken by spermatozoa. Sci Rep 10(1):1–13. https://doi.org/10.1038/s41598-020-65517-9

Nadkarni S, Smith J, Sferruzzi-Perri AN, Ledwozyw A, Kishore M, Haas R, Mauro C, Williams DJ, Farsky SHP, Marelli-Berg FM, Perretti M (2016) Neutrophils induce proangiogenic T cells with a regulatory phenotype in pregnancy. Proc Natl Acad Sci USA 113(52):E8415–E8424. https://doi.org/10.1073/pnas.1611944114

Naji A, Suganuma N, Espagnolle N, Yagyu K, Baba N, Sensebé L, Deschaseaux F (2017) Rationale for determining the functional potency of mesenchymal stem cells in preventing regulated cell death for therapeutic use. Stem Cells Transl Med 6(3):713–719. https://doi.org/10.5966/sctm.2016-0289

Nakano S, Yamamoto S, Okada A, Nakajima T, Sato M, Takagi T, Tomooka Y (2017) Role of extracellular vesicles in the interaction between epithelial and mesenchymal cells during oviductal ciliogenesis. Biochem Biophys Res Commun 483(1):245–251. https://doi.org/10.1016/j.bbrc.2016.12.158

Neuer A, Spandorfer SD, Giraldo P, Dieterle S, Rosenwaks Z, Witkin SS (2000) The role of heat shock proteins in reproduction. Hum Reprod Update 6(2):149–159

Ng YH, Rome S, Jalabert A, Forterre A, Singh H, Hincks CL, Salamonsen LA (2013) Endometrial exosomes/microvesicles in the uterine microenvironment: A new paradigm for embryo-endometrial Cross talk at implantation. PLoS One 8(3):1–13. https://doi.org/10.1371/journal.pone.0058502

Nguyen HPT, Simpson RJ, Salamonsen LA, Greening DW (2016) Extracellular vesicles in the intrauterine environment: challenges and potential functions. Biol Reprod 95(5):1–12. https://doi.org/10.1095/biolreprod.116.143503

Nollen E, Morimoto R (2002) Chaperoning signaling pathways: molecularchaperones as stress-sensing 'heat shock' proteins. J Cell Sci 155:2809–20816. https://doi.org/10.1242/jcs.115.14.2809

OCDE-FAO (2023) Food and agriculture Organization of the United Nations. In: OCDE FAO agricultural outlook 2021–2030, 1st edn. OECD Publising, pp 181–196. https://doi.org/10.1787/agr-outl-data-en

Ogawa Y, Miura Y, Harazono A, Kanai-Azuma M, Akimoto Y, Kawakami H, Yamaguchi T, Toda T, Endo T, Tsubuki M, Yanoshita R (2011) Human whole saliva is an aqueous complex mixture of proteins and minerals. Biol Pharm Bull 34(1):13–13. http://www.matrix-science.com/

O'Leary S, Jasper MJ, Robertson SA, Armstrong DT (2006) Seminal plasma regulates ovarian progesterone production, leukocyte recruitment and follicular cell responses in the pig. Reproduction 132(1):147–158. https://doi.org/10.1530/rep.1.01119

O'Leary S, Jasper MJ, Warnes GM, Armstrong DT, Robertson SA (2004) Seminal plasma regulates endometrial cytokine expression, leukocyte recruitment and embryo development in the pig. Reproduction 128(2):237–247. https://doi.org/10.1530/rep.1.00160

O'Neil EV, Burns GW, Spencer TE (2020) Extracellular vesicles: novel regulators of conceptus-uterine interactions? Theriogenology 150:106–112. https://doi.org/10.1016/j.theriogenology.2020.01.083

Parks JC, Mccallie BR, Patton AL, Al-Safi ZA, Polotsky AJ, Griffin DK, Schoolcraft WB, Katz-Jaffe MG (2018) The impact of infertility diagnosis on embryo-endometrial dialogue. Reproduction 155:543–552. https://doi.org/10.1530/REP

Parrilla I, Martinez EA, Gil MA, Cuello C, Roca J, Rodriguez-Martinez H, Martinez CA (2020) Boar seminal plasma: current insights on its potential role for assisted reproductive technologies in swine. Anim Reprod 17(3):1–20. https://doi.org/10.1590/1984-3143-AR2020-0022

Parrilla I, Perez-Patiño C, Li J, Barranco I, Padilla L, Rodriguez-Martinez H, Martinez EA, Roca J (2019) Boar semen proteomics and sperm preservation. Theriogenology 137:23–29. https://doi.org/10.1016/j.theriogenology.2019.05.033

Piehl LL, Fischman ML, Hellman U, Cisale H, Miranda PV (2013) Boar seminal plasma exosomes: effect on sperm function and protein identification by sequencing. Theriogenology 79(7):1071–1082. https://doi.org/10.1016/j.theriogenology.2013.01.028

Piibor J, Waldmann A, Dissanayake K, Andronowska A, Ivask M, Prasadani M, Kavak A, Kodithuwakku S, Fazeli A (2023) Uterine fluid extracellular vesicles proteome is altered during the estrous cycle. Mol Cell Proteomics 22(11):1–16. https://doi.org/10.1016/j.mcpro.2023.100642

Pisitkun T, Shen R-F, Knepper MA (2004) Identification and proteomic profiling of exosomes in human urine. PNAS 101(36):13368–13373. https://www.pnas.org

Pohler KG, Green JA, Moley LA, Gunewardena S, Hung WT, Payton RR, Hong X, Christenson LK, Geary TW, Smith MF (2017) Circulating microRNA as candidates for early embryonic viability in cattle. Mol Reprod Dev 84(8):731–743. https://doi.org/10.1002/mrd.22856

Pons-Rejraji H, Artonne C, Sion B, Brugnon F, Canis M, Janny L, Grizard G (2011) Prostasomes: inhibitors of capacitation and modulators of cellular signalling in human sperm. Int J Androl 34(6 PART 1):568–580. https://doi.org/10.1111/j.1365-2605.2010.01116.x

Qu P, Qing S, Liu R, Qin H, Wang W, Qiao F, Ge H, Liu J, Zhang Y, Cui W, Wang Y (2017) Effects of embryo-derived exosomes on the development of bovine cloned embryos. PLoS One 12(3):1–17. https://doi.org/10.1371/journal.pone.0174535

Qu P, Zhao J, Hu H, Cao W, Zhang Y, Qi J, Meng B, Zhao J, Liu S, Ding C, Wu Y, Liu E (2022) Loss of renewal of extracellular vesicles: harmful effects on embryo development in vitro. Int J Nanomedicine 17:2301–2318. https://doi.org/10.2147/IJN.S354003

Qu P, Zhao Y, Wang R, Zhang Y, Li L, Fan J, Liu E (2019) Extracellular vesicles derived from donor oviduct fluid improved birth rates after embryo transfer in mice. Reprod Fertil Dev 31(2):324–332. https://doi.org/10.1071/RD18203

Rana S, Yue S, Stadel D, Zöller M (2012) Toward tailored exosomes: the exosomal tetraspanin web contributes to target cell selection. Int J Biochem Cell Biol 44(9):1574–1584. https://doi.org/10.1016/j.biocel.2012.06.018

Raposo G, Nijman HW, Stoorvogel W, Leijendekker R, Hardingfl CV, Cornelis JM, Melief, Geuze HJ (1996) B lymphocytes secrete antigen-presenting vesicles. J Exp Med 183:1161–1172. http://rupress.org/jem/article-pdf/183/3/1161/1678390/1161.pdf

Raposo G, Stoorvogel W (2013) Extracellular vesicles: exosomes, microvesicles, and friends. J Cell Biol 200(4):373–383. https://doi.org/10.1083/jcb.201211138

Rath D, Knorr C, Taylor U (2016) Communication requested: boar semen transport through the uterus and possible consequences for insemination. Theriogenology 85(1):94–104. https://doi.org/10.1016/j.theriogenology.2015.09.016

Rejraji H, Sion B, Prensier G, Carreras M, Motta C, Frenoux JM, Vericel E, Grizard G, Vernet P, Drevet JR (2006) Lipid remodeling of murine epididymosomes and spermatozoa during epididymal maturation. Biol Reprod 74(6):1104–1113. https://doi.org/10.1095/biolreprod.105.049304

Rhodes M, Brendemuhl JH, Hansen PJ (2006) Litter characteristics of gilts artificially insemi-nated with transforming growth factor-β. Am J Reprod Immunol 56(3):153–156. https://doi.org/10.1111/j.1600-0897.2006.00423.x

Riesenbeck A (2011) Review on international trade with boar semen. Reprod Domest Anim 46(SUPPL. 2):1–3. https://doi.org/10.1111/j.1439-0531.2011.01869.x

Riesenbeck A, Schulze M, Rüdiger K, Henning H, Waberski D (2015) Quality control of boar sperm processing: implications from European AI Centres and two Spermatology reference laboratories. Reprod Domest Anim 50(Suppl. 2):1–4. https://doi.org/10.1111/RDA.12573

Robertson SA (2007) Seminal fluid signaling in the female reproductive tract: lessons from rodents and pigs. J Anim Sci 85(13 Suppl):E36. https://doi.org/10.2527/jas.2006-578

Robertson SA, Ingman WV, O'leary S, Sharkey DJ, Tremellen KP (2002) Transforming growth factor b-a mediator of immune deviation in seminal plasma. J Reprod Immunol 57:109–128. www.elsevier.com/locate/jreprimm

Robertson SA, Sharkey DJ (2016) Seminal fluid and fertility in women. Fertil Steril 106(3):511–519. https://doi.org/10.1016/j.fertnstert.2016.07.1101

Robinson DG, Ding Y, Jiang L (2016) Unconventional protein secretion in plants: a critical assess-ment. Protoplasma 253(1):31–43. https://doi.org/10.1007/s00709-015-0887-1

Roca J, Vázquez JM, Gil MA, Cuello C, Parrilla I, Martínez EA (2006) Challenges in pig artificial insemination. Reprod Domest Anim 41(Suppl. 2):43–53

Rodríguez-Martínez H, Kvist U, Ernerudh J, Sanz L, Calvete JJ (2011) Seminal plasma pro-teins: what role do they play? Am J Reprod Immunol 66(SUPPL. 1):11–22. https://doi.org/10.1111/j.1600-0897.2011.01033.x

Rodríguez-Martínez, H., Kvist, U., Saravia, F., Wallgren, M., Johannisson, A., Sanz, L., Peña, F. J., Martínez, E. A., Roca, J., Vázquez, J. M., & Calvete, J. J. (2009). The physiological roles of the boar ejaculate

Rodriguez-Martinez H, Martinez EA, Calvete JJ, Peña Vega FJ, Roca J (2021) Seminal plasma: relevant for fertility? Int J Mol Sci 22(9). https://doi.org/10.3390/ijms22094368

Rodríguez-Martínez H, Saravia F, Wallgren M, Tienthai P, Johannisson A, Vázquez JM, Martínez E, Roca J, Sanz L, Calvete JJ (2005) Boar spermatozoa in the oviduct. Theriogenology 63:514–535. https://doi.org/10.1016/j.theriogenology.2004.09.028

Rodriguez-Martinez H, Tienthai P, Suzuki K, Funahashi H, Ekwall H, Johannisson A (2001) Involvement of oviduct in sperm capacitation andd oocyte development in pigs. Reprod Suppl 58:129–145

Romar R, Cánovas S, Matás C, Gadea J, Coy P (2019) Pig in vitro fertilization: where are we and where do we go? Theriogenology 137:113–121. https://doi.org/10.1016/j.theriogenology.2019.05.045

Ronquist G (2012) Prostasomes are mediators of intercellular communication: from basic research to clinical implications. J Intern Med 271(4):400–413. https://doi.org/10.1111/j.1365-2796.2011.02487.x

Ronquist G, Hedstrom M (1977) Restoration of detergent-inactivated adenosine tri-phosphatase activity of human prostatic fluid with Cocavalin A. Biochim Biophys Acta 483:483–486

Ronquist G, Nilsson B, Hjerten S (1990) Interaction between prostasomes and spermatozoa from human semen. Arch Androl 24:147–157

Rosenbluth EM, Shelton DN, Wells LM, Sparks AET, Van Voorhis BJ (2014) Human embryos secrete microRNAs into culture media–A potential biomarker for implantation. Fertil Steril 101(5):1493–1500. https://doi.org/10.1016/j.fertnstert.2014.01.058

Rozeboom KJ, Troedsson MHT, Crabo BG (1998) Characterization of uterine leukocyte infiltra-tion in gilts after artificial insemination. J Reprod Fertil 114:195–199

Saadeldin IM, Kim SJ, Choi YB, Lee BC (2014) Improvement of cloned embryos development by co-culturing with parthenotes: A possible role of exosomes/microvesicles for embryos para-crine communication. Cell Reprogram 16(3):223–234. https://doi.org/10.1089/cell.2014.0003

Saadeldin IM, Oh HJ, Lee BC (2015) Embryonic–maternal cross-talk via exosomes: potential implications. Stem Cells Cloning 8:103–107. https://doi.org/10.2147/SCCAA.S84991

Sabapatha A, Gercel-taylor C, Taylor DD (2006) Specific isolation of placenta-derived exosomes from the circulation of pregnant women and their immunoregulatory consequences. Am J Reprod Immunol 56(5–6):345–355. https://doi.org/10.1111/j.1600-0897.2006.00435.x

Saez F, Frenette G, Sullivan R (2003) Epididymosomes and prostasomes: their roles in posttesticular maturation of the sperm cells. J Androl 24(2):149–154. https://doi.org/10.1002/j.1939-4640.2003.tb02653.x

Saez F, Motta C, Boucher D, Grizard G (2000) Prostasomes inhibit the NADPH oxidase activity of human neutrophils. Mol Hum Reprod 6(10):883–891

Samanta L, Parida R, Dias TR, Agarwal A (2018) The enigmatic seminal plasma: A proteomics insight from ejaculation to fertilization. Reprod Biol Endocrinol 16(1):1–11. https://doi.org/10.1186/s12958-018-0358-6

Santonocito M, Vento M, Guglielmino MR, Battaglia R, Wahlgren J, Ragusa M, Barbagallo D, Borzì P, Rizzari S, Maugeri M, Scollo P, Tatone C, Valadi H, Purrello M, Di Pietro C (2014) Molecular characterization of exosomes and their microRNA cargo in human follicular fluid: Bioinformatic analysis reveals that exosomal microRNAs control pathways involved in follicular maturation. Fertil Steril 102(6):1751–1761. https://doi.org/10.1016/j.fertnstert.2014.08.005

Saravia F, Wallgren M, Johannisson A, Calvete JJ, Sanz L, Peña FJ, Roca J, Rodríguez-Martínez H (2009) Exposure to the seminal plasma of different portions of the boar ejaculate modulates the survival of spermatozoa cryopreserved in MiniFlatPacks. Theriogenology 71(4):662–675. https://doi.org/10.1016/j.theriogenology.2008.09.037

Schauer SN, Sontakke SD, Watson ED, Esteves CL, Donadeu FX (2013) Involvement of miR-NAs in equine follicle development. Reproduction 146(3):273–282. https://doi.org/10.1530/REP-13-0107

Schjenken JE, Robertson SA (2014) Seminal fluid and immune adaptation for pregnancy–comparative biology in mammalian species. Reprod Domest Anim 49:27–36. https://doi.org/10.1111/rda.12383

Schjenken JE, Robertson SA (2020) The female response to seminal fluid. Physiol Rev 100(3):1077–1117. https://doi.org/10.1152/physrev.00013.2018

Schorey JS, Cheng Y, Singh PP, Smith VL (2015) Exosomes and other extracellular vesicles in host–pathogen interactions. EMBO Rep 16(1):24–43. https://doi.org/10.15252/embr.201439363

Schuster A, Tang C, Xie Y, Ortogero N, Yuan S, Yan W (2016) SpermBase: A database for sperm-borne RNA contents. Biol Reprod 95(5):1–12. https://doi.org/10.1095/biolreprod.116.142190

Seytanoglu A, Stephen Georgiou A, Sostaric E, Watson PF, Holt WV, Fazeli A (2008) Oviductal cell proteome alterations during the reproductive cycle in pigs. J Proteome Res 7(7):2825–2833. https://doi.org/10.1021/pr8000095

Siciliano L, Marcianò V, Carpino A (2008) Prostasome-like vesicles stimulate acrosome reaction of pig spermatozoa. Reprod Biol Endocrinol 6:1–7. https://doi.org/10.1186/1477-7827-6-5

Sims PJ, Faioni EM, Wiedmer T, Shattil SJ (1988) Complement proteins C5b-9 cause release of membrane vesicles from the platelet surface that are enriched in the membrane receptor for coagulation factor Va and express Prothrombinase activity*. J Biol Chem 263(34):18205–18212

Skotland T, Hessvik NP, Sandvig K, Llorente A (2019) Exosomal lipid composition and the role of ether lipids and phosphoinositides in exosome biology. J Lipid Res 60(1):9–18. https://doi.org/10.1194/jlr.R084343

Skotland T, Sagini K, Sandvig K, Llorente A (2020) An emerging focus on lipids in extracellular vesicles. Adv Drug Deliv Rev 159:308–321. https://doi.org/10.1016/j.addr.2020.03.002

Soleilhavoup C, Riou C, Tsikis G, Labas V, Harichaux G, Kohnke P, Reynaud K, De Graaf SP, Gerard N, Druart X (2016) Proteomes of the female genital tract during the oestrous cycle. Mol Cell Proteomics 15(1):93–108. https://doi.org/10.1074/mcp.M115.052332

Soriano-Úbeda C, García-Vázquez FA, Romero-Aguirregomezcorta J, Matás C (2017) Improving porcine in vitro fertilization output by simulating the oviductal environment. Sci Rep 7(1):1–12. https://doi.org/10.1038/srep43616

Soriano-Úbeda C, Romero-Aguirregomezcorta J, Matás C, Visconti PE, García-Vázquez FA (2019) Manipulation of bicarbonate concentration in sperm capacitation media improvesin vitro fertilisation output in porcine species. J Anim Sci Biotechnol 10(1):1–15. https://doi.org/10.1186/s40104-019-0324-y

Steverink DWB, Soede NM, Bouwman EG, Kemp B (1998) Semen backflow after insemination and its effect on fertilisation results in sows. Anim Reprod Sci 54:109–119

Strzeżek J, Lecewicz M (1999) A note on antioxidant capacity of boar seminal plasma. Anim Sci Rep 17(4):181–188. https://www.researchgate.net/publication/285322216

Sullivan R (2015) Epididymosomes: A heterogeneous population of microvesicles with multiple functions in sperm maturation and storage. Asian J Androl 17(5):726–729. https://doi.org/10.4103/1008-682X.155255

Sullivan R, Saez F (2013) Epididymosomes, prostasomes, and liposomes: their roles in mammalian male reproductive physiology. Reproduction 146(1):R21–R35. https://doi.org/10.1530/REP-13-0058

Sutovsky P, Moreno R, Ramalho-Santos J, Dominko T, Thompson WE, Schatten G (2001) A putative, ubiquitin-dependent mechanism for therecognition and elimination of defective spermatozoain the mammalian epididymis. J Cell Sci 114(9):1665–1675

Tamai K, Tanaka N, Nakano T, Kakazu E, Kondo Y, Inoue J, Shiina M, Fukushima K, Hoshino T, Sano K, Ueno Y, Shimosegawa T, Sugamura K (2010) Exosome secretion of dendritic cells is regulated by Hrs, an ESCRT-0 protein. Biochem Biophys Res Commun 399(3):384–390. https://doi.org/10.1016/j.bbrc.2010.07.083

Tan Q, Shi S, Liang J, Zhang X, Cao D, Wang Z (2020) MicroRNAs in small extracellular vesicles indicate successful embryo implantation during early pregnancy. Cells 9(645):1–17. https://doi.org/10.3390/cells9030645

Taylor U, Rath D, Zerbe H, Schuberth HJ (2008) Interaction of intact porcine spermatozoa with epithelial cells and neutrophilic granulocytes during uterine passage. Reprod Domest Anim 43(2):166–175. https://doi.org/10.1111/j.1439-0531.2007.00872.x

Théry C, Ostrowski M, Segura E (2009) Membrane vesicles as conveyors of immune responses. Nat Rev Immunol 9(8):581–593. https://doi.org/10.1038/nri2567

Thimon V, Frenette G, Saez F, Thabet M, Sullivan R (2008) Protein composition of human epididymosomes collected during surgical vasectomy reversal: A proteomic and genomic approach. Hum Reprod 23(8):1698–1707. https://doi.org/10.1093/humrep/den181

Tienthai P, Kjellén L, Pertoft H, Suzuki K, Rodriguez-Martinez H (2000) Localization and quantitation of hyaluronan and sulfated glycosaminoglycans in the tissues and intraluminal fluid of the pig oviduct. Reprod Fertil Dev 12(3–4):173–182. https://doi.org/10.1071/rd00034

Toledo-Guardiola SM, Luongo C, Abril-Parreño L, Soriano-Úbeda C, Matás C (2023) Different seminal ejaculated fractions in artificial insemination condition the protein cargo of oviductal and uterine extracellular vesicles in pig. Front Cell Dev Biol 11:1–15. https://doi.org/10.3389/fcell.2023.1231755

Toledo-Guardiola SM, Martínez-Díaz P, Martínez-Núñez R, Navarro-Serna S, Soriano-Úbeda C, Romero-Aguirregomezcorta J, Matás C (2024a) Sperm functionality is differentially regulated by porcine oviductal extracellular vesicles from the distinct phases of the estrous cycle. Reprod Fertil Dev 36(8):1–14. https://doi.org/10.1071/RD23239

Toledo-Guardiola SM, Párraga-Ros E, Seva J, Luongo C, García-Vázquez FA, Soriano-Úbeda C, Matás C (2024b) Artificial insemination of all ejaculated sperm fractions accelerates embryo development and increases the uterine vascularity in the pig. Theriogenology 219:32–38. https://doi.org/10.1016/j.theriogenology.2024.02.017

Tomlinson MJ, White A, Barratt CLR, Bolton AE, Cooke ID (1992) The removal of morphologically abnormal sperm forms by phagocytes: a positive role for seminal leukocytes? Hum Reprod 7:517–522

Van Niel G, D'Angelo G, Raposo G (2018) Shedding light on the cell biology of extracellular vesicles. In nature reviews molecular cell biology (Vol. 19, issue 4, pp. 213–228). Nature publishing group. https://doi.org/10.1038/nrm.2017.125

Vargas A, Zhou S, Ethier-Chiasson M, Flipo D, Lafond J, Gilbert C, Barbeau B (2014) Syncytin proteins incorporated in placenta exosomes are important for cell uptake and show variation in abundance in serum exosomes from patients with preeclampsia. FASEB J 28(8):1–17. https:// doi.org/10.1096/fj.13-239053

Visconti PE (2009) Understanding the molecular basis of sperm capacitation through kinase design. PNAS 106(3):667–668

Waberski D, Claassen R, Hahn T, Jungblut PW, Parvizi N, Kallweit E, Weitze F (1997) LH profile and advancement of ovulation after transcervical infusion of seminal plasma at different stages of oestrus in gilts. J Reprod Fertil 109:29–34

Waberski D, Riesenbeck A, Schulze M, Weitze KF, Johnson L (2019) Application of preserved boar semen for artificial insemination: past, present and future challenges. Theriogenology 137:2–7. https://doi.org/10.1016/j.theriogenology.2019.05.030

Wallgren M, Saravia F, Rodríguez-Martínez H (2010) The vanguard sperm cohort of the boar ejaculate is overrepresented in the tubal sperm reservoir in vivo. J Reprod Dev 56(1):9–125

Walzel H, Brock J, Pöhland R, Vanselow J, Tomek W, Schneider F, Tiemann U (2004) Effects of galectin-1 on regulation of progesterone production in granulosa cells from pig ovaries in vitro. Glycobiology 14(10):871–881. https://doi.org/10.1093/glycob/cwh101

Waqas MY, Zhang Q, Ahmed N, Yang P, Xing G, Akhtar M, Basit A, Liu T, Hong C, Arshad M, Rahman HMSU, Chen Q (2017) Cellular evidence of exosomes in the reproductive tract of Chinese soft-shelled turtle Pelodiscus sinensis. J Exp Zool A Ecol Integr Physiol 327(1):18–27. https://doi.org/10.1002/jez.2065

Watkins AJ, Dias I, Tsuro H, Allen D, Emes RD, Moreton J, Wilson R, Ingram RJM, Sinclair KD (2018) Paternal diet programs offspring health through sperm- and seminal plasma-specific pathways in mice. Proc Natl Acad Sci USA 115(40):10064–10069. https://doi.org/10.1073/ pnas.1806333115

Watkins AJ, Sinclair KD, Watkins AJ (2014) Paternal low protein diet affects adult offspring cardiovascular and metabolic function in mice. Am J Physiol Heart Circ Physiol 306:1444–1452. https://doi.org/10.1152/ajpheart.00981.2013.-Although

Watson AL, Carlson DF, Largaespada DA, Hackett PB, Fahrenkrug SC (2016) Engineered swine models of cancer. Front Genet 7(78):1–16. https://doi.org/10.3389/fgene.2016.00078

Winuthayanon W, Li S (2018) Fallopian tube/oviduct: structure and cell biology. In: Encyclopedia of reproduction, vol 2. Elsevier, pp 282–290. https://doi.org/10.1016/B978-0-12-801238-3.64401-X

Wolf P (1967) The nature and significance of platelet products in human plasma. Br J Haematol 13:269–288

Yanagimachi R, Kamiguchi Y, Mikamo K, Suzuki F, Yanagimachi H (1985) Maturation of spermatozoa in the epididymis of the Chinese hamster. Am J Anat 172:317–330

Yao Y, Chen R, Wang G, Zhang Y, Liu F (2019) Exosomes derived from mesenchymal stem cells reverse EMT via TGF-β1/Smad pathway and promote repair of damaged endometrium. Stem Cell Res Ther 10(1):1–17. https://doi.org/10.1186/s13287-019-1332-8

Yu C, Li M, Wang Y, Liu Y, Yan C, Pan J, Liu J, Cui S (2017) miR-375 mediates CRH signaling pathway in inhibiting E2 synthesis in porcine ovary. Reproduction 153(1):63–73. https://doi. org/10.1530/REP-16-0323

Zhang H, Martin-DeLeon PA (2003) Mouse Spam1 (PH-20) is a multifunctional protein: evidence for its expression in the female reproductive tract. Biol Reprod 69(2):446–454. https://doi. org/10.1095/biolreprod.102.013854

Zhang T, Hu R, Wang Y, Guo S, Wu Z, Liu J, Han C, Qiu C, Deng G (2023) Extracellular matrix stiffness mediates uterine repair via the Rap1a/ARHGAP35/RhoA/F-actin/YAP axis. Cell Commun Signal 21(1):1–16. https://doi.org/10.1186/s12964-022-01018-8

Zhao G, Yang C, Yang J, Liu P, Jiang K, Shaukat A, Wu H, Deng G (2018) Placental exosome-mediated Bta-miR-499-Lin28B/let-7 axis regulates inflammatory bias during early pregnancy. Cell Death Dis 9(6):1–18. https://doi.org/10.1038/s41419-018-0713-8

Zhou C, Cai G, Meng F, Xu Z, He Y, Hu Q, Zheng E, Huang S, Xu Z, Gu T, Hu B, Wu Z, Hong L (2020) Deep-sequencing identification of MicroRNA biomarkers in serum exosomes for early pig pregnancy. Front Genet 11. https://doi.org/10.3389/fgene.2020.00536

Zhou W, De Iuliis GN, Dun MD, Nixon B (2018) Characteristics of the epididymal luminal environment responsible for sperm maturation and storage. Front Endocrinol 9(59):1–13. https://doi.org/10.3389/fendo.2018.00059

Zigo M, Jonakova V, Manaskova-Postlerova P, Kerns K, Sutovsky P (2019) Ubiquitin-proteasome system participates in the de-aggregation of spermadhesins and DQH protein during boar sperm capacitation. Reproduction 157:283–295. https://doi.org/10.1530/REP

Zigo M, Kerns K, Sutovsky P (2023) The ubiquitin-proteasome system participates in sperm surface subproteome remodeling during boar sperm capacitation. Biomol Ther 13(6):1–27. https://doi.org/10.3390/biom13060996

Zitvogel L, Armelle R, Lozier A, Wolfers J, Flament C, Tenza D, Ricciardi-Castagnoli P, Rapaso G, Amigorena S (1998) Erradication of established murine tumors using a novel cell-free vaccine: dentritic cell-derived exosomes. Nat Med 4(5):594–600

Early Embryonic Development in the Mare: From Fertilization to Implantation

Juan Carlos Gardón ⓘ, María Gemma Velasco-Martínez ⓘ, and Katy Satué ⓘ

Abstract

The development of the equine embryo is a complex process that is critical to the establishment of pregnancy. It begins with fertilization, where the spermatozoa penetrate the oocyte and a series of intricate molecular and cellular events unfold. After fertilization, the zygote undergoes cleavage divisions, forming a multicellular embryo. These divisions are accompanied by changes in gene expression and cellular differentiation, leading to the formation of the blastocyst. During blastocyst formation, cells differentiate into distinct lineages, including the inner cell mass (ICM) and the trophectoderm (TE). Implantation is the process by which the embryo attaches to the uterine wall. The success of this process is contingent upon intricate interactions between the embryo and the maternal environment, involving signaling molecules, adhesion molecules, and immune responses. Throughout early development, the embryo is highly susceptible to environmental influences, including maternal health, nutrition, and stress. Disruptions during this period can lead to developmental abnormalities or pregnancy loss. Advances in assisted reproductive technologies have enabled the manipulation of early equine embryos for research purposes and assisted breeding programs. In vitro fertilization (IVF) offers opportunities for studying early development and enhancing breeding outcomes. Understanding the mechanisms

J. C. Gardón (✉)
Department of Animal Medicine and Surgery, Faculty of Veterinary and Experimental Sciences, Catholic University of Valencia-San Vicente Mártir, Valencia, Spain
e-mail: jc.gardon@ucv.es

M. G. Velasco-Martínez · K. Satué
Department of Animal Medicine and Surgery, CEU Cardenal Herrera University, Valencia, Spain
e-mail: mariagemma.velasco@alumnos.uchceu.es; ksatue@uchceu.es

© The Author(s), under exclusive license to Springer Nature Switzerland AG 2024
J. C. Gardón, K. Satué Ambrojo (eds.), *Assisted Reproductive Technologies in Animals Volume 1*, https://doi.org/10.1007/978-3-031-73079-5_14

underlying equine early embryo development is essential for improving reproductive success and advancing our knowledge of mammalian embryogenesis.

Keywords

Equine · Embryo development · In vitro · In vivo · Pregnancy recognition

1   Introduction

The development of the equine embryo, from fertilization to implantation, is a meticulously orchestrated process essential for the continuation of the species. Upon successful fertilization of the mare's egg by the stallion's sperm, a series of intricate cellular events ensue, initiating embryonic development. This process commences with the formation of the zygote, the single-cell entity resulting from the fusion of the egg and sperm. It then progresses through stages of cleavage, blastocyst formation, and gastrulation, ultimately leading to the formation of distinct embryonic structures (Betteridge 2000).

During cleavage, the zygote undergoes rapid cell division, dividing into smaller and smaller cells called blastomeres. These blastomeres continue to divide and form a compact ball of cells known as the morula, which eventually differentiates into a blastocyst. The blastocyst consists of two distinct cell populations: the inner cell mass (ICM), which will develop into the embryo itself, and the outer trophoblast cells, which will contribute to the formation of the placenta (Betteridge 2007). Following the formation of the blastocyst, the embryo enters the uterus and undergoes implantation, a critical step in which it attaches to the uterine wall. Implantation involves complex interactions between the embryonic trophoblast cells and the maternal endometrium, facilitated by factors such as adhesion molecules and cytokines. Successful implantation ensures proper nourishment and support for the developing embryo throughout gestation (Allen and Wilsher 2009).

From the moment of conception, the mare's uterus provides a nurturing environment for the developing embryo, facilitating implantation and establishing the foundation for placental development. Maternal factors, such as uterine histotroph composition and endometrial receptivity, influence the embryo's ability to attach to the uterine wall and initiate the formation of vital placental structures. (Aurich and Budik 2015). The development of equine embryos is profoundly dependent on intricate maternal interactions, which mark a pivotal relationship essential for successful reproduction in horses. Throughout gestation, the maternal environment plays a crucial role in nurturing the developing embryo and shaping its growth and development. This dynamic interplay encompasses a myriad of physiological processes within the mare's reproductive system, ensuring optimal conditions for embryonic survival and fetal well-being (Bauersachs and Almiñana 2020).

The in vitro development of equine embryos represents a significant advancement in assisted reproductive technologies, offering potential solutions to various reproductive challenges in the equine industry. In vitro fertilization (IVF) and embryo culture techniques permit researchers and practitioners to manipulate and monitor embryonic development outside the mare's reproductive tract. This approach provides valuable insights into early embryo physiology and allows for the production of embryos independent of traditional breeding constraints (Hinrichs 2010). During in vitro embryo culture, a variety of parameters, including pH, temperature, gas composition, and nutrient concentrations, are meticulously regulated to closely mimic the physiological conditions of the mare's reproductive tract. This optimized environment is conducive to embryo viability and developmental competence, resulting in the production of high-quality blastocysts suitable for transfer or cryopreservation (Galli et al. 2001).

Similarly, equine chorionic gonadotropin (eCG), also known as pregnant mare serum gonadotropin (PMSG), plays a pivotal role in equine reproduction, particularly during the early stages of pregnancy. This glycoprotein hormone, primarily produced by the endometrial cups of the developing placenta, exerts profound effects on the mare's reproductive physiology, influencing ovarian function and maintaining pregnancy. An understanding of the production and functions of eCG is essential for the development of optimal reproductive management strategies in horses (Murphy 2012). In summary, eCG production plays a pivotal role in equine reproduction, exerting multifaceted effects on ovarian function, pregnancy maintenance, and maternal physiology. An understanding of the regulation and functions of eCG is fundamental for the optimization of breeding management practices and the assurance of successful outcomes in equine reproduction (Wildman et al. 2006; Mess and Carter 2007; Galet et al. 2009).

2   Equine Embryonic Development in the Oviduct

Following ovulation, the mare's oocyte is released into the oviductal fimbria and travels to the ampullary region, where fertilization occurs in the presence of capacitated sperm (Hunter 2012). Following fertilization, the initial cleavage of the equine zygote is observed approximately 20 h post-ovulation, with subsequent divisions occurring at intervals of approximately 12 h, leading to the formation of a 12–36 cell stage by day 4 of pregnancy (Webel et al. 1977). From the moment of fertilization, the newly formed zygote is endowed with all the genetic information necessary to progress through the early stages of embryonic development, initiating a sequence of cell divisions that culminate in the formation of a morula (Gilbert 2000).

Palmer et al. (1997) demonstrated that while follicular fluid may be beneficial, it is not indispensable for conveying the oocyte to the ampulla. Nevertheless, fibroblasts are known to enter the oviduct at ovulation and produce collagen masses in its lumen, which may impact early development (Lantz et al. 1998). Given the differential transport of fertilized and unfertilized equine eggs in the mare, it appears unlikely that follicular contents play a significant role in embryo transport, contrary to proposals for cattle by Wijayagunawardane et al. (1999).

The cleavage process in horse eggs is comparable to that observed in other domestic mammals, with variations in cleavage rate potentially influencing the initial development of the embryo and resulting in embryos of varying sizes at the same developmental stage (Gaivão and Stout 2012). The early stages of equine embryogenesis present several unique features, including the polarity and opacity of the oocyte and zygote. Furthermore, distinctive aspects include the contents of the perivitelline space, the often-ellipsoidal shape of cleaving eggs, and the acquisition of additional coats on the outside of the zona pellucida by cleaving embryos, a process known as deutoplasmolysis, which is particularly observed in embryos with fewer than 16 cells (Betteridge 2000).

Early embryonic development in the horse is characterized by several peculiarities (Betteridge 2007). First, the equine embryo does not exit the oviduct via the prominent uterotubal papilla until as late as 144–156h after ovulation (Freeman et al. 1991; Smits et al. 2012). The duration for the cleaving embryo to traverse the oviduct is estimated to be 5.5 to 6.0 days, with the complete journey taking 6.0 to 6.5 days (Betteridge 2011). The final segment of this journey, through the isthmus, occurs relatively swiftly (Weber et al. 1996).

It was described that the communication between the embryo and the mare's reproductive tract likely initiates early in the oviduct. The characteristics of the oviductal epithelium vary depending on the reproductive status and location within the tract (Ball et al. 1997). It is therefore important to note that the equine embryo predominantly resides near the ampullary–isthmic junction of the oviduct during its 6-day journey (Weber et al. 1996). Although oviductal secretions are abundant, their composition does not appear to vary significantly with the mare's reproductive status or region (McDowell et al. 1993).

The oviductal epithelium exhibits varying histochemical characteristics depending on reproductive status and location. It is crucial to recognize that the embryo is just one participant in the complex physiological dialogue with its environment. Unlike in cattle, where follicular contents may influence embryo transport, fertilized and unfertilized equine eggs are transported differently in the mare (Kölle et al. 2020).

The reciprocal interaction between the embryo and the oviduct is notably evidenced by the differential transport of fertilized and unfertilized eggs to the uterus. Initially, Van Niekerk and Gerneke 1996, and subsequent studies (Freeman et al. 1992; Weber and Woods 1993; Weber et al. 1991a, b, 1995), have highlighted the involvement of prostaglandin E2 produced by horse embryos in this phenomenon.

This understanding of horses has prompted similar investigations across species, with observations made in bats, hamsters, and rats (Rasweiler 1979; Ortiz et al. 1989; Ortiz et al. 1986; Velasquez et al. 1995, 1997). Other studies in hamsters have indicated the necessity to investigate additional embryonic products, such as platelet-activating factor, in horses (Velasquez et al. 1995, 1997).

It is important to consider the activation of the embryonic genome while still in the oviduct, which occurs gradually at the four- to eight-cell stage (Ball et al. 1993; Grøndahl et al. 1993; Brinsko et al. 1995). Further insights have been gained from mouse studies, which have enhanced our comprehension of the influence of the interaction between developing embryos and the oviduct on subsequent interactions with the uterus and pregnancy rates. Research has demonstrated that the structure of the oviduct wall can be locally influenced by the presence of oocytes shortly after ovulation in mice (Sato et al. 1995).

The equine embryo primarily resides near the ampullary–isthmic junction of the oviduct during its 6-day journey. The functional aspects of the oviduct influence embryo survival and exhibit age-dependent effects. In vitro studies have demonstrated that co-culturing cleaving embryos with oviductal epithelial cells has beneficial effects on embryo development (Morris 2018).

Nevertheless, the equine endosalpinx is not considered essential for normal development, as viable horse embryos can develop in the oviduct of nonpregnant ewes (Ball and Miller 1992; Carnevale et al. 1993; Weber et al. 1996; Wijayagunawardane et al. 1999; Betteridge 2000; Raeside et al. 2004).

The oviduct exerts a beneficial influence on equine embryo development, as evidenced by the accelerated progression and enhanced probability of reaching the blastocyst stage in 2–4 cell embryos transferred into the oviduct of a recipient mare shortly after intracytoplasmic sperm injection (ICSI) compared to embryos maintained in vitro (Choi et al. 2004b; Rader et al. 2016).

The presence of an embryo in the mare's oviduct appears to induce alterations in the gene and protein expression profiles of oviductal epithelial cells, as evidenced by studies conducted by Smits et al. (2012, 2016). Nevertheless, it does not appear to impede the mare's capacity to support pregnancy, as evidenced by the consistently high pregnancy rates observed in well-synchronized embryo transfer recipient mares, as demonstrated by Cuervo-Arango et al. (2019).

During its residence within the oviduct, the embryo's genetic material becomes active, advancing it to the late morula phase of development. Despite the lack of comprehensive understanding regarding the interaction between embryos and the oviduct, the environment within the oviductal ampulla actively supports early development (Stout 2020). Embryos produced in vitro (IVP) reach the blastocyst stage at a similar time to those in vivo (7–8 days post-fertilization) but have a significantly lower number of cells at this stage, with a higher proportion undergoing apoptosis (Tremoleda et al. 2003a; Choy, et al. 2004a; Pomar et al. 2005). Upon transfer to a recipient mare, IVP embryos exhibit a developmental lag of 2–3 days and are more susceptible to early embryonic death compared to in vivo embryos (Morris 2018).

Selective oviductal transport marks the conclusion of the equine embryo's journey in the oviduct. This process is facilitated by an enigmatic embryonic signaling mechanism known as selective oviductal transport. This ensures that only developing embryos progress into the uterus, while unfertilized oocytes remain confined within the oviductal ampulla (Betteridge and Mitchell 1974). Prostaglandins, particularly prostaglandin E2 (PGE2), serve as the key mediators of this selective transport. Viable embryos initiate secretion of PGE2 approximately 4 days after fertilization (Weber et al. 1991a, b). Initially, the early postfertilization embryo is ensnared within the ampulla of the oviduct due to the tightly sealed ampullary–isthmic junction. However, the PGE2 released by the developing morula triggers the relaxation of the circular smooth muscle in the isthmus, prompting its dilation and facilitating the passage of the embryo into the uterus (Stout 2020). Notably, the absence of viable embryos failing to initiate oviductal transport suggests that tubal pregnancy is not a complication observed in mares (Stout 2020).

The functional role of the oviduct, as evidenced by its effects on embryo survival, appears to change with age (Carnevale et al. 1993). Notably, in vitro studies have illustrated the positive impact of co-culturing cleaving embryos with oviductal epithelial cells (Ball and Miller 1992), indicating the influence of the oviductal epithelium on embryo development. It is, however, important to note that the equine endosalpinx may not be essential for normal development, as viable horse embryos can develop in the oviduct of nonpregnant ewes (Fehilly and Willadsen 1986; Meinecke and Meinecke-Tillmann 2023).

Following the release of an equine oocyte from the follicle, it might be expected that the final preparations for fertilization would be significantly influenced by follicular fluid. However, Townson and Ginther (1989) observed no corresponding distension of the oviduct immediately after ovulation.

3   Equine Embryonic Migration from the Oviduct to the Uterus

Equine embryos in the compact morula or early blastocyst stage enter the uterus 144 to 156 h (day 6 or 6.5) after ovulation (Battut et al. 1997; Allen 2001). One of the characteristics that differentiates the mare from other domestic species regarding the beginning of gestation is the ability of the embryo to move in the uterine lumen (Ginther 2021). Such ability to move occurs similarly in the mare (Ginther 1983, 1984, 1995; Leith and Ginther 1984), jennies (Meira et al. 1998; Gastal et al. 1993) and mules carrying horse embryos (Camargo et al. 2020). From its entry into the uterus, the equine embryo is mobile until the moment of its fixation in the caudal uterine horn around days 16 (Ginther 1983) or 17 (Allen and Wilsher 2009), moving from the tips of the horns to the junction between the body and the cervix (Ginther 2021).

Both the endometrium (Watson and Sertich 1989) and the myometrium (Piotrowska-Tomala et al. 2020) and the early equine embryo (Klein 2016a) produce both prostaglandin F2α (PGF2α) and PGE2 in vitro. Myometrial peristaltic contractions induced by the rhythmic synthesis of PGF2α (Vegas et al. 2021) and PGE2 (Allen 2001; Klein 2016a; Cuervo-Arango et al. 2018) induce uterine migration of the concept. Specifically, PGE2 hormone acts locally by relaxing the smooth muscle fibers of the circular layer of the wall of the uterine tube, allowing the passage of the embryo to the uterus after 24 h (Fig. 1) (Allen 2001). So, the embryonic vesicle undergoes periodic compressions that lead to changing the initial spherical shape to a wider dimension during the period of maximum mobility (days 12 to 15 or 16) (Ginther 1983). These morphological changes in the embryo vesicle can be detected by ultrasound in a longitudinal direction in the uterine lumen, in which the longitudinal arrangement of the endometrial folds facilitates progression (Ginther 1983; Leith and Ginther 1984; Ginther 2021).

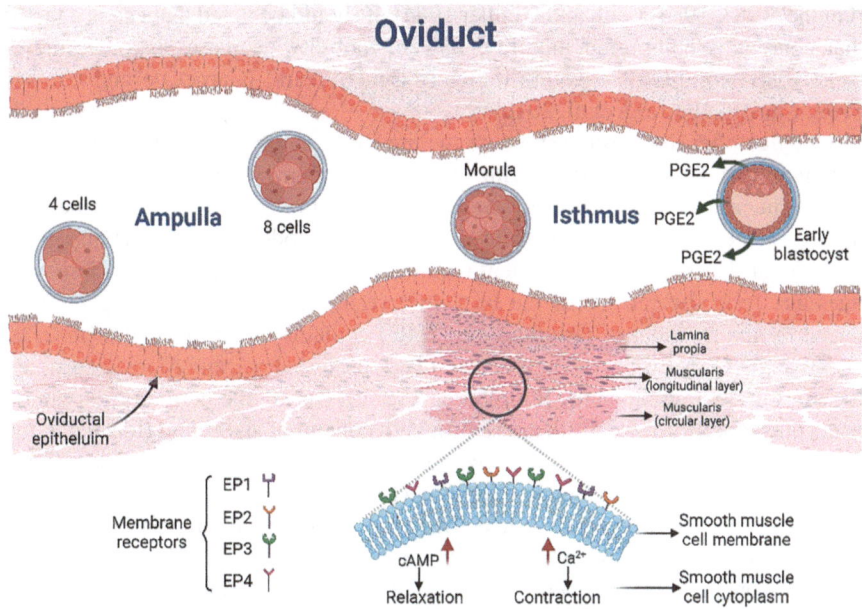

Fig. 1 Twenty-four hours after fertilization, the zygote has divided by mitosis into two cells and continues to divide within the zona pellucida into 4, 16, and 32 cells until it reaches the morula stage after 4 days. The passage of the embryo through the oviduct is accomplished by the secretion of appreciable amounts of PGE2 produced by the embryo itself. PGE2 acts locally on the smooth and circular muscle fibers of the oviductal wall and causes the embryo to advance with the help of rhythmic movements of the cilia until it enters the uterus on days 6 to 7. The biological effects of prostaglandin E2 (PGE2) on the oviduct are mediated through four different receptors: EP1, EP2, EP3, and EP4. EP1 and EP3 receptors induce contraction of the smooth muscle, whereas EP2 and EP4 receptors induce relaxation of the same tissue. Image of own origin made with biorender.com

The synthesis of PG by enzymes of the embryonic vesicle during mobility and the uterus is receiving attention. Studies of the mechanisms that underlie embryo mobility have now reached the level of enzymes and genes using polymerase chain reaction (PCR) to amplify certain segments of DNA (Klohonatz et al. 2015; Kalpokas et al. 2021). These studies have involved the embryo or uterus. It has been reported that PGE2 synthesis is restricted to the pole opposite to the embryonic disc and converted to PGF2α, which accumulates at the pole with the disc (Budik et al. 2021). It was suggested that a different PG on opposite poles might be crucial for intrauterine embryo movement in each direction (Ginther 2021).

The 2 to 3 mm diameter embryo first arrives at the uterine body on approximately day 6 (Griffin et al. 1993), remaining more than half of its time in the uterine body on days 9 and 10 (Leith and Ginther 1984). When the embryo vesicle reaches a diameter of 9 mm around day 12, the number of entrances to the uterine horn increases, beginning its period of maximum mobility until reaching 21 mm on the day of fixation on day 16. It has been estimated that the movements occur 0.5 to 0.9 times per hour, equivalent to 12 to 22 times per day or 100 times/mobility period

(Ginther 1983, 1984; Leith and Ginther 1984). It should be noted that age is a factor that conditions embryonic mobility. Indeed, mobility decreases in older mares ($\geq$15 years) compared to young mares (mean; 6 years) (Carnevale and Ginther 1992; Camargo Ferreira et al. 2019).

One day after the blastocyst enters the uterus on day 6 (Jacob et al. 2012) and specifically at 12 h coinciding with the moment of blastulation, high molecular weight mucin-like glycoproteins with a high content of threonine and serine residues are secreted by the outer palisade layer of trophectoderm cells of the blastocyst (Oriol et al. 1993a, b). The persistent remains of the zona pellucida trap these residues and form the capsule of the blastocyst (Betteridge 1989). The capsule produced by the trophoblast (Albihn et al. 2003) contains mucin-like glycoproteins, and glycine (Gly), serine (Ser), threonine (Thr), proline (Pro), and Alanine (Ala) are the most present amino acids (Oriol et al. 1993b). During the period of maximum conceptus mobility, the capsule presents a high content of sialic acid (located in the sugar side chains of the capsule glycoproteins), providing the fetus with anti-adhesive properties that facilitate migration (Oriol et al. 1993a; Ginther 2021).

Around day 16, glycoproteins lose their sialic acid content, decreasing the anti-adhesion ability. This happens at the same time when the equine conceptus finishes its migration and attaches itself to the base of one of the uterine horns. It has been demonstrated that the embryo itself regulates the sialic acid content of its capsule by the expression of neuraminidase 2 (NEU2), an enzyme that removes sialic acid residues from glycoproteins (Klein and Troedsson 2012). The capsule begins to decrease by day 18 and disappears around day 23 (Oriol et al. 1993a). The capsule is a thin, elastic, and tough 3 μm covering that encapsulates the embryo for the next 20 to 25 days (Betteridge 1989) necessary for the hatching and survival of the embryo (Stout et al. 2005).

The capsule provides internal turgor pressure which prevents trophoblast elongation and ensures that the embryo maintains a spherical configuration to allow its propulsion in the uterus (Ginther 1983). Additionally, the capsule provides tensile strength to allow the embryo to resist myometrial compressive forces (Stout et al. 2005). Additionally, the high concentration of negatively charged sialic acid residues in glycoproteins (Oriol et al. 1993b) not only regulates the intrauterine movement of the embryo but also allows the absorption of exocrine secretions from the endometrial glands ("uterine milk"). These secretions contain transport proteins such as uterocalin (Stewart et al. 1995) and uteroglobulin (McDowell et al. 1990) among other substances and are the main source of nutrients for the rapidly expanding, unattached embryo during the first 40 days. Embryo transfer to synchronous recipient mares after capsule removal drastically reduces pregnancy rates (Stout et al. 2005). Endometrial uterocalin is functionally correlated with capsule formation and persistence (Crossett et al. 1996; Tremoleda et al. 2003a). Although uterocalin is produced by the endometrium (Crossett et al. 1996), its mRNA has been localized in the glandular and luminal epithelium (Stewart et al. 2000; Ellenberger et al. 2008; Hoffmann et al. 2009). In addition to its nutritional function, uterocalin is a component that improves capsule formation (Smits et al. 2012).

Endometrial secretion of uterocalin appears to depend on progesterone (P4) since its concentration in the uterine fluid is correlated with P4 levels in the peripheral blood both during the estrous cycle and early gestation (Stewart et al. 1995). Nevertheless, uterocalin addition to culture media does not lead to the physiological formation of a capsule in equine embryos produced in vitro (Smits et al. 2012). However, there is controversy in this issue, as some authors suggest that the expression of this protein is regulated by other factors than P4 (Klein 2016a; Bastos et al. 2019).

Insulin-like growth factor binding protein 3 (IGFBP3) has also been identified in the capsule. Herrler et al. (2000) proved that IGFBP3 is secreted by the equine conceptus from day 10 post-ovulation and may promote embryonic development by the regulation of the action of maternal IGFs (especially IGF1) on the conceptus. Furthermore, these researchers suggested a protective function of IGFBP3 due to the properties of the cleaved form of IGFBP3, which is the predominant form of IGFBP3 bound to the capsule.

Therefore, being in contact with the uterine environment seems to be essential for capsule formation. In addition, by transcriptomics, de Castro et al. (2024) identified increased gene expression (until 30 genes) in the epithelium of the uterine horn that supported the embryo. These genes have been related to the regulation of vascular growth factors and the immune system during the period of embryonic mobility.

The embryo progresses through the uterus due to increased uterine contractility (Griffin and Griffin and Ginther 1990), tone (Bonafos et al. 1994), vascularization (Silva et al. 2005; Silva and Ginther 2006), and endometrial edema (Griffin et al. 1993) during the period of mobility. These uterine mechanisms, except for vascularization (Silva et al. 2005), bind the horn containing the embryo to prevent it from entering the body or the other uterine horn (Griffin et al. 1993). Therefore, the uterine horn with the confined embryo but not the remaining uterus shows increased edema, tone, and mobility. Endometrial vascular perfusion increases at 7 min around 12 to 15 days after entry of the embryo into the uterine horn (Silva et al. 2005) that contains the embryo (Ginther 1985) versus the opposite horn. This increase in vascular perfusion is faster than that of the stimulation of uterine contractions.

4   Equine Embryo Development in the Uterus

The developmental characteristics of the equine embryo in utero differ significantly from those of other domestic species (Betteridge 2007). Upon arrival in utero, the embryo usually transitions to the late morula or early blastocyst stage (Betteridge 2011). At this stage, the late morula is composed primarily of two cell groups: a smaller inner cell group surrounded by a larger peripheral cell mass. Most of the peripheral cells will contribute to the trophoblast, which will form the chorion, the outer layer of the fetal portion of the placenta. Initially, the morula lacks an internal cavity. However, through a process called cavitation, the trophoblast cells secrete fluid to create small lacunae, which eventually fuse to form a single blastocele (Gilbert 2000).

Although initial blastocyst cavity formation has been observed as early as 5.5 days post-ovulation, blastocysts from days 6.5 to early day 7 often exhibit an incompletely segregated ICM. It is not until late day 7 that equine blastocysts develop a well-defined ICM, accompanied by dehiscence of the zona pellucida to expose the newly formed blastocyst capsule (Betteridge et al. 1982). By day 7 post-ovulation, the ICM begins to consolidate on one side of the blastocyst, forming a discontinuous layer of cells beneath it. Isolated cells of the primitive endoderm overlie the mural trophoblast. Rauber's layer, which lines the epiblast cells of the ICM, remains intact but diminishes later, disappearing completely between days 10 and 12. As early as day 8, primitive endoderm cells form in the blastocyst. As early as day 8, the cells of the primitive endoderm begin to coalesce into a continuous layer, forming the bilaminar blastocyst. This process, in which scattered cells give rise to a layer of primitive endoderm, is characteristic and deviates markedly from the pattern observed in other mammals (Enders et al. 1993).

During this period, a layer of endodermal cells designated the yolk sac endoderm or primitive endoderm emerges from the ICM to create a second cell layer situated between the blastocyst cavity and the trophoblast (Enders et al. 1988). The mechanisms of blastocyst expansion in horses appear to diverge markedly from those observed in other species. The fluid within the expanding yolk sac is significantly hypotonic (120 mosM/kg) until approximately day 16, gradually increasing to 250 mosM/kg by day 22 (Betteridge 2007).

From this point onward, the yolk sac becomes trilaminar at the embryonic pole and bilaminar at the embryonic pole, while exhibiting a discernible thickening in the future neural folds. Within the developing mesoderm layer, blood islands are formed that subsequently fuse to establish a continuous vascular network, thus forming a primitive vitello-embryonic circulatory system (Sharp 2000). At the time of attachment (days 16–17), less than half of the yolk sac possesses a layer of mesoderm, and the developing blood vessels are concentrated around the embryo proper. By day 18, the blood vessels have interconnected to form a network between the endoderm and trophectoderm, as well as adjacent to the embryo. In addition, a cardiac chamber and somites have formed, and the cephalic fold is present in the mesoderm lateral to the neural tube (Ginther 1998, 2022).

The equine embryo, fetal membranes, and circulatory system undergo substantial development even before the first signs of a functional yolk sac emerge around day 22 (Enders et al. 1993), followed by the formation of the allantochorionic placenta at approximately 40 days post-ovulation. In addition, the endodermal wall of the yolk sac is composed of cells that are unusually dispersed within the blastocyst and exhibit distinct functional properties (Enders et al. 1993). Unlike other species, the horse sheds the zona pellucida of the blastocyst and replaces it with a glycoprotein capsule of remarkable functional importance between about days 7 and 21. As the encapsulated conceptus develops, it undergoes two distinct phases of interaction with its environment. It displays remarkable mobility until approximately days 16–17, after which it becomes immobilized at the eventual site of placentation.

Throughout this process, the capsule maintains the spherical shape of the conceptus, especially before fixation. In addition, intact conceptuses can be readily retrieved transcervically from the standing mare until approximately day 35, which is a crucial feature for experimental investigations (Fig. 2) (Enders et al. 1993; Ginther 2022).

During the second and third weeks of gestation, significant structural transformations occur in the conceptus, as extensively documented by Betteridge et al. (1982), Enders and Liu (1991), and Enders et al. (1988, 1993). An understanding of these changes is crucial for interpreting concurrent functional modifications, particularly elucidated for cellular tissues by Enders et al. (1993). Crossett et al. (1998) and Herrler and Beier (2000) have highlighted the role of the acellular capsule as a conduit for the exchange of proteins and, potentially, other molecules between the conceptus and the endometrium. Furthermore, the timing of capsule development at the time of blastocyst entry into the uterus is of practical importance, as it profoundly affects the efficacy of equine blastocyst cryopreservation. This is probably due to its impact on the permeation of cryoprotective agents, as previously noted by Bruyas et al. (2000) and De Coster et al. (2020). Furthermore, it presents challenges for embryo micromanipulation, as observed by Huhtinen et al. (1997).

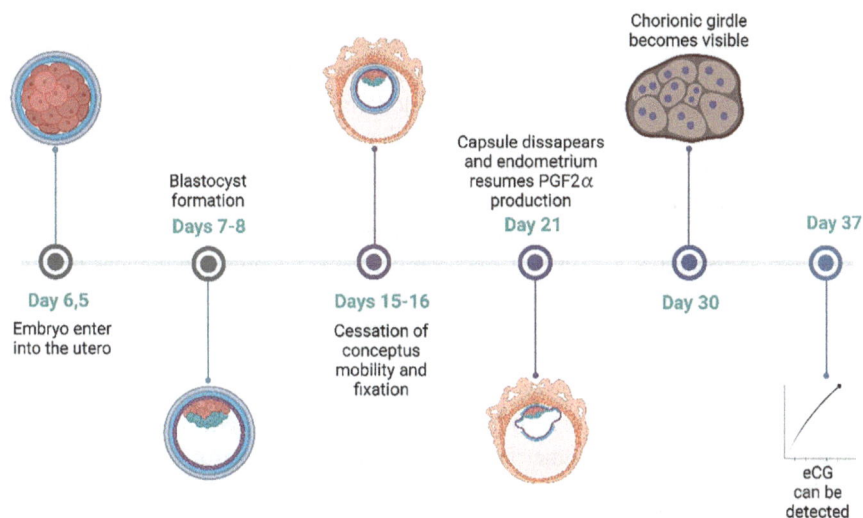

Timeline of the equine embryo in the uterus

Chorionic girdle becomes visible

Blastocyst formation
Days 7-8

Capsule dissapears and endometrium resumes PGF2α production
Day 21

Day 37

Day 6,5
Embryo enter into the utero

Days 15-16
Cessation of conceptus mobility and fixation

Day 30

eCG can be detected

Fig. 2 The equine embryo enters the uterus on days 6–7 after ovulation at the late morula or early blastocyst stage of development. The capsule is formed between the trophoblast and the zona pellucida. The blastocyst has a central cavity and the inner cell mass (ICM), which is established at one of the poles. The embryo remains mobile for approximately 2 weeks in the uterus until fixation, which occurs on day 15 in ponies and day 16 in mares. From the third week, the capsule disappears, and on day 30, the chorionic girdle becomes visible. As a consequence, the eCG hormone shows significant levels in plasma, allowing it to be detected. Image of own origin made with biorender.com

Two fundamental approaches to studying the secretory substances of the conceptus are likely to be of great importance in understanding their interaction with the mare (Swegen et al. 2017). The first is to hypothesize the presence of a substance and then look for it in the blastocyst fluid or culture medium used for conceptus tissues. The second approach is to employ various analytical techniques to detect a series of products, followed by the identification of specific members of this series and speculation about their functions. While these approaches complement each other, the emergence of gene banks is particularly enhancing the second approach (Simpson et al. 1999), which has the potential to lead to rapid advances in our understanding of the interactions between the conceptus and the endometrium. The main categories of secretory products identified so far include steroids, eicosanoids, proteins, and peptides. Any of these substances discovered in blastocyst fluid could have been transported there by carrier proteins such as uterocalin (Crossett et al. 1998). Therefore, to determine the source of a putative product, in vitro culture and/ or tissue analysis using immunological or molecular probes is necessary.

Since the initial demonstrations of estrogen and androgen production by the equine conceptus in the 1970s (Flood et al. 1979; Zavy et al. 1979), the most extensive research on steroidogenesis in these tissues has been conducted by Goff and colleagues. The research team has demonstrated that between days 7 and 14 of pregnancy, 17α-hydroxyprogesterone, and not estradiol as previously assumed, is the primary steroid synthesized by the equine conceptus. Furthermore, they have shown that this steroid undergoes downstream metabolism to an unidentified steroid by the endometrium, suggesting that these steroids may influence conceptus development or play a role in preventing luteolysis (Goff et al. 1993). Before this, the same research group had indicated that endoderm and trophectoderm cells of the conceptus exhibit distinct steroidogenic profiles in vitro, suggesting that these two tissues must cooperate for pregnancy to progress (Marsan et al. 1987).

Concerning eicosanoids, the equine embryo continues to synthesize PGE2 following its entry into the uterus, reaching its maximum production between days 11 and 15 (Vanderwall et al. 1993). Concurrently, the conceptus commences production of PGF2α by day 14 (Watson and Sertich 1989). A multitude of studies have focused on the production of proteins and peptides by the conceptus. McDowell et al. (1990) demonstrated that the protein profile of conceptuses before day 14 differed from that of older conceptuses, which exhibited a pattern characteristic of yolk sac tissue. Sharp et al. (1989) investigated a secretory product of the conceptus capable of suppressing endometrial PGF2α production in vitro. The investigators employed dialysis membranes with varying molecular exclusion properties to conclude that a molecule with a molecular weight greater than 1000 and less than 6000 was responsible. Simpson et al. (1999) employed suppression subtractive hybridization to isolate transcripts that are more abundantly expressed in day 15 conceptuses compared to day 12 conceptuses. The cDNA clones for a calcium-binding protein (calcyclin) and phospholipase A2 (involved in eicosanoid metabolism) were identified as differentially expressed during these periods. Green et al. (1999) identified a novel aspartic proteinase produced by the equine trophoblast as early as day 25, which appears to be related to gestation-associated glycoproteins.

5 The Capsule: Formation and Functions

In most large domestic animals, the zona pellucida envelops the developing oocyte and conceptus before implantation or formation of the placenta proper. However, in horses, a second acellular glycoprotein membrane, known as the "capsule," forms between the trophoblast and zona pellucida shortly before zona loss, which continues to envelop the developing conceptus during the second and third weeks of gestation (Fig. 3) (Stout et al. 2005).

The formation of the capsule occurs between the trophectoderm and zona pellucida shortly after the embryo enters the uterus, coinciding with the development of the blastocyst (Flood et al. 1979). Initially, the capsule's mucin-like glycoproteins are secreted by the trophectoderm (Oriol et al. 1993a; Albihim et al. 2003). Nevertheless, research conducted in vitro indicates that these glycoproteins alone are insufficient to create a cohesive structure through cross-linking (Tremoleda et al. 2003a). It is therefore probable that the formation of the capsule requires the involvement of the endometrium, as evidenced by the increased production of capsular glycoproteins in IVF embryos when exposed to uterocalin (Smits et al. 2012), a P4-dependent endometrial protein strongly linked to early uterine conceptus capsules (Stewart et al. 1995). The properties of the capsule change as conceptus development progresses. These changes include alterations in glycosylation characteristics around day 9, which coincides with the loss of the zona pellucida. In addition, a decrease in sialic acid content is observed from day 16, which correlates with the

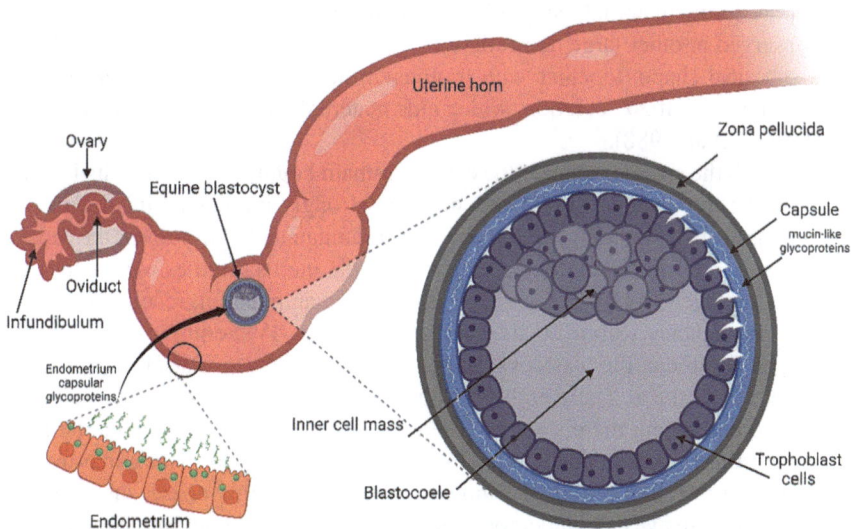

Fig. 3 The formation of the capsule occurs between the trophectoderm and zona pellucida coinciding with the development of the blastocyst. The capsule's mucin-like glycoproteins are secreted by the trophectoderm. Also it is probable that the formation of the capsule requires the involvement of the endometrium. Image of own origin made with biorender.com

attachment of the conceptus to the uterus (Wilsher et al. 2013). The disappearance of the capsule begins around day 22.5 and is complete by day 24.5 of gestation (Oriol et al. 1993b). However, the precise role of the primary uterine contribution to capsule formation remains uncertain. It is unclear whether this contribution is primarily to supply structural components or to create a microenvironment conducive to the cross-linking and assembly of trophectoderm-derived glycoproteins into a cohesive structure (Stout 2020).

Following the completion of capsule formation, typically occurring around day 7, the embryo exits its zona pellucida. Nevertheless, the newly formed capsule persists as a physical barrier between the trophectoderm and the endometrium. After this, the capsule undergoes gradual thickening until approximately day 18, acquiring a bilaminar structure that may indicate an additional source of structural components, likely from the endometrium (Oriol et al. 1993a; Stewart et al. 1995). Once the embryo attaches, marking the end of migration, the dissolution of the capsule commences. This dissolution is characterized by a rapid decrease in dry weight and a loss of continuity, typically occurring between days 20 and 23; Chu et al. 1997), with complete disintegration of the capsule by day 30 (Enders and Liu 1991; Oriol et al. 1993a). The mechanism of capsule dissolution remains unknown. However, both trophectodermal and endometrial enzymes are likely involved. Furthermore, P4 appears to play a central role in stimulating enzymatic dissolution. Induction of luteolysis shortly before conceptus fixation has been demonstrated to prevent the dissolution, desialylation, and degradation of capsule-associated proteins linked to normal degradation (Chu et al. 1997; Quinn et al. 2007).

This capsule likely plays a role in maintaining the spherical shape of the equine conceptus during maternal recognition of pregnancy, in contrast to the rapid elongation observed in other large domestic species during this phase (Kölle et al. 2007). This spherical shape, together with prolonged uterine migration, is linked to the suppression of PGF2α secretion, which aids maternal recognition of the conceptus (McDowell et al. 1988).

Although the specific roles of the capsule remain largely speculative, its location between the trophectoderm and endometrium strongly implies its involvement in facilitating embryo-maternal interaction and communication. Moreover, the capsule is indispensable for the conceptus's survival in the uterus. Studies, where the capsule was removed from day 7 embryos via micromanipulation, resulted in unsuccessful pregnancies (Stout et al. 2005). In contrast, IVP embryos lack a fully developed capsule at the blastocyst stage yet accumulate capsular glycoproteins in the perivitelline space. These embryos subsequently develop a normal capsule after transfer to a recipient mare's uterus. The capsule is structurally resilient and flexible, suggesting its role in maintaining the blastocyst's spherical shape and providing mechanical protection during the mobile phase. During this phase, the relatively delicate conceptus vesicle undergoes compression around the uterine lumen due to myometrial contractions. This compression is thought to be a factor in the development of the maternal recognition of pregnancy signaling (Stout and Allen 2001).

Additionally, capsular glycoproteins contain a substantial number of negatively charged sialic acid residues, which are believed to confer anti-adhesive properties, facilitating conceptus migration (Chu et al. 1997). Indeed, the conclusion of the mobile phase of the conceptus coincides with extensive desialylation of the capsule (Arar et al. 2007). Therefore, it is probable that the capsule indirectly contributes to maternal recognition of pregnancy signaling by promoting migration. Moreover, it has been postulated that the capsule actively engages in embryo-maternal communication to store, modify, or transfer endometrial proteins involved in nutrient transport or stimulation of trophectoderm cell proliferation (Oriol et al. 1993a; Stewart et al. 1995; Crossett et al. 1998; Herrler and Beier 2000; Quinn et al. 2007; Arar et al. 2007; Allen and Wilsher 2009). Similarly, the capsule may serve as a reservoir for trophectodermal proteins that regulate endometrial function or act in a paracrine manner to control trophectodermal cell migration or proliferation (Herrler and Beier 2000).

The capsule also acts as a solid barrier that protects the embryo from myometrial contractions and facilitates its movement within the uterine lumen before fixation. Studies on equine blastocysts showed that, when the capsule was removed, no regeneration occurred, leading to the disappearance of the embryo, underlining its essential role in maintaining pregnancy (Betteridge 1989; Stout et al. 2005).

The location of the capsule suggests that it may be involved in maternal-fetal communication for the establishment and maintenance of pregnancy. It may serve as a temporary reservoir for crucial substances, such as IGF-1 and IGFBP-3, which are essential for embryonic development (Pereira et al. 2013). In addition, endometrial lipocalin P19 (Stewart et al. 1995; Crossett et al. 1998) is secreted into the uterine lumen and is dependent on P4. In addition, the yolk sac wall is involved in steroid synthesis and metabolism, as well as in the production of prostaglandins PGF2α and PGE2 (Betteridge 2007; Klein and Troedsson 2011).

During fixation, the capsule loses sialic acid, a phenomenon that can be prevented by treatment with PGF2α around day 14 post-ovulation (Chu et al. 1997). This loss of galactose and N-acetyl galactose residues from the main type 1 O-glycan core may result in alterations in capsule permeability, which could influence the attachment of the conceptus to the endometrium. Subsequently, capsule production decreases until its enzymatic rupture on day 21, although the underlying mechanisms remain unclear (Denker 2000).

Most of the duration of the capsule is characterized by resilience, facilitating extensive migration of the conceptus into the uterus until approximately days 16–17. This marks the cessation of migration and fixation of the conceptus for subsequent implantation (Ginther 1983, 1985; Gastal et al. 1996). Pre-fixation mobility is essential for sustaining pregnancy (McDowell et al. 1988), enabling communication between the conceptus and the entire endometrium. The capsule is essential for early embryo survival, shielding it during its mobile phase (Stout et al. 2005) and acting as a regulatory interface between the trophoblast and the uterine luminal epithelium (Quinn et al. 2007).

Failures in pregnancy due to embryonic loss impose significant economic burdens on horse breeding, with a considerable portion of losses occurring during the presence of the embryonic capsule (Ball 1988, Carnevale et al. 2000, Morris and Allen 2002). Morris and Allen (2002) indicate that approximately 17% of pregnancies diagnosed by ultrasound around day 15 experience subsequent losses, with 60% of these losses occurring between days 15 and 35.

The embryonic capsule is believed to regulate the transfer of uterine substances to the equine embryo. There is evidence that specific proteins bind to the capsule, which allows essential substances to be delivered to the embryo (Herrler et al. 2000). For example, maternal P19/uterocalin binds to the capsule and is primarily involved in lipid delivery to the equine conceptus (Stewart et al. 1995, 2000; Crossett et al. 1996, 1998; Suire et al. 2001). Moreover, equine yolk sac wall tissues express considerable levels of GM2 activator protein (GM2AP) before attachment. This protein plays a role in glycolipid transport and breakdown, as well as the production of insulin-like growth factor binding protein 3, which is essential for embryonic development (Herrler et al. 2000; Quinn et al. 2007).

The capsule, which is primarily composed of polysaccharides, may act as a selectively permeable barrier, protecting embryonic antigens from immune rejection before the establishment of immunoregulatory mechanisms (Betteridge 1989; Hunt et al. 2005; Trowsdale and Betz 2006). Alterations in both the capsule and the endometrium likely play a functional role in the fixation process or its failure. During the fixation process, the capsule becomes lax and undergoes desialylation, which may affect the binding of proteins to exposed galactose or mannose residues (Oriol et al. 1993a, b; Chu et al. 1997; Gray et al. 2005). Several proteins present in and around the early equine conceptus have been identified in uterine fluid, the zona pellucida, the capsule, and yolk sac fluid (McDowell et al. 1990, Miller et al. 1992; Crossett et al. 1996; Herrler & Beier 2000, Herrler et al. 2000).

Legrand et al. (2001) employed a conventional freeze/thaw method to investigate equine capsule permeability. Their findings indicated that capsule thickness was directly correlated with the rate of cell death after thawing. Embryos with thicker capsules showed higher rates of cell damage compared to embryos with thinner or absent capsules. Similar studies (Legrand et al. 2002) have also shown a negative correlation between capsule thickness and the ability of large embryos to withstand freezing, a fact that suggests that the capsule may impede the penetration of cryoprotectants into the embryo itself.

A notable abnormality in IVP is the absence of a normal capsule. Despite the in vitro production of mucin-like capsular glycoproteins, these proteins fail to assemble into the capsule as do naturally conceived equine embryos from the early blastocyst stage. The exact cause of this failure of capsular glycoprotein coalescence under in vitro conditions remains unclear. It is postulated that the glycoproteins may disperse in the culture medium, preventing the concentration necessary for capsule assembly. Another possibility is that there is a lack of hydration and cross-linking of the capsular glycoproteins in the absence of specific uterine components. In both cases, the participation of the mare's uterus is crucial for the capsule formation process (Tremoleda et al. 2003a).

6   Equine Embryonic Mobility and Maternal Recognition of Pregnancy

The complete and uninterrupted interaction between the uterus and the conceptus guarantees the establishment and maintenance of pregnancy (Swegen 2021). Maternal recognition of pregnancy (MRP) refers to a sequence of events by which embryonic signals prolong luteal function beyond its lifespan during a natural estrous cycle by ensuring continuous secretion of progesterone (P4) by the corpus luteum. (CL) (Allen and Stewart 2001; Klein 2016b) since the primary CL is the only source of P4 during the first 40 days of gestation (Squires and Ginther 1975). The presence of P4 is a necessary condition for the mobility of the conceptus, its fixation at the base of a uterine horn, and the orientation within the uterus (Kastelic et al. 1987). Although the expression of P4 receptors in the trophoblast exerts P4 effects on the fetus (Rambags et al. 2008; Willmann et al. 2011), its main effect is to prepare the endometrium for pregnancy. This requires a negative regulation of P4 receptors in the endometrial epithelium to increase the expression of proteins associated with pregnancy (Spencer and Bazer 2002). P4 receptors are not present in the endometrial epithelium from day 20 onwards, although they are abundant in stromal cells (Wilsher et al. 2011). Indeed, the synthetic progestin in mares starting on day 5 after ovulation results in a downregulation of P4 receptors in the endometrium on day 11 (Willmann et al. 2011).

Compared to the normal estrous cycle, PGF2α is reduced in the uterine vein between days 10 and 14 of gestation (Douglas and Ginther 1976). However, by day 18, or the time of luteolysis in cyclical mares, concentrations are similar in both pregnant and nonpregnant mares (Stout and Allen 2002; Stout 2016). It is hypothesized that the embryo delays, rather than prevents, the release of endometrial PGF2α during MRP. The signal produced by the equine embryo is primarily antiluteolytic in nature, rather than luteoprotective (or inhibitory of PGF2α action) or luteotropic (or supportive of CL function) (Klein and Troedsson 2011). The concentration of PGF2α reaches its highest level on day 20, declining to undetectable levels by day 30. The mechanism underlying the failure of PGF2α secretion to induce CL lysis remains unclear (Klein, 2016a, b). However, the observed capacity of CLs to bind to PGF2α (Vernon et al. 1981) may offer a potential explanation for this seemingly contradictory phenomenon. Although Sharp et al. (1989) showed an antiluteolytic agent with a molecular weight between 1000 and 6000 D secreted by the equine conceptus, molecules that fit the molecular weight such as PGE2 or insulin do not prolong the useful life of the CL when they are infused into the uterine lumen in cyclic mares (Rambags et al. 2008).

The MRP signal is unknown in the mare, although it is believed that it is a low molecular weight molecule (3–10 kDa) that would begin to be secreted 10 days after ovulation (Klein 2016b). The uterine response to oxytocin (OXT) is also altered during pregnancy in the mare. Administration of OXT in pregnant mares 14 days after ovulation decreases PGF2α (Goff et al. 1987). The embryo-maternal interaction is responsible for the low concentration of endometrial OXT receptors during the first weeks post-ovulation and, consequently, for the lower production of

PGF2α in the gravid uterus (Sharp et al. 1997; Starbuck et al. 1998). The main function of the embryonic vesicle in its uterine mobilization is to signal endometrial cells to reduce the production of PGF2α (Klohonatz et al. 2015). This reduction in the response to OXT in pregnant mares coincides with a decrease in the binding capacity of OXT to its receptors (OXTR) (Starbuck et al. 1998; Boerboom et al. 2004). However, the capacity to respond to endometrial OXT develops around day 11 post-ovulation in cyclic mares, so the embryonic signal that would suppress this increased sensitivity to OXT would begin to be released from day 10 (Goff et al. 1987). This inhibitory effect is maintained until day 18 of gestation when the binding of OXT to the receptor (OXTR) in the endometrium causes the pulsatile liberation of PGF2α (Starbuck et al. 1998).

Although endometrial and luteal OXT concentrations change with gestation status and the day after ovulation (Diel de Amorin et al. 2002), OXTR transcriptional levels are similar between pregnant and nonpregnant mares at the time of MRP (Klein 2016a; de Ruijter-Villani et al. 2015). However, OXTR levels are decreased in pregnant animals (Starbuck et al. 1998; Diel de Amorin et al. 2002), suggesting that OXTRs are modulated at the posttranscriptional level and not at the transcriptional level. Versus pregnant mares, in cyclic mares, OXTR concentrations increment approximately threefold on day 14 in comparison to days 10 and 18 (Starbuck et al. 1998), suggesting that suppression of OXTR expression must be initiated earlier from day 14 in pregnant mares.

In cyclical mares, P4 concentrations start to decrease approximately 14 days post-ovulation, coinciding with peak levels of PGF2α, being assessed by the presence of its metabolite (PGF metabolite, PGFM) in the vein and bloodstream uterine lumen as well as in the endometrium (Stout and Allen 2002; Ginther et al. 2019). Furthermore, the enzyme that is responsible for the arachidonic acid release, and therefore initiator of PGF2α release, is higher in nonpregnant mares at day 14 (Ababneh and Troedsson 2013). On the other hand, the endometrial release of PGF2α is attenuated in pregnant mares at the time of luteolysis (Stout and Allen 2002; Stout 2016).

As previously described, embryonic mobility increases on day 10, reaching a maximum on days 11 to 14 and ceases on day 16 at the time of fixation (Ginther 1983, 1985). The embryo stimulates uterine contractions that propel it into the uterine lumen (Klein 2016b). thanks to the production of PGE2 (Weber et al. 1991a, b; Vanderwall et al. 1993) and PGF2α (Watson and Sertich 1989), paradoxically the same substance that causes luteolysis and whose pulsations should be avoided during the early stages of pregnancy (Allen and Stewart 2001; Klein 2016a, b). However, there is evidence that PGE2 would predominantly cause embryonic motility (Klein 2016a). Stout and Allen (2002) proposed that this feature compensates for the relatively small surface area of the trophoblast in this species. Both mobility and adequate conceptus size are essential requirements for MRP (Ginther 1983, 1985), while delayed growth and inadequate development lead to early pregnancy loss in mares (Vanderwall 2008; Wilsher et al. 2011). The goal of extensive embryonic migration is for the distribution of antiluteolytic factors throughout the endometrial surface (McDowell et al. 1988). Since endometrial PGF2α reaches the CL through systemic circulation, the embryonic RMP signal must be transported to the entire endometrial surface (Klein 2016b). Restriction of embryonic movement to less than 2/3 of the uterine surface luteolysis appears with the same systemic pathways induced in nonpregnant mares (Piotrowska-Tomala et al. 2020) resulting in pregnancy loss. However, these events can be prevented by administration of P4, indicating that CL lysis is the cause of embryonic loss when embryonic movement is restricted (McDowell et al. 1988).

The regulation of endometrial expression of the PG-endoperoxide synthase 2 enzyme that is involved in the transformation of arachidonic acid to PG H2 was considered a key event in MRP in horses (Boerboom et al. 2004; Ealy et al. 2010). The suppression of this enzyme occurs until day 13 of pregnancy (Boerboom et al. 2004), so the time for the release of the MRP signal occurs between days 12 and 14. Recently, Wilsher et al. (2022) proved that manual reduction of day 11 embryos 24 h or 12 h after the transference to recipient mares on days 10, 11, 12, or 13 post-ovulation results in 93% beginning a period of luteostasis. Wilsher et al. (2022) also proved that in vitro rupture of a day 11 embryo recovered in vivo and its transfer together with the medium where it was ruptured led to a significantly increased number of day 12 recipients entering the embryo in a luteostasis period in comparison to day 11 recipients. These studies suggest that MRP signaling can occur in a relatively short period and seem to confirm that day 12 is the optimal time to suppress endometrial PGF2α release. In addition, Wilsher et al. (2022) and Newcombe et al. (2022) also suggest that the reduction of a single embryo on day 12 or later leads to most mares beginning a period of luteostasis.

Although MRP in equids occurs around day 12 after ovulation (Klein 2016a), other studies found that the timing of MRP signal delivery seemed to be variable but repeatable in individual mares during a post-ovulation window of ≤24 h. for each individual (Klein 2016a; Wilsher et al. 2022). Newcombe et al. (2023) showed that the timing of the MRP signal fluctuated up to 72 h on an individual basis, as reductions of embryos before a certain time after ovulation lead to luteolysis, while reductions afterward result in luteolysis.

The persistence of the CL is observed in a determined percentage of nonpregnant mares after the introduction of a glass marble (Nie et al. 2003) or a rubber ball filled with liquid (Rivera del Alamo et al. 2008) into the uterine lumen during the first days post-ovulation. It has been suggested that these spherical intrauterine devices resemble conceptus presence by producing pressure or contact on the uterine wall (Ribera del Alamo et al. 2008). This effect appears to depend on adequate endometrial perfusion and drainage, being less effective in older mares (Katila 2015). This evidence suggests that the embryonic signal for MRP could at least partially be of mechanical rather than secretory origin. This was supported by the regulation of PG production and the increased CL lifespan described after intrauterine administration of various vegetable oils in the luteal phase (Wilsher and Allen 2011b). Although these researchers did not exclude that physical interference with the endometrium could be involved, luteolysis was not prevented by intrauterine administration of mineral oil. In another research, an intrauterine device (IUD) was placed in the uterus of mares between 2 and 4 days post-ovulation, preventing luteolysis on day

14 and extending the luteal phase (Rivera del Álamo et al. 2008). PG levels of mares that showed an increased luteal phase were basal from day 12 post-ovulation, suggesting that the endometrial pulsatile liberation of PGF2α is prevented by the mechanical stimulation of an IUD. Nevertheless, other research indicates that it is preventable. The surface and secretions released by the conceptus are necessary for the attenuation of the luteolytic liberation of PGF2α of endometrial origin, rather than merely the mechanical stimulation of the moving conceptus or the IUD (Klohonatz et al. 2019).

Although it has not been proven, it is thought that when the conceptus meets the endometrium, it could induce local endometrial secretion of PG under the secretion of estrogens of embryonic origin, which begin to be noticeable 6–10 days post-ovulation (Zavy et al. 1979). In vitro (Vernon et al. 1981) and in vivo (Goff et al. 1993) studies demonstrated the stimulatory effect of estrogens on endometrial PGF2α secretion. While levels of uterine fluid do not differ (Watson 1989), myometrial contractions differ completely between pregnant and cyclic mares. Goff et al. (1993) demonstrated that although hormones such as testosterone, P4, androstenedione, 17αhydroxyprogesterone, and 20α-hydroxyprogesterone are secreted by the equine conceptus, these do not change at the time of MRP. Of these steroids, 17α-hydroxyprogesterone (Budik et al. 2021) and even estrogens (Raeside et al. 2004) are the majority. In the equine concept, steroidogenesis begins around day 6–10 (Paulo and Tischner 1985), reaching considerable levels by day 12 post-ovulation (Flood et al. 1979; Zavy et al. 1984; Choi et al. 1997).

Although interferon delta (IFN-delta) expression (EqIFN-delta 1 and EqIFN-delta 2) occurs between days 16 and 22 post-ovulation (Cochet et al. 2009), they have not been identified at earlier stages in conception. Since no antiviral activity could also be identified in the uterine fluid of pregnant mares (Sharp et al. 1989), it has been suggested that IFNs do not have an important function in MRP in horses.

Current evidence has revealed transcriptomic and proteomic regulations associated with embryonic growth. The identification of the regulation of proteins, mRNAs, and miRNAs on days 12 and 13 of gestation indicates that they are essential for embryonic development, as well as they are involved in MRP and the establishment of gestation (Rudolf Vegas et al. 2021; Vegas et al. 2021; de Castro et al. 2024). Indeed, de Castro et al. (2024) identified increased gene expression (until 30 genes) in the epithelium of the uterine horn that supported the embryo. These genes have been related to the transcriptomic regulation of vascular growth factors and the immune system during the period of embryonic mobility.

The termination of the classic RMG or successful prolongation of the life of the primary CL is temporally coincident with the end of embryo migration on days 16–17 post-ovulation, at which time fixation appears to result from a continuous increase in the embryo diameter in combination with a higher uterine tone, the embryo being lodged at the base of one of the uterine horns. Beyond the fixation period, but before the formation of the endometrial cups around day 35 of gestation, the ability to secrete PGF2α in response to OXT stimulation is regained by the uterus.

7 In Vitro Equine Embryo Development

7.1 Equine Early Cleavage Stages

The initial processes linked to fertilization do not typically induce observable morphological changes in vivo (Bezard et al. 1989). However, the zygote often undergoes alterations in size (reduction) and shape (cleavage or loss of spherical form) as cleavage approaches. In the majority of equine zygotes (80%) derived from oocytes retrieved from preovulatory follicles, pronuclei are present 8 h after the intracytoplasmatic sperm injection (ICSI). However, pronucleus formation may be delayed when oocytes mature in vitro, with 80% of zygotes lacking pronuclei until 12 h after ICSI (Ruggeri et al. 2017). The assessment of male and female pronuclei is not typically conducted using standard microscopy for equine zygotes. However, they are consistently observable in zygotes of certain species, such as humans (Gardner and Balaban 2016). To date, there have been no reports of real-time identification and quality assessment of pronuclei in equine zygotes. Cleavage, which entails division into two blastomeres, represents the primary developmental milestone following ICSI that can be visually monitored. Indicators signaling impending cleavage can be detected before the initial cell division (Carnevale and Metcalf 2019).

In horses, small anuclear fragments of membrane and extruded cytoplasm are frequently observed. Fragmentation is not exclusive to any particular species and is linked to the progression of meiosis or mitosis in human cleavage-stage embryos before the activation of the embryonic genome at the 4- to 8-cell stages (Stensen et al. 2015). In horses, fragmentation was initially documented in 1945 (Hamilton and Day 1945) and was termed deutoplasmolysis. At the time, it was believed to be associated with the high-fat content in equine oocytes. During the preparation of the zygote for cleavage, the membrane may appear distorted or cleaved, while the cytoplasm may elongate or display polarization with opposing dark areas that are dense in organelles and/or lipids. The precise timeframe from fertilization to cleavage in vivo is often difficult to ascertain due to the uncertainty surrounding the precise time of fertilization. However, oviductal collections performed 24 to 48 h after ovulation from inseminated mares yielded embryos with an average of three to four cells (Carnevale et al. 1993). Following in vivo fertilization, collections resulted in one-cell structures at 6 or 12 h, with division into two cells occurring at 24 h (Bezard et al. 1989). Other researchers (Betteridge et al. 1982) observed that oviductal collections at 24 or 36 h after fertilization yielded embryos with three or four cells. Consequently, the initial cleavage division appears to occur 24 h after fertilization in vivo, followed by a second division the subsequent day.

In the case of in vivo matured oocytes fertilized by ICSI, the division into two cells or indications of imminent division, such as membrane fragmentation and/or cleavage, typically occurs around 20–24 h after ICSI. However, the time to division may vary when conventional ICSI is conducted on in vitro matured oocytes. The interval between the initial cell divisions has been demonstrated to predict subsequent embryo quality in humans and other mammals, particularly in non-human primates (Van Royen et al. 1999; Daughtry and Chavez 2018). When transferred into recipient oviducts, early-stage equine embryos with fewer culture hours per resulting blastomere were notably more likely to achieve pregnancy, indicating that the timing of cell divisions reflects developmental potential (Frank et al. 2019). Ultimately, time-lapse imaging is expected to offer the most comprehensive insights into the timing of cell divisions in early equine embryos. If sperm-injected, uncleaved oocytes are left in culture for several days, some may undergo fragmentation and exhibit characteristics that resemble cleavage-stage embryos or arrested early-stage embryos. Cleavage-stage equine embryos have been observed to exhibit a lower number of nuclei than anticipated based on morphology (Lewis et al. 2016). Consequently, the presence of anuclear blastomeres or large cytoplasmic fragments may complicate the initial cleavage assessment in certain embryos.

The timing for assessing cleavage and subsequent embryo development is often determined by the specific laboratory protocols and the method of ICSI used for fertilization. The two most employed methods for equine ICSI are the piezoelectric micropipette and the conventional ICSI needle. While cleavage and blastocyst rates do not differ significantly between these methods, there may be delays in oocyte activation and sperm processing with conventional ICSI compared to piezo-driven ICSI (Salgado et al. 2018), which could influence the optimal timing for assessing cleavage. With conventional ICSI and time-lapse imaging, scoring of the first three mitotic divisions is feasible, as optimal timing for this assessment has been linked to embryo health and development in various mammalian species, including horses (Burruel et al. 2014; Daughtry et al. 2016; Weinerman et al. 2016; Sugimura et al. 2017; Daughtry and Chavez 2018). The utilization of time-lapse imaging or the evaluation of cleavage status at 1 to 2 days after ICSI facilitates the determination of accurate cleavage rates. Oocytes that fail to cleave can be removed from the culture, thus preventing uncertainties regarding embryo development in subsequent assessments. However, the evaluation of cleavage timing varies among laboratories. Limiting the number of observations or employing time-lapse microscopy for early developmental stage observations helps minimize embryonic stress associated with incubator opening and plate manipulation (Carnevale and Metcalf 2019).

The timing of early cleavage divisions may occur synchronously or asynchronously, resulting in the formation of even or odd numbers of cells, respectively. It is optimal for blastomeres to undergo synchronous division, resulting in consecutive divisions and an even number of cells at 24-h intervals. Nevertheless, a certain degree of asynchrony in cell divisions may result in the formation of odd numbers of blastomeres, with larger cells that have not yet undergone cleavage (Prados et al. 2012). On rare occasions, a blastomere may be undergoing division and may appear elongated or have an indentation within the cell. The presence of a large extruded cytoplasm fragment or anucleate blastomere can present challenges in accurately determining the number of cells (Lewis et al. 2016). Blastomeres in the cleavage phase are typically large and occupy a significant portion of the space within the zona pellucida (ZP). As successive divisions occur, individual cells will become relatively smaller. The number of cells can be determined by focusing on different planes of the embryo. It is expected that individual blastomeres will exhibit similar size, texture, and hue.

The occurrence of fragmentation in equine embryos is not uncommon and can be quantified as a percentage of the total blastomere volume. Clinical assessments of human embryos indicate that fragmentation is deemed significant if it accounts for 0.25% of the total cell volume, which has been demonstrated to have a negative correlation with pregnancy outcomes (Racowsky et al. 2003; Prados et al. 2012). In the evaluation of equine embryos, fragments typically appear clustered within the ZP cavity, although occasionally they may be dispersed (Carnevale and Metcalf 2019). Both clustered and dispersed types of fragmentation have been reported in human embryos. Dispersed fragments are more commonly associated with chromosomal abnormalities (Magli et al. 2007; Prados et al. 2012).

The assessment of fragmentation, including membrane-bound anuclear fragments in early embryos, has been linked to pregnancy success rates in human fertility clinics. A variety of forms of fragmentation have been identified, including large fragments that may be mistaken for blastomeres. This phenomenon has also been observed in equine embryos (Lewis et al. 2016; Frank et al. 2019). Fragmentation in human embryos has been classified into two categories: clustered or dispersed among cells. The former is further subdivided into definitive fragmentation and pseudo-fragmentation. The latter is transient and associated with cell excision (Prados et al. 2012). Similar types of fragmentation have been observed in in vitro-produced equine embryos, with clustered and small fragments being the most prevalent and observed to varying extents (Frank et al. 2019).

Fragmentation is not exclusive to in vitro-produced equine embryos; even early-stage embryos generated in vivo display fragmentation (Hamilton and Day 1945; Betteridge et al. 1982; Bezard et al. 1989), with a higher prevalence noted in lower-quality oviductal embryos from older mares compared to younger ones (Carnevale et al. 1993). In the assessment of early human embryos, less than 20% fragmentation has been regarded as indicative of optimal morphology (Gardner and Balaban 2016). Mild fragmentation is not associated with adverse fertility outcomes. However, more severe fragmentation at specific developmental stages has been linked to diminished outcomes and meiotic delay (Prados et al. 2012; Stensen et al. 2015). As equine embryos progress in development, fragmentation and extruded or degenerated cells often coalesce into a dark "debris" zone, a phenomenon observed in in vivo-collected embryos (McCue 2021). Furthermore, it has been observed that embryos produced by intracytoplasmic sperm injection (ICSI) exhibited elevated percentages of apoptotic cells (10% vs. 0.3% in vivo embryos), an altered pattern of microfilament distribution, and irregularities in cell size and shape (Tremoleda et al. 2003a).

Embryos containing 16 or more cells begin to undergo compaction approximately 4 to 5 days post-ICSI, forming a loosely cohesive mass with distinguishable cell borders (Carnevale and Metcalf 2019). This aligns with the developmental timeline observed in vivo, where embryos collected 5 or 5.5 days after ovulation typically exhibit 10–16 cells or a compacting cell mass (Betteridge et al. 1982). Similarly, Carnevale and Metcalf (2019) observed that the most significant developmental delays and subsequent embryo losses occur between cleavage and Day 5. During this developmental window, the embryonic genome is activated (Brinsko et al. 1995), and the embryo relies on mitochondria derived from the oocyte (Hendriks et al. 2019). The culture of ICSI-fertilized oocytes within mare oviducts resulted in embryos with twice as many cells per embryo at 96 h compared to those cultured in vitro (mean 16 vs. 8 cells). This suggests that culture conditions may impact the rate of embryo development (Choi et al. 2002). The presence of day 5 embryos with a well-defined cell mass of at least 16 cells indicates the potential for further development. Nevertheless, at this stage, equine embryos may exhibit disorganization or signs of degeneration, with viable cells surrounded by debris. Consequently, it is challenging to ascertain the viability of embryos during this period. In such cases, embryos may require further culture for additional evaluations (Carnevale and Metcalf 2019).

Between days 5 and 6, the compactness of the morula serves as a positive indicator for subsequent development. This is characterized by blastomeres forming a dark mass of tightly adherent cells, with less noticeable borders compared to earlier stages. In general, the cell mass of the compact morula is smaller than that observed in earlier developmental phases. This developmental timing and appearance of the morula align with observations from embryos collected from oviducts, as reported by Hamilton and Day (1945), Betteridge et al. (1982), Bezard et al. (1989), and Carnevale and Metcalf (2019).

Following compaction, the morula typically experiences a period of size enlargement on a subsequent day, although there is potential for variation over time. It is possible to discern multiple cells, which results in a lobulated appearance of the outer rim with slight variations in texture and hue. As the blastocyst cavity begins to form, the central area lacks distinct cell borders, although trophoblast cells can be identified at the outer margins (Carnevale and Metcalf 2019). The embryo begins to enlarge, filling the inner space of the ZP. Extruded cells and fragmentation within the ZP cavity often aggregate into a dark mass of debris, which is comparable to observations made in embryos collected in vivo (McCue 2021). Consequently, as the embryo progresses in development, the assessment of embryo quality may improve, as the discarded debris constitutes a smaller portion of the embryo. This is evidenced by the observation that on Day 7, the proportion of discarded debris is less than that observed on Day 5 (Carnevale and Metcalf 2019). Until the blastocyst stage of development, the outer diameter of the ZP of the early embryo remains similar to that of the oocyte. As the blastocyst forms and the embryo exerts pressure against the inner zona pellucida, some growth of the embryo and thinning of the zona pellucida may be observed, although this is less pronounced than in vivo (Carnevale and Metcalf 2019).

In vitro-produced embryos undergo assessment for the ZP size and abnormalities, which mirrors the evaluation conducted on in vivo-produced embryos. ZP morphology provides insights into oocyte development and potential viability. It has been demonstrated that excessive internal ZP space is associated with a reduced developmental potential of oocytes following ICSI (Altermatt et al. 2009). Furthermore, ZP size can also be indicative of the developmental capacity of the originating oocyte, even in the early stages of embryonic development. Longitudinal clefts may occasionally be present in the ZP of certain mares, although their impact remains unclear. The ZP should not be cracked or broken, as embryos lacking a ZP or capsule exhibit poor survival rates post-transfer to recipient uteri (Stout et al. 2005). Following the formation of the blastocyst, the blastocyst must occupy the ZP. If a blastocyst forms but markedly shrinks away from the ZP, this suggests an unfavorable or hypertonic environment or diminishing viability. Furthermore, blastocyst collapse can also occur in in vivo-produced embryos (Vanderwall 1996).

7.2   In Vitro Development of Equine Blastocysts

The efficacy of blastocyst production from zygotes has been constrained regardless of the utilization of co-culture (Li et al. 2001), conditioned medium (Choi et al. 2001), or synthetic oviductal fluid (Dell'Aquila et al. 2001; Galli et al. 2002), indicating that the conditions for in vitro culture were not optimal (Galli et al. 2014). Some environmental factors can impact embryo development in culture. It has been proposed that elevated rates of apoptosis with cell fragmentation might be associated with incorrect embryo-to-medium ratios, thermal stress, and excessive ROS generation (Tremoleda et al. 2003b; Mortensen et al. 2010). For instance, the arrangement of actin filaments in ICSI-generated embryos could be influenced by factors such as temperature, pH, and ion concentration (Tremoleda et al. 2003b). Additionally, embryos produced in vitro typically exhibit reduced cell numbers, less distinct inner cell masses, smaller blastocoels, and greater irregularity compared to their in vivo counterparts (Tremoleda et al. 2003a, b). These morphological disparities in quality also render in vitro embryos more challenging to assess before transfer. To mitigate the risk of misinterpreting embryo quality, it is advised to remove uncleaved embryos from the culture, which additionally helps in reducing ROS production in small volumes of medium (Lewis et al. 2016).

The accurate identification of the blastocyst is of paramount importance, as it frequently marks the culmination of laboratory procedures for both research and clinical purposes. However, due to variations in imaging systems and expertise levels, there is a risk of misidentifying other structures as blastocysts, as demonstrated in previous studies employing nuclear staining techniques (Lewis et al. 2016). The establishment of cleavage within 2 days post-ICSI and the observation of a morula provide initial indicators of the potential for blastocyst formation. Oocytes or embryos that have aged or degenerated may resemble blastocysts. In some cases, the aged zona pellucida may appear thin and enlarged, despite the absence of a blastocyst (Lewis et al. 2016). The most distinctive feature of the blastocyst is the presence of an outer layer composed of trophoblast cells, which can be observed under a microscope by adjusting the focal plane to capture the outer cell layer's textured or cobblestone-like appearance. Additionally, stereoscopic imaging can assist in the identification of blastocysts. It should be noted that in vitro-produced blastocysts tend to be smaller and darker in appearance compared to those collected in vivo. Time-lapse microscopy offers the advantage of monitoring embryonic development at frequent intervals, allowing for the visualization of blastocoel formation concurrent with blastocyst expansion in viable embryos. Conversely, embryos that cease development exhibit slowing or cessation of movement and blastomere splitting (Carnevale and Metcalf 2019).

While early blastocysts exhibit similar morphological features both in vivo and in vitro, discrepancies become more noticeable as development progresses in culture compared to the natural uterine environment. Blastocysts collected from mare uteri typically display a distinct, well-defined trophoblast layer and an organized ICM (Vanderwall 1996; McCue et al. 2009). In contrast, ICM cells in blastocysts produced via ICSI are not organized and are distributed widely (Tremoleda et al. 2003a). Initial attempts to selectively stain ICM cells in equine blastocysts produced by conventional IVF were unsuccessful in delineating the dispersed ICM cells (Choi et al. 2009). One proposed explanation for the difficulty in visualizing and defining ICM cells is the overlapping of the ICM endoderm with trophoblast cells during the blastocyst's multicentric expansion, a phenomenon unique to horses (Choi et al. 2015). Similarly, IVF horse blastocysts are reported to exhibit a more dispersed epiblast (EPI) compared to those produced via in vitro embryo production (IVEP), with a mixture of EPI and primitive endoderm (PE) cells. This increased dispersion may be attributed to developmental delays, particularly evident in slowly developing IVP horse blastocysts by day 9. Nevertheless, the EPI of IVEP horse blastocysts undergoes rapid compaction upon transfer to the uterus of a recipient mare, suggesting that the uterine environment may mitigate developmental disparities in IVP embryos (Umair et al. 2023).

Research indicates that the transfer of in vitro-produced embryos into a mare's uterus results in the formation of both the embryo capsule and a distinguishable ICM, as reported by Choi et al. (2009). This indicates that the uterus provides essential factors necessary for ICM and capsule development in horses. During this period, the ZP undergoes thinning from approximately 17 to 3 mm, before appearing to dehisce. This is observed by Betteridge et al. (1982) to occur concurrently with the expansion in diameter of the embryo. However, in contrast to embryos retrieved from mare uteri, the expansion of the ZP and embryonic growth are constrained in embryos produced in vitro, as observed by Carnevale and Metcalf (2019). It has been documented that monozygotic twinning can occur following the transfer of a single ICSI-produced embryo. This phenomenon is potentially attributed to the dissociation of trophoblast extrusions through the ZP, as indicated by Roberts et al. (2015) and Dijkstra et al. (2020). Moreover, the presence of scattered ICM cells within IVP embryos may contribute to the formation of two ICMs (Dijkstra et al. 2020).

The formation of the capsule, a glycoprotein layer situated between the trophectoderm and the ZP, has been documented by numerous researchers, including Betteridge (1989), Oriol et al. (1993a), Vanderwall (1996), Tremoleda et al. (2003a), and McCue et al. (2009). During in vitro culture, the developing blastocyst secretes capsular glycoproteins; however, these proteins fail to aggregate to form a distinct structure, as outlined by Tremoleda et al. (2003a). Upon transfer to the uterus, these glycoproteins bind together to form the capsule around the zona pellucida-enclosed embryo. Capsule formation is of paramount importance, as it is a vital step in the survival of the equine embryo in utero, as highlighted by Stout et al. (2005).

7.3   Progression Through the Stages of Development

It is of paramount importance to ensure synchronization across various developmental stages when producing embryos via ICSI. Embryos produced in vivo may
exhibit different developmental stages despite being of similar age. A delay in
embryonic development has been linked to a decreased pregnancy rate following
embryo transfer (Carnevale et al. 2000). In the context of in vitro embryo production, it is possible to evaluate intervals between specific developmental stages,
which serve as potential indicators of developmental potential. Nevertheless, the
timing of embryo development may vary depending on factors such as the laboratory, the type of oocytes used, and the culture systems employed. In their 2019
publication, Carnevale and Metcalf outlined the ideal progression from ICSI (Day
0) as follows: On the first day, the embryo cleaves into two cells or exhibits signs of
impending cleavage. On the second day, the embryo reaches the four-cell stage. On
the fifth day, the embryo reaches the 16-cell stage with signs of compaction. On the
sixth day, the embryo reaches the compact morula or early blastocyst stage. On the
seventh day, the embryo reaches the early blastocyst or blastocyst stage. This
sequence of events is consistent with the timeline of equine embryo development
observed in vivo (Betteridge et al. 1982; Bezard et al. 1989; Carnevale et al. 1993;
Vanderwall 1996). The compactness of the developing morula is a positive indicator
of embryo health, as it signifies the initiation of tight junction formation between
individual blastomeres (Betteridge et al. 1982; Vanderwall 1996).

The progression of embryonic development can vary significantly between different laboratories, with several factors contributing to this discrepancy. These
include specific laboratory procedures, the source of oocytes, the method of oocyte
maturation, and the techniques used for embryo culture (Choi et al. 2004a; Salgado
et al. 2018). Consequently, individual laboratories may establish their own developmental parameters that they consider to be indicative of embryonic developmental
potential, which may differ from the standardized system outlined above. For example, research has demonstrated that the choice between conventional ICSI and
Piezo-drill techniques can result in blastocysts with significantly different numbers
of nuclei, reflecting variations in cell numbers during blastocyst production (Salgado
et al. 2018). Furthermore, studies have demonstrated that factors such as the initial
morphology of the oocyte cumulus and the glucose concentration of the medium
can influence cleavage and the number of nuclei in early embryos (Choi et al. 2004b).

In contrast to the single evaluation typically conducted for in vivo-collected
embryos, assessing the quality of in vitro-produced embryos involves multiple evaluations and monitoring of development over time. Temporal markers of quality may
include the duration between cleavage divisions and intervals between specific
developmental milestones. In humans, high-quality embryos are characterized by
having four to five blastomeres on day 2 and seven or more blastomeres on day 3,
with less than 20% fragmentation (Van Royen et al. 1999; Rienzi et al. 2005;
Gardner and Balaban 2016). Species-specific cleavage timings for horses are best
determined using time-lapse imaging. Nevertheless, optimal pregnancy outcomes
have been associated with sperm-injected oocytes that divide within approximately

24 h and have an average of three cells by 32 h (Frank et al. 2019). It is also important to note that the timing of assessments and, to some extent, the intervals between developmental stages may vary among different programs.

8   The Endometrial Cups

In 1897, J. Cossar Ewart described a band approximately 0.5 cm wide comprising numerous folds separated by deep grooves, which facilitated the attachment of the horse embryo to the uterus. Subsequently, in 1912, Wilhelm Schauder published findings on distinctive structures in the endometrium of pregnant mares, marking the initial documentation of endometrial cups. These cups are circular, raised protrusions with ulcers situated at the base of the gravid horn in the equine uterus during early gestation (Schauder 1912). Each protrusion contains a central depression filled with yellow exocrine secretion, which Schauder hypothesized to be a significant component of fetal histotrophs. Ewart's "chorionic girdle," previously identified as a distinct thickened band of specialized trophoblast cells that proliferate rapidly between days 25 and 35 of gestation, is now recognized as the precursor of gonadotropin-secreting endometrial cups in mares. This connection was established six decades later (Hamilton et al. 1973).

The chorionic girdle initially appears around day 25 as shallow ripples in the chorion. Over the following 10 days, it deepens significantly, becoming elongated and finger-like, with hairy ridges due to rapid trophoblast cell hyperplasia at their apex. These ridges eventually fold and flatten under the pressure of the uterus and the expansion of the conceptus, with the spaces between them assuming a glandular appearance and function. Subsequently, around day 36, although with some variability among individual mares, the entire girdle detaches from the fetal membranes, and the now binucleated girdle cells invade maternal tissue. These binucleated trophoblast cells traverse the epithelial cells of the endometrial lumen, at times penetrating them directly to reach the underlying basement membrane. Subsequently, they descend through the endometrial glands, shedding epithelial cells from the lining as they progress, before crossing the basement membranes and entering the endometrial stroma between days 38 and 40. This intricate process involves the rearrangement of placental and maternal cells, reflecting complex interactions between the uterus and conceptus (Wooding et al. 2001).

A microscopic examination reveals that each cup is composed of a densely packed cluster of large binucleated epithelioid-like cells, interspersed with occasional blood vessels and the dilated fundic portions of the endometrial glands. The apical regions and outgrowths of these glands were obliterated during the initial invasion of the chorionic girdle around day 38, as reported by Daels et al. (1991). Subsequently, substantial lymphatic sinuses develop in the stroma beneath each cup, accompanied by an increasing presence of maternal immune cells, including CD41 and CD81 lymphocytes, plasma cells, macrophages, and eosinophils, which accumulate in the stroma at the periphery (Fig. 4) (Daels et al. 1991; Haig 1993).

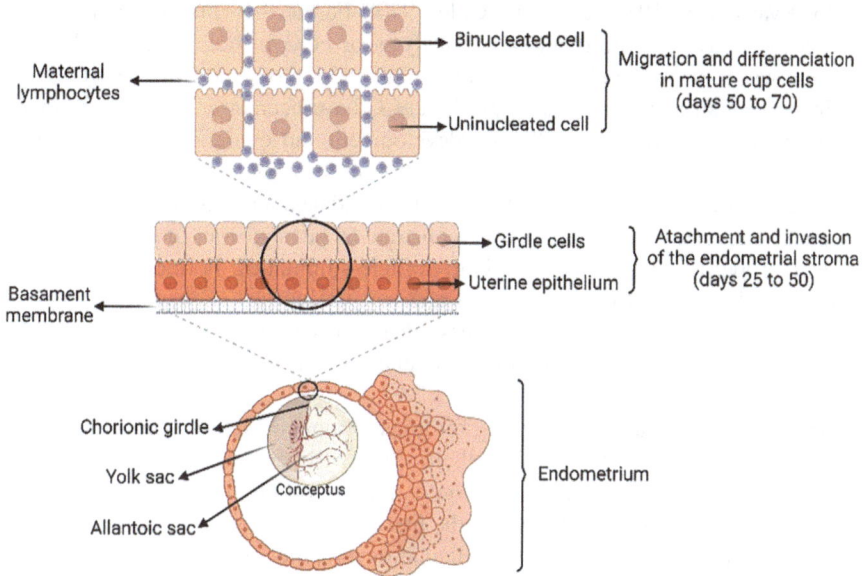

Fig. 4 Formation of endometrial cups from cells of the chorionic girdle. The chorionic girdle appears around day 25 as shallow ripples in the chorion. The bi-nucleated trophoblast cells traverse the epithelial cells of the endometrial lumen and penetrate them directly to reach the underlying basement membrane. Subsequently, substantial lymphatic sinuses develop in the stroma beneath each cup, accompanied by an increasing presence of maternal immune cells, including lymphocytes, CD41 and CD81, plasma cells, macrophages, and eosinophils. Image of own origin made with biorender.com

It took 31 years after Schauder's initial description for endometrial cups to be identified as the origin of high concentrations of eCG in pregnant mares' blood (Cole and Goss 1943). Similarly, Cole and Hart (1930) and Day and Rowlands (1940) demonstrated that small amounts of serum recovered from mares between 40 and 150 days of gestation, but not before or after these gestational stages, induced significant ovarian and uterine enlargement and follicular growth when injected into sexually immature rats.

The discovery of elevated gonadotropin levels in the bloodstream of pregnant mares prompted a decade-long program of meticulous experimental inquiry by the Davis group. The studies revealed that gonadotrophic activity first became noticeable in the mare's blood serum between 37 and 41 days after mating. Subsequently, hormone levels exhibited a sharp surge, reaching an individually varying peak between 60 and 75 days (20–300 IU/ml). Subsequently, a gradual decline was observed until the hormone activity completely disappeared between days 120 and 150 (Cole and Hart 1930; Evans et al. 1933; Cole 1938; Policastro et al. 1983, 1986). The biological characteristics of eCG in laboratory rodents and farm animals were studied by Cole et al. (1940), and its prolonged biological lifespan in serum was determined (Catchpole et al. 1935). The isolation of eCG from mare serum was achieved, enabling the investigation of its chemical properties (Saunders and Cole

1935; Evans et al. 1936; Goss and Cole 1940). Furthermore, the development of biological assays for the quantitative analysis of eCG in both serum and saline extracts was improved and refined (Evans et al. 1936; Goss and Cole 1940).

Increasing amounts of partially purified eCG have been employed in veterinary, with applications including follicular development and ovulation in sheep, cows, and noncycling pigs, as well as superovulation in these and laboratory species for embryo transfer (Rowson 1971). Despite its abundance and ease of isolation from pregnant mares' blood serum, eCG demonstrated a peculiar trait of possessing biological properties that resembled both follicle-stimulating hormone (FSH) and luteinizing hormone (LH) in a ratio of approximately 1.4:1 (Schams and Papkoff 1972; Stewart et al. 1976). Nevertheless, uncertainty remained regarding the relationship between the chorionic girdle and endometrial vessels, as well as the true source of eCG (Antczak et al. 2013).

The hormone has a limited affinity for gonadotropin receptors in equine gonadal tissues (Bousfield et al. 1987). Nevertheless, its LH-like component either triggers ovulation or induces luteinization of the dominant follicle in successive waves of follicles stimulated for growth during the first half of gestation by continuous pituitary FSH release, which regulates follicular development during the estrous cycle (Murphy and Martinuk 1991; Galet et al. 2009). Consequently, secondary corpora lutea begin to accumulate in maternal ovaries from the onset of eCG appearance in maternal blood, approximately day 38 after ovulation, leading to elevated maternal serum progesterone concentrations with the formation of each additional luteal structure (Stewart and Maher 1991; Sherman et al. 1992; Wildman et al. 2006). Furthermore, the onset of estrogen-secreting endometrial cups in pregnant mares is accompanied by a rapid increase in peripheral serum levels of conjugated estrogens (Mess and Carter 2007; Galet et al. 2009). These estrogens, which originate from the ovaries (Mess and Carter 2007), are produced by primary and/or secondary corpora lutea, rather than Graffian follicles, in direct response to eCG's gonadotrophic effects (Wildman et al. 2006).

The irregular shapes and sizes of the individual cups, ranging from a few millimeters to several centimeters, are attributed to variations in the contact of the girdle cells across the irregular terrain of the endometrial folds. The similar weight of the cups in pony and horse mares, which is approximately 10 grams, is explained by the comparable diameters of the conceptus at day 36 of gestation. This similarity contributes to higher eCG concentrations in ponies because of their smaller blood volume (Ginther 1992). Each cup comprises adjacent cells with distinct maternal and fetal genotypes, essentially functioning as fetal allografts. Because the fetal antigens are foreign to the mare, a remarkable influx of immune lymphocytes occurs, constituting one of the most significant examples of a cell-mediated immune response to pregnancy in all species. Although the vessels initially resist the lymphocytic attack, they eventually succumb to it, indicating lymphocyte-mediated death (Allen and Wilsher 2009).

Around the time that eCG becomes detectable in the bloodstream, the primary corpus luteum (CL) begins to increase in size, boost progesterone (P4) production, and initiate estradiol synthesis, marking the second luteal response to pregnancy

(Bergfelt and Ginther 1996). At the same time, the formation of supplemental corpora lutea is closely correlated with circulating levels of eCG, representing the third luteal response to pregnancy. As the fetoplacental source of P4 and estrogen becomes predominant, the role of eCG shifts to stimulation of a temporary steroid source until the fetoplacental axis is fully functional. In addition, it has been proposed that the cups immunologically prepare the uterine environment for subsequent allantochorionic placental attachment (Ginther 1992).

Several factors have been demonstrated to influence blood eCG levels in mares, including mare size (Allen et al. 2002), parity (Day and Rowlands 1947), the paternity of the conceptus (Manning et al. 1987), twin gestation (Rowlands 1949), and the timing of conception relative to the first postpartum estrus (Bell and Bristol 1991). Recent studies, which controlled for mare size, age, parity, and conceptus paternity, have also found significant impacts of body condition and exercise on maternal blood eCG levels (Wilsher and Allen 2011a). In a study involving 61 Thoroughbred mares bred to the same stallion, peak eCG concentrations occurred between days 50 and 85 after ovulation (mean ± s.e. 62.4 ± 1.0 days), with a range of 14.5 to 126 iu/ml (64.5 ± 3.7 iu/ml). eCG became undetectable in circulation by day 134.1 ± 1.7 (range, 105–150 days), taking a mean of 71.7 ± 1.4 days (range, 50–95 days) to return to basal levels after the peak concentration (Wilsher and Allen 2011a, b). It is noteworthy that body condition scores influenced eCG levels in this investigation. Specifically, mares with higher (fatter) scores exhibited lower circulating eCG levels compared to mares with lower (thinner) scores. Furthermore, the effect of exercise was observed in non-exercised mares exhibited higher eCG levels starting 60 days after ovulation than those subjected to mild exercise (Wilsher and Allen 2011a, b).

The most significant influences on eCG levels are those resulting from the fetal genotype. Bielanski et al. (1955) observed lower eCG concentrations in the serum of mares carrying hybrid mule fetuses (female horse × male donkey) compared with those carrying normal intraspecies horse fetuses. Subsequently, Clegg et al. (1962) and Allen (1969) corroborated and extended these findings by evaluating eCG levels in groups of mares (*Equus caballus*, 2n = 64) and donkeys (*Equus asinus*, 2n = 62) with normal intraspecific gestations or interspecific mule (donkey-horse) or donkey (horse-horse) gestations. These authors observed that the genotype of the sire was a determining factor in the maternal serum eCG level. These observations were later attributed to consistent differences in the amount of endometrial tissue formed in the four types of gestation (Allen 1969).

9   Conclusion

The equine pregnancy process is complex and fascinating. It begins with fertilization and culminates in implantation. The journey starts when a viable sperm fertilizes an ovum in the mare's oviduct. This fertilized egg, now a zygote, undergoes a series of cellular divisions as it travels down the oviduct, becoming a morula and eventually a blastocyst.

During these early stages, the equine embryo is protected by a unique glycoprotein capsule, which plays a crucial role in its development and protection. The embryo migrates through the uterus, a critical process that ensures proper maternal recognition of pregnancy and prevents luteolysis, thereby maintaining the necessary hormonal environment for pregnancy continuation.

Implantation in mares is noninvasive and occurs around days 40 to 45 of gestation. The embryo attaches to the uterine lining through a series of complex interactions between the trophoblast cells of the embryo and the endometrial cells of the uterus. This period is vital for establishing a stable and efficient nutrient exchange system between the mare and the developing fetus.

Throughout these stages, the mare's reproductive physiology is intricately regulated by hormonal signals, ensuring that each step occurs in a precisely timed manner to support a successful pregnancy. It is crucial to understand these processes to improve reproductive management and outcomes in equine breeding programs.

In conclusion, equine pregnancy is a meticulously regulated process involving complex biological and physiological mechanisms. From the moment of fertilization, through the critical stages of early embryonic development and implantation, to the final preparations for birth, each phase is vital for the successful development and delivery of a healthy foal. A clear understanding of these processes not only enhances reproductive management in equine breeding but also contributes to the overall health and well-being of both the mare and her offspring.

References

Ababneh MM, Troedsson MH (2013) Endometrial phospholipase A2 activity during the oestrous cycle and early pregnancy in mares. Reprod Domest Anim 48:46–52. https://doi.org/10.1111/j.1439-0531.2012.02023.x

Albihn A, Waelchli RO, Samper J, Oriol JG, Croy BA, Betteridge KJ (2003) Production of capsular material by equine trophoblast transplanted into immunodeficient mice. Reproduction 125:855–863. https://doi.org/10.1530/rep.0.1250855

Allen WR (1969) Factors influencing pregnant mare serum gonadotrophin production. Nature 223:64–65. https://doi.org/10.1038/223064a0

Allen WR (2001) Fetomaternal interactions and influences during equine pregnancy. Reproduction 121:513–527. https://doi.org/10.1530/rep.0.1210513

Allen WR, Stewart F (2001) Equine placentation. Reprod Fertil Dev 13(8):623–634. https://doi.org/10.1071/RD01063

Allen WR, Wilsher S, Stewart F, Ousey J, Fowden A (2002) The influence of maternal size on placental, fetal and postnatal growth in the horse. II. Endocrinology of pregnancy. J Endocrinol 172(2):237–246. https://doi.org/10.1677/joe.0.1720237

Allen WR, Wilsher S (2009) A review of implantation and early placentation in the Mare. Placenta 30(12):1005–1015. https://doi.org/10.1016/j.placenta.2009.09.007

Altermatt JL, Suh TK, Stokes JE, Carnevale EM (2009) Effects of age and equine follicle-stimulating hormone (eFSH) on collection and viability of equine oocytes assessed by morphology and developmental competency after intracytoplasmic sperm injection (ICSI). Reprod Fertil Dev 21:615–623. https://doi.org/10.1071/RD08210

Antczak DF, de Mestre AM, Wilsher S, Allen WR (2013) The equine endometrial cup reaction: a fetomaternal signal of significance. Ann Rev Anim Biosci 1:419–442. https://doi.org/10.1146/annurev-animal-031412-103703

Arar S, Chan KH, Quinn BA, Waelchli RO, Hayes MA, Betteridge KJ, Monteiro MA (2007) Desialylation of core type 1 Oeglycan in the equine embryonic capsule coincides with immobilization of the conceptus in the uterus. Carbohydr Res 342:1110e5. https://doi.org/10.1016/j. carres.2007.02.016

Aurich C, Budik S (2015) Early pregnancy in the horse revisited–does exception prove the rule? J Anim Sci Biotechnol 6:50. https://doi.org/10.1186/s40104-015-0048-6

Ball BA (1988) Embryonic loss in mares. Incidence, possible causes, and diagnostic considerations. Veterinary clinics North America: equine. Practice 4(2):263–290. https://doi.org/10.1016/ S0749-0739(17)30641-7

Ball BA, Miller PG (1992) Survival of equine embryos co-cultured with equine oviductal epithelium from the four- to eight-cell to the blastocyst stage after transfer to synchronous recipient mares. Theriogenology 37:979–991. https://doi.org/10.1016/0093-691X(92)90097-B

Ball BA, Ignotz GG, Brinsko SP, Thomas PGA, Miller PG, Ellington JE, Currie WB (1993) The in-vitro block to development and initiation of transcription in early equine embryos. Equine Vet J Suppl 15:87–90. https://doi.org/10.1111/j.2042-3306.1993.tb04835.x

Ball BA, Dobrinski I, Fagnan MS, Thomas PG (1997) Distribution of glycoconjugates in the uterine tube oviduct. Of horses. Am J Vet Res 58:816–822. https://doi.org/10.2460/ajvr.1997.58.08.816

Bastos HBA, Martinez MN, Camozzato GC, Estradé MJ, Barros E, Vital CE, Vidigal PMP, Meikle A, Jobim MIM, Gregory RM, Mattos RC (2019) Proteomic profile of histotroph during early embryo development in mares. Theriogenology 125:224–235. https://doi.org/10.1016/j. theriogenology.2018.11.002

Battut I, Colchen S, Fieni F, Tainturier D, Bruyas JF (1997) Success rates when attempting to nonsurgically collect equine embryos at 144, 156 or 168 hours after ovulation. Equine Vet J 29:60–62. https://doi.org/10.1111/j.2042-3306.1997.tb05102.x

Bauersachs S, Almiñana C (2020) Embryo-maternal interactions underlying reproduction in mammals. International journal of molecular. Science 10;21(14):4872. https://doi.org/10.3390/ ijms21144872

Bell RJ, Bristol F (1991) Equine chorionic gonadotrophin in mares that conceive at foal oestrus. J Reprod Fertil Suppl 44:719–721

Bergfelt DR, Ginther OJ (1996) Ovarian, uterine and embryo dynamics in horses versus ponies. J Equine Vet Sci 16:66–72. https://doi.org/10.1016/S0737-0806(96)80158-4

Betteridge KJ, Mitchell D (1974) Direct evidence of retention of unfertilized ova in the oviduct of the mare. J Reprod Fertil 39:145–148. https://doi.org/10.1530/jrf.0.0390145

Betteridge KJ, Eaglesome MD, Mitchell D, Flood PF, Beriault R (1982) Development of horse embryos up to twenty two days after ovulation: observations on fresh specimens. J Anat 135:191–209. PMID: 7130052

Betteridge KJ (1989) The structure and function of the equine capsule in relation to embryo manipulation and transfer. Equine Vet J Suppl 8:92–100. https://doi.org/10.1111/j.2042-3306.1989. tb04690.x

Betteridge KJ (2000) Comparative aspects of equine embryonic development. Anim Reprod Sci 2(60–61):691–702. https://doi.org/10.1016/S0378-4320(00)00075-0

Betteridge KJ (2007) Equine embryology: an inventory of unanswered questions. Theriogenology 1(68):S9–S21. https://doi.org/10.1016/j.theriogenology.2007.04.037

Betteridge KJ (2011) Embryo morphology, growth and development. In: McKinnon AO, Squires EL, Vaala WE, Varner DD (eds) Equine Reproduction. Willey-Blackwell, pp 2167–2186

Bezard J, Magistrini M, Duchamp G, Palmer E (1989) Chronology of equine fertilization and embryonic development in vivo and in vitro. Equine Vet J 21:105–110. https://doi. org/10.1111/j.2042-3306.1989.tb04692.x

Bielanski W, Ewy Z, Pigoniowa H (1955) Preliminary comparative investigations on endocrine secretion in mares mated with stallions and donkeys. Folia Biol 3:19–30. PMID: 14391422

Boerboom D, Brown KA, Vaillancourt D, Poitras P, Goff AK, Watanabe K, Sirois J (2004) Expression of key prostaglandin synthases in equine endometrium during late diestrus and early pregnancy. Biol Reprod 70(2):391–399. https://doi.org/10.1095/biolreprod.103.020800

Bonafos LD, Carnevale EM, Smith CA, Ginther OJ (1994) Development of uterine tone in nonbred and pregnant mares. Theriogenology 42:1247–1255. https://doi.org/10.1016/0093-691X(94)90244-D

Bousfield GR, Liu WK, Sugino H, Ward DN (1987) Structural studies on equine glycoprotein hormones. Amino acid sequence of equine lutropin β-subunit. J Biol Chem 262:8610–8620. PMID: 3298239

Brinsko SP, Ball BA, Ignotz GG, Thomas PGA, Currie WB, Ellington JE (1995) Initiation of transcription and nucleologenesis in equine embryos. Mol Reprod Dev 42:298–302. https://doi.org/10.1002/mrd.1080420306

Bruyas JF, Sanson JP, Battut I, Fiéni F, Tainturier D (2000) Comparison of the cryoprotectant properties of glycerol and ethylene glycol for early (day 6) equine embryos. J Reprod Fertil Suppl 2000:549–560

Budik S, Walter I, Leitner MC, Ertl R, Aurich C (2021) Expression of enzymes associated with prostaglandin synthesis in equine conceptuses. Anim (Basel) 11:1180. https://doi.org/10.3390/ani11041180

Burruel V, Klooster K, Barker CM, Riejo Pera R, Meyers S (2014) Abnormal early cleavage events predict early embryo demise: sperm oxidative stress and early abnormal cleavage. Sci Rep 4:6598. https://doi.org/10.1038/srep06598

Camargo Ferreira J, Linhares Boakari Y, Sousa Rocha N, Saules Ignácio F, Barbosa da Costa G, de Meira C (2019) Luteal vascularity and embryo dynamics in mares during early gestation: effect of age and endometrial degeneration. Reprod Domest Anim 54(3):571–579. https://doi.org/10.1111/rda.13396

Camargo CE, Rechsteiner SF, Macan RC, Kozicki LE, Gastal MO, Gastal EL (2020) The mule (Equus mulus) as a recipient of horse (Equus caballus) embryos: comparative aspects of early pregnancy with mares. Theriogenology 145:217–225. https://doi.org/10.1016/j.theriogenology.2019.10.029

Carnevale EM, Ginther OJ (1992) Relationships of age to uterine function and reproductive efficiency in mares. Theriogenology 37:1101–1115. https://doi.org/10.1016/0093-691X(92)90108-4

Carnevale EM, Griffin PG, Ginther OJ (1993) Age-associated subfertility before entry of embryos into the uterus of the mare. Equine Vet J Suppl 15:31–35. https://doi.org/10.1111/j.2042-3306.1993.tb04820.x

Carnevale EM, Ramirez RJ, Squires EL, Alvarenga MA, Vanderwall DK, McCue PM (2000) Factors affecting pregnancy rates and early embryonic death after equine embryo transfer. Theriogenology 54:965–979. https://doi.org/10.1016/S0093-691X(00)00405-2

Carnevale EM, Metcalf ES (2019) Morphology, developmental stages and quality parameters of in vitro-produced equine embryos. Reprod Fertil Dev 31(12):1758–1770. https://doi.org/10.1071/RD19257

Catchpole HR, Cole HH, Pearson PB (1935) Studies on the rate of disappearance and fate of mare gonadotrophic hormone following intravenous injection. Am J Physiol 1935(112):21–26. https://doi.org/10.1152/ajplegacy.1935.112.1.21

Choi SJ, Anderson G, Roser J (1997) Production of free estrogens and estrogen conjugates by the preimplantation equine embryo. Theriogenology 47:457–466. https://doi.org/10.1016/S0093-691X(97)00004-6

Choi YH, Chung YG, Seidel GE Jr, Squires EL (2001) Developmental capacity of equine oocytes matured and cultured in equine trophoblast-conditioned media. Theriogenology 56:329–339. https://doi.org/10.1016/S0093-691X(01)00567-2

Choi YH, Love CC, Love LB, Varner DD, Brinsko S, Hinrichs K (2002) Developmental competence in vivo and in vitro of in vitro-matured equine oocytes fertilized by intracytoplasmic sperm injection with fresh or frozen–thawed spermatozoa. Reproduction 123:455–465. https://doi.org/10.1530/rep.0.1230455

Choi YH, Roasa LM, Love CC, Varner DD, Brinsko SP, Hinrichs K (2004a) Blastocyst formation rates in vivo and in vitro of in vitro-matured equine oocytes fertilized by intracytoplasmic sperm injection. Biol Reprod 70(5):1231–1238. https://doi.org/10.1095/biolreprod.103.023903

Choi YH, Love LB, Varner DD, Hinrichs K (2004b) Factors affecting developmental competence of equine oocytes after intracytoplasmic sperm injection. Reproduction 127:187–194. https://doi.org/10.1530/rep.1.00087

Choi YH, Harding HD, Hartman DL, Obermiller AD, Kurosaka S, McLaughlin KJ, Hinrichs K (2009) The uterine environment modulates trophectodermal POU5F1 levels in equine blastocysts. Reproduction 138:589–599. https://doi.org/10.1530/REP-08-0394

Choi YH, Ross P, Velez IC, Macias-Garcia B, Riera FL, Hinrichs K (2015) Cell lineage allocation in equine blastocysts produced in vitro under varying glucose concentrations. Reproduction 150:31–41. https://doi.org/10.1530/REP-14-0662

Chu JW, Sharom FJ, Oriol JG, Betteridge KJ, Cleaver BD, Sharp DC (1997) Biochemical changes in the equine capsule following prostaglandin-induced pregnancy failure. Mol Reprod Dev 46:286–295. https://doi.org/10.1002/(SICI)1098-2795(199703)46:3<286::AID-MRD7>3.0.CO;2-L

Clegg MT, Cole HH, Howard CB, Pigon H (1962) The influence of foetal genotype on equine gonadotrophin secretion. J Endocrinol 25:245–248. https://doi.org/10.1677/joe.0.0250245

Cochet M, Vaiman D, Lefèvre F (2009) Novel interferon delta genes in mammals: cloning of one gene from the sheep, two genes expressed by the horse conceptus and discovery of related sequences in several taxa by genomic database screening. Gene 15, 433(1-2):88–99. https://doi.org/10.1016/j.gene.2008.11.026

Cole HH, Hart GH (1930) The potency of blood serum of mares in progressive stages of pregnancy in effecting the sexual maturity of the immature rat. Am J Physiol 93:57–68. https://doi.org/10.1152/ajplegacy.1930.93.1.57

Cole HH (1938) High gonadotrophic hormone concentration in pregnant ponies. Proc Soc Exp Biol Med 38:193–194. https://doi.org/10.3181/00379727-38-9788P

Cole HH, Pencharz RI, Goss H (1940) On the biological properties of highly purified gonadotropin from pregnant mare serum. Endocrinology 27:548–553. https://doi.org/10.1210/endo-27-4-548

Cole HH, Goss H (1943) Essays in biology in honor of Herbert M Evans. Univ. Calif. Press, Berkeley. The source of equine gonadotropin, pp 107–119

Crossett B, Allen WR, Stewart F (1996) A 19 kDa protein secreted by the endometrium of the mare is a novel member of the lipocalin family. Biochem J 320:137–143. https://doi.org/10.1042/bj3200137

Crossett B, Suire S, Herrler A, Allen WR, Stewart F (1998) Transfer of a uterine lipocalin from the endometrium of the mare to the developing equine conceptus. Biol Reprod 59(3):483–490. https://doi.org/10.1095/biolreprod59.3.483

Cuervo-Arango J, Claes AN, Ruijter-Villani MM, Stout TA (2018) Likelihood of pregnancy after embryo transfer is reduced in recipient mares with a short preceding oestrus. Equine Vet J 50:386–390. https://doi.org/10.1111/evj.12739

Cuervo-Arango J, Claes AN, Stout TA (2019) A retrospective comparison of the efficiency of different assisted reproductive techniques in the horse, emphasizing the impact of maternal age. Theriogenology 132:36–44. https://doi.org/10.1016/j.theriogenology.2019.04.010

Daels PF, DeMoraes JJ, Stabenfeldt GH, Hughes JP, Lasley BL (1991) The corpus luteum: source of oestrogen during early pregnancy in the mare. J Reprod Fertil Suppl 44:501–508. PMID: 1665517

Daughtry BL, Masterson KR, Metcalf ES, Battaglia D, Fei SS, Carbone L, Beck R, Cook N, Chavez SL (2016) Combining time-lapse imaging and next generation RNA-sequencing to assess equine embryo developmental potential. J Equine Vet Sci 41:80–81. https://doi.org/10.1016/j.jevs.2016.04.081

Daughtry BL, Chavez SL (2018) Time-lapse imaging for detection of chromosomal abnormalities in primate preimplantation embryos. Method Mol Biol 1769:293–317. https://doi.org/10.1007/978-1-4939-7780-2_19

Day FT, Rowlands IW (1940) The time and rate of appearance of gonadotrophin in the serum of pregnant mares. J Endocrinol 2:255–261. https://doi.org/10.1677/joe.0.0020255

Day FT, Rowlands IW (1947) Serum gonadotrophin in welsh and Shetland ponies. J Endocrinol 5:1–8. https://doi.org/10.1677/joe.0.0050001

de Castro T, van Heule M, Domingues RR, Jacob JCF, Daels PF, Meyers SA, Conley AJ, Dini P (2024) Embryo-endometrial interaction associated with the location of the embryo during the mobility phase in mares. Sci Rep 714(1):3151. https://doi.org/10.1038/s41598-024-53578-z

de Ruijter-Villani M, van Tol HT, Stout TA (2015) Effect of pregnancy on endometrial expression of luteolytic pathway components in the mare. Reprod Fertil Dev 27:834–845. https://doi.org/10.1071/RD13381

Dell'Aquila ME, Masterson M, Maritato F, Hinrichs K (2001) Influence of oocyte collection technique on initial chromatin configuration, meiotic competence, and male pronucleus formation after intracytoplasmic sperm injection (ICSI) of equine oocytes. Mol Reprod Dev 60:79–88. https://doi.org/10.1002/mrd.1064

Denker HW (2000) Structural dynamics and function of early embryonic coats. Cells Tissues Organs 166(2):180–207. https://doi.org/10.1159/000016736

De Coster T, Velez DA, Van Soom A, Woelders H, Smits K (2020) Cryopreservation of equine oocytes: looking into the crystal ball. Reprod Fertil Dev 32(5):453–467. https://doi.org/10.1071/RD19229

Diel de Amorin M, Bramer SA, Rajamnickam GD, Klein C, Card C (2002) Endometrial and luteal gene expression of putative gene regulators of equine maternal recognition of pregnancy. Anim Reprod Sci 245:107064. https://doi.org/10.1016/j.anireprosci.2022.107064

Dijkstra A, Cuervo-Arango J, Stout TAE, Claes A (2020) Monozygotic multiple pregnancies after transfer of single in vitro produced equine embryos. Equine Vet J 52(2):258–261. https://doi.org/10.1111/evj.13146

Douglas RH, Ginther OJ (1976) Concentration of prostaglandins F in uterine venous plasma of anesthetized mares during the estrous cycle and early pregnancy. Prostaglandins 11(2):251–260. https://doi.org/10.1016/0090-6980(76)90148-9

Ealy AD, Eroh ML, Sharp DC (2010) Prostaglandin H synthase type 2 is differentially expressed in endometrium based on pregnancy status in pony mares and response to oxytocin and conceptus secretions in explant culture. Anim Reprod Sci 117:99–105. https://doi.org/10.1016/j.anireprosci.2009.03.014

Ellenberger C, Wilsher S, Allen WR, Hoffmann C, Kölling M, Bazer FW, Klug J, Schoon D, Schoon HA (2008) Immunolocalisation of the uterine secretory proteins uterocalin, uteroferrin and uteroglobin in the mare's uterus and placenta throughout pregnancy. Theriogenology 70(5):746–757. https://doi.org/10.1016/j.theriogenology.2008.04.050

Enders AC, Lantz KC, Liu IKM, Schlafke S (1988) Loss of polar trophoblast during differentiation of the blastocyst of the horse. J Reprod Fertil 83(1):447–460. https://doi.org/10.1530/jrf.0.0830447

Enders AC, Liu IK (1991) Lodgement of the equine blastocyst in the uterus from fixation through endometrial cup formation. J Reprod Fertil Suppl 44:427–438. PMID: 1795287

Enders AC, Schlafke S, Lantz KC, Liu IK (1993) Endoderm cells of the equine yolk sac from day 7 until formation of the definitive yolk sac placenta. Equine Vet J 25(S15):3–9. https://doi.org/10.1111/j.2042-3306.1993.tb04814.x

Evans HM, Korpi KJ, Simpson ME, Pencharz RI (1936) Fractionation of gonadotrophic hormones in pregnant mare serum by means of ammonium sulphate. University of California Publications in Anatomy 1:275–281. https://www.cabidigitallibrary.org/doi/full/10.5555/19370101851

Evans HM, Gustus EL, Simpson ME (1933) Concentration of the gonadotropic hormone pregnant mare's serum. J Exp Med 58:569–574. https://doi.org/10.1084/jem.58.5.569

Ewart JC (ed) (1987) A critical period in the development of the horse. Adam and Charles Black, London

Fehilly CB, Willadsen SM (1986) Embryo manipulation in farm animals. Oxf Rev Reprod Biol 8:379–413. PMID:3540807

Flood P, Betteridge K, Irvine D (1979) Oestrogens and androgens in blastocoelic fluid and cultures of cells from equine conceptuses of 10-22 days gestation. J Reprod Fertil Suppl 27:413–420. PMID: 289818

Frank BL, Doddman CD, Stokes JE, Carnevale EM (2019) Association of equine oocyte and cleavage stage embryo morphology with maternal age and pregnancy after intracytoplasmic sperm injection. Reprod Fertil Dev 12:1812–1822. https://doi.org/10.1071/RD19250

Freeman DA, Woods GL, Vanderwall DK, Weber JA (1992) Embryo-initiated oviductal transport in the mare. J Reprod Fertil 95:535–538. https://doi.org/10.1530/jrf.0.0950535

Freeman DA, Weber JA, Geary T, Woods GL (1991) Time of embryo transport through the mare oviduct. Theriogenology 36:823–830. https://doi.org/10.1016/0093-691X(91)90348-H

Gaivão M, Stout T (2012) Equine conceptus development—a mini review. Revista Lusófona de Ciência e Medicina Veterinária 5:64–72

Galli C, Crottii G, Notari C, Turini P, Lazzari G (2001) Embryo production by ovum pick up from live donors. Theriogenology 55(6):1341–1357. https://doi.org/10.1016/S0093-691X(01)00486-1

Galet C, Guillou F, Foulon-Gauze F, Combarnous Y, Chopineau M (2009) The β104–109 sequence is essential for the secretion of correctly folded single-chain βα horse LH/CG and for its FSH activity. J Endocrinol 203:167–174. https://doi.org/10.1677/JOE-09-0141

Galli C, Crotti G, Turini P, Duchi R, Mari G, Zavaglia G, Duchamp G, Daels P, Lazzari G (2002) Frozen–thawed embryos produced by ovum pick up of immature oocytes and ICSI are capable to establish pregnancies in the horse. Theriogenology 58:705–708. https://doi.org/10.1016/S0093-691X(02)00771-9

Galli C, Duchi R, Colleoni S, Lagutina I, Lazzari G (2014) Ovum. Pick up, intracytoplasmic sperm injection and somatic cell nuclear transfer in cattle, buffalo and horses: from the research laboratory to clinical practice. Theriogenology 81:138–151. https://doi.org/10.1016/j.theriogenology.2013.09.008

Gardner DK, Balaban B (2016) Assessment of human embryo development using morphological criteria in an era of time-lapse, alogorithms and 'OMICS': is looking good still important? Mol Hum Reprod 22:704–718. https://doi.org/10.1093/molehr/gaw057

Gastal EL, Santos GF, Henry M, Piedade HM (1993) Embryonic and early foetal development in donkeys. Equine Vet J 25:10–13. https://doi.org/10.1111/j.2042-3306.1993.tb04815.x

Gastal MO, Gastal EL, Kot K, Ginther OJ (1996) Factors related to the time of fixation of the conceptus in mares. Theriogenology 46:1171–1180. https://doi.org/10.1016/S0093-691X(96)00288-9

Gilbert SF (2000) Early embryonic development. In: Gilbert FL (ed) Developmental Biology. Sinauer Associates, Sunderland, pp 185–373

Ginther OJ (1983) Mobility of the early equine conceptus. Theriogenology 19:603–611. https://doi.org/10.1016/0093-691X(83)90180-2

Ginther OJ (1984) Intrauterine movement of the early conceptus in barren and postpartum mares. Theriogenology 21:633–644. https://doi.org/10.1016/0093-691X(84)90448-5

Ginther OJ (1985) Dynamic physical interactions between the equine embryo and uterus. Equine Vet J 17(S3):41–47. https://doi.org/10.1111/j.2042-3306.1985.tb04592.x

Ginther OJ (ed) (1992) Reproductive biology of the mare: basic and applied aspects, 2nd edn. Equiservices Publishing, Cross Plains, WI

Ginther OJ (ed) (1995) Ultrasonic imaging and animal reproduction: horses. Book 2. Cross. Equiservices Publishing, Plains, WI

Ginther OJ (1998) Equine pregnancy: physical interactions between the uterus and conceptus. Proc Am Assoc Equine Pract 44:73–105. https://api.semanticscholar.org/CorpusID:22186126

Ginther OJ, Domingues RR, Kennedy VC, Dangudubiyyam SV (2019) Endogenous and exogenous effects of PGF2α during luteolysis in mares. Theriogenology 132:45–52. https://doi.org/10.1016/j.theriogenology.2019.04.004

Ginther OJ (2021) Equine embryo mobility. A friend of Theriogenologists. J Equine Vet Sci 106:103747. https://doi.org/10.1016/j.jevs.2021.103747

Ginther OJ (2022) The dynamic equine embryo from Postfixation (Day 17) to the end of the embryo stage (Day 40). J Equine Vet Sci 108:103808. https://doi.org/10.1016/j.jevs.2021.103808

Goff AK, Leduc S, Poitras P, Vaillancourt D (1993) Steroid synthesis by equine conceptuses between days 7 and 14 and endometrial steroid metabolism. Domest Anim Endocrinol 10:229–236. https://doi.org/10.1016/0739-7240(93)90027-9

Goff AK, Pontbriand D, Sirois J (1987) Oxytocin stimulation of plasma 15- keto-13, 14-dihydro prostaglandin F-2 alpha during the oestrous cycle and early pregnancy in the mare. J Reprod Fertil Suppl 35:253–260. PMID: 3479581

Goss H, Cole HH (1940) Further studies on the purification of mare gonadotropic hormone. Endocrinology 26:244–249. https://doi.org/10.1210/endo-26-2-244

Gray CA, Dunlap KA, Burghardt RC, Spencer TE (2005) Galectin-15 in ovine uteroplacental tissues. Reproduction 130:231–240. https://doi.org/10.1530/rep.1.00637

Green JA, Xie S, Szafranska B, Gan X, Newman AG, McDowell K, Roberts RM (1999) Identification of a new aspartic proteinase expressed by the outer chorionic cell layer of the equine placenta. Biol Reprod 60(5):1069–1077. https://doi.org/10.1095/biolreprod60.5.1069

Griffin PG, Ginther OJ (1990) Uterine contractile activity in mares during the estrous cycle and early pregnancy. Theriogenology 34:47–56. https://doi.org/10.1016/0093-691X(90)90576-F

Griffin PG, Carnevale EM, Ginther OJ (1993) Effects of the embryo on uterine morphology and function in mares. Anim Reprod Sci 31:311–329. https://doi.org/10.1016/0378-4320(93)90015-J

Grøndahl C, Grøndahl Nielsen C, Eriksen T, Greve T, Hyttel P (1993) In-vivo fertilisation and initial embryogenesis in the mare. Equine Vet J Suppl 15:79–83. https://doi.org/10.1111/j.2042-3306.1993.tb04833.x

Haig D (1993) Genetic conflicts in human pregnancy. Q Rev Biol 68:495–532. https://doi.org/10.1086/418300

Hamilton WJ, Day FT (1945) Cleavage stages of the ova of the horse, with notes on ovulation. J Anat 79(127–130):3

Hamilton DW, Allen WR, Moor RM (1973) The origin of equine endometrial cups. Light and electron microscopic study of fully developed equine endometrial cups. Anat Rec 177:503–517. https://doi.org/10.1002/ar.1091770404

Hendriks WK, Colleoni S, Galli G, Paris DBBP, Colenbrander B, Stout TAE (2019) Mitochondrial DNA replication is initiated at blastocyst formation in equine embryos. Reprod Fertil Dev 31:570–578. https://doi.org/10.1071/RD17387

Herrler A, Beier HM (2000) Early embryonic coats: morphology, function, practical applications. Cells Tissues Organs 166:233–246. https://doi.org/10.1159/000016736

Herrler A, Pell JM, Allen WR, Beier HM, Stewart F (2000) Horse conceptuses secrete insulin-like growth factor-binding protein 3. Biol Reprod 62:1804–1811. https://doi.org/10.1095/biolreprod62.6.1804

Hinrichs K (2010) In vitro production of equine embryos: state of the art. Reprod Domestic Anim Supp 2:3–8. https://doi.org/10.1111/j.1439-0531.2010.01624.x

Hoffmann C, Bazer FW, Klug J, Aupperle H, Ellenberger C, Schoon HA (2009) Immunohistochemical and histochemical identification of proteins and carbohydrates in the equine endometrium expression patterns for mares suffering from endometrosis. Theriogenology 71(2):64–74. https://doi.org/10.1016/j.theriogenology.2008.07.008

Hunt JS, Petroff MG, McIntire H, Ober C (2005) HLA-G and immune tolerance in pregnancy. FASEB J 19:681–693. https://doi.org/10.1096/fj.04-2078rev

Huhtinen M, Peippo J, Bredbacka P (1997) Successful transfer of biopsied equine embryos. Theriogenology 48:361–367. https://doi.org/10.1016/S0093-691X(97)00247-1

Hunter RHF (2012) Components of oviduct physiology in eutherian mam-mals. Biol Rev 87(1):244–255. https://doi.org/10.1111/j.1469-185X.2011.00196.x

Jacob JC, Haag KT, Santos GO, Oliveira JP, Gastal MO, Gastal EL (2012) Effect of embryo age and recipient asynchrony on pregnancy rates in a commercial equine embryo transfer program. Theriogenology 77:1159–1166. https://doi.org/10.1016/j.theriogenology.2011.10.022

Kalpokas I, Martínez MN, Cavestany D, Perdigon F, Mattos RC, Meikle A (2021) Equine early pregnancy endocrine profiles and ipsilateral endometrial immune cell, gene expression and protein localization response. Reprod Fertil Dev 33:410–426. https://doi.org/10.1071/RD21001

Kastelic JP, Adams GP, Ginther OJ (1987) Role of progesterone in mobility, fixation, orientation, and survival of the equine embryonic vesicle. Theriogenology 27:655–663. https://doi.org/10.1016/0093-691X(87)90059-8

Katila T (2015) Clinical commentary: techniques to suppress oestrus in mares. Equine Vet Edu 27:344–345. https://doi.org/10.1111/eve.12324

Klein C, Troedsson MHT (2011) Transcriptional profiling of equine conceptuses reveals new aspects of embryo-maternal communication in the horse. Biol Reprod 84:872–885. https://doi.org/10.1095/biolreprod.110.088732

Klein C, Troedsson M (2012) Equine pre-implantation conceptuses express neuraminidase 2--a potential mechanism for desialylation of the equine capsule. Reprod Domest Anim 47(3):449–454. https://doi.org/10.1111/j.1439-0531.2011.01901.x

Klein C (2016a) Early pregnancy in the mare: old concepts revisited. Domest Anim Endocrinol 56:S212–S217. https://doi.org/10.1016/j.domaniend.2016.03.006

Klein C (2016b) Maternal recognition of pregnancy in the context of equine embryo transfer. J Equine Vet Sci 41:22–28. https://doi.org/10.1016/j.jevs.2016.04.001

Klohonatz KM, Hess AM, Hansen TR, Squires EL, Bouma GJ, Bruemmer JE (2015) Equine endometrial gene expression changes during and after maternal recognition of pregnancy. J Anim Sci 93:3364–3376. https://doi.org/10.2527/jas.2014-8826

Klohonatz KM, Nulton LC, Hess AM, Bouma GJ, Bruemmer JE (2019) The role of embryo contacts and focal adhesions during maternal recognition of pregnancy. PLoS One 5;14(3):e0213322. https://doi.org/10.1371/journal.pone.0213322

Kölle S, Dubois CS, Caillaud M, Lahuec C, Sinowatz F, Goudet G (2007) Equine zona protein synthesis and ZP structure during folliculogenesis, oocyte maturation, and embryogenesis. Mol Reprod Dev 74(7):851–859. https://doi.org/10.1002/mrd.20501

Kölle S, Hughes B, Steele H (2020) Early embryo-maternal communication in the oviduct: A review. Mol Reprod Dev 87(6):650–662. https://doi.org/10.1002/mrd.23352

Lantz KC, Enders AC, Liu IK (1998) Possible significance of cells within intraluminal collagen masses in equine oviducts. Anat Rec 252:568–579

Leith GS, Ginther OJ (1984) Characterization of intrauterine mobility of the early equine conceptus. Theriogenology 22:401–408. https://doi.org/10.1016/0093-691X(84)90460-6

Legrand E, Krawiecki JM, Tainturier D, Bruyas JF (2001) Does the embryonic capsule impede freezing of equine embryos?. In: Proceedings of the 5th Internation Symposium on Equine Embryo Transfer, pp. 62-65.

Legrand E, Bencharif D, Barrier-Battut I, Delajarraud H, Cornière P, Fiéni F, Tainturier D, Bruyas JF (2002) Comparison of pregnancy rates for days 7-8 equine embryos frozen in glycerol with or without previous enzymatic treatment of their capsule. Theriogenology 58:721–723. https://doi.org/10.1016/S0093-691X(02)00891-9

Lewis N, Hinrichs K, Schnauffer K, Morganti M, McG. Argo, C. (2016) Effect of oocyte source and transport time on rates of equine oocyte. Maturation and cleavage after fertilization by ICSI, with a note on the validation of equine embryo morphological classification. Clin Theriogenol 8:25–39. https://doi.org/10.1038/s41598-020-60624-z

Li X, Morris LH, Allen WR (2001) Influence of co-culture during maturation on the developmental potential of equine oocytes fertilized by intracytoplasmic sperm injection (ICSI). Reproduction 121:925–932. https://doi.org/10.1530/rep.0.1210925

Magli MC, Gianaroli L, Ferraretti AP, Lappi M, Ruberti A, Farfalli V (2007) Embryo morphology and development are dependent on the chromosomal complement. Fertil Steril 87:534–541. https://doi.org/10.1016/j.fertnstert.2006.07.1512

Manning AW, Rajkumar K, Bristol F, Flood PF, Murphy BD (1987) Genetic and temporal variation in serum concentrations and biological activity of horse chorionic gonadotrophin. J Reprod Fertil Suppl 35:389–397

Marsan C, Goff AK, Sirois J, Betteridge KJ (1987) Steroid secretion by different cell types of the horse conceptus. J Reprod Fertil Suppl 35:363–369

McCue PM (2021) Embryo evaluation. In: Dascanio J, McCue P (eds) Equine reproductive procedures, second edition, vol 63. John Wiley & Sons, Inc, pp 231–234. https://doi.org/10.1002/9781119556015.ch63

McCue PM, DeLuca CA, Ferris RA, Wall JJ (2009) How to evaluate equine embryos. AAEP Proc 55:252–256

McDowell KJ, Sharp DC, Grubaugh W, Thatcher WW, Wilcox CJ (1988) Restricted conceptus mobility results in failure of pregnancy maintenance in mares. Biol Reprod 39(2):340–348. https://doi.org/10.1095/biolreprod39.2.340

McDowell KJ, Sharp DC, Fazleabas AT, Roberts RM (1990) Synthesis and release of proteins by endometrium from pregnant and non-pregnant mares, and by conceptus membranes: characterization by two-dimensional gel electrophoresis. J Reprod Fertil 89:107–115. https://doi.org/10.1530/jrf.0.0890107

McDowell KJ, Adams MH, Williams NM (1993) Characterization of equine oviductal proteins synthesized and released at estrus and at day 4 after ovulation inbred and nonbred mares. J Exp Zool 267:217–224. https://doi.org/10.1002/jez.1402670215

Meinecke B, Meinecke-Tillmann S (2023) Lab partners: oocytes, embryos and company. A personal view on aspects of oocyte maturation and the development of monozygotic twins. Anim Reprod 20(2):e20230049. https://doi.org/10.1590/1984-3143-ar2023-0049

Meira C, Ferreira JC, Papa FO, Henry M (1998) Ultrasonographic evaluation of the conceptus from days 10 to 60 of pregnancy in jennies. Theriogenology 49(8):1475–1482. https://doi.org/10.1016/S0093-691X(98)00093-4

Mess A, Carter AM (2007) Evolution of the placenta during the early radiation of placental mammals. Comp Biochem Physiol A Mol Integr Physiol 1(148):769–779. https://doi.org/10.1016/j.cbpa.2007.01.029

Miller CC, Fayrer-Hosken RA, Timmons TM, Lee VH, Caudle AB, Dunbar BS (1992) Characterization of equine zona pellucida glycoproteins by polyacrylamide gel electrophoresis and immunological techniques. J Reprod Fertil 96:815–825. https://doi.org/10.1530/jrf.0.0960815

Morris LH, Allen WR (2002) Reproductive efficiency of intensively managed thoroughbred mares in Newmarket. Equine Vet J 34:51–60. https://doi.org/10.2746/042516402776181222

Morris LHA (2018) The development of in vitro embryo production in the horse. Equine Vet J 59:712–720. https://doi.org/10.1111/evj.12839

Mortensen CJ, Choi YH, Ing NH, Kraemer DC, Vogelsang MM, Hinrichs K (2010) Heat shock protein 70 gene expression in equine blastocysts after exposure of oocytes to high temperatures in vitro or in vivo after exercise of donor mares. Theriogenology 74(3):374–383. https://doi.org/10.1016/j.theriogenology.2010.02.020

Murphy BD, Martinuk SD (1991) Equine chorionic gonadotropin. Endocr Rev 12:27–44. https://doi.org/10.1210/edrv-12-1-27

Murphy BD (2012) Equine chorionic gonadotropin: an enigmatic but essential tool. Anim Reprod 9(3):223–223. https://www.animal-reproduction.org/article/5b5a6057f7783717068b46e2

Newcombe JR, Wilsher S, Cuervo-Arango J (2022) The timing of the maternal recognition signal is specific to individual mares. J Equine Vet Sci 113:103977. https://doi.org/10.1016/j.jevs.2022.103977

Newcombe JR, Cuervo-Arango J, Wilsher S (2023) The timing of the maternal recognition of pregnancy is specific to individual mares. Animals (Basel) 13(10):1718. https://doi.org/10.3390/ani13101718

Nie GJ, Johnson KE, Braden TD, Wenzel JGW (2003) Use of an intra-uterine glass ball protocol to extend luteal function in mares. J Equine Vet Sci 23(6):266–273. https://doi.org/10.1053/jevs.2003.75

Oriol JG, Sharom FJ, Betteridge KJ (1993a) Developmentally regulated changes in the glycoproteins of the equine embryonic capsule. J Reprod Fertil 99(2):653–664. https://doi.org/10.1530/jrf.0.0990653

Oriol JG, Betteridge KJ, Clarke AJ, Sharom FJ (1993b) Mucin-like glycoproteins in the equine embryonic capsule. Mol Reprod Dev 34:255–265. https://doi.org/10.1002/mrd.1080340305

Ortiz ME, Llados C, Croxatto HB (1989) Embryos of different ages transferred to the rat oviduct enter the uterus at different times. Biol Reprod 41:381–384. https://doi.org/10.1095/biolreprod41.3.381

Palmer E, Duchamp G, Cribiu EP, Mahla R, Boyazoglu S, Bézard J (1997) Follicular fluid is not a compulsory carrier of the oocyte at ovulation in the mare. Equine Vet J Suppl 25:22–24. https:// doi.org/10.1111/j.2042-3306.1997.tb05094.x

Paulo E, Tischner M (1985) Activity of delta(5)3beta-hydroxysteroid dehydrogenase and steroid hormones content in early preimplantation horse embryos. Folia Histochem Cytobiol 23(1–2):81–84

Pereira GR, Lorenzo PL, Carneiro GF, Ball BA, Pegoraro LM, Pimentel CA, Liu IK (2013) Influence of equine growth hormone, insulin-like growth factor-I and its interaction with gonadotropins on in vitro maturation and cytoskeleton morphology in equine oocytes. Animal 7(9):1493–1499. https://doi.org/10.1017/S175173111300116X

Piotrowska-Tomala KK, Jonczyk AW, Skarzynski DJ, Szóstek-Mioduchowska AZ (2020) Luteinizing hormone and ovarian steroids affect in vitro prostaglandin production in the equine myometrium and endometrium. Theriogenology 1(153):1–8. https://doi.org/10.1016/j. theriogenology.2020.04.039

Policastro P, Ovitt CE, Hoshina M, Fukuoka H, Boothby MR, Boime I (1983) The β subunit of human chorionic gonadotropin is encoded by multiple genes. J Biol Chem 258:11492–11499

Policastro PF, Daniels-McQueen S, Carle G, Boime I (1986) A map of the hCGβ-LHβ gene cluster. J Biol Chem 261:5907–5916

Pomar FJ, Teerds KJ, Kidson A, Colenbrander B, Tharasanit T, Aguilar B, Roelen BA (2005) Differences in the incidence of apoptosis between in vivo and in vitro produced blastocysts of farm animal species: a comparative study. Theriogenology 63:2254–2268. https://doi. org/10.1016/j.theriogenology.2004.10.015

Prados FJ, Debrock S, Lemmen JG, Agerholm I (2012) The cleavage stage embryo. Hum Reprod 27:i50–i71. https://doi.org/10.1093/humrep/des224

Quinn BA, Hayes MA, Waelchli RO, Kennedy MW, Betteridge KJ (2007) Changes in major proteins in the embryonic capsule during immobilization (fixation) of the conceptus in the third week of pregnancy in the mare. Reproduction 134(1):161–170. https://doi.org/10.1530/ REP-06-0241

Racowsky C, Combelles CMH, Nureddin A, Pan Y, Finn A, Miles L, Gale S, O'Leary T, Jackson KV (2003) Day 3 and Day 5 morphology predictors of embryo viability. Reprod Biomed Online 6:323–331. https://doi.org/10.1016/S1472-6483(10)61852-4

Rader K, Choi YH, Hinrichs K (2016) Intracytoplasmic sperm. Injection, embryo culture, and transfer of in vitro-produced blastocysts. Veterinary clinics of North America. Equine Pract 32:401–413. https://doi.org/10.1016/j.cveq.2016.07.003

Raeside JI, Christie HL, Renaud RL, Waelchli RO, Betteridge KJ (2004) Estrogen metabolism in the equine conceptus and endometrium during early pregnancy in relation to estrogen concentrations in yolk-sac fluid. Biol Reprod 71:1120–1127. https://doi.org/10.1095/ biolreprod.104.028712

Rambags BP, van Rossem AW, Blok EE, de Graaf-Roelfsema E, Kindahl H, van der Kolk JH, Stout TAE (2008) Effects of exogenous insulin on luteolysis and reproductive cyclicity in the mare. Reprod Domest Anim 43:422–428. https://doi.org/10.1111/j.1439-0531.2007.00929.x

Rasweiler JJIV (1979) Differential transport of embryos and degenerating ova by the oviducts of the long-tongued bat, Glossophaga soricina. J Reprod Fertil 55:329–334. https://doi. org/10.1530/jrf.0.0550329

Rienzi L, Ubaldi F, Iacobelli M, Romano S, Giulia Minasi M, Ferrero S, Sapienza F, Baroni E, Greco E (2005) Significance of morphological attributes of the early embryo. Reprod Biomed Online 10:669–681. https://doi.org/10.1016/S1472-6483(10)61676-8

Rivera del Alamo MM, Reilas T, Kindahl H, Katila T (2008) Mechanisms behind intrauterine device-induced luteal persistence in mares. Anim Reprod Sci 107(1–2):94–106. https://doi. org/10.1016/j.anireprosci.2007.06.010

Rivera del Alamo MM, Reilas T, Kindahl H, Katila T (2008) Mechanisms behind intrauterine device-induced luteal persistence in mares. Anim Reprod Sci 107(1-2):94–106. https://doi. org/10.1016/j.anireprosci.2007.06.010

Roberts MA, London K, Campos-Chillon LF, Altermatt JL (2015) Presumed monozygotic twins develop following transfer of an in vitro-produced equine embryo. J Equine Sci 26:89–94. https://doi.org/10.1294/JES.26.89

Rowlands IW (1949) Serum gonadotrophin and ovarian activity in the pregnant mare. J Endocrinol 6:184–191. https://doi.org/10.1677/joe.0.0060184

Rowson LE (1971) The second Hammond memorial lecture: the role of reproductive research in animal production. J Reprod Fertil 26:113–126. https://doi.org/10.1530/jrf.0.0260113

Rudolf Vegas A, Podico G, Canisso IF, Bollwein H, Almiñana C, Bauersachs S (2021) Spatiotemporal endometrial transcriptome analysis revealed the luminal epithelium as key player during initial maternal recognition of pregnancy in the mare. Sci Rep 11(1):22293. https://doi.org/10.1038/s41598-021-01785-3

Ruggeri E, DeLuca KF, Galli C, Lazzari G, DeLuca JG, Stokes JE, Carnevale EM (2017) Use of confocal microscopy to evaluate equine zygote development after sperm injection of oocytes matured in vivo or in vitro. Microsc Microanal 23:1197–1206. https://doi.org/10.1294/JES.26.89

Salgado RM, Brom-de-Luna JG, Resende HL, Canesin HS, Hinrichs K (2018) Lower blastocyst quality after conventional vs. piezo ICSI in the horse reflects delayed sperm component remodeling and oocyte activation. J Assist Reprod Genet 35:825–840. https://doi.org/10.1007/S10815-018-1174-9

Sato E, Ando N, Takahashi Y, Miyamoto H, Toyoda Y (1995) Structural changes in the oviductal wall during the passage of unfertilized cumulus-oocyte complexes in mice. Anat Rec 241:363–368. https://doi.org/10.1002/ar.1092410207

Saunders FJ, Cole HH (1935) Two gonadotropic substances in mare serum. Proc Soc Exp Biol Med 32:1476–1478. https://doi.org/10.3181/00379727-32-81

Sharp DC, McDowell KJ, Weithenauer J, Thatcher WW (1989) The continuum of events leading to maternal recognition of pregnancy in mares. J Reprod Fertil Suppl 37:101–107

Sharp DC (2000) The early fetal life of the equine conceptus. Anim Reprod Sci 60–61:679–689. https://doi.org/10.1016/s0378-4320(00)00138-x

Sharp DC, Thatcher MJ, Salute ME, Fuchs AR (1997) Relationship between endometrial oxytocin receptors and oxytocin-induced prostaglandin F2 alpha release during the oestrous cycle and early pregnancy in pony mares. J Reprod Fertil 109(1):137–144. https://doi.org/10.1530/jrf.0.1090137

Schams D, Papkoff H (1972) Chemical and immunological studies on pregnant mare serum gonadotropin. Biochim Biophys Acta 263:139–148. https://doi.org/10.1016/0005-2795(72)90168-7

Schauder W (1912) Untersuchungen uber die eithaute und Embryotrophe des pferdes. Arch Anat Phys 192:259–302

Sherman GB, Wolfe MW, Farmerie TA, Clay CM, Threadgill DS, Sharp DC, Nilson JH (1992) A single gene encodes the β-subunits of equine luteinizing hormone and chorionic gonadotropin. Mol Endocrinol 6:951–959. https://doi.org/10.1210/mend.6.6.1379674

Silva LA, Gastal EL, Beg MA, Ginther OJ (2005) Changes in vascular perfusion of the endometrium in association with changes in location of the embryonic vesicle in mares. Biol Reprod 72:755–761. https://doi.org/10.1095/biolreprod.104.036384

Silva LA, Ginther OJ (2006) An early endometrial vascular indicator of completed orientation of the embryo and the role of dorsal endometrial encroachment in mares. Biol Reprod 74(2):337–343. https://doi.org/10.1095/biolreprod.105.047621

Simpson KS, Adams MH, Behrendt-Adam CY, Baker CB, McDowell KJ (1999) Identification and initial characterization of calcyclin and phospholipase A2 in equine conceptuses. Mol Reprod Dev 53:179–187. https://doi.org/10.1002/(SICI)1098-2795(199906)53:2%3C179::AIDMRD7%3E3.0.CO;2-P

Smits K, Govaere J, Peelman LJ, Goossens K, de Graaf DC, Vercauteren D, Vandaele L, Hoogewijs M, Wydooghe E, Stout T, Van Soom A (2012) Influence of the uterine environment on the development of in vitro-produced equine embryos. Reproduction 143(2):173–181. https://doi.org/10.1530/REP-11-0217

Smits K, De Coninck DI, Van Nieuwerburgh F, Govaere J, Van Poucke M, Peelman L, Deforce D, Van Soom A (2016) The equine embryo influences immune-related gene expression in the oviduct. Biol Reprod 94(2):36. https://doi.org/10.1095/biolreprod.115.136432

Spencer TE, Bazer FW (2002) Biology of progesterone action during pregnancy recognition and maintenance of pregnancy. Front Biosci 7(4):1879–1898. https://doi.org/10.2741/spencer

Squires EL, Ginther OJ (1975) Follicular and luteal development in pregnant mares. J Reprod Fertil Suppl 23:429–433

Starbuck GR, Stout TA, Lamming GE, Allen WR, Flint AP (1998) Endometrial oxytocin receptor and uterine prostaglandin secretion in mares during the oestrous cycle and early pregnancy. J Reprod Fertil 113(2):173–179. https://doi.org/10.1530/jrf.0.1130173

Stensen MH, Tanbo TG, Storeng R, Abyholm T, Fedorcsak P (2015) Fragmentation of human cleavage-stage embryos is related to the progression through meiotic and mitotic cell cycles. Fertil Steril 103:374–381.e4. https://doi.org/10.1016/j.fertnstert.2014.10.031

Stewart F, Allen WR, Moor RM (1976) Pregnant mare serum gonadotrophin: ratio of follicle-stimulating hormone and luteinizing hormone activities measured by radioreceptor assay. J Endocrinol 71:471–482. https://doi.org/10.1677/joe.0.0710371

Stewart F, Maher JK (1991) Analysis of horse and donkey gonadotrophin genes using southern blotting and DNA hybridization techniques. J Reprod Fertil Suppl 44:19–25

Stewart F, Charleston B, Crossett B, Barker PJ, Allen WR (1995) A novel uterine protein that associates with the embryonic capsule in equids. Reproduction 105:65. https://doi.org/10.1530/jrf.0.1050065

Stewart F, Gerstenberg C, Suire S, Allen WR (2000) Immunolocalization of a novel protein (P19) in the endometrium of fertile and subfertile mares. J Reprod Fertil Suppl 56:593–599

Stout TAE, Allen WR (2001) Role of prostaglandins in intrauterine migration of the equine conceptus. Reproduction 121:771–775

Stout TAE, Allen WR (2002) Prostaglandin E(2) and F(2 alpha) production by equine conceptuses and concentrations in conceptus fluids and uterine flushings recovered from early pregnant and dioestrous mares. Reproduction 123:261–268. https://doi.org/10.1530/rep.0.1230261

Stout TAE, Meadows S, Allen WR (2005) Stage-specific formation of the equine blastocyst capsule is instrumental to hatching and to embryonic survival in vivo. Anim Reprod Sci 87(3):269–281. https://doi.org/10.1016/j.anireprosci.2004.11.009

Stout TA (2016) Embryo-maternal communication during the first 4 weeks of equine pregnancy. Theriogenology 86(1):349–354. https://doi.org/10.1016/j.theriogenology.2016.04.048

Stout T (2020) Embryo-maternal communication during the establishment of equine pregnancy. Clin Theriogenol 12(4):536–542. https://doi.org/10.58292/ct.v12.9455

Sugimura S, Akai T, Imai K (2017) Selection of viable in vitrofertilized bovine embryos using time-lapse monitoring in microwell culture dishes. J Reprod Dev 63:353–357. https://doi.org/10.1262/JRD.2017-041

Suire S, Stewart F, Beauchamp J, Kennedy MW (2001) Uterocalin, a lipocalin provisioning the preattachment equine conceptus: fa.tty acid and retinol binding properties, and structural characterization. Biochem J 356(2):369–376. https://doi.org/10.1042/bj3560369

Swegen A, Grupen CG, Gibb Z, Baker MA, de Ruijter-Villani M, Smith ND, Stout TAE, Aitken RJ (2017) From peptide masses to pregnancy maintenance: A comprehensive proteomic analysis of the early equine embryo Secretome, blastocoel fluid, and capsule. Proteomics 17:17–18. https://doi.org/10.1002/pmic.201600433

Swegen A (2021) Maternal recognition of pregnancy in the mare: does it exist and why do we care? Reproduction 161(6):R139–R155. https://doi.org/10.1530/REP-20-0437

Townson DH, Ginther OJ (1989) Ultrasonic characterization of follicular evacuation during ovulation and fate of the discharged follicular fluid in mares. Anim Reprod Sci 20:131–141. https://doi.org/10.1016/0378-4320(89)90070-5

Tremoleda JL, Stout TAE, Lagutina I, Lazzari G, Bevers MM, Colebrander B, Galli C (2003a) Effects of in vitro production on horse embryo morphology, cytoskeletal characteristics, and blastocyst capsule formation. Biol Reprod 69:1895–1906. https://doi.org/10.1095/BIOLREPROD.103.018515

Tremoleda JL, Tharasanit T, Van Tol HT, Stout TA, Colenbrander B, Bevers MM (2003b) Effects of follicular cells and FSH on the resumption of meiosis in equine oocytes matured in vitro. Reproduction 125:565–577

Trowsdale J, Betz AG (2006) Mother's little helpers: mechanisms of maternal–fetal tolerance. Nat Immunol 7:241–246. https://doi.org/10.1038/ni1317

Umair M, Scheeren VFDC, Beitsma MM, Colleoni S, Galli C, Lazzari G, de Ruijter-Villani M, Stout TAE, Claes A (2023) In vitro-produced equine blastocysts exhibit greater dispersal and intermingling of inner cell mass cells than in vivo embryos. Int J Mol Sci 24(11):9619. https://doi.org/10.3390/ijms24119619

Van Niekerk CH, Gerneke WH (1996) Persistence and parthenogenetic cleavage of tubal ova in the mare. Onderstepoort J Vet Res 31:195–232

Van Royen E, Mangelschots K, De Neubourg D, Valkenburg M, Van de Meerssche M, Rychaert G, Eestermans W, Gerris J (1999) Characterization of a top quality embryo, as step towards single-embryo transfer. Hum Reprod 14:2345–2349. https://doi.org/10.1093/HUMREP/14.9.2345

Vanderwall DK, Woods G, Weber J, Lichtenwalner A (1993) PGE2 secretion by the conceptus and binding by non-pregnant endometrium in the horse. Equine Vet J 25:24–27. https://doi.org/10.1111/j.2042-3306.1993.tb04818.x

Vanderwall DK (1996) Early embryonic development and evaluation of equine embryo viability. Veterinary clinics of North America. Equine Pract 12:61–83. https://doi.org/10.1016/s0749-0739(17)30295-x

Vanderwall DK (2008) Early embryonic loss in the Mare. J Equine Vet Sci 28(11):691–702. https://doi.org/10.1016/j.jevs.2008.10.001

Vegas AR, Podico G, Canisso IF, Bollwein H, Fröhlich T, Bauersachs S, Almiñana C (2021) Dynamic regulation of the transcriptome and proteome of the equine embryo during maternal recognition of pregnancy. FASEB Bioadvances 4(12):775–797. https://doi.org/10.1096/fba.2022-00063

Velasquez LA, Aguilera JG, Croxatto HB (1995) Possible role of platelet-activating factor in embryonic signaling during oviductal transport in the hamster. Biol Reprod 52:1302–1306. https://doi.org/10.1095/biolreprod52.6.1302

Velasquez LA, Ojeda SR, Croxatto HB (1997) Expression of platelet-activating factor receptor in the hamster oviduct: localization to the endosalpinx. J Reprod Fertil 109:349–354. https://doi.org/10.1530/jrf.0.1090349

Vernon MW, Zavy MT, Asquith RL, Sharp DC (1981) Prostaglandin F2alpha in the equine endometrium: steroid modulation and production capacities during the estrous cycle and early pregnancy. Biol Reprod 25:581–589. https://doi.org/10.1095/biolreprod25.3.581

Watson ED (1989) Release of immunoreactive arachidonate metabolites by equine endometrium in vitro. Am J Vet Res 50(8):1207–1209

Watson ED, Sertich PL (1989) Prostaglandin production by horse embryos and the effect of co-culture of embryos with endometrium from pregnant mares. J Reprod Fertil 87(1):331–336. https://doi.org/10.1530/jrf.0.0870331

Webel SK, Franklin V, Hardland B, Dziuk PJ (1977) Fertility, ovulation and maturation of eggs in mares injected with HCG. J Reprod Fertil 51:337–341. https://doi.org/10.1530/JRF.0.0510337

Weber JA, Freeman DA, Vanderwall DK, Woods GL (1991a) Prostaglandin E2 secretion by oviductal transport-stage equine embryos. Biol Reprod 45:540–543. https://doi.org/10.1095/biolreprod.45.4.540

Weber JA, Freeman DA, Vanderwall DK, Woods GL (1991b) Prostaglandin E2 hastens oviductal transport of equine embryos. Biol Reprod 45:544–546. https://doi.org/10.1095/biolreprod.45.4.544

Weber JA, Woods GL (1993) Influence of embryonic secretory chemicals on selective oviductal transport in mares. Equine Vet J Suppl 15:36–38. https://doi.org/10.1111/j.2042-3306.1993.tb04821.x

Weber JA, Woods GL, Lichtenwalder AB (1995) Relaxatory effects of prostaglandin E2 on circular smooth muscle isolated from the equine oviductal isthmus. Biol Reprod 1:125–130. https://doi.org/10.1093/biolreprod/52.monograph_series1.125

Weber JA, Woods GL, Aigular JJ (1996) Location of equine oviductal embryos on day 5 post ovulation and oviductal transport time of day 5 embryos autotransferred to the contralateral oviduct. Theriogenology 46:1477–1483. https://doi.org/10.1016/S0093-691X(96)00325-1

Weinerman R, Feng R, Ord TS, Schultz RM, Bartolomei MS, Coutifaris C, Mainigi M (2016) Morphokinetic evaluation of embryo development in a mouse model: functional and molecular correlates. Biol Reprod 94:84. https://doi.org/10.1095/BIOLREPROD.115.134080

Wijayagunawardane MP, Choi YH, Miyamoto A, Kamishita H, Fujimoto S, Takagi M, Sato K (1999) Effect of ovarian steroids and oxytocin on the production of prostaglandin E2, prostaglandin F2alpha and endothelin-1 from cow oviductal epithelial cell monolayers in vitro. Anim Reprod Sci 56:11–17. https://doi.org/10.1016/s0378-4320(99)00021-4

Wildman DE, Chen C, Erez O, Grossman LI, Goodman M, Romero R (2006) Evolution of the mammalian placenta revealed by phylogenetic analysis. Proc Natl Acad Sci 103:3203–3208. https://doi.org/10.1073/pnas.0511344110

Willmann C, Budik S, Walter I, Aurich C (2011) Influences of treatment of early pregnant mares with the progestin altrenogest on embryonic development and gene expression in the endometrium and conceptus. Theriogenology 76:61–73. https://doi.org/10.1016/j.theriogenology.2011.01.018

Wilsher S, Allen WR (2011a) Factors influencing equine chorionic gonadotrophin production in the mare. Equine Vet J 43:430–438. https://doi.org/10.1111/j.2042-3306.2010.00309.x

Wilsher S, Allen WR (2011b) Intrauterine administration of plant oils inhibits luteolysis in the mare. Equine Vet J 43:99–105. https://doi.org/10.1111/j.2042-3306.2010.00131.x

Wilsher S, Gower S, Allen WR (2011) Immunohistolocalisation of progesterone and oestrogen receptors at the placental interface in mares during early pregnancy. Anim Reprod Sci 129:200–208. https://doi.org/10.1016/j.anireprosci.2011.11.004

Wilsher S, Gower S, Allen WR (2013) Persistence of an immunoreactive MUC1 protein at the feto-maternal interface throughout pregnancy in the mare. Reprod Fertil Dev 25:753–761. https://doi.org/10.1071/RD12152

Wilsher S, Newcombe JR, Ismer A, Duarte A, Kovacsy S, Allen WR (2022) The effect of embryo reduction and transfer on luteostasis in the mare. Anim Reprod Sci 242:107002. https://doi.org/10.1016/j.anireprosci.2022.107002

Wooding FB, Morgan G, Fowden AL, Allen WR (2001) A structural and immunological study of chorionic gonadotrophin production by equine trophoblast girdle and cup cells. Placenta 22:749–767. https://doi.org/10.1053/plac.2001.0707

Zavy MT, Mayer R, Vernon MW, Bazer FW, Sharp DC (1979) An investigation of the uterine-luminal environment of non-pregnant and pregnant pony mares. J Reprod Fertil Suppl 27:403–411. PMID: 289817

Zavy MT, Vernon MW, Sharp DC III, Bazer FW (1984) Endocrine aspects of early pregnancy in pony mares: a comparison of uterine luminal and peripheral plasma levels of steroids during the estrous cycle and early pregnancy. Endocrinology 115:214–219. https://doi.org/10.1210/endo-115-1-214

The Role of the Oviduct in Gameto-Maternal Interaction and Embryonic Development

Sabine Kölle (iD)

Abstract

The oviduct is a key player for successful establishment of pregnancy. It is pivotal for oocyte and sperm transport, formation of the sperm reservoir, fertilization, and early embryonic development and orchestrates a comprehensive set of gameto-maternal interactions. These include the communication between the cumulus-oocyte complex and tubal epithelium, as well as the binding of the sperm to the oviductal ciliated cells and the formation of the sperm reservoir, which maintains sperm fertilizing capacity for days, weeks, or even months. After fertilization, the oviduct regulates the proper transport and nutrition of the early embryo during cleavage and migration toward the uterus. Most importantly, the fallopian tube is the site where the first embryo-maternal communication occurs ensuring the survival and continuous development of the early embryo. This article provides an overview of the findings to date highlighting the morphological, functional, and molecular changes induced by gameto-maternal interactions and early embryo-maternal communication using cutting-edge molecular and imaging technologies.

Keywords

Oviduct · Oocyte · Sperm · Embryo

S. Kölle (✉)
School of Medicine, Health Sciences Centre, University College Dublin (UCD),
Dublin, Ireland
e-mail: sabine.koelle@ucd.ie

© The Author(s), under exclusive license to Springer Nature
Switzerland AG 2024
J. C. Gardón, K. Satué Ambrojo (eds.), *Assisted Reproductive Technologies in Animals Volume 1*, https://doi.org/10.1007/978-3-031-73079-5_15

473

1   Introduction

Ovulation, fertilization, and successful embryonic development are vitally dependent on the timely preparation of the female genital tract by systemic hormonal actions (Kolle et al. 2020). Additionally, precisely timed local interactions between the oviductal epithelium and the gametes and active communication between the maternal tubal epithelium and the early embryo are essential for enabling successful fertilization and for supporting embryo survival during cleavage and migration toward the uterus (Kolle et al. 2009; Camara Pirez et al. 2020).

With regard to gameto-maternal interactions, the first step is the attachment of the cumulus-oocyte complex (COC) to the ampullar epithelium after ovulation (Kolle 2012; Kolle et al. 2009). Only COCs that are capable of attaching will maintain the ability to be fertilized (Kolle et al. 2009; Kolle 2012). Similarly, when looking at spermatozoa, only those ones that can bind to the tubal epithelium will survive and maintain fertilizing ability (Camara Pirez et al. 2020; Kolle 2022). Successful fertilization will only occur if the oviduct is able to preserve the vitality of both the spermatozoa and the COC and to orchestrate a meeting in the right place and at the right time, i.e., in the tubal ampulla within 12–24 hs after ovulation (Talbot et al. 2003). After fertilization, the early embryo is continuously cleaving and continues its migration towards the uterus (Kolle et al. 2010). During this time the oviduct plays an essential role in managing the timely transport as well as in providing nutrition produced by the tubal secretory cells. The transport of the oocyte and the embryo is achieved by a) ciliary beating of the tubal epithelial cells, b) contraction of the tubal smooth muscle, and c) the resulting movement of tubal fluid (Croxatto 2002). With regard to the role of ciliary beating, it is known that after excision and reverse reimplantation of a part of the ampulla in the reverse direction, pregnancy sporadically occurs (McComb et al. 1980). Similarly, women with Kartagener Syndrome who have ciliary dyskinesia can be fertile (McComb et al. 1986). Regarding smooth muscle contraction, it is regulated both systemically (prostaglandin F2alpha increases smooth muscle contraction, whereas progesterone decreases it) as well as by endocrine, autocrine, and paracrine factors secreted locally by the oocyte and embryo (Croxatto 2002; Kolle 2015).

Besides providing transport and nutrition, the oviduct is the first site of communication between the embryo and the mother. Successful pregnancy only occurs if the embryo is able to signal to the mother that it is here and if the mother can answer. Consequently, a failure in communication between the embryo and the mother is the major cause of embryonic loss before implantation. The precise percentage of early embryonic loss is unknown, but it is estimated that up to 30–43% of pregnancies are terminated in ruminants before implantation (Diskin and Morris 2008).

Overall, the importance of the oviduct for gamete transport sperm and oocyte survival, fertilization, embryonic development, and early embryo-maternal communication has been underestimated for a long time. Firstly, this is because the fallopian tube is well hidden in the abdominal cavity and there are very few imaging technologies to visualize it. Clinically, it can only be examined by surgical procedures such as endoscopy (Marano 1998). Secondly, most studies on the oviduct have been conducted in in vitro tubal epithelial cell cultures and oviductal explants (Walter 1995). These experiments have provided essential information on hormonal regulation, gene expression, and changes in the oviductal environment due to hormonal actions (Kodithuwakku et al. 2007; Yeste et al. 2014). However, in these experiments, the comprehensive interactions between the tubal epithelium, the underlying connective tissue, and the smooth muscle layers are missing. Consequently, the paracrine signaling between cells is decreased. To overcome these limitations and to gain a comprehensive understanding of the dynamic interplay between tubal epithelium, oocyte, sperm, and embryo, novel cutting-edge technologies have been developed in recent years. Thus, digital live cell imaging (LCI) within the oviduct in real-time, confocal fluorescence microscopy using a fiber-optic probe and optical coherence tomography providing 3D high resolution have been successfully applied to shed light on the complex role of the oviduct in gameto-maternal and embryo-maternal interactions (Druart et al. 2009; Kolle et al. 2009; Kolle 2012, 2015; Camara Pirez et al. 2020).

Overall, the knowledge of the comprehensive interactions between the oviduct on the one side and the gametes and the early embryo on the other side hold promise to establish novel diagnostic and therapeutic strategies for the treatment of subfertility and infertility in husbandry and to maximize economic success in cattle breeding. With regard to in vitro embryo production, a deep understanding of gameto- and embryo-maternal communication provides the basis for the development of novel IVF concepts across all species. As such, the supplementation of signaling molecules involved in embryo-maternal communication during the in vitro culture of early embryos might contribute to further improving the outcome of assisted reproductive technologies (ART) in both animals and humans.

2   The Oviductal Epithelium Interacts with the Cumulus-Oocyte Complex after Ovulation

When ovulation occurs, the cumulus-oocyte complex (COC) adheres to the fimbriae of the infundibulum and then slides over the inner surface in the direction of the tubal ostium (Talbot et al. 1999). The COC is then transported to the ampulla by ciliary beating and smooth muscle contraction (Talbot et al. 2003). As soon as it has reached the ampulla, the COC settles in the depth between the folds and firmly attaches to the inner surface by mechanical interaction between the tubal epithelial cells and the cumulus cells (Fig. 1) (Kolle et al. 2009). The capacity of the COC to settle in between the folds is due to a specific transport mechanism in the ampulla, which can be visualized using dynabeads, small spherical polymer particles. In the ampulla, the dynabeads settle between the folds and stay there, whereas in the isthmus the particles are rapidly transported to the apical ridges of the fold moving in a guided stream (Kolle et al. 2009). The firm contact between cumulus cells and the oviductal epithelium is only lost when the cumulus cells are destroyed. Importantly, if the oocyte is of low quality and/or degenerating, which is seen by an irregular

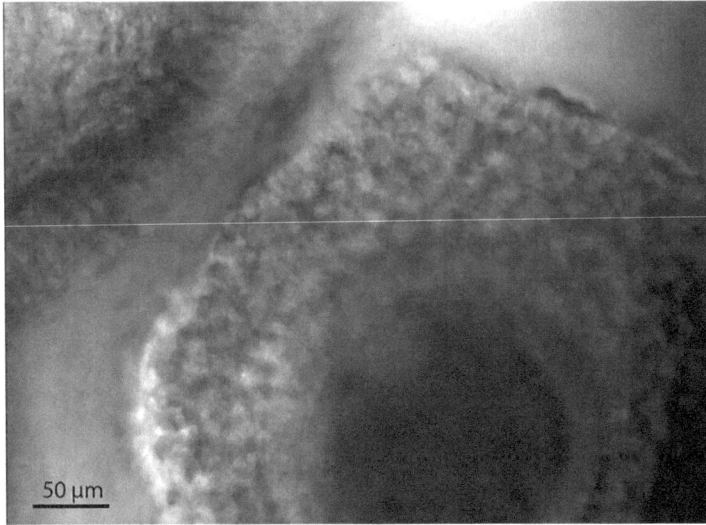

Fig. 1 The bovine cumulus-oocyte complex attaches to the tubal epithelium after ovulation

form of the nucleus and the dark granular appearance of the cytoplasm, attachment does not occur (Kolle et al. 2009). If there is no attachment, the oocyte rapidly dies. Experiments using denuded oocytes and zona-free oocytes have shown that the physical and molecular interactions between the tubal cells and the oocyte are mediated by the cumulus cells and their intercellular matrix. In the bovine, E-cadherin has been documented to be involved in somatic cell adhesion (Caballero et al. 2014). Additionally, extracellular matrix proteins such as fibronectin and tenascin-c have been hypothesized to be involved in the dialogue between COC and tubal epithelium (Relucenti et al. 2005). However, the precise molecular interactions and signaling pathways between cumulus cells, oocytes, and tubal epithelium are unknown to date and remain to be elucidated.

3 The Oviduct Maintains Sperm Fertilizing Ability by Formation of a Sperm Reservoir

After insemination, sperm transport in the female genital tract is achieved by the motility of spermatozoa and smooth muscle contraction of the uterus and oviduct (Kolle et al. 2009; Kolle 2015). Whereas sperm motility ensures that sperm move in the middle of the lumen and do not get stuck to the inner surface of the female genital tract, smooth muscle contraction in the oviduct is pivotal for sperm being able to successfully move against the strong current of the fluid resulting from ciliary beating (Kolle et al. 2009). As soon as the sperm enter the oviduct, they bind with their head to the ciliated cells of the tubal epithelium at a tangential angle (Fig. 2). The cilia embrace the sperm head and get in close mechanical contact with the plasma

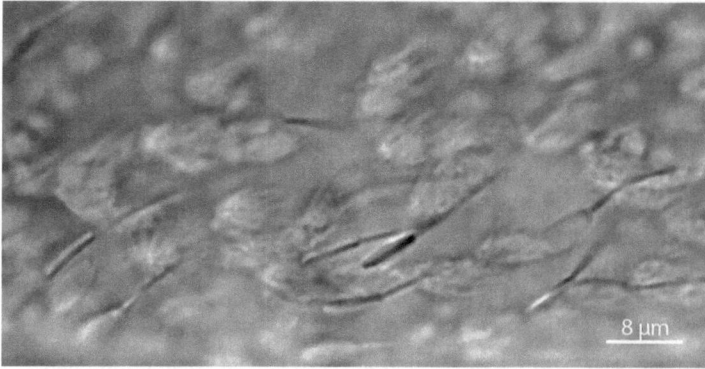

Fig. 2 Live cell imaging of the sperm reservoir in the bovine oviduct

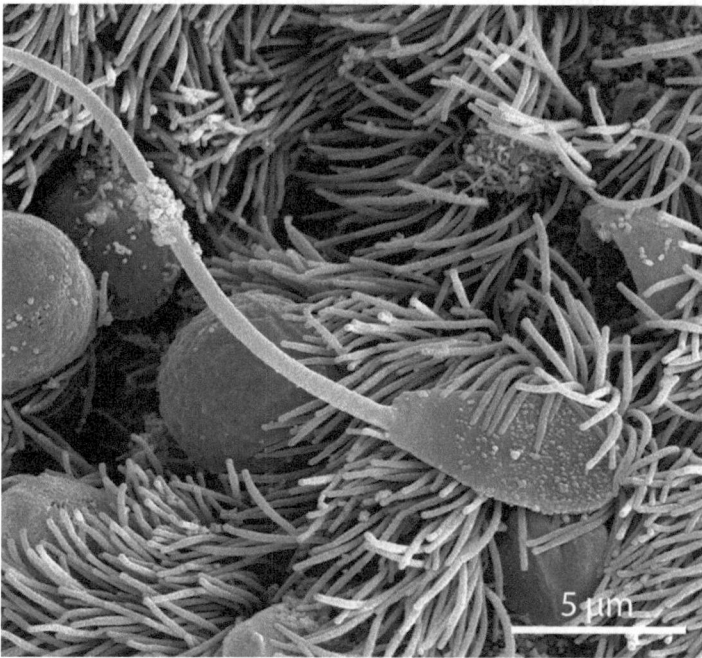

Fig. 3 Scanning electron micrograph of a bovine spermatozoon bound to the ciliary cells of the tubal epithelium

membrane (Fig. 3). As spermatozoa first enter the isthmus, most of the sperm are found in the isthmus; however, sperm also bind to the ampullar epithelium (Suarez and Pacey 2006; Kolle et al. 2009). It is important that sperm bound in the ampulla have a higher probability of fertilizing as they will be the first to reach the oocyte after ovulation (Suarez and Pacey 2006). The formation of a sperm reservoir has been confirmed in birds and reptiles as well as in rodents, pigs, cows, and horses

(Ellington et al. 1999; Suarez and Pacey 2006; Hunter 2008; Suarez 2008b). It enables the maintenance of motility and fertilizing capacity of spermatozoa for 3–4 days (most mammals), weeks (birds), months (bats), or even years (reptiles) (Suarez 2008b). Binding of the sperm head to the cilia is mediated by species-specific carbohydrate moieties on the cilia, such as fucose in the bovine, mannose/Lewis X trisaccharide in the porcine, and galactose in the equine (DeMott et al. 1995; Dobrinski et al. 1996; Lefebvre et al. 1997; Green et al. 2001; Machado et al. 2014). In bovine and ovine, the carbohydrate moieties on the cilia have been shown to bind to sperm binder proteins (BSPs) on the sperm head (Hung and Suarez 2012; Pini et al. 2018). After sperm binding, a series of molecular signaling pathways are activated. Consequently, tubal gene and protein expression are changed.

(Bergqvist et al. 2005; Georgiou et al. 2007). This change modulates hormone and enzyme synthesis as well as the production of mucopolysaccharides, glycoproteins, and hyaluronan, thus affecting the composition of the tubal fluid (Sostaric et al. 2008). Additionally, the ciliary beat frequency within the bovine oviduct is increased after sperm binding (Camara Pirez et al. 2020). In the porcine, gene expression patterns after sperm binding have been shown to be different in the presence of an X or Y chromosome with transcripts involved in signal transduction and immune response being mostly affected (Alminana et al. 2014).

Analyses of sperm behavior using live cell imaging in the bovine oviduct allowed us to characterize sperm motility and sperm behavior within the sperm reservoir (Camara Pirez et al. 2020). Overall, immotile sperm, which are present after freezing and thawing, are not able to bind and are flushed away by the current of the tubal fluid which is a result of ciliary beating. Motile sperm immediately bind when they reach the oviduct. Importantly, sperm stay motile after binding to the oviductal epithelium (Camara Pirez et al. 2020). As shown in the bovine, sperm binding in the reservoir results in increased activity of the oviductal endoplasmic reticula, which plays an important role in calcium storage and protein synthesis and ensures nutrition (Camara Pirez et al. 2020). The patterns of sperm movement and sperm behavior are not affected by the site (ampulla or isthmus) or by the cycle stage (Camara Pirez et al. 2020). The majority of sperm bound in the reservoir are agile sperm which are characterized by a binding angle of 30 degrees and have an actively beating, undulating tail. 5–15% of the bound sperm show hyperactivation, which is characterized by high amplitude and whip-like beating of the tail and a turning head. Over time, 10–20% of the bound sperm turn into lagging spermatozoa, which bind at a decreased angle with a merely slightly moving tail (Camara Pirez et al. 2020). As time further progresses, they become immotile and lie flat on the epithelium. These spermatozoa are unable to detach from the tubal epithelium in the presence of the oocyte and will not reach the oocyte for fertilization (Camara Pirez et al. 2020). Sex-sorting, which is associated with mechanical, chemical, and time stress, leads to reduced binding capacity of sperm in the reservoir (Camara Pirez et al. 2020). Additionally, single-sex-sorted sperm show a unique motility pattern within the reservoir which is characterized by a 360-degree rotational movement of the head and tail (Camara Pirez et al. 2020). When there is inflammation in the female genital tract, increased secretion is present which covers many sperm in the sperm reservoir, so that they lie flat and are not able to detach and find their way to the oocyte after ovulation (Owhor et al. 2019).

As soon as the COC is in the ampulla, spermatozoa detach from the tubal epithelium and leave the sperm reservoir. This is achieved by sperm hyperactivation, which is characterized by vigorous high-amplitude, asymmetrical beating of the tail (Suarez 2008a) and which is triggered by increased Ca^{2+} influx via sperm-specific cation channels (CatSper) (Suarez 2008a). Additionally, capacitation in the oviduct supports detachment as the binding affinity to the tubal epithelium is reduced (Lefebvre and Suarez 1996). With regard to systemic hormonal actions, progesterone has been shown to support the release of sperm from the reservoir by affecting sperm proteomics and lipidomics (Mirihagalle et al. 2022).

It is important to highlight that the interaction between COC and oviductal epithelium might be pivotal for sperm detachment and for guiding the sperm to the COC after ovulation. This is supported by the fact that, as seen in the daily routine in the IVF lab, the COC alone is not able to attract spermatozoa. Recently, it has been shown in mice that oocyte-dependent and cumulus cell-derived transforming growth factor B1 (TGFB1) supports the synthesis of natriuretic peptide type C (NPPC) in the ampulla, thus guiding the sperm in the oviduct (Marchais et al. 2022). However, the majority of local paracrine signaling pathways and molecules directing the sperm are unknown to date.

4   The Oviduct Acts as a Major Site of Sperm Selection

Although millions or even billions of sperm are inseminated, only 10–50 sperm reach the site of fertilization (Kolle et al. 2009). It is known that up to 70% of sperm are selected out in the folds and the mucus of the cervix, which acts as the first and major barrier. The second selection site is the utero-tubal junction (UTJ) which is a barrier due to its narrow lumen and its wounded course (Suarez 2008b). When sperm have successfully passed these anatomical barriers, the oviduct acts as a third, most sensitive, and most effective organ for sperm selection. Only sperm that are able to bind and maintain motility will survive and maintain fertilizing ability (Kolle 2022; Kolle et al. 2009). As such, any morphological alteration such as deviations in the form of the head or tail, or any molecular alterations such as an altered sperm plasma membrane (which often occurs after freezing and thawing), or any reduced function of important organelles such as endoplasmic reticula and mitochondria will negatively affect sperm binding capacity and will subsequently lead to inability to hyperactivate and to detach from the oviductal epithelium in the presence of the oocyte (Kolle et al. 2009). Consequently, all those sperm are selected out and eliminated from the fertilization process. In summary, the fallopian tube is able to identify and eliminate sperm with functional impairments that are not recognized during the routine investigations of ejaculates. Consequently, molecular alterations on the sperm head or tail might be the cause of impaired fertility in individuals with a good spermiogram.

5   The Oviduct Manages Embryo Transport and Nutrition

The oviduct creates a perfect microenvironment for the developing embryo by ensuring precise and timely transport and by providing nutrition. Overall, the transport of the early embryo is achieved by ciliary beating, tubal fluid flow, and smooth muscle contraction (Li and Winuthayanon 2017). Interestingly, if the cilia are immotile (ciliary dyskinesia, Kartagener Syndrome) or if smooth muscle contractions are inhibited (Halbert et al. 1976), pregnancy may occur, indicating the high adaptive capacity of the female genital tract. Importantly, successful pregnancy only occurs if the transport of the embryo takes place in a precise and timely manner (Akira et al. 1993) The average time for the embryo traveling through the oviduct is species-specific and takes 1–10 days (e.g., bovine: 4–5 days) (Kolle 2022).

The histomorphological equivalents managing the transport are the tubal ciliated cells and smooth muscle cells. Further to that, nutrition for the early embryo is provided by secretory cells of the tubal epithelium (Kolle et al. 2020). These cells might secrete small vesicles, known as extracellular vesicles (EVs), or they leave the epithelium as a whole cell and then present as protruding cells. Secretory cells that have already been secreted are named peg cells. After successful fertilization, a distinct increase of secretory and protruding cells is visible in the bovine ampulla which provides nutrition for the developing embryo during its migration down the oviduct (Kolle et al. 2009). The secretions also maintain an optimal pH required for embryonic development and to promote embryo cleavage. Thus, co-cultures of embryos and tubal cells have been shown to improve embryonic development in sheep, mice, and cattle (Gandolfi and Moor 1987; Eyestone and First 1989; Sakkas and Trounson 1990). Importantly, comparative co-culture experiments in sheep with fibroblasts and oviductal cells have confirmed that, specifically, the presence of tubal cells and not just any type of somatic cells plays a vital role in early embryonic development (Rizos et al. 2002).

The secretory cells secrete glycoproteins, glucose, lactate, pyruvate, transferrin, and amino acids (Leese 1988). Additionally, immunoglobulins, cytokines, and growth factors such as epidermal growth factor (EGF), fibroblast growth factor (FGF), and transforming growth factor (TGF) (Leese 1988; Li and Winuthayanon 2017) are secreted into the tubal fluid. The main energy sources for the early embryo are pyruvate and lactate as well as Acetyl-CoA in the porcine and bovine (Sturmey et al. 2009). Importantly, the early embryo changes from an oxidative metabolism to a glycolic metabolism as soon as the mitochondria mature in the morula or blastocyst stage (Sathananthan and Trounson 2000).

As the composition of the tubal fluid changes according to the actual needs of the developing early embryo, it might be hypothesized that the tubal fluid not only provides nutrition but also prepares and supports the metabolic transit of the early embryo.

6   The Oviduct Is the First Site of Embryo-Maternal Communication

Recent research has provided morphological, functional, and molecular proof that the oviduct is the first site of embryo-maternal communication.

Regarding morphological signs of embryo-maternal interaction, the best approach is to investigate the functional anatomy and histomorphology in the same individual, comparing the oviduct containing the embryo(s) (ipsilateral tube) with that without an embryo (contralateral tube). In the bovine, already at day 2.5 after fertilization, a change in the course of the tubal artery is visible (Kolle et al. 2009). Whereas this vessel runs straight and parallel to the oviduct on the contralateral side, it is strongly wound in the ipsilateral tube. Further, the ipsilateral tube shows increased thickness, edema, and increased transparency when compared with the contralateral tube (Kolle et al. 2009). Besides affecting the vascularization of the oviduct, the early embryo is also able to change the composition of the oviductal epithelium on the inner surface. Thus, during its migration in the oviduct, the embryo is able to initiate the transformation of ciliated cells to secretory cells, thereby locally modulating its microenvironment and ensuring nutrition (Kolle et al. 2009). These changes are already visible within the first 2 days after fertilization pointing to the initiation of rapid molecular signaling pathways, which are induced by the early embryo and which are acting both on morphology and function of the oviduct.

Another effect of the early embryo acting on the function of the oviduct includes the regulation of its own speed of transport. Thus, the early bovine embryo induces the downregulation of its own transport speed in the oviduct already on days 2–3.5 of pregnancy (Kolle et al. 2009). This allows the embryo to get in close contact with the tubal epithelium by moving in the depth between the primary and secondary ampullar folds and getting into close contact with the tubal epithelium (Fig. 4).

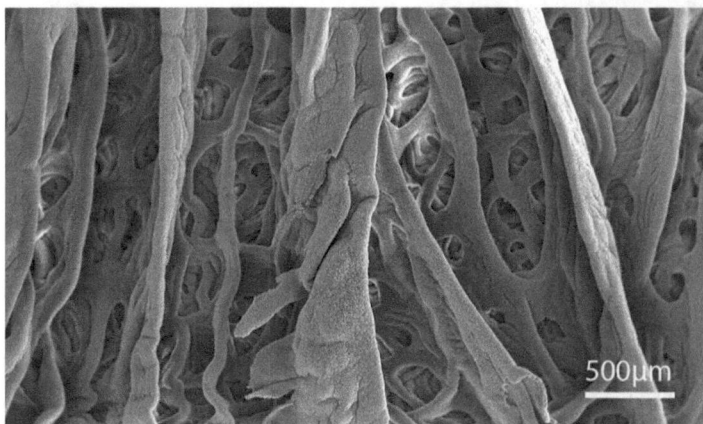

Fig. 4 Scanning electron micrograph of the inner surface of the bovine tubal ampulla revealing the primary, secondary, and tertiary folds

Interestingly, this downregulation of transport speed only occurs in the ipsilateral oviduct, but not in the contralateral oviduct highlighting the importance of local signals of the embryo within the female genital tract. Similarly, in mares, embryo-induced tubal transport has been reported, which has accounted for the increased production of prostaglandin E2 in the embryo (Weber et al. 1992). When the behavior of the early bovine embryo within the oviduct is observed under near in vivo conditions, it becomes obvious that the early embryo is not only transported in the oviduct but gets in close contact with the epithelium (Fig. 5) (Kolle et al. 2009).

Besides these effects of the embryo on the oviduct, it has also been confirmed that the tubal cells exert positive effects on the development of the early embryos in pigs, sheep, cattle, and mice (Hamdi et al. 2018). The in vitro experiments using epithelial cell cultures or oviductal explants provide evidence that proteins and growth factors, which are secreted by the epithelial cells, promote the development of the early embryo. Thus, 5–10% oviductal fluid (OF) supplemented with the culture medium has been shown to support embryonic developmental competence, cryotolerance, cell number, and gene expression in the early bovine embryo (Hamdi et al. 2018). Specifically, epidermal growth factor (EGF) (Adachi et al. 1995), fibroblast growth factor (FGF) (Wollenhaupt et al. 2004), transforming growth factor (TGF) (Wollenhaupt et al. 2004), platelet-activating factor (Roberts et al. 1993) (PAF), and leptins (Kawamura et al. 2002) have been shown to actively promote early embryonic development. An important proof of embryo-maternal interaction is the fact that, when specific proteins are synthesized in the oviductal epithelium, the respective receptors are found in the early embryo. Thus, the receptors for insulin growth factor (IGF), GH, EGF, and cytokines have been found in the early

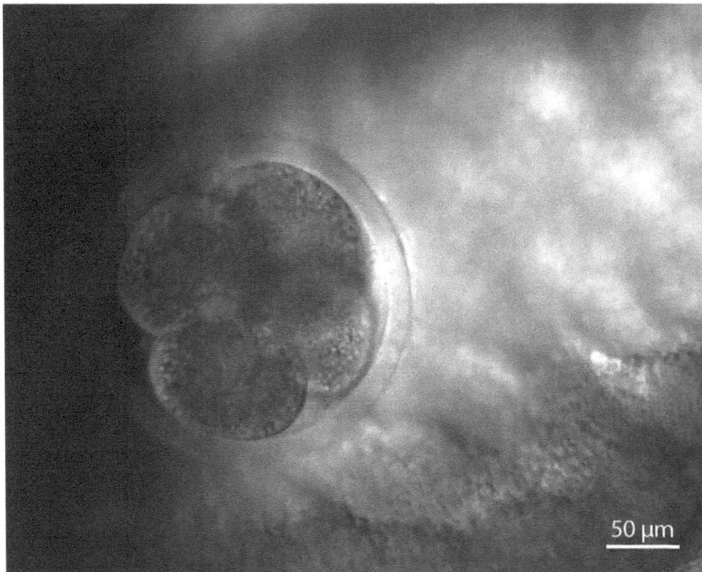

Fig. 5 Bovine four-cell embryo in close contact with the tubal epithelium

preimplantation embryo in mice and bovine (Kaye and Harvey 1995; Chow et al. 2001). Vice versa, the receptors for the embryo-derived platelet-activating factor (PAF) are localized in the oviduct (Tiemann et al. 2001). PAF is known to regulate embryo transport through the oviduct by affecting ciliary beating and the intracellular calcium concentration (Tiemann et al. 1996) and to modulate the electrophysiology of the oviductal epithelium (Downing et al. 2002).

Apart from these specific molecules, extracellular vesicles (EVs), which transfer bioactive proteins, messenger RNAs, micro RNA, and lipids, have been shown to play a pivotal role in embryo-maternal communication and the activity of tubal cells. In these EVs, oviductin (OVGP1) has been described to be an important modulator for gameto-maternal and embryo-maternal communication (Alminana et al. 2018). Recently, it has been shown that not only the tubal cells but also the early embryo is able to secrete EVs (Dissanayake et al. 2021).

Finally, the presence of an embryo-maternal crosstalk in the oviduct becomes evident in numerous molecular analyses. Those studies compare gene expression patterns in the ipsi- and contralateral oviduct before and after embryo transfer in mice, bovine, porcine, and equine using SSH, reverse dot blot analysis, or a combination of subtracted complementary cDNA libraries and cDNA array hybridization (Lee and Lie 2012; Maillo et al. 2015). The differently expressed genes are species-specific (Gonzalez-Brusi et al. 2020). In the ipsilateral oviduct, the encoded gene products include cell-surface proteins, cell-cell interaction proteins enzymes, and immune-related proteins (Bauersachs et al. 2003). However, a lot of differently expressed genes have not been identified yet. Alternatively, gene expression patterns in the oviducts of pregnant heifers on day 2.5 of embryogenesis have been compared to cyclic heifers (Maillo et al. 2016). Different gene expression patterns were specifically identified in the area where the embryo was located compared to areas where the embryo had passed (Maillo et al. 2015).

7   Conclusions

The oviduct is the key organ for keeping the COC vital after ovulation. It is critical for maintaining sperm fertilizing ability in the sperm reservoir and for sperm selection, thus enabling successful fertilization. To fulfill these roles, numerous mechanical and molecular gameto-maternal interactions take place. The precise molecular signaling pathways between COC and tubal cells, as well as between spermatozoa and oviductal epithelium, still need to be elucidated. However, if this knowledge is established, it holds promise to create novel therapeutic strategies for the treatment of infertility and subfertility and for further improving the outcome of assisted reproductive technologies (ART), e.g., by developing matrices with supplements guiding the sperm to the oocyte.

After fertilization, the embryo spends the first 3–4 days of its life in the oviduct. During this time, nutrition, embryonic cell growth and differentiation, and the precisely timed transport are orchestrated by systemic hormonal stimuli as well as by a local, active, and continuous embryo-maternal dialogue. This dialogue is initiated by the first mechanical contract between the embryo and the oviductal inner surface shortly after fertilization and is the most critical time in early embryonic development. Only if the embryo is able to signal its presence to the mother—and if the mother is able to respond—a successful pregnancy will be established.

Conflict of Interest The author declares no conflict of interest.

References

Adachi K, Kurachi H, Homma H, Adachi H, Imai T, Sakata M, Higashiguchi O, Yamaguchi M, Morishige K, Sakoyama Y et al (1995) Estrogen induces epidermal growth factor (EGF) receptor and its ligands in human fallopian tube: involvement of EGF but not transforming growth factor-alpha in estrogen-induced tubal cell growth in vitro. Endocrinology 136(5):2110–2119. https://doi.org/10.1210/endo.136.5.7720660

Akira S, Sanbuissho A, Lin YC, Araki T (1993) Acceleration of embryo transport in superovulated adult rats. Life Sci 53(15):1243–1251. https://doi.org/10.1016/0024-3205(93)90543-c

Alminana C, Caballero I, Heath PR, Maleki-Dizaji S, Parrilla I, Cuello C, Gil MA, Vazquez JL, Vazquez JM, Roca J, Martinez EA, Holt WV, Fazeli A (2014) The battle of the sexes starts in the oviduct: modulation of oviductal transcriptome by X and Y-bearing spermatozoa. BMC Genomics 15(1):293. https://doi.org/10.1186/1471-2164-15-293

Alminana C, Tsikis G, Labas V, Uzbekov R, da Silveira JC, Bauersachs S, Mermillod P (2018) Deciphering the oviductal extracellular vesicles content across the estrous cycle: implications for the gametes-oviduct interactions and the environment of the potential embryo. BMC Genomics 19(1):622. https://doi.org/10.1186/s12864-018-4982-5

Bauersachs S, Blum H, Mallok S, Wenigerkind H, Rief S, Prelle K, Wolf E (2003) Regulation of ipsilateral and contralateral bovine oviduct epithelial cell function in the postovulation period: a transcriptomics approach. Biol Reprod 68(4):1170–1177. https://doi.org/10.1095/biolreprod.102.010660

Bergqvist AS, Yokoo M, Heldin P, Frendin J, Sato E, Rodriguez-Martinez H (2005) Hyaluronan and its binding proteins in the epithelium and intraluminal fluid of the bovine oviduct. Zygote 13(3):207–218. https://doi.org/10.1017/s0967199405003266

Caballero JN, Gervasi MG, Veiga MF, Dalvit GC, Perez-Martinez S, Cetica PD, Vazquez-Levin MH (2014) Epithelial cadherin is present in bovine oviduct epithelial cells and gametes, and is involved in fertilization-related events. Theriogenology 81(9):1189–1206. https://doi.org/10.1016/j.theriogenology.2014.01.028

Camara Pirez M, Steele H, Reese S, Kolle S (2020) Bovine sperm-oviduct interactions are characterized by specific sperm behaviour, ultrastructure and tubal reactions which are impacted by sex sorting. Sci Rep 10(1):16522. https://doi.org/10.1038/s41598-020-73592-1

Chow JF, Lee KF, Chan ST, Yeung WS (2001) Quantification of transforming growth factor beta1 (TGFbeta1) mRNA expression in mouse preimplantation embryos and determination of TGFbeta receptor (type I and type II) expression in mouse embryos and reproductive tract. Mol Hum Reprod 7(11):1047–1056. https://doi.org/10.1093/molehr/7.11.1047

Croxatto HB (2002) Physiology of gamete and embryo transport through the fallopian tube. Reprod Biomed Online 4(2):160–169. https://doi.org/10.1016/s1472-6483(10)61935-9

DeMott RP, Lefebvre R, Suarez SS (1995) Carbohydrates mediate the adherence of hamster sperm to oviductal epithelium. Biol Reprod 52(6):1395–1403. https://doi.org/10.1095/biolreprod52.6.1395

Diskin MG, Morris DG (2008) Embryonic and early foetal losses in cattle and other ruminants. Reprod Domest Anim 43(Suppl 2):260–267. https://doi.org/10.1111/j.1439-0531.2008.01171.x

Dissanayake K, Nomm M, Lattekivi F, Ord J, Ressaissi Y, Godakumara K, Reshi QUA, Viil J, Jaager K, Velthut-Meikas A, Salumets A, Jaakma U, Fazeli A (2021) Oviduct as a sensor of embryo quality: deciphering the extracellular vesicle (EV)-mediated embryo-maternal dia-logue. J Mol Med (Berl) 99(5):685–697. https://doi.org/10.1007/s00109-021-02042-w

Dobrinski I, Ignotz GG, Thomas PG, Ball BA (1996) Role of carbohydrates in the attachment of equine spermatozoa to uterine tubal (oviductal) epithelial cells in vitro. Am J Vet Res 57(11):1635–1639. https://www.ncbi.nlm.nih.gov/pubmed/8915444

Downing SJ, Maguiness SD, Tay JI, Watson A, Leese HJ (2002) Effect of platelet-activating fac-tor on the electrophysiology of the human fallopian tube: early mediation of embryo-maternal dialogue? Reproduction 124(4):523–529. https://doi.org/10.1530/rep.0.1240523

Druart X, Cognie J, Baril G, Clement F, Dacheux JL, Gatti JL (2009) In vivo imaging of in situ motility of fresh and liquid stored ram spermatozoa in the ewe genital tract. Reproduction 138(1):45–53. https://doi.org/10.1530/REP-09-0108

Ellington JE, Samper JC, Jones AE, Oliver SA, Burnett KM, Wright RW (1999) In vitro interactions of cryopreserved stallion spermatozoa and oviduct (uterine tube) epithelial cells or their secre-tory products. Anim Reprod Sci 56(1):51–65. https://doi.org/10.1016/s0378-4320(99)00030-5

Eyestone WH, First NL (1989) Co-culture of early cattle embryos to the blastocyst stage with ovi-ducal tissue or in conditioned medium. J Reprod Fertil 85(2):715–720. https://doi.org/10.1530/jrf.0.0850715

Gandolfi F, Moor RM (1987) Stimulation of early embryonic development in the sheep by co-culture with oviduct epithelial cells. J Reprod Fertil 81(1):23–28. https://doi.org/10.1530/jrf.0.0810023

Georgiou AS, Snijders AP, Sostaric E, Aflatoonian R, Vazquez JL, Vazquez JM, Roca J, Martinez EA, Wright PC, Fazeli A (2007) Modulation of the oviductal environment by gametes. J Proteome Res 6(12):4656–4666. https://doi.org/10.1021/pr070349m

Gonzalez-Brusi L, Algarra B, Moros-Nicolas C, Izquierdo-Rico MJ, Aviles M, Jimenez-Movilla M (2020) A comparative view on the Oviductal environment during the Periconception period. Biomol Ther 10(12). https://doi.org/10.3390/biom10121690

Green CE, Bredl J, Holt WV, Watson PF, Fazeli A (2001) Carbohydrate mediation of boar sperm binding to oviductal epithelial cells in vitro. Reproduction 122(2):305–315. https://doi.org/10.1530/rep.0.1220305

Halbert SA, Tam PY, Blandau RJ (1976) Egg transport in the rabbit oviduct: the roles of cilia and muscle. Science 191(4231):1052–1053. https://doi.org/10.1126/science.1251215

Hamdi M, Lopera-Vasquez R, Maillo V, Sanchez-Calabuig MJ, Nunez C, Gutierrez-Adan A, Rizos D (2018) Bovine oviductal and uterine fluid support in vitro embryo development. Reprod Fertil Dev 30(7):935–945. https://doi.org/10.1071/RD17286

Hung PH, Suarez SS (2012) Alterations to the bull sperm surface proteins that bind sperm to ovi-ductal epithelium. Biol Reprod 87(4):88. https://doi.org/10.1095/biolreprod.112.099721

Hunter RH (2008) Sperm release from oviduct epithelial binding is controlled hormonally by peri-ovulatory graafian follicles. Mol Reprod Dev 75(1):167–174. https://doi.org/10.1002/mrd.20776

Kawamura K, Sato N, Fukuda J, Kodama H, Kumagai J, Tanikawa H, Nakamura A, Tanaka T (2002) Leptin promotes the development of mouse preimplantation embryos in vitro. Endocrinology 143(5):1922–1931. https://doi.org/10.1210/endo.143.5.8818

Kaye PL, Harvey MB (1995) The role of growth factors in preimplantation development. Prog Growth Factor Res 6(1):1–24. https://doi.org/10.1016/0955-2235(95)00001-1

Kodithuwakku SP, Miyamoto A, Wijayagunawardane MP (2007) Spermatozoa stimulate prostaglandin synthesis and secretion in bovine oviductal epithelial cells. Reproduction 133(6):1087–1094. https://doi.org/10.1530/REP-06-0201

Kolle S (2012) Live cell imaging of the oviduct. Method Enzymol 506:415–423. https://doi.org/10.1016/B978-0-12-391856-7.00045-7

Kolle S (2015) Transport, distribution and elimination of mammalian sperm following natural mating and insemination. Reprod Domest Anim 50(Suppl 3):2–6. https://doi.org/10.1111/rda.12576

Kolle S (2022) Sperm-oviduct interactions: key factors for sperm survival and maintenance of sperm fertilizing capacity. Andrology 10(5):837–843. https://doi.org/10.1111/andr.13179

Kolle S, Dubielzig S, Reese S, Wehrend A, Konig P, Kummer W (2009) Ciliary transport, gamete interaction, and effects of the early embryo in the oviduct: ex vivo analyses using a new digital videomicroscopic system in the cow. Biol Reprod 81(2):267–274. https://doi.org/10.1095/biolreprod.108.073874

Kolle S, Hughes B, Steele H (2020) Early embryo-maternal communication in the oviduct: a review. Mol Reprod Dev 87(6):650–662. https://doi.org/10.1002/mrd.23352

Kolle S, Reese S, Kummer W (2010) New aspects of gamete transport, fertilization, and embryonic development in the oviduct gained by means of live cell imaging. Theriogenology 73(6):786–795. https://doi.org/10.1016/j.theriogenology.2009.11.002

Lee CS, Lie AT (2012) Successful pregnancy outcome following gamete intra-fallopian transfer in a patient with Mullerian dysgenesis. Reprod Biomed Online 24(5):547–549. https://doi.org/10.1016/j.rbmo.2012.01.021

Leese HJ (1988) The formation and function of oviduct fluid. J Reprod Fertil 82(2):843–856. https://doi.org/10.1530/jrf.0.0820843

Lefebvre R, Lo MC, Suarez SS (1997) Bovine sperm binding to oviductal epithelium involves fucose recognition. Biol Reprod 56(5):1198–1204. https://doi.org/10.1095/biolreprod56.5.1198

Lefebvre R, Suarez SS (1996) Effect of capacitation on bull sperm binding to homologous oviductal epithelium. Biol Reprod 54(3):575–582. https://doi.org/10.1095/biolreprod54.3.575

Li S, Winuthayanon W (2017) Oviduct: roles in fertilization and early embryo development. J Endocrinol 232(1):R1–R26. https://doi.org/10.1530/JOE-16-0302

Machado SA, Kadirvel G, Daigneault BW, Korneli C, Miller P, Bovin N, Miller DJ (2014) LewisX-containing glycans on the porcine oviductal epithelium contribute to formation of the sperm reservoir. Biol Reprod 91(6):140. https://doi.org/10.1095/biolreprod.114.119503

Maillo V, Gaora PO, Forde N, Besenfelder U, Havlicek V, Burns GW, Spencer TE, Gutierrez-Adan A, Lonergan P, Rizos D (2015) Oviduct-embryo interactions in cattle: two-way traffic or a one-way street? Biol Reprod 92(6):144. https://doi.org/10.1095/biolreprod.115.127969

Maillo V, Lopera-Vasquez R, Hamdi M, Gutierrez-Adan A, Lonergan P, Rizos D (2016) Maternal-embryo interaction in the bovine oviduct: evidence from in vivo and in vitro studies. Theriogenology 86(1):443–450. https://doi.org/10.1016/j.theriogenology.2016.04.060

Marano P (1998) Female infertility: radiology and endoscopy in morphofunctional imaging. Rays 23(4):597–599. https://www.ncbi.nlm.nih.gov/pubmed/10191653

Marchais M, Gilbert I, Bastien A, Macaulay A, Robert C (2022) Mammalian cumulus-oocyte complex communication: a dialog through long and short distance messaging. J Assist Reprod Genet 39(5):1011–1025. https://doi.org/10.1007/s10815-022-02438-8

McComb PF, Halbert SA, Gomel V (1980) Pregnancy, ciliary transport, and the reversed ampullary segment of the rabbit fallopian tube. Fertil Steril 34(4):386–390. https://www.ncbi.nlm.nih.gov/pubmed/7418893

McComb P, Langley L, Villalon M, Verdugo P (1986) The oviductal cilia and Kartagener's syndrome. Fertil Steril 46(3):412–416. https://www.ncbi.nlm.nih.gov/pubmed/3488922

Mirihagalle S, Hughes JR, Miller DJ (2022) Progesterone-induced sperm release from the oviduct sperm reservoir. Cells 11(10). https://doi.org/10.3390/cells11101622

Owhor LE, Reese S, Kolle S (2019) Salpingitis impairs bovine tubal function and sperm-oviduct interaction. Sci Rep 9(1):10893. https://doi.org/10.1038/s41598-019-47431-x

Pini T, de Graaf SP, Druart X, Tsikis G, Labas V, Teixeira-Gomes AP, Gadella BM, Leahy T (2018) Binder of sperm proteins 1 and 5 have contrasting effects on the capacitation of ram spermatozoa. Biol Reprod 98(6):765–775. https://doi.org/10.1093/biolre/ioy032

Relucenti M, Heyn R, Correr S, Familiari G (2005) Cumulus oophorus extracellular matrix in the human oocyte: a role for adhesive proteins. Ital J Anat Embryol 110(2 Suppl 1):219–224. https://www.ncbi.nlm.nih.gov/pubmed/16101041

Rizos D, Ward F, Duffy P, Boland MP, Lonergan P (2002) Consequences of bovine oocyte maturation, fertilization or early embryo development in vitro versus in vivo: implications for blastocyst yield and blastocyst quality. Mol Reprod Dev 61(2):234–248. https://doi.org/10.1002/mrd.1153

Roberts C, O'Neill C, Wright L (1993) Platelet activating factor (PAF) enhances mitosis in pre-implantation mouse embryos. Reprod Fertil Dev 5(3):271–279. https://doi.org/10.1071/rd9930271

Sakkas D, Trounson AO (1990) Co-culture of mouse embryos with oviduct and uterine cells prepared from mice at different days of pseudopregnancy. J Reprod Fertil 90(1):109–118. https://doi.org/10.1530/jrf.0.0900109

Sathananthan AH, Trounson AO (2000) Mitochondrial morphology during preimplantational human embryogenesis. Hum Reprod 15(Suppl 2):148–159. https://doi.org/10.1093/humrep/15.suppl_2.148

Sostaric E, Dieleman SJ, van de Lest CH, Colenbrander B, Vos PL, Garcia-Gil N, Gadella BM (2008) Sperm binding properties and secretory activity of the bovine oviduct immediately before and after ovulation. Mol Reprod Dev 75(1):60–74. https://doi.org/10.1002/mrd.20766

Sturmey RG, Reis A, Leese HJ, McEvoy TG (2009) Role of fatty acids in energy provision during oocyte maturation and early embryo development. Reprod Domest Anim 44(Suppl 3):50–58. https://doi.org/10.1111/j.1439-0531.2009.01402.x

Suarez SS (2008a) Control of hyperactivation in sperm. Hum Reprod Update 14(6):647–657. https://doi.org/10.1093/humupd/dmn029

Suarez SS (2008b) Regulation of sperm storage and movement in the mammalian oviduct. Int J Dev Biol 52(5–6):455–462. https://doi.org/10.1387/ijdb.072527ss

Suarez SS, Pacey AA (2006) Sperm transport in the female reproductive tract. Hum Reprod Update 12(1):23–37. https://doi.org/10.1093/humupd/dmi047

Talbot P, Geiske C, Knoll M (1999) Oocyte pickup by the mammalian oviduct. Mol Biol Cell 10(1):5–8. https://doi.org/10.1091/mbc.10.1.5

Talbot P, Shur BD, Myles DG (2003) Cell adhesion and fertilization: steps in oocyte transport, sperm-zona pellucida interactions, and sperm-egg fusion. Biol Reprod 68(1):1–9. https://doi.org/10.1095/biolreprod.102.007856

Tiemann U, Neels P, Kuchenmeister U, Walzel H, Spitschak M (1996) Effect of ATP and platelet-activating factor on intracellular calcium concentrations of cultured oviductal cells from cows. J Reprod Fertil 108(1):1–9. https://doi.org/10.1530/jrf.0.1080001

Tiemann U, Tomek W, Schneider F, Wollenhaupt K, Kanitz W, Becker F, Pohland R, Alm H (2001) Platelet-activating factor (PAF)-like activity, localization of PAF receptor (PAF-R) and PAF-acetylhydrolase (PAF-AH) activity in bovine endometrium at different stages of the estrous cycle and early pregnancy. Prostaglandins Other Lipid Mediat 65(2–3):125–141. https://doi.org/10.1016/s0090-6980(01)00130-7

Walter I (1995) Culture of bovine oviduct epithelial cells (BOEC). Anat Rec 243(3):347–356. https://doi.org/10.1002/ar.1092430309

Weber JA, Woods GL, Freeman DA, Vanderwall DK (1992) Prostaglandin E2-specific binding to the equine oviduct. Prostaglandins 43(1):61–65. https://doi.org/10.1016/0090-6980(92)90065-2

Wollenhaupt K, Welter H, Einspanier R, Manabe N, Brussow KP (2004) Expression of epidermal growth factor receptor (EGF-R), vascular endothelial growth factor receptor (VEGF-R) and fibroblast growth factor receptor (FGF-R) systems in porcine oviduct and endometrium during the time of implantation. J Reprod Dev 50(3):269–278. https://doi.org/10.1262/jrd.50.269

Yeste M, Holt WV, Bonet S, Rodriguez-Gil JE, Lloyd RE (2014) Viable and morphologically normal boar spermatozoa alter the expression of heat-shock protein genes in oviductal epithelial cells during co-culture in vitro. Mol Reprod Dev 81(9):805–819. https://doi.org/10.1002/mrd.22350